W9-DHM-370

Complete Guide to

Understanding Electronics Diagrams

Edward A. Lacy

Prentice Hall, Englewood Cliffs, New Jersey 07632

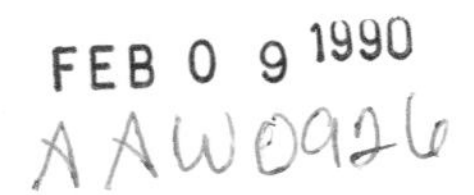

Library of Congress Cataloging-in-Publication Data

Lacy, Edward A.,
Complete guide to understanding electronics diagrams / Edward A.
Lacy.
p. cm.
Includes index.
ISBN 0-13-160920-3
1. Electronics--Charts, diagrams, etc. I. Title.
TK7866.L33 1989
621.381'014--dc19 88-8030
CIP

Editorial/production supervision
and interior design: Elaine Lynch
Cover design: Edsal Enterprises
Manufacturing buyer: Robert Anderson

The publisher offers discounts on this book when ordered
in bulk quantities. For more information, write:
Special Sales/College Marketing
Prentice-Hall, Inc.
College Technical and Reference Division
Englewood Cliffs, NJ 07632

Printed in the United States of America
10 9 8 7 6 5 4 3 2 1

ISBN 0-13-160920-3

PRENTICE-HALL INTERNATIONAL (UK) LIMITED, *London*
PRENTICE-HALL OF AUSTRALIA PTY. LIMITED, *Sydney*
PRENTICE-HALL CANADA INC., *Toronto*
PRENTICE-HALL HISPANOAMERICANA, S.A., *Mexico*
PRENTICE-HALL OF INDIA PRIVATE LIMITED, *New Delhi*
PRENTICE-HALL OF JAPAN, INC., *Tokyo*
PRENTICE-HALL OF SOUTHEAST ASIA PTE. LTD., *Singapore*
EDITORA PRENTICE-HALL DO BRASIL, LTDA., *Rio de Janeiro*

Dedicated

to

Lee and Dorie Lacy

Contents

Preface

Several excellent books show how to read or interpret schematics and other electronics diagrams, but we believe this is the first text to help the engineer, technician, or hobbyist to interpret logic diagrams as well as other electronics diagrams.

In a sense this book is the reverse of electronics drafting textbooks and standards. It tells you how to interpret what you see, not how to draw a circuit. To do this, it includes *non*standard symbols and drawing practices because such exist in the real world. Despite numerous military and civilian drawing specifications, different electronics companies have their own variations of these specifications.

Even within the same company, the engineers and drafters may differ on symbols and drawing practices because they are not aware of the latest standards or they disagree with the value of the newest symbols or practices.

Although only one chapter is devoted specifically to interpreting logic diagrams, the other chapters are essential because logic diagrams are frequently combined with elements of other drawings—block diagrams, schematics, connection diagrams—so that interpretation or understanding must go beyond an understanding of logic circuits. In addition, it is necessary at times to be able to interpret mechanical drawings and electrical diagrams in order to figure out logic diagrams.

Within electronics there is a good bit of looseness in describing or defining illustrations. For example, the terms "drawings" and "diagrams" are often used interchangeably. (In this book we use "diagram" as a term referring to a drawing that explains.) "Schematic diagram" is frequently shortened to "schematic."

One engineer or drafter may call a diagram with four blocks, a meter, and two potentiometers a ''schematic,'' while another engineer may call it a ''block diagram.'' Despite the looseness of these terms, it is soon apparent which type of diagram is being discussed.

Manufacturing drawings are not covered here, as they are not needed for circuit test, installation, or maintenance (and therefore are not likely to be supplied to the technician, operator, or other end user.)

A fundamental knowledge of electronics and digital logic circuits is assumed; short explanations are placed near some circuits and components as reminders, not for tutorial purposes. Electron tube circuits are included because they are still in use despite the proliferation of semiconductor circuits.

Acknowledgments

This book has been made possible through the kind assistance of Thelma Braswell, the Institute of Electrical and Electronics Engineers, Inc., and Cynthia Clark, the American Society of Mechanical Engineers. Material labeled ANSI/IEEE Std 91-1984 has been extracted from IEEE Standard Graphic Symbols for Logic Functions, copyright 1984 by the IEEE; ANSI/IEEE Std 991-1986 from IEEE Standard for Logic Circuit Diagrams, copyright 1986; USAS Y14.15-1966 from USA Standard Drafting Practices—Electrical and Electronics Diagrams, copyright 1986 by the American Society of Mechanical Engineers. The symbols in Chapter 4 have been extracted from ANSI Y32.2-1975 and ANSI/IEEE Std 315A-1986.

Thanks also to the following companies for the use of copyrighted materials:

American Radio Relay League
Eaton Corporation
EDN
EEM 86/87
GE/RCA Corporation
Hewlett-Packard Company
Howard W. Sams & Co.
MSN & Communications Technology
Prentice Hall
QST

TAB Books, Inc.
Texas Instruments Incorporated

Thanks, too, to the following individuals for providing information:

Earnest W. Thompson, Siemens Components, Inc.

Ms. Lucy A. Tarnell, Standard Microsystems Corporation

Joel Weeks, Advanced Micro Devices, Inc.

Noelle Greene, Gould, Inc.

Peter H. Kowalchuk, United Technologies Microelectronic Center, Inc.

Ms. Denise M. Olds, Commodore Semiconductor Group

Jack Galpin, RCA Service Company

And thanks to my wife, Rita, for her support in ways too numerous to mention.

Chapter 1

Introduction

Without the use of numerous types of diagrams or drawings, it would be most difficult, if not impossible, to build and maintain modern-day electronic equipment. By representing three-dimensional electronic components with two-dimensional graphic symbols, however, it is a straightforward technique to show the relationships and interconnections between these components on various types of drawings. Before we consider methods of interpreting diagrams, let's look at same basic types.

Mechanical drawings such as top assembly drawings, panel drawings, and pictorial drawings are perhaps the easiest to understand because they most nearly resemble the equipment they depict. To interpret them it is not necessary to know or recognize dozens of different symbols.

The *top assembly* drawing (not shown here because of its size) is used primarily for military electronic equipment and complicated civilian electronic gear. Before the equipment is built, the top assembly drawing gives the technician or other interested parties an idea of how the finished item will appear.

It combines outline drawings for subassemblies and some components and a parts location drawing. It also lists the subassemblies and components (parts) and numbers for key related drawings such as schematics, logic, and panel drawings. With this drawing you can track down all the other drawings related to this equipment. Once the equipment is built, the primary value of the drawing is its list of parts. However, it can be modified to be a parts identification drawing.

Like the top assembly drawing, the *panel* drawing (Fig. 1.1) gives a preview of the front panel of the equipment, showing switches, controls, and meters and their names. For the technical writer preparing instructional material or the train-

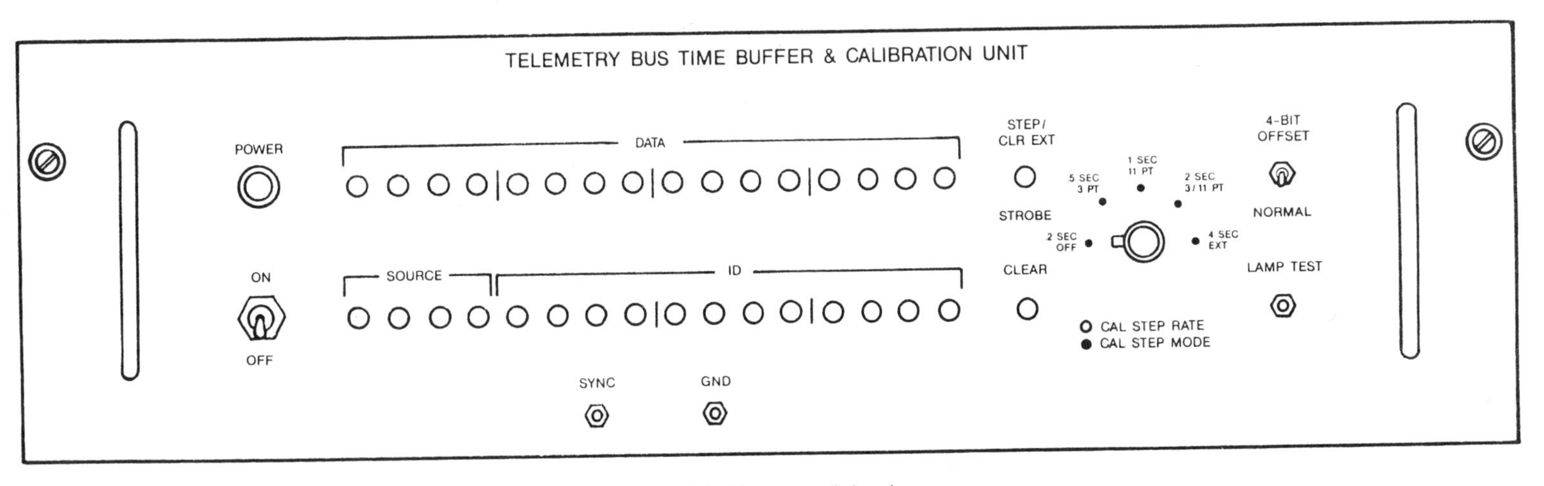

Figure 1.1 Front-panel drawing.

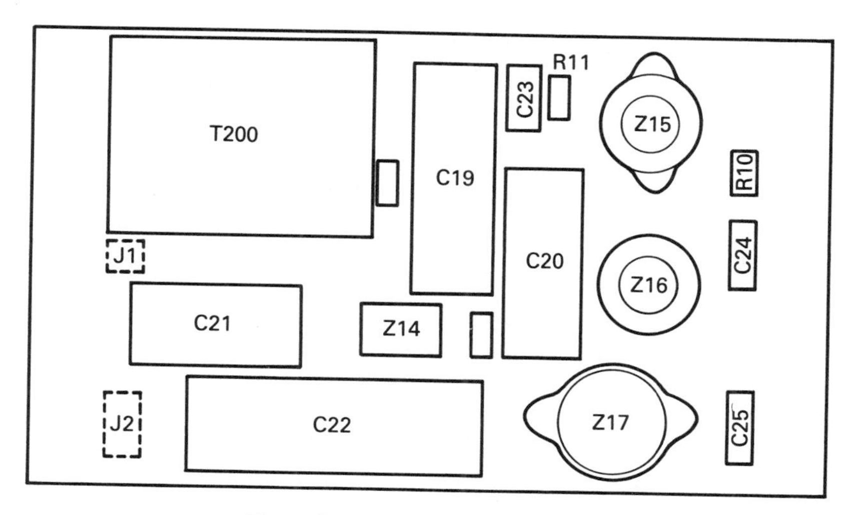

Figure 1.2 Parts location diagram.

ing department putting together a course on this equipment, this drawing allows them to proceed without waiting for the panel to be manufactured.

The *component location* (or parts arrangement) drawing (see Fig. 1.2) is frequently available as part of the service literature for that equipment. It is an essential drawing for the technician who has determined, from a schematic or logic diagram (to be discussed), that component "X" is probably defective and who must now find that component in order to test and replace it.

Although the information on a component location diagram may be stamped or engraved on the equipment chassis, the diagram may still be useful for those cases where the information on the chassis has been obliterated through dirt and grime, wear, and even charring. It may also show test points.

The *pictorial* drawing (see Fig. 1.3), a form of line drawing, shows the

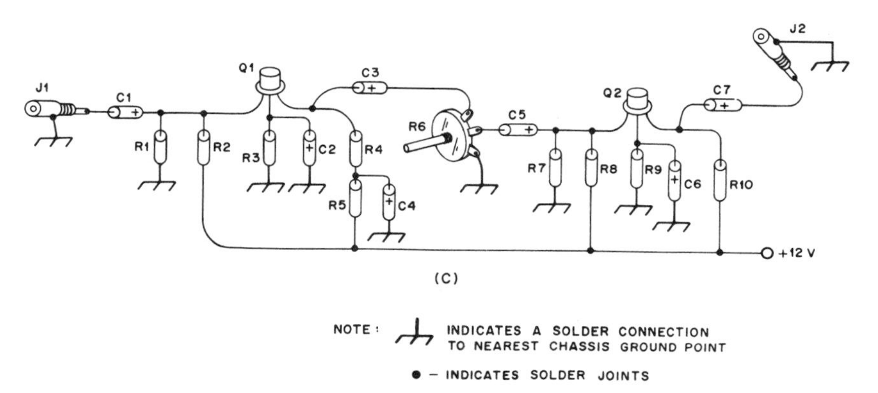

Figure 1.3 Pictorial wiring diagram—component to component. (Copyright 1984 by QST.)

physical relationship of components, subassemblies, and assemblies. It is much like a photo except that in some cases it may reveal more details than a photo. Photos may lose details in shadows and through limitations in the printing process. While often used in instructions for constructing electronic gear from kits, pictorial drawings are also used to show connections between units of equipment (Fig. 1.4). As the interconnections become complicated, the pictorial interconnection is replaced with a wiring-type interconnection. A pictorial drawing may be combined as shown in Fig. 1.5 with other drawings to form *hybrid* drawings.

Wiring diagrams are either connection or interconnection types. A *connection* diagram (Fig. 1.6) shows the point-to-point wiring within one chassis. An *interconnection* diagram shows the wiring, frequently in cables, between assemblies, units, or racks of equipment (depending on the complexity of the equipment). When an interconnection diagram becomes too complicated, it may be replaced with a *wire list,* which indicates the source ("from") and destination ("to") points for each wire in a code made of letters and numbers. These diagrams are discussed in more detail in Chapter 6.

Electronics drawings include block, flow, schematic, logic, state, timing, and printed-wiring diagrams. Not all of these will be encountered in any given service or instruction manual, but you should know how to interpret them whenever they are encountered.

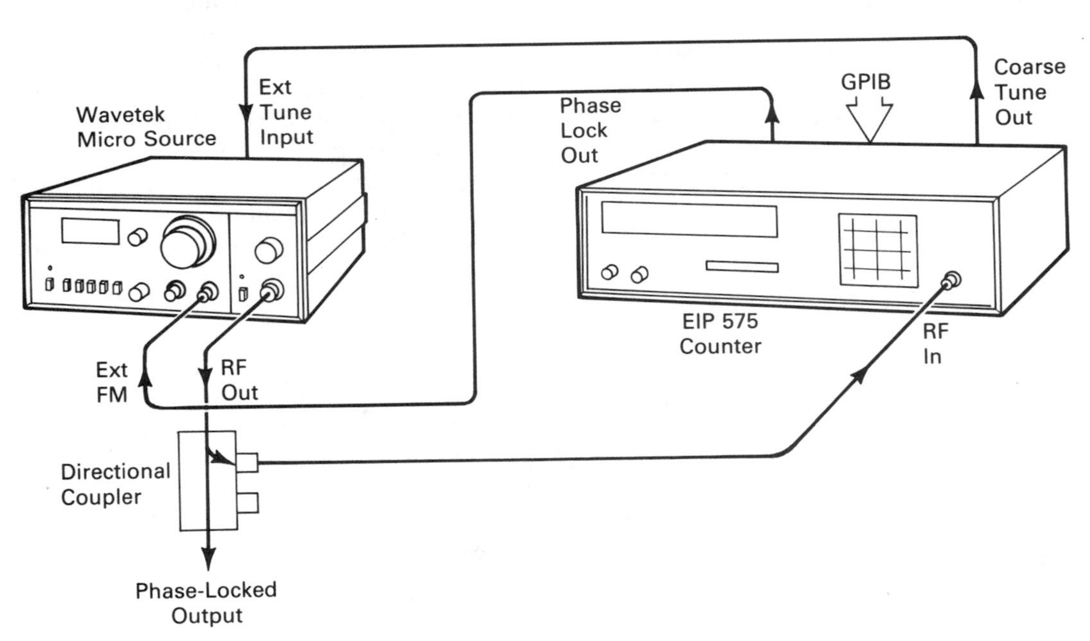

Figure 1.4 Pictorial wiring diagram—equipment to equipment. (Copyright 1986 by EW Communications.)

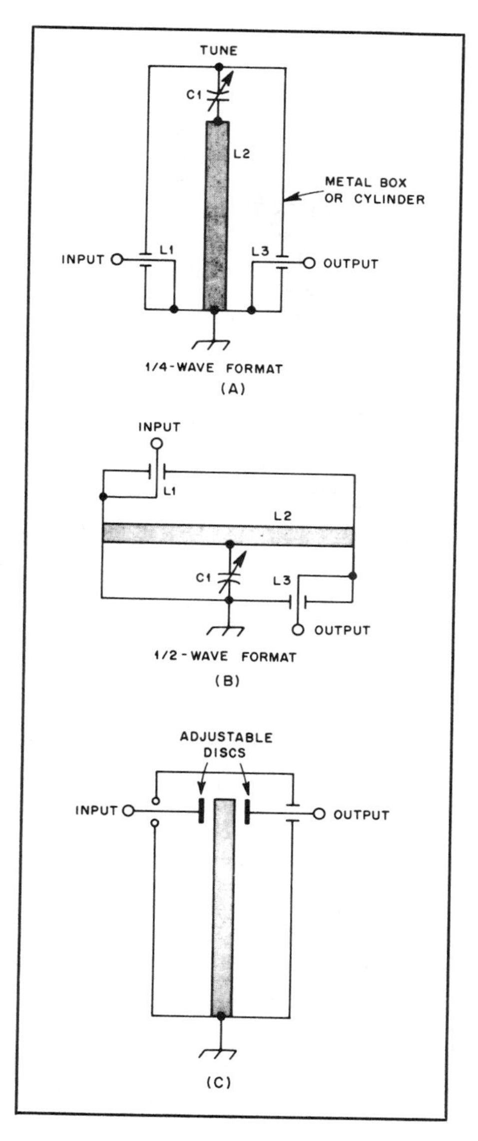

Figure 1.5 Hybrid pictorial/schematic diagram. (Copyright 1984 by QST.)

The *block* diagram (Fig. 1.7) is an uncluttered drawing that uses blocks and similar shapes to represent entire circuits, assemblies, or equipment and to show the path of the signal. It is a versatile drawing because complex circuits of any size can be reduced to a simple block for ease of understanding the relationships between stages or higher-level assemblies. Interconnecting flow lines between the blocks indicate the relationship of the blocks. Notice that the block diagram is a form of a single-line diagram (used for microwave circuits), as it uses single lines

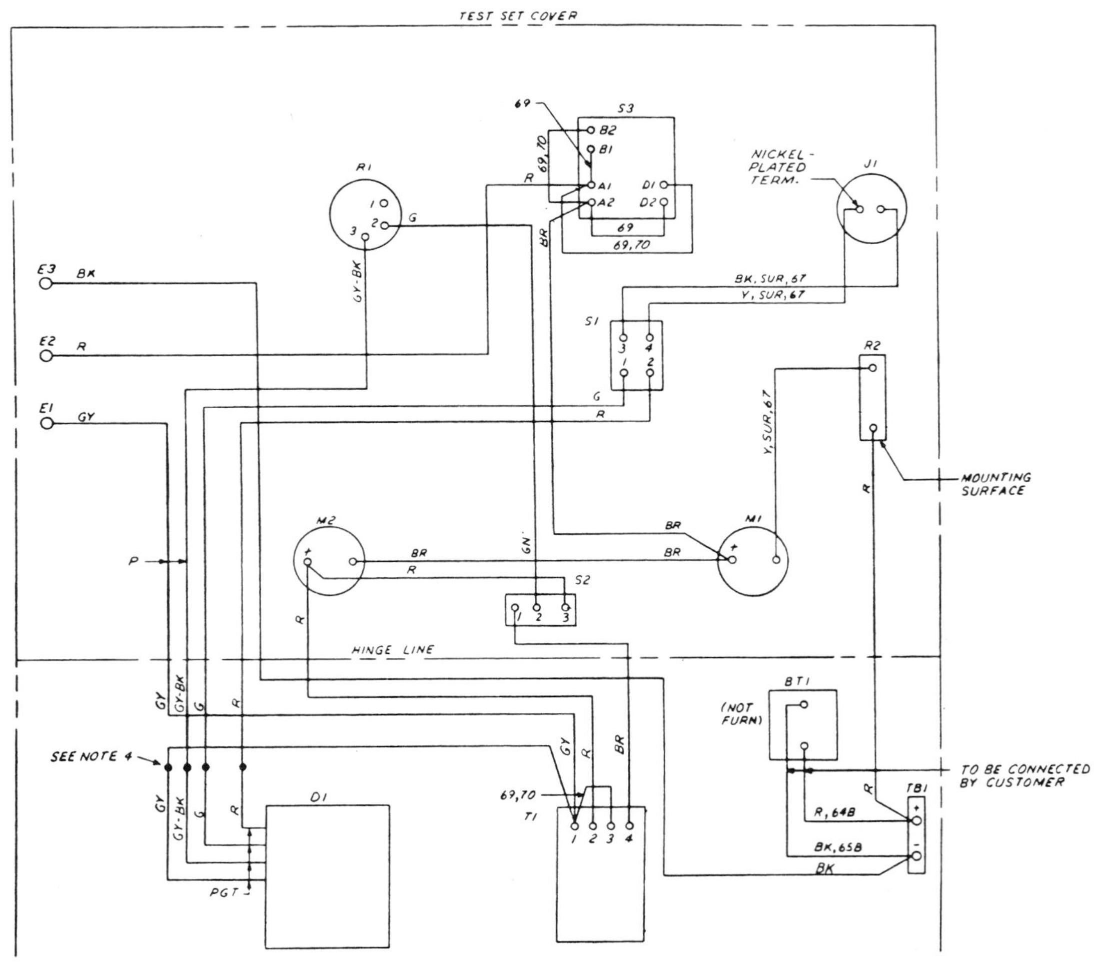

NOTES:

1. UNLESS OTHERWISE SPECIFIED, ALL WIRES ARE INCLUDED IN THE CABLE ASSEMBLY XXXXX.
2. ITEM NUMBERS REFERRED TO ARE SHOWN IN PARTS LIST OF ASSEMBLY DRAWING XXXXX.
3. ALL SOLDERING SHALL BE IN ACCORDANCE WITH QQ-S-524 METHOD C.
4. SPLICE AND SOLDER AND WRAP WITH ONE LAYER OF TAPE ITEM 58 AND TWO LAYERS OF TAPE ITEM 60.
5. SUR-WIRING-WIRE TO BE DRESSED BACK AND RUN ALONG THE MOUNTING SURFACES IN THE MOST CONVENIENT MANNER.
6. PGT - LEADS FURNISHED WITH PART.

Figure 1.6 Point-to-point connection diagram. (From USAS Y14.15-1966.)

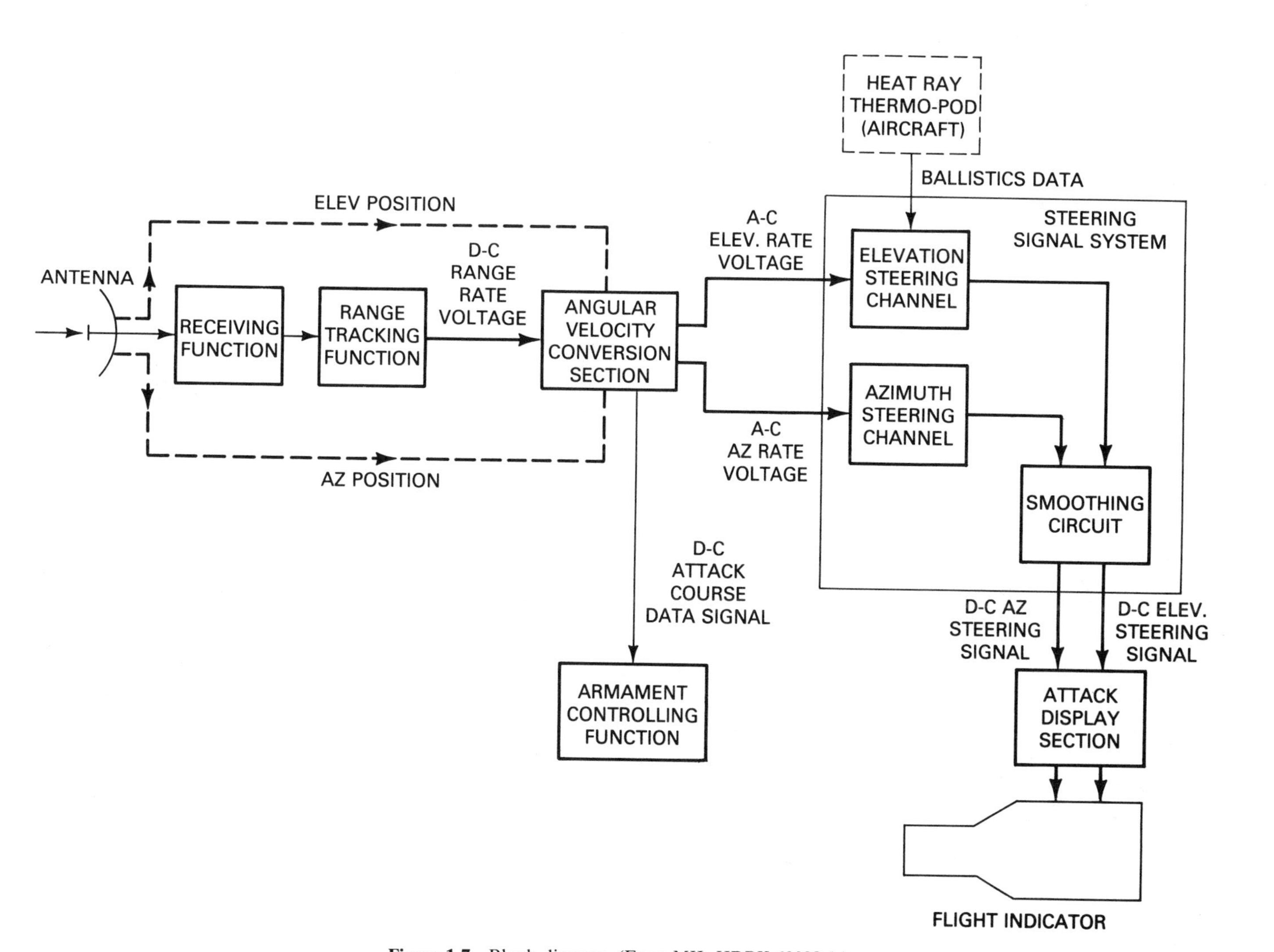

Figure 1.7 Block diagram. (From MIL-HDBK-63038-1.)

to show information or control signal flow, not electron flow. Block diagrams are discussed further in Chapter 3.

Schematic diagrams (see Fig. 1.8) have long been the most popular and useful drawings in electronics because they show all the components and electrical connections within a unit. Graphic symbols in the form of various solid and broken lines, circles, squares, arcs, and combinations thereof show the function of components without regard to their physical shape or size. Reference labels on these components show part values. Lines between the symbols indicate wires that allow electron flow. Expected voltages and currents may be shown at critical points in the circuit represented by the schematic. The arrangement of these symbols on the schematic has nothing to do with the position of the components in the equipment. In some cases, the schematic may be the only type of drawing available to the technician or engineer. To try to repair modern electronic equipment without a schematic would be unthinkable. In Chapter 5 we present a further discussion of schematics.

The *printed wiring master* (Fig. 1.9) shows the exact position and size of printed wiring between components. It is an essential drawing for manufacturing printed circuits but has little value for understanding or repairing a circuit. When combined with a parts placement diagram, as in Fig. 1.10, it becomes a more useful drawing.

The *logic* diagram (Fig. 1.11) is a particular type of schematic that uses logic symbols instead of conventional electronic symbols to show the operation of digital electronic equipment. It is the only practical drawing for digital circuits, as each of its logic symbols may be the equivalent of many conventional electronic components, from a dozen to many thousands. Without these symbols, diagrams for digital circuits would be hopelessly large and complex. In a sense, logic sym-

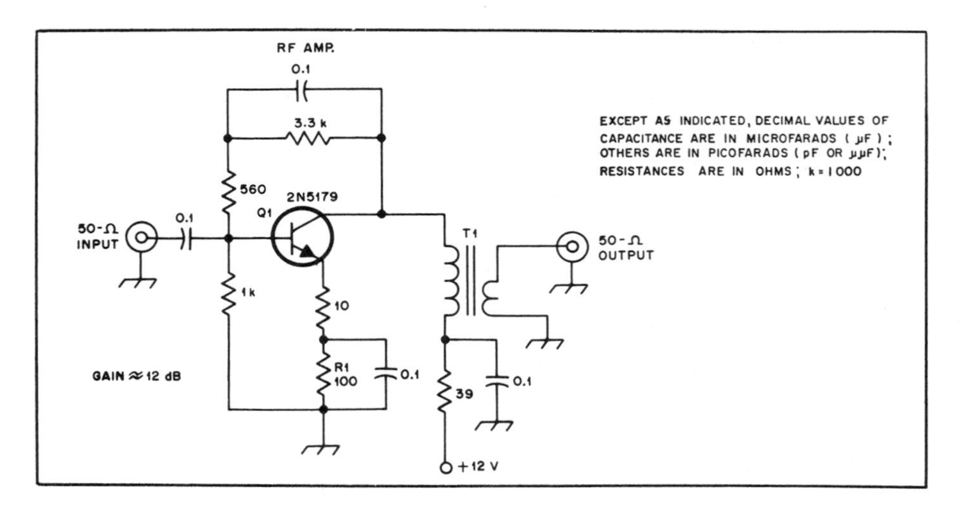

Figure 1.8 Schematic. (Copyright 1984 by QST.)

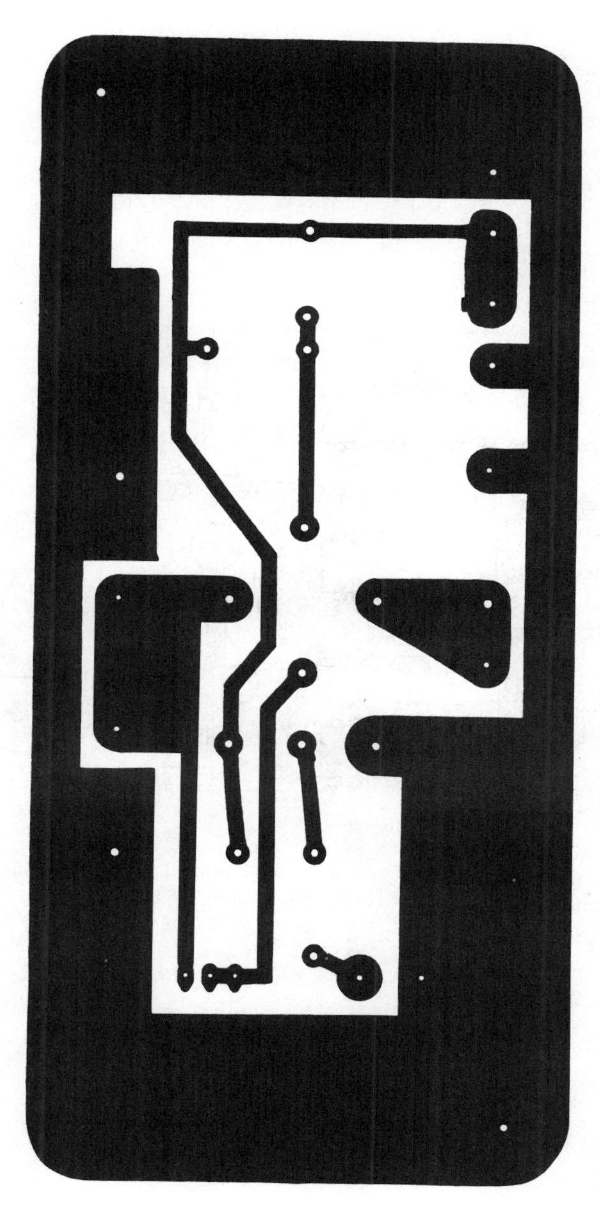

Figure 1.9 Printed-circuit master.

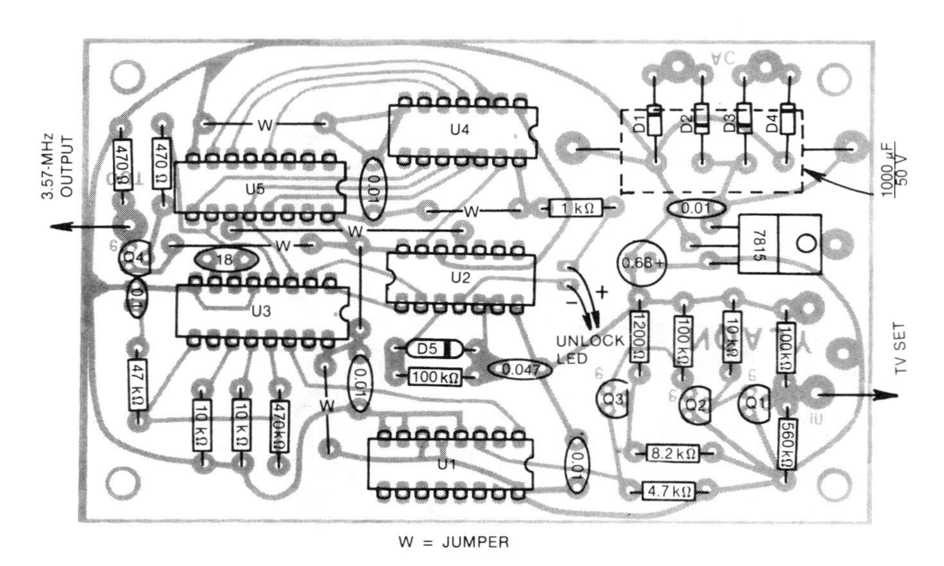

Figure 1.10 Parts-placement diagram combined with printed wiring diagram (gray areas represent unetched copper). (Copyright 1983 by QST.)

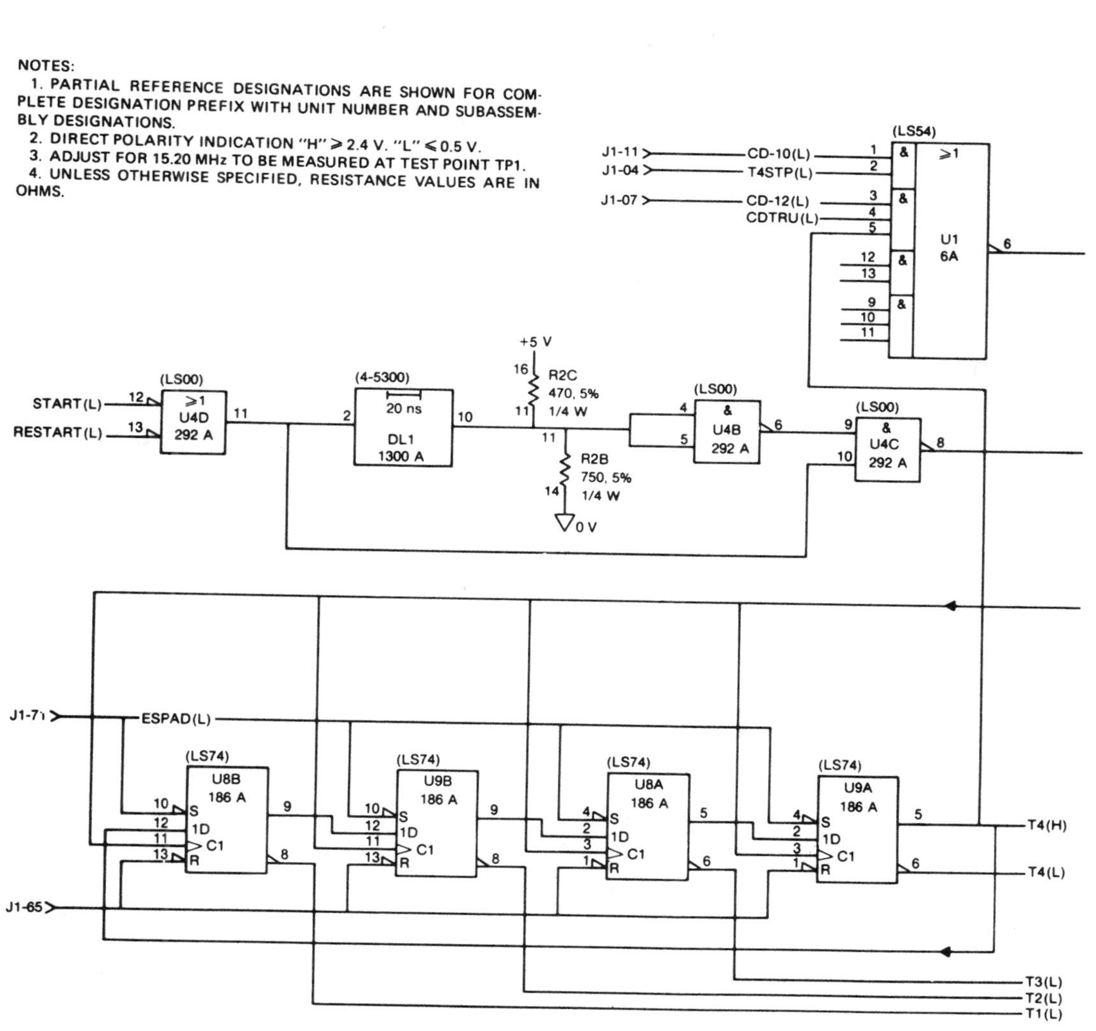

Figure 1.11 Detailed logic diagram. (From ANSI/IEEE Std 991-1986.)

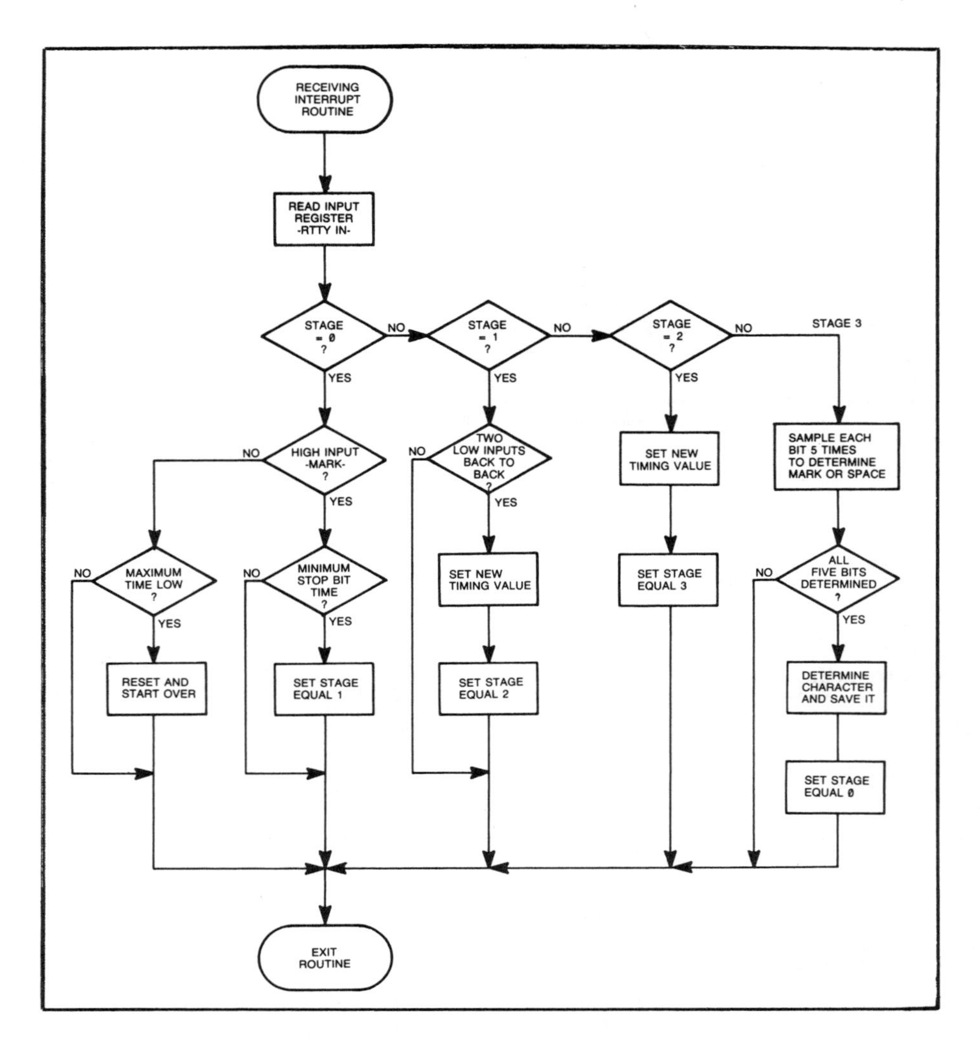

Figure 1.12 Flowchart. (Copyright 1987 by QST.)

bols are ''black boxes'' whose detailed inner workings are irrelevant as long as the user knows what comes out of the ''box'' when certain signals go in. The symbol shows *what* the circuit does, not *how* it does it. Logic diagrams may also show discrete components. Logic symbols are discussed in Chapters 7 and 8; logic diagrams are covered in Chapter 9.

Flowcharts (Fig. 1.12) show the sequence of steps in a microprocessor. The flowchart is a variation of the block diagram. The symbols in the flowchart, however, represent actions, not electronic parts. The arrangement of the symbols indicates the time sequence of the actions that are performed.

Chapter 2

Abbreviations and Acronyms

In reading electronics diagrams, the first step is to interpret the words, letters, abbreviations, and acronyms. Words on diagrams can often be interpreted with the aid of an electronics dictionary or the frequent glossaries that appear in the IEEE's *Spectrum* magazine.

On the other hand, letters, abbreviations, and acronyms may present trouble, as no single dictionary is likely to list all that you will encounter. This chapter will assist you in deciphering abbreviations in any type of electronics diagram: schematic, logic, block, or wiring.

By *abbreviation* we mean a letter or short combination of letters that have an alphabetical similarity to the parent word or phrases. It is used to save space on a diagram and to avoid needlessly spelling out repetitious words and phrases in text. In diagrams, abbreviations are most often all capital letters.

An *acronym* is similar to an abbreviation except that it forms a pronounceable word spelled in all capital letters. If the acronym becomes popular, it loses its identity as an acronym and is typed and treated like a word. For instance, at one time RADAR was an acronym for "radio detecting and ranging." Now it has become a word with its own meaning and is usually spelled as "radar."

A *reference designation* has the appearance (for example, R10, C24, U6) of an abbreviation and may match them in a few cases, but generally a reference designation has no meaning out of its system.

The distinction between abbreviations and acronyms is rapidly being lost so that the two words are being used interchangeably. Because of this, we will use the term "abbreviation" to include abbreviations and acronyms in the following discussion.

2.1 ABBREVIATIONS

Where abbreviations were at one time used mainly for mathematical terms and units of measurement, they are now being used in all fields, not just technical. In the field of electronics alone there are now more than 10,000 abbreviations. "If the trend continues, there will eventually be no plain English at all; one will simply string together the initial letters of the words," laments the author of *Jane's Aerospace Dictionary* [1].

Not only are abbreviations being coined at an amazing rate, they are acquiring more than one meaning, which makes interpretation even more difficult. HVAC, for example, is an abbreviation of "high-voltage alternating current," but it is also the abbreviation for "heating, ventilation, and air-conditioning." Only the context can tell you which meaning of such abbreviations is intended. The profusion of abbreviations has resulted in what some call "technospeak," an incomprehensible language for the nontechnical world [2].

Table 2.1 gives an abbreviated list of common abbreviations and acronyms used in electronics. A more comprehensive list is given in Appendix A. Some of the abbreviations in Appendix A have been invented by electronics companies for use in their advertisements and technical data sheets. In many cases they are not likely to survive, because of their limited usefulness. However, some of them have (or will) become so useful that they have been adopted for industry-wide usage. As we cannot predict which of these abbreviations will pass into common usage, we are including as many of these company abbreviations as we could find.

TABLE 2.1 COMMON ABBREVIATIONS AND ACRONYMS

A	ampere
AC	alternating current
A/D	analog to digital
ADC	analog-to-digital converter
ALU	arithmetic-logic unit
AM	amplitude modulation
AMP	ampere
AND	logic element
AOI	AND-OR-INVERT
B	bit
BCD	binary-coded decimal
BER	bit error rate
BiMOS	bipolar plus MOS
BO	blocking oscillator
BP	bandpass
BPS	bits per second
B/S	bits per second
C	capacitor
CLK	clock
CMOS	complementary metal-oxide semiconductor
CMRR	common-mode rejection ratio
CPU	central processing unit

TABLE 2.1 (continued)

CS	chip select
D/A	digital to analog
DAC	digital-to-analog converter
dB	decibel
DC	direct current
DIP	dual-in-line package
DMM	digital multimeter
DMOS	diffused MOS
DTL	diode–transistor logic
DVM	digital voltmeter
ECL	emitter-coupled logic
EEPROM	electrically erasable programmable read-only memory
EMF	electromotive force
F	farad, frequency
FET	field-effect transistor
FF	flip-flop
FIFO	first in, first out
FM	frequency modulation
FSK	frequency-shift keying
GND	ground
GPIB	general-purpose interface bus
H	henry
HEX	hexadecimal
HP	high pass
Hz	hertz
I	current
IC	integrated circuit
I/O	input/output
JFET	junction field-effect transistor
JK	type of flip-flop
JMOS	junction MOS
L	inductor
LAN	local-area network
LP	low pass
LSB	least-significant bit
LSI	large-scale integration
mA	milliampere
MOS	metal-oxide semiconductor
MOSFET	metal-oxide semiconductor field-effect transistor
MPU	microprocessor unit
MPX	multiplex
MSB	most significant bit
MSI	medium-scale integration
NAND	logic element
NC	no connection
nF	nanofarad
NF	noise figure
opamp	operational amplifier
OR	logic element
OS	one-shot
PCB	printed-circuit board

TABLE 2.1 (continued)

PLL	phase-locked loop
p-p	peak to peak
PROM	programmable read-only memory
Q	quality factor
R	resistor
RAM	random-access memory
RDY	ready
REF	reference
RF	radio frequency
RISC	reduced instruction set computer
rms	root mean square
ROM	read-only memory
SCR	silicon-controlled rectifier
SEL	selector
S/N	signal-to-noise ratio
SPDT	single-pole, double-throw switch
SR	shift register
SSB	single sideband
SW	switch
TTL	transistor–transistor logic
UART	universal asynchronous receiver–transmitter
UHF	ultrahigh frequency
V	volts
V_{CC}	collector power supply voltage
VCO	voltage-controlled oscillator
V_{DD}	drain power supply voltage
V/F	voltage to frequency
VHSIC	very high speed IC
VOM	volt-ohmmeter
WR	write
X	multiplier
XOR	logical element

An additional source of pertinent abbreviations is given in American National Standard ANSI Y1.1-1972, Abbreviations for Use on Drawings and in Text. A second major reference is Military Standard 12D (MIL-STD-12D), Abbreviations for Use on Drawings and in Specifications, Standards and Technical Documents. It lists abbreviations in two ways: by terms and their corresponding abbreviations and by abbreviations and their corresponding terms.

No matter how extensive any list of abbreviations used nationally or internationally is, there will often be locally coined abbreviations. By "local" we mean in any manufacturing or test operation. To learn the meaning of these local abbreviations, start with the pertinent project engineer. If he or she is not available, hunt for the nearest old-timer. From there you can try the publications department or your company library.

TABLE 2.2 ABBREVIATED SEMICONDUCTOR SYMBOL LIST

Field-effect transfer symbols		Bipolar transistor symbols		Bipolar transistor symbols	
A	voltage amplification	C_{ibo}	input capacitance, open circuit (common base)	I_{CEO}	collector-cutoff current, base open
C_c	intrinsic channel capacitance	C_{ieo}	input capacitance, open circuit (common emitter)	I_E	emitter current
C_{ds}	drain-to-source capacitance (includes approximately 1-pF drain-to-case and interlead capacitance)	C_{obo}	output capacitance, open circuit (common base)	MAG	maximum available amplifier gain
C_{gd}	gate-to-drain capacitance (includes 0.1-pF interlead capacitance)	C_{oeo}	output capacitance, open circuit (common emitter)	P_{CE}	total dc or average power input to collector (common emitter)
C_{gs}	gate-to-source interlead and case capacitance	f_c	cutoff frequency	P_{OE}	large-signal output power (common emitter)
C_{iss}	small-signal input capacitance, short circuit	f_T	gain-bandwidth product (frequency at which small-signal forward current-transfer ratio, common emitter, is unity or 1)	R_L	load resistance
C_{rss}	small-signal reverse transfer capacitance, short circuit	g_{me}	small-signal transconductance (common emitter)	R_s	source resistance
g_{fs}	forward transconductance	h_{FB}	static forward-current transfer ratio (common base)	V_{BB}	base-supply voltage
g_{is}	input conductance	h_{fb}	small-signal forward-current transfer ratio, short circuit (common base)	V_{BC}	base-to-collector voltage
g_{os}	output conductance	h_{FE}	static forward-current transfer ratio (common emitter)	V_{BE}	base-to-emitter voltage
I_D	dc drain current	h_{fe}	small-signal forward-current transfer ratio, short circuit (common emitter)	V_{CB}	collector-to-base voltage
$I_{DS(OFF)}$	drain-to-source OFF current	h_{IE}	static input resistance (common emitter)	V_{CBO}	collector-to-base (emitter open)
I_{GSS}	gate leakage current	h_{ie}	small-signal input impedance, short circuit (common emitter)	V_{CC}	collector-supply voltage
r_c	effective gate series resistance	I_b	base current	V_{CE}	collector-to-emitter voltage
$r_{DS(ON)}$	drain-to-source ON resistance	I_c	collector current	V_{CEO}	collector-to-emitter voltage (base open)
r_{gd}	gate-to-drain leakage resistance	I_{CBO}	collector-cutoff current, emitter open	$V_{CE(sat)}$	collector-to-emitter saturation voltage
r_{gs}	gate-to-source leakage resistance			V_{EB}	emitter-to-base voltage
V_{DB}	drain-to-substrate voltage			V_{EBO}	emitter-to-base voltage (collector open)
V_{DS}	drain-to-source voltage			V_{EE}	emitter-supply voltage
V_{GB}	dc gate-to-substrate voltage			Y_{fe}	forward transconductance
V_{GB}	peak gate-to-substrate voltage			Y_{ie}	input admittance
V_{GS}	dc gate-to-source voltage			Y_{oe}	output admittance
V_{GS}	peak gate-to-source voltage				
$V_{GS(OFF)}$	gate-to-source cutoff voltage				
Y_{fs}	forward transadmittance $\approx g_{fs}$				
Y_L	load admittance				
Y_{os}	output admittance				

Source: Copyright 1979 by the American Radio Relay League.

In deciphering abbreviations, note that there is a trend to deleting punctuation, such as periods, from abbreviations. Also, in some cases the same abbreviation will be used for both the singular and plural forms without an "s" to indicate plural form.

If an abbreviation on a drawing is followed by an asterisk (*) or is underlined (__), it should be read as a lowercase letter. For example, A* = A = a. Letter symbols are similar to abbreviations and are defined in Tables 2.2, 2.3, and 2.4 and Appendix B.

Once you figure out what the letters in an abbreviation stand for, of course you may still have a problem: What do the words mean [3]? Although this does happen, we believe that most readers know more words than abbreviations. For example, an electronics worker is likely to learn "digital multimeter" before he or she learns its abbreviation—DMM.

TABLE 2.3 TTL SYMBOLS, TERMS, AND DEFINITIONS

Symbol	Definition
f_{max}	Maximum clock frequency: the highest rate at which the clock input of a bistable circuit can be driven through its required sequence while maintaining stable transitions of logic level at the output with input conditions established that should cause changes of output logic level in accordance with the specification
I_{CC}	Supply current: the current into[a] the V_{CC} supply terminal of an integrated circuit
I_{CCH}	Supply current, outputs high: the current into[a] the V_{CC} supply terminal of an integrated circuit when all (or a specified number) of the outputs are at the high level
I_{CCL}	Supply current, outputs low: the current into[a] the V_{CC} supply terminal of an integrated circuit when all (or a specified number) of the outputs are at the low level
I_{IH}	High-level input current: the current into[a] an input when a high-level voltage is applied to that input
I_{IL}	Low-level input current: the current into[a] an input when a low-level voltage is applied to that input
I_{OH}	High-level output current: the current into[a] an output with input conditions applied that, according to the product specification, will establish a high level at the output
I_{OL}	Low-level output current: the current into[a] an output with input conditions applied that, according to the product specification, will establish a low level at the output
I_{OS}	Short-circuit output current: the current into[a] an output when that output is short-circuited to ground (or other specified potential) with input conditions applied to establish the output logic level farthest from ground potential (or other specified potential)
I_{OZH}	Off-state (high-impedance-state) output current (of a three-state output) with high-level voltage applied: the current flowing into[a] an output having three-state capability with input conditions established that, according to the product specification, will establish the high-impedance state at the output and with a high-level voltage applied to the output *Note:* This parameter is measured with other input conditions established that would cause the output to be at a low level if it were enabled.

TABLE 2.3 (continued)

I_{OZL}	Off-state (high-impedance-state) output current (of a three-state output) with low-level voltage applied: the current flowing into[a] an output having three-state capability with input conditions established that, according to the product specification, will establish the high-impedance state at the output and with a low-level voltage applied to the output *Note:* This parameter is measured with other input conditions established that would cause the output to be at a high level if it were enabled.
V_{IH}	High-level input voltage: an input voltage within the more positive (less negative) of the two ranges of values used to represent the binary variables *Note:* A minimum is specified that is the least-positive value of high-level input voltage for which operation of the logic element within specification limits is guaranteed.
V_{IK}	Input clamp voltage: an input voltage in a region of relatively low differential resistance that serves to limit the input voltage swing
V_{IL}	Low-level input voltage: an input voltage level within the less positive (more negative) of the two ranges of values used to represent the binary variables *Note:* A maximum is specified that is the most-positive value of low-level input voltage for which operation of the logic element within specification limits is guaranteed.
V_{OH}	High-level output voltage: the voltage at an output terminal with input conditions applied that, according to the product specification, will establish a high level at the output
V_{OL}	Low-level output voltage: the voltage at an output terminal with input conditions applied that, according to the product specification, will establish a low level at the output
t_a	Access time: the time interval between the application of a specific input pulse and the availability of valid signals at an output
t_{dis}	Disable time (of a three-state output): the time interval between the specified reference points on the input and output voltage waveforms, with the three-state output changing from either of the defined active levels (high or low) to a high-impedance (off) state ($t_{dis} = t_{PHZ}$ or t_{PLZ})
t_{en}	Enable time (of a three-state output): the time interval between the specified reference points on the input and output voltage waveforms, with the three-state output changing from a high-impedance (off) state to either of the defined active levels (high or low) ($t_{en} = t_{PZH}$ or t_{PZL})
t_h	Hold time: the time interval during which a signal is retained at a specified input terminal after an active transition occurs at another specified input terminal *Notes:* 1. The hold time is the actual time interval between two signal events and is determined by the system in which the digital circuit operates. A minimum value is specified that is the shortest interval for which correct operation of the digital circuit is guaranteed. 2. The hold time may have a negative value in which case the minimum limit defines the longest interval (between the release of the signal and the active transition) for which correct operation of the digital circuit is guaranteed.
t_{pd}	Propagation delay time: the time between the specified reference points on the input and output voltage waveforms with the output changing from one defined level (high or low) to the other defined level ($t_{pd} = t_{PHL}$ or t_{PLH})

TABLE 2.3 (continued)

t_{PHL}	Propagation delay time, high-to-low-level output: the time between the specified reference points on the input and output voltage waveforms with the output changing from the defined high level to the defined low level
t_{PHZ}	Disable time (of a three-state output) from high level: the time interval between the specified reference points on the input and output voltage waveforms with the three-state output changing from the defined high level to a high-impedance (off) state
t_{PLH}	Propagation delay time, low-to-high-level output: the time between the specified reference points on the input and output voltage waveforms with the output changing from the defined low level to the defined high level
t_{PLZ}	Disable time (of a three-state output) from low level: the time interval between the specified reference points on the input and output voltage waveforms with the three-state output changing from the defined low level to a high-impedance (off) state
t_{PZH}	Enable time (of a three-state output) to high level: the time interval between the specified reference points on the input and output voltage waveforms with the three-state output changing from a high-impedance (off) state to the defined high level
t_{PZL}	Enable time (of a three-state output) to low level: the time interval between the specified reference points on the input and output voltage waveforms with the three-state output changing from a high-impedance (off) state to the defined low level
t_{sr}	Sense recovery time: the time interval needed to switch a memory from a write mode to a read mode and to obtain valid data signals at the output
t_{su}	Setup time: the time interval between the application of a signal at a specified input terminal and a subsequent active transition at another specified input terminal *Notes:* 1. The setup time is the actual time interval between two signal events and is determined by the system in which the digital circuit operates. A minimum value is specified that is the shortest interval for which correct operation of the digital circuit is guaranteed. 2. The setup time may have a negative value in which case the minimum limit defines the longest interval (between the active transition and the application of the other signal) for which correct operation of the digital circuit is guaranteed.
t_w	Pulse duration (width): the time interval between specified reference points on the leading and trailing edges of the pulse waveform

Source: Courtesy of Texas Instruments Incorporated.

[a] Current out of a terminal is given as a negative value.

TABLE 2.4 CMOS SYMBOLS AND DEFINITIONS

C_{IN}	input capacitance per unit load
I_{DD}	quiescent device current
I_{OL}	output low (sink) current
I_{OH}	output high (source) current
I_{IN}	input current
$I_{OUT}max$	three-state output leakage current
	propagation delay:
t_{PHL}	outputs going high to low
t_{PLH}	outputs going low to high
	output transition time:
t_{THL}	outputs going high to low
t_{TLH}	outputs going low to high
t_{WL} or t_{WH}	pulse width: set, reset, preset, enable, disable, strobe, clock
f_{CL}	clock input frequency
t_{rCL}, t_{fCL}	clock input rise and fall time
t_{SU}	setup time
t_H	hold time
t_{REM}	removal time: set, reset, preset-enable
	three-state disable delay times:
t_{PHZ}	high level to high impedance
t_{PLZ}	low level to high impedance
t_{PZH}	high impedance to high level
t_{PZL}	high impedance to low level
V_{OL}	low-level output voltage
V_{OH}	high-level output voltage
V_{IL}	input low voltage
V_{IH}	input high voltage

Source: Copyright 1983 by GE/RCA Corporation.

2.2 REFERENCE DESIGNATIONS

A *reference designation* is a unique combination of letters and numbers placed next to a component or subassembly on a schematic or single-line drawing. The usual combination is one or two capital letters followed by one or more digits: for example, R23.

The letter or letters identify a particular type or class of component: for example, S for switch, T for transformer. Table 2.5 gives a list of the most common reference designations.

The number part of the designation is called a *component sequence number*. These numbers are frequently assigned according to some specific pattern. For example, R11, R12, and R13 will be in proximity, either horizontally or vertically, on the diagram.

Within each class, each component on a given schematic will have a unique number. For example, only one resistor will be labeled R1.

TABLE 2.5 REFERENCE DESIGNATIONS

A	separable or repairable assembly or subassembly
	amplifier
	computer
	electronic divider
	electronic function generator
	electronic multiplier
	facsimile set
	integrator
	modulator
	phototransistor (isolator)
	recording unit
	sensor (transducer to electrical power)
	positional servomechanism
	teleprinter
AR	amplifier
AT	attenuator (fixed or variable)
	pad
	terminating resistor
B	blower
	fan
	motor
	resolver
BT	battery
	battery cell
	solar cell
C	capacitor
	capacitor bushing
CB	circuit breaker
CP	connector adapter
	coupling (aperture, loop, or probe)
	junction (coaxial or waveguide)
CR	semiconductor diode
	capacitive diode
	current regulator (semiconductor device)
	backward diode
	photodiode
	semiconductor rectifier
	stabistor
	varactor
	asymmetrical varistor
D	backward diode
	light-emitting diode
	varactor
	varistor, asymmetrical
DC	directional coupler
DL	delay function
	delay line
DS	lamp
	indicator (except meter or thermometer)
	audible signaling device (alarm)

TABLE 2.5 (continued)

	bell
	buzzer
	light-emitting diode
	annunciator
	signal light
DT	gas discharge tube
E	antenna
	lightning arrester
	electrical contact brush
	electrical contact
	ferrite bead rings
	gap (horn, protective, or sphere)
	Hall element
	insulator
	loop antenna
	miscellaneous electrical part
	rotary joint (microwave)
	electrical shield
	optical shield
	electrolytic cell
EQ	equalizer
F	fuse
FL	filter
G	generator
	electronic chopper
	rotating amplifier
	interrupter vibrator
H	hardware (common fasteners, etc.)
HP	hydraulic part
HR	heater
HS	handset
	electrical headset
HT	telephone receiver
HY	hybrid junction (magic tee)
	equalizing network
IC	integrated circuit
J	jack
	disconnecting device (connector, receptacle)
	receptacle (connector, stationary portion)
	waveguide flange (choke)
K	relay
	contactor, electrically operated
	contactor, magnetically operated
L	inductor
	choke coil
	coil (all not classified as transformers)
	radio-frequency coil [radio-frequency choke (RFC)]
	reactor
	electrical solenoid
	winding

TABLE 2.5 (continued)

LS	loudspeaker
	electrical horn
	speaker
M	instrument
	meter
	oscillograph
	oscilloscope
	elapsed-time recorder
MG	motor–generator
	dynamotor
MK	microphone
MP	mechanical interlock
	mechanical part
MT	mode transducer
N	equipment subdivision
P	disconnecting device (connector, plug)
	electrical plug (connector, movable portion)
	waveguide flange (plain)
PS	power supply
	rectifier (complete power supply assembly)
PU	head (with various modifier)
Q	transistor
	semiconductor-controlled rectifier
	semiconductor-controlled switch
	field-effect transistor
	gated switch triac
	thyristor triode
R	resistor
	potentiometer
	rheostat
RE	radio receiver
RT	ballast tube or lamp
	resistance lamp
	current-regulating resistor
	thermal resistor
RV	voltage-sensitive resistor
	symmetrical varistor
S	switch
	disconnecting device (switch)
	interlock, safety, electrical
	electric squib
	contactor (manually, mechanically, or thermally operated)
SQ	explosive squib
SR	slip ring (ring, electrical contact)
T	transformer
	autotransformer
	repeating coil
	taper, coaxial or waveguide
TB	connecting block
	terminal board or strip
TR	radio transmitter

TABLE 2.5 (continued)

U	integrated-circuit package
	microcircuit
	micromodule
	assembly, inseparable or nonrepairable
	phototransistor (isolator)
V	electron tube
	barrier photocell
	blocking layer cell
	phototube, photoelectric cell
	voltage regulator
VR	breakdown diode
	voltage regulator (semiconductor device)
W	bus bar
	cable, cable assembly (with connectors)
	strip-type transmission line
	transmission path
	waveguide
X	fuseholder
	lampholder
	socket
Y	piezoelectric, crystal unit, quartz crystal
	magnetostriction oscillator
Z	tuned cavity
	discontinuity (usually coaxial or waveguide transmission)
	artificial line
	mode suppressor
	network
	phase-changing network
	resonator (tuned cavity)
	phase shifter
	E-H tuner

Reference designations allow a common reference to a part on different drawings: parts lists, wiring diagram, schematic, and so on.

2.3 SIGNAL MNEMONICS

A *signal mnemonic* is an abbreviation of a signal name used on a logic diagram. Typically, it includes the first letter or letters of each word in the name: for example, DB for "data bus," CB for "control bit." Once one mnemonic can be interpreted, it may be easier to define similar words. CONWDRDY, for example, could be used for *CON*TROL *W*OR*D* *R*EA*DY*. Knowing this, you can see how CONWACK could be used for CONTROL WORD ACKNOWLEDGE, even if WORD was abbreviated as WD in the first case but as W in the second case.

The meaning of these mnemonics is of great importance in interpreting a

TABLE 2.6 SAMPLE TABLE OF SIGNAL MNEMONICS

Mnemonic	Description
A0, A1	Address bits. In conjunction with MEMR and MEMW, these bits select the appropriate U26 port to be used.
D0–D7	Bidirectional data bus to 1640 microprocessor. This bus passes data to and from the selected port of U26.
HABR	High Abort Receive. When high, RX data has been aborted.
HABT	High Abort Transmit. When high, TX data has been aborted.
HERR	High Error Receive. When high, an error has been detected in the RX data.
HERT	High Error Transmit. When high, an error has been detected in the TX data.
HFGR	High Flag Receive. When high, a flag has been detected in the incoming RX data.
HFGT	High Flag Transmit. When high, a flag has been detected in the incoming TX data.
HRF	High Receive Flag. True when RX character is a flag.
HTF	High Transmit Flag. True when transmitted character is a flag.
LOPPI	Low Option Board Programmable Peripheral Interface. Active low.
LZI	Low Zero Insert. When low, a 0 is inserted in the data bit stream.
MEMR	Memory Read. When low, the data bus (D0–D7) is routed to the SDLC board via the port specified by A0 and A1 when U26 is enabled.

Source: Copyright 1981 by Hewlett-Packard Company.

logic diagram. Check the technical manual for a table that defines the mnemonics and gives, in some cases, a description of the signal, as in Table 2.6.

Signal names are discussed more fully in Section 9.9. The illustrations there may help you to interpret the mnemonics if you do not have access to a table of mnemonics for your diagram.

2.4 IC PIN NAMES

No standard exists for abbreviations for pin numbers on integrated circuits. Each manufacturer seems to have its own. The following table gives a few of the abbreviations you may see on IC pins:

A	address inputs
CE	chip enable
CLK	clock
CP	clock pulse
CS	chip select
D	data inputs
INT	interrupt
O	data outputs
P	parallel data inputs
Q	flip-flop outputs
R/W	read/write
Vcc	positive power supply

REFERENCES

1. *Jane's Aerospace Dictionary*, Jane's Publishing Co. Ltd., London.
2. "The Dangers of 'Technospeak,' " *Telephony*, December 22, 1986, p. 27.
3. Chuck Cozette, "The ABCs of AD—Acronym Dropping," *Infosystems*, August 1985, p. 38.

Chapter 3

Block Diagrams

One of the most common and elementary drawings in electronics is the *block diagram,* which is a relatively simple line drawing (Fig. 3.1) showing how the basic elements of an electronic system are connected in a functional relationship. Depending on the purposes of the drawing and the complexity of the system it represents, the symbols may represent simple functional stages (such as a single transistor used as a preamplifier in a stereo receiver) up through complex racks of equipment. In a sense, a block diagram is a greatly simplified schematic diagram. It may be confined to one $8\frac{1}{2}$- by 11-inch page or may comprise foldout sheets a yard long.

The most common or predominant symbol on a block diagram is a rectangle or square, but it is possible to have a block diagram that has neither squares nor rectangles. Other symbols, such as circles, triangles, and conventional graphic symbols (instead of blocks), are used along with the blocks, as shown in Fig. 3.2. The graphic symbols include switches, connectors, resistors, capacitors, loudspeaker, earphones, current generators, batteries, meters, and others.

Flow lines connecting these symbols show how the signal or data flows between stages or equipment. Control points may be shown. Input and output lines may be so labeled.

Thus a block diagram gives a quick overview to any system; it lets you understand its operation more easily and provides a handy way to diagnose and isolate equipment faults. It also aids one in understanding the theory of operation of complex equipment before proceeding to an examination of detailed schematics.

In the following paragraphs we explain how to interpret block diagrams.

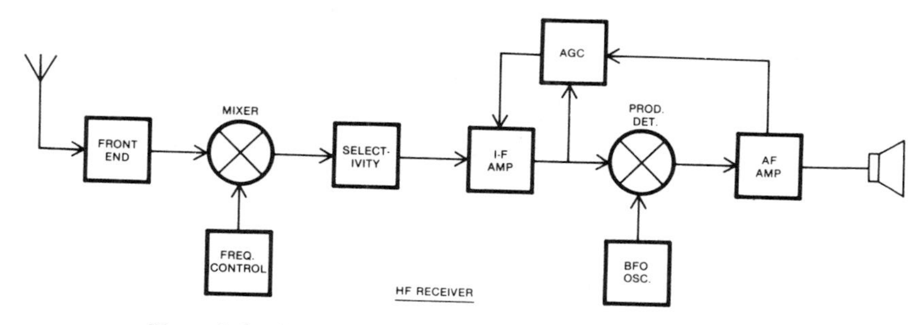

Figure 3.1 Simple block diagram. (Copyright 1983 by QST.)

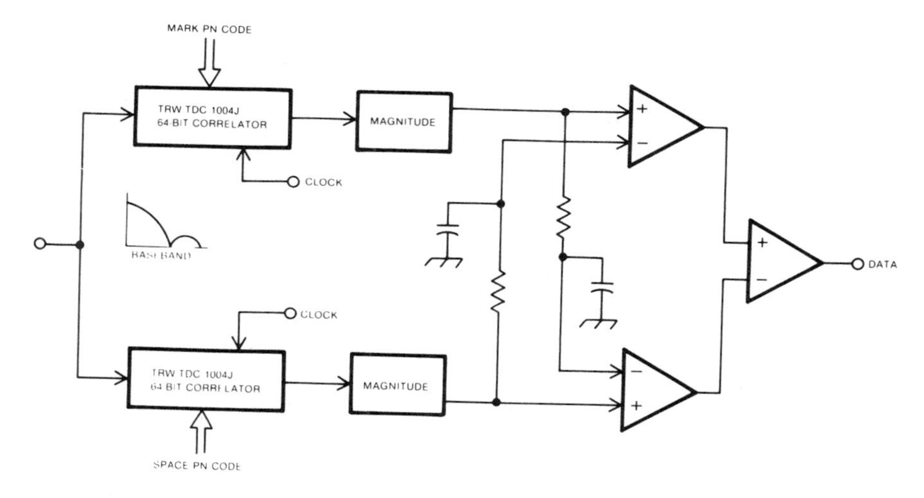

Figure 3.2 Block diagram with conventional electronic symbols. (Copyright 1983 by QST.)

3.1 SYMBOLS

Rectangle symbols are most common on block diagrams, but square symbols can be used for the same function. The proportion of the symbols is irrelevant—it is strictly up to the drafter. The orientation of the symbols may follow some convention, but it is not critical for the diagram if they do not.

The blocks are usually all the same size, but some may be larger either to indicate greater importance or to allow flow lines to be drawn with fewer bends when a block has several inputs or outputs. Some of the symbols will carry word or abbreviation identifiers; if their meaning is not obvious, refer back to Chapter 2. Symbols that lack identification can be interpreted by their unusual shapes, as shown in Fig. 3.3. A complete list of symbols is included in Chapter 4.

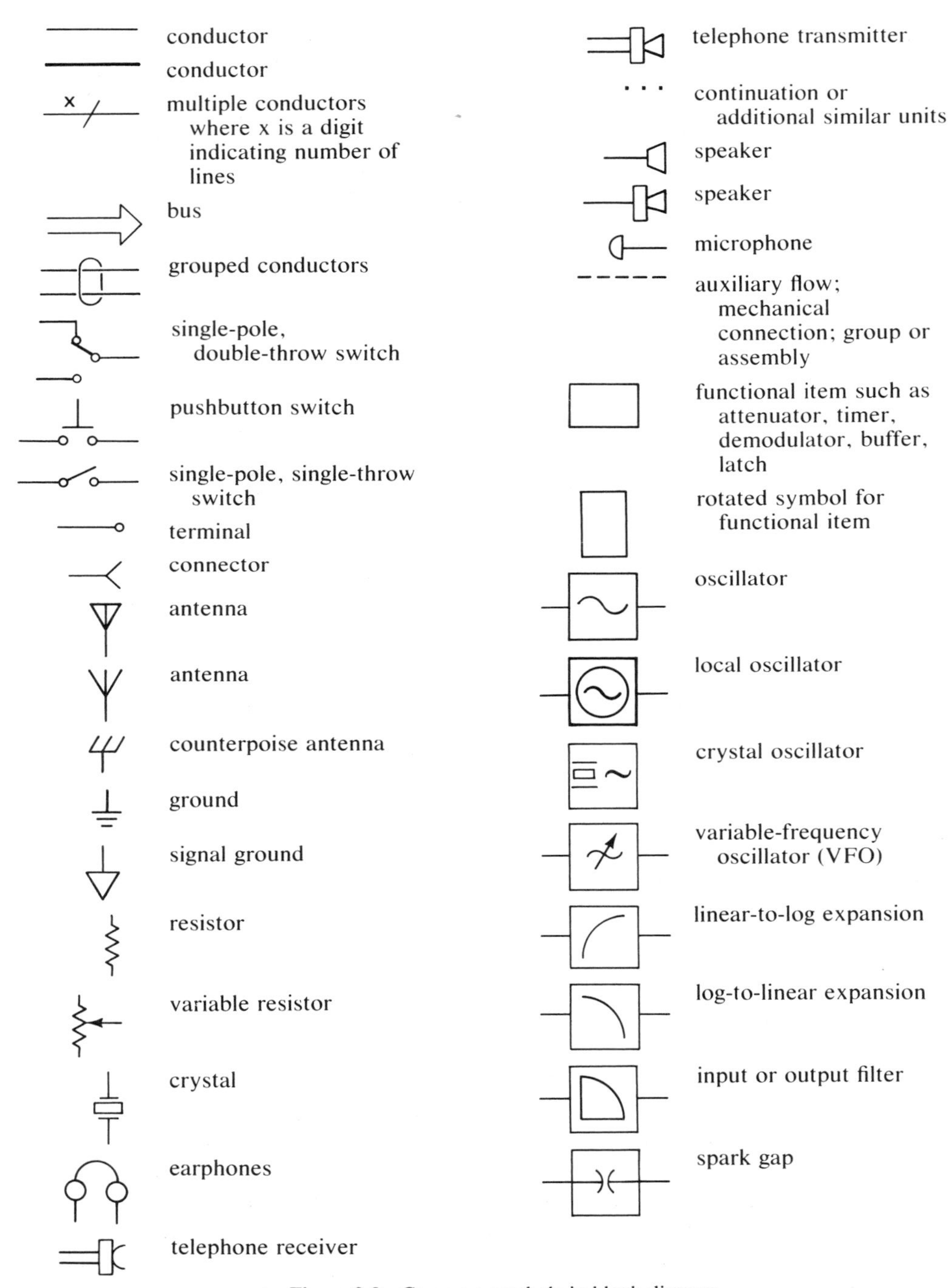

Figure 3.3 Common symbols in block diagram.

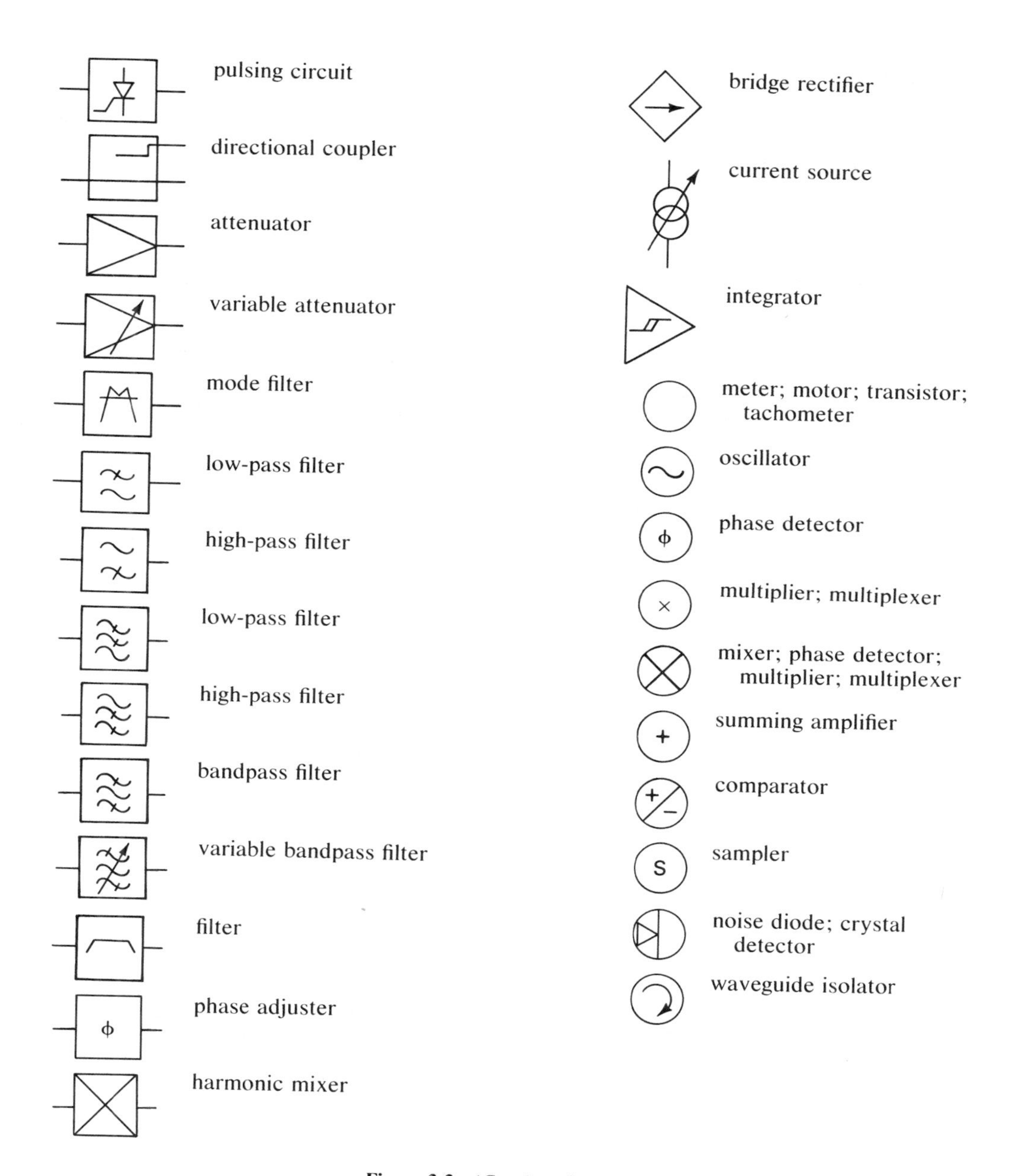

Figure 3.3 (*Continued*)

3.2 FLOW LINES

Flow lines are vertical, horizontal, or slanted lines that show the path of various signals. On any block diagram there will be only one main flow line but sometimes several minor lines.

Arrowheads on flow lines indicate the direction of information flow, not necessarily the direction of current flow. If the arrowheads are missing (as in Fig. 3.4), the flow may be considered to be from left to right, or from top to bottom. The arrowheads may be in the middle of the flow lines, as shown in Fig. 3.5, or at the end of the lines, as shown in Fig. 3.6. They may be open (→) or solid (—▶).

The flow lines are generally solid but may be broken, as shown in Fig. 3.7, to indicate an auxiliary circuit. Very large flow lines (Fig. 3.8) indicate buses. Bold lines, as in Fig. 3.9, may be used to emphasize the most important signal flow. A combined bold and special broken flow line is used to indicate a feedback path as

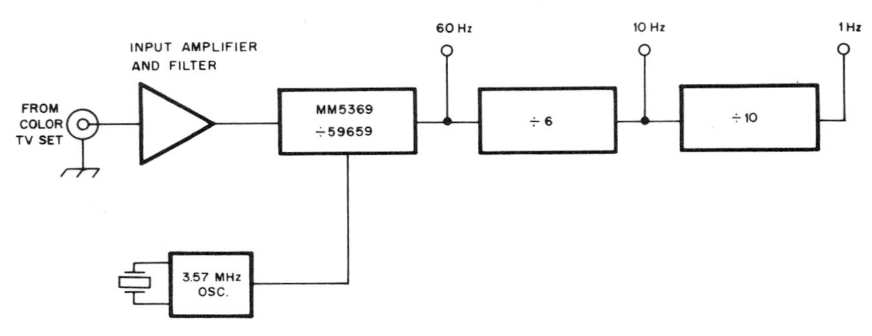

Figure 3.4 Flow lines without arrows. (Copyright 1983 by QST.)

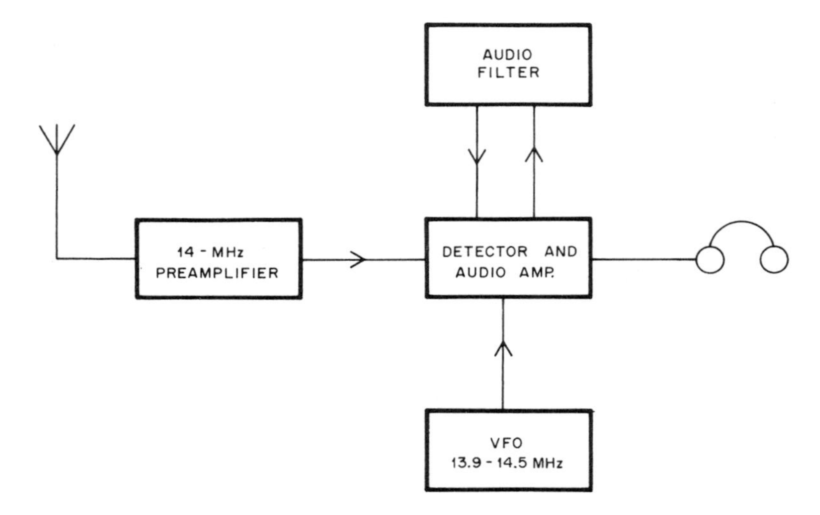

Figure 3.5 Flow lines with arrows in middle of line. (Copyright 1983 by QST.)

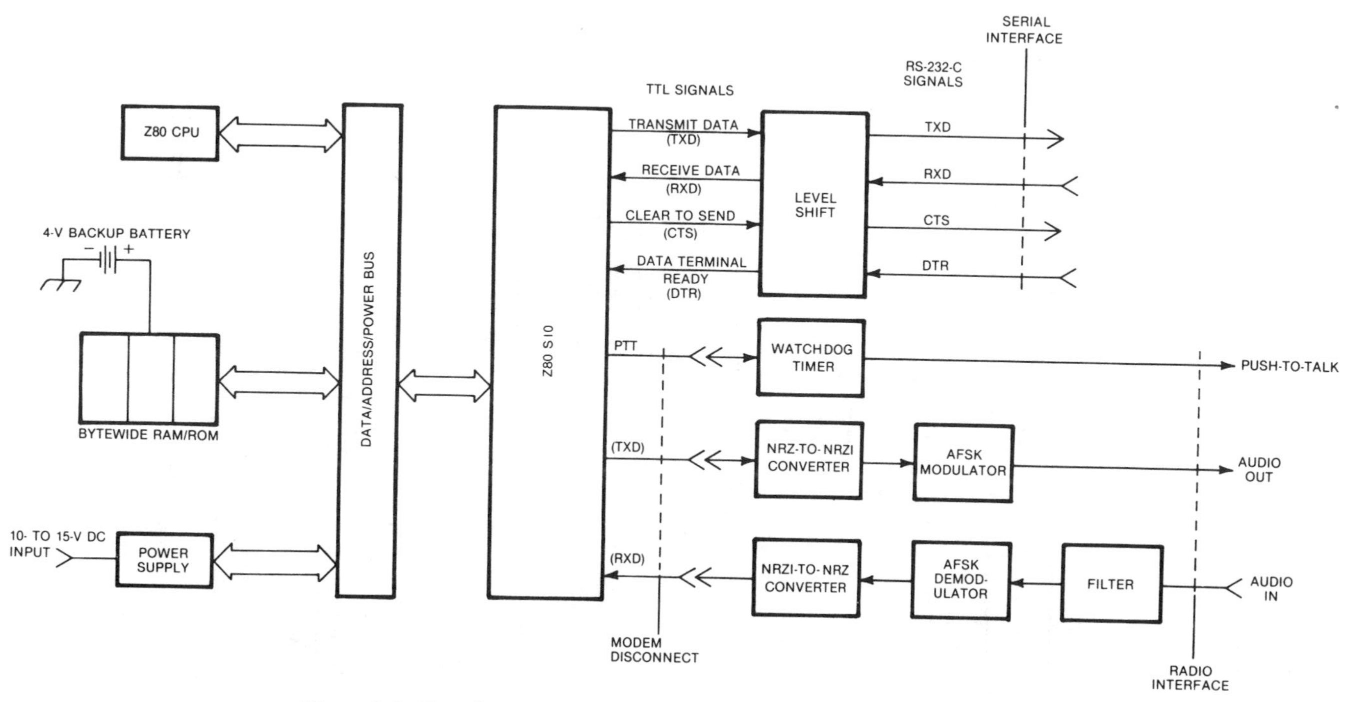

Figure 3.6 Flow lines with arrows at ends of lines. (Copyright 1985 by QST.)

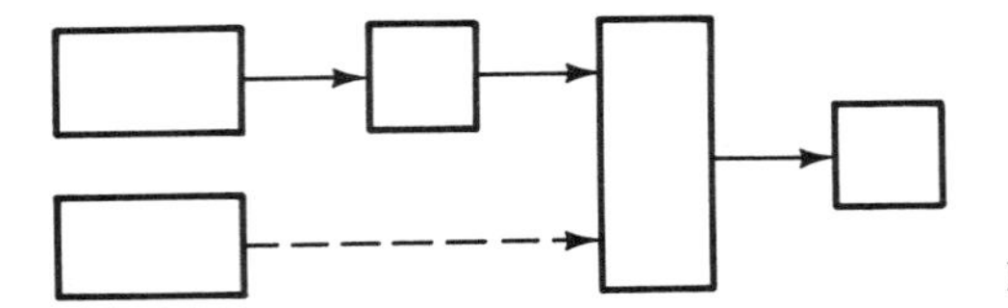

Figure 3.7 Broken flow line.

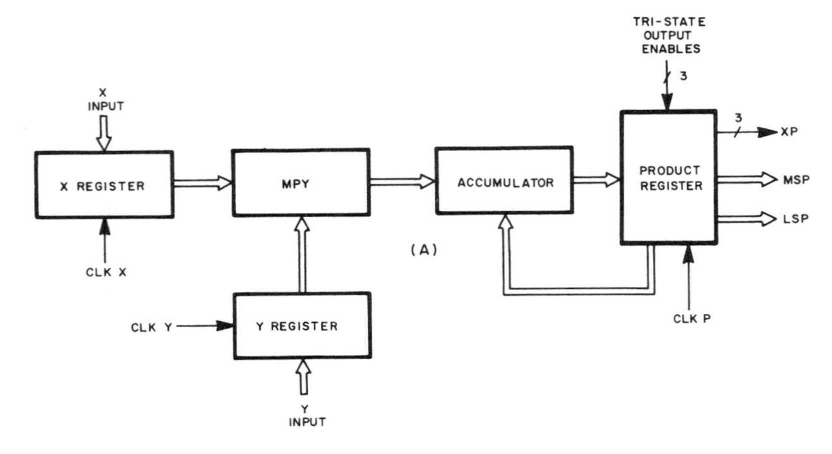

Figure 3.8 Large flow lines. (Copyright 1984 by QST.)

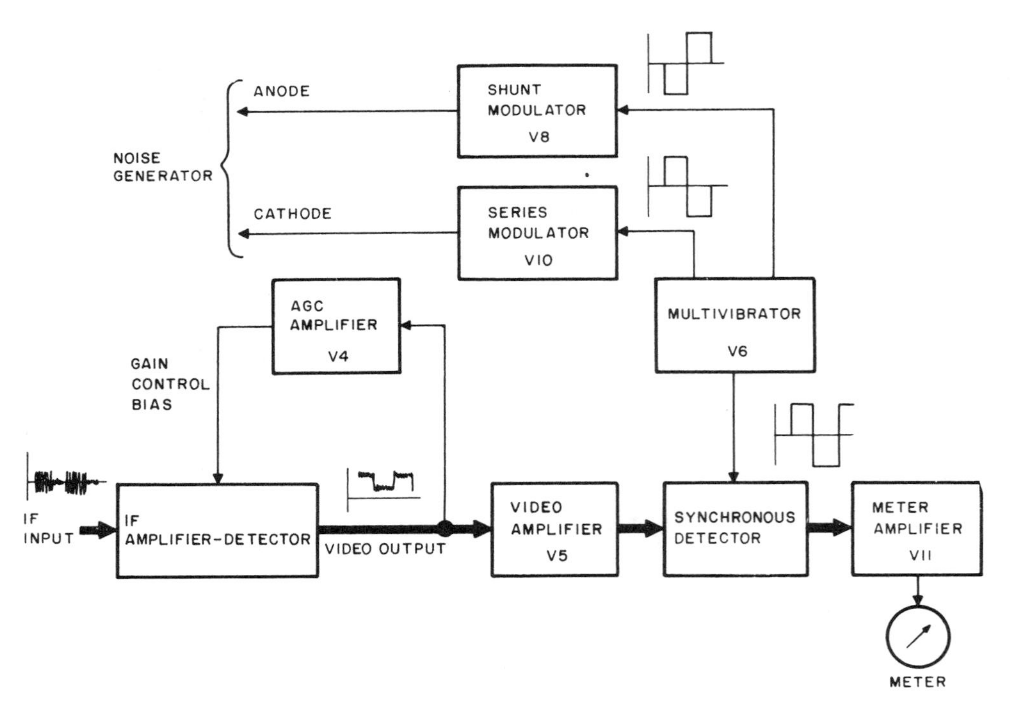

Figure 3.9 Bold flow lines. (Courtesy of Eaton Corporation.)

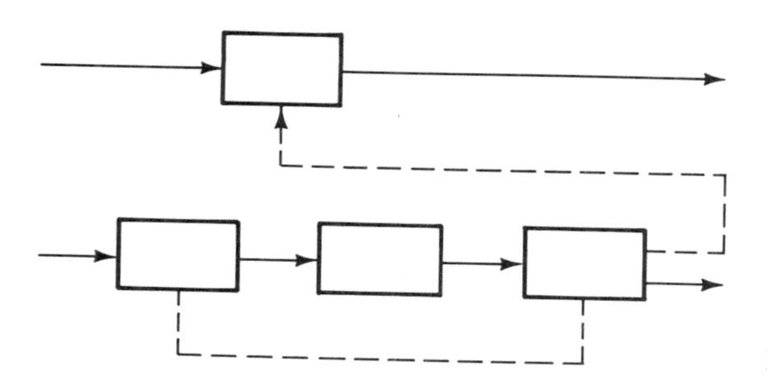

Figure 3.10 Feedback lines.

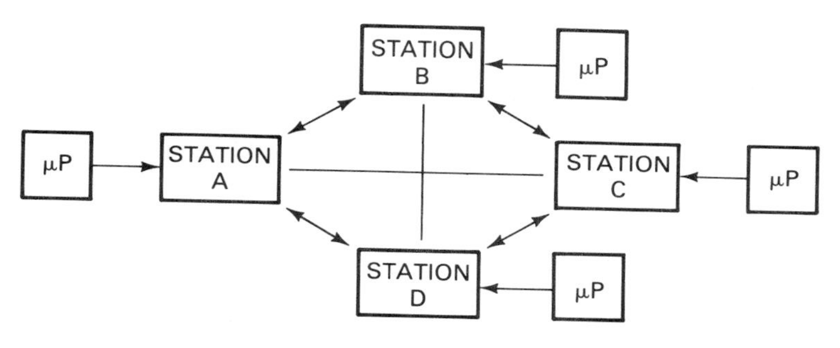

Figure 3.11 Oblique flow lines. (Copyright 1983 by QST.)

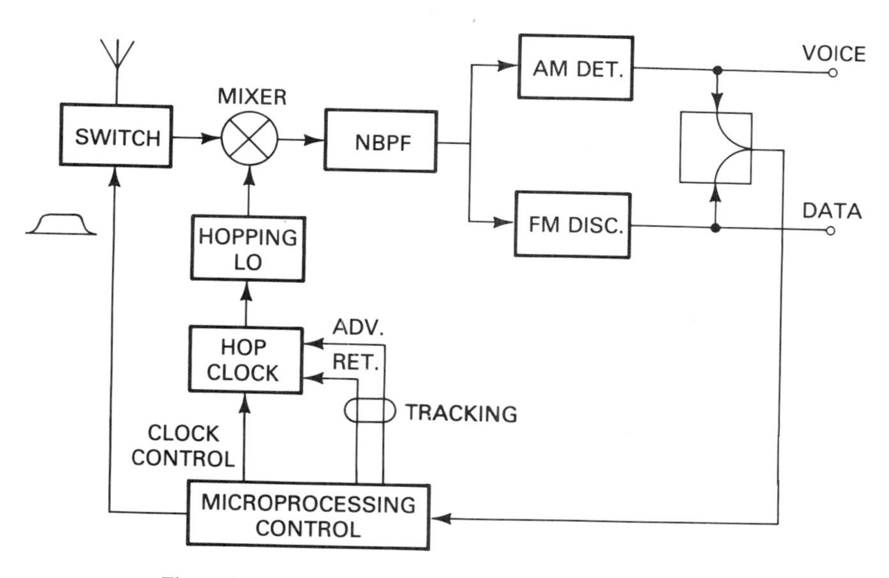

Figure 3.12 Circled flow lines. (Copyright 1983 by QST.)

in Fig. 3.10. If bold lines are used, the medium-weight lines may indicate control circuits. Notice in Fig. 3.11 that flow lines can be oblique instead of the normal horizontal and vertical. Flow lines may be grouped by a circle as indicated in Fig. 3.12. To reduce clutter on block diagrams, multiple flow lines may be combined into one flow line, as we see in Fig. 3.13. Callouts (see Fig. 3.14) on flow lines identify the data on the flow lines; take care that you do not mistake them for flow lines. Waveforms next to flow lines, as in Fig. 3.15, help to explain the signals flowing on the flow lines.

Just as in schematics and logic diagrams, flow lines may cross over each other or may be connected as in Fig. 3.16.

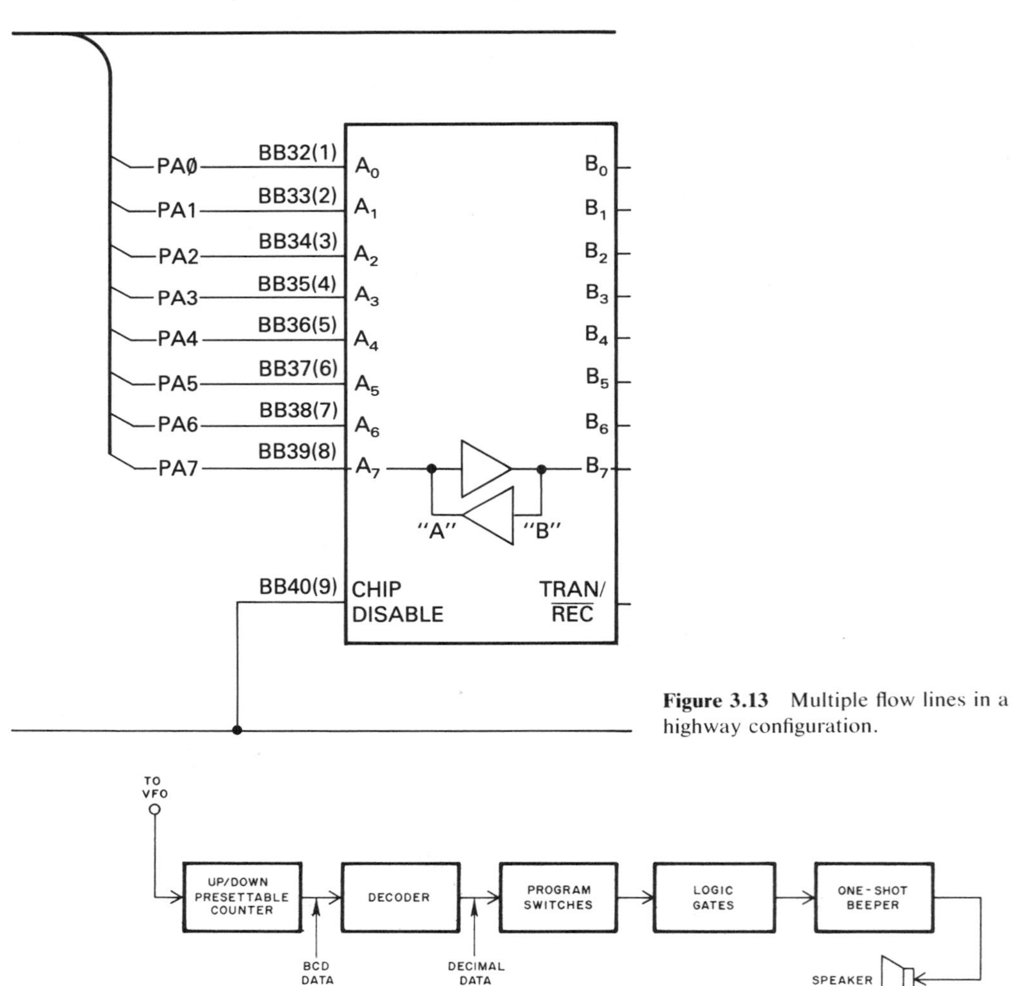

Figure 3.13 Multiple flow lines in a highway configuration.

Figure 3.14 Callouts for flow lines. (Copyright 1983 by QST.)

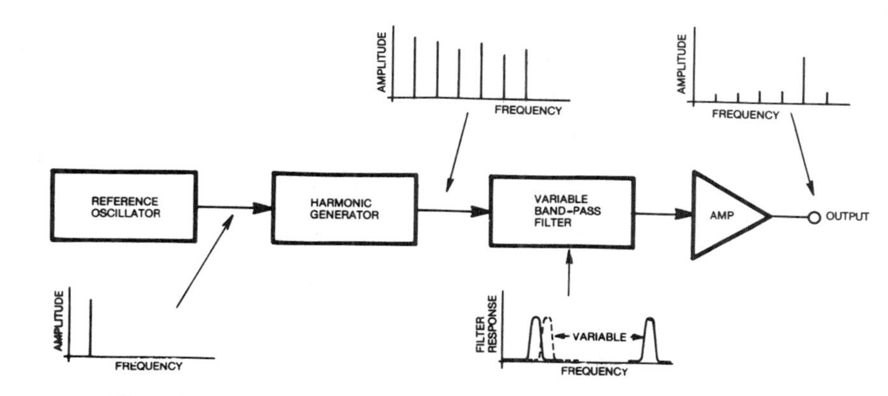

Figure 3.15 Illustration of data on flow lines. (Copyright 1984 by QST.)

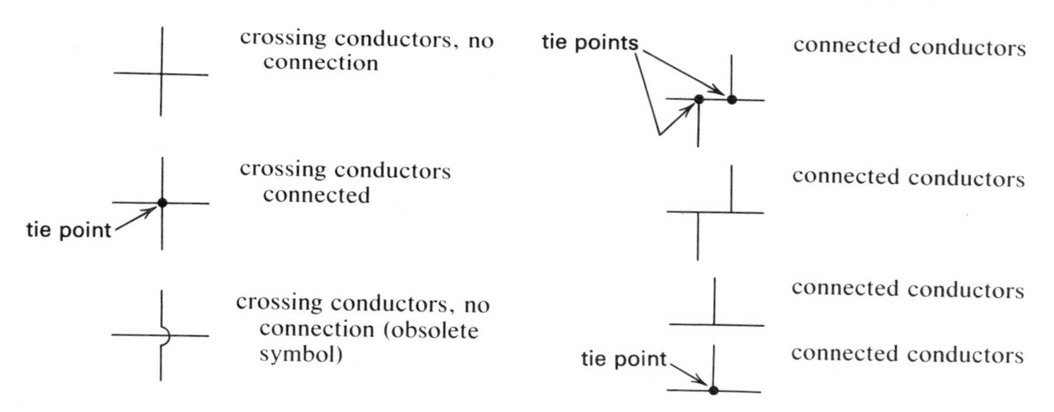

Figure 3.16 Flow path line crossovers and connections.

3.3 ARRANGEMENT OF SYMBOLS

Blocks may be shown within larger solid-line blocks (Fig. 3.17) or within broken-line blocks (Fig. 3.18) to indicate a functional relationship.

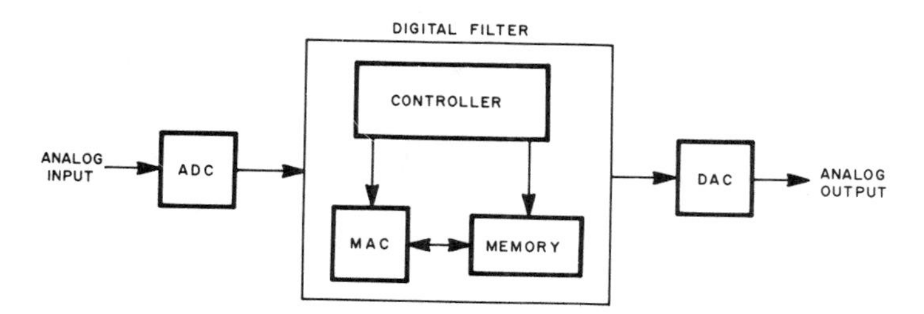

Figure 3.17 Blocks within solid-line blocks. (Copyright 1984 by QST.)

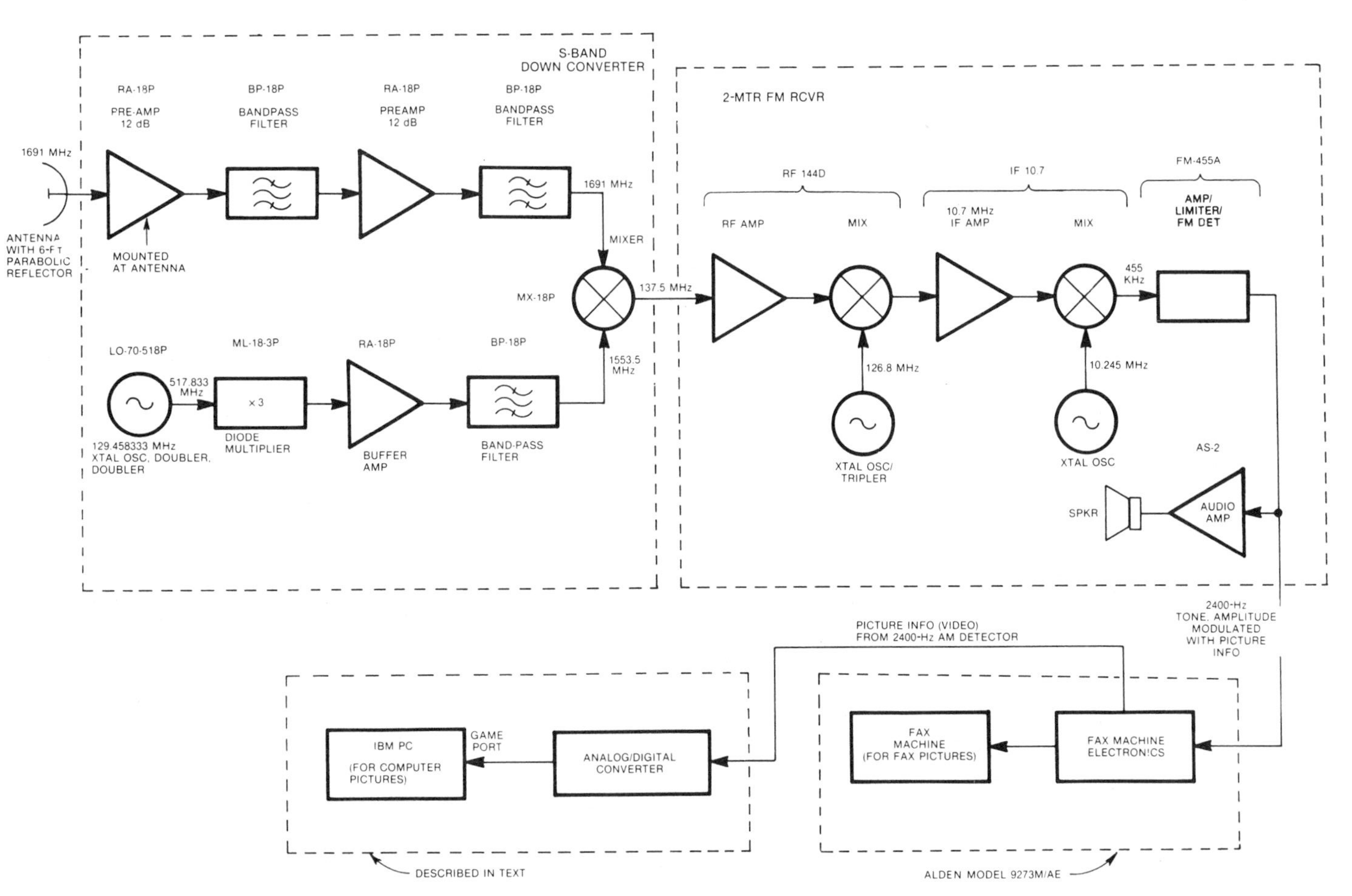

Figure 3.18 Blocks within broken-line blocks. (Copyright 1985 by QST.)

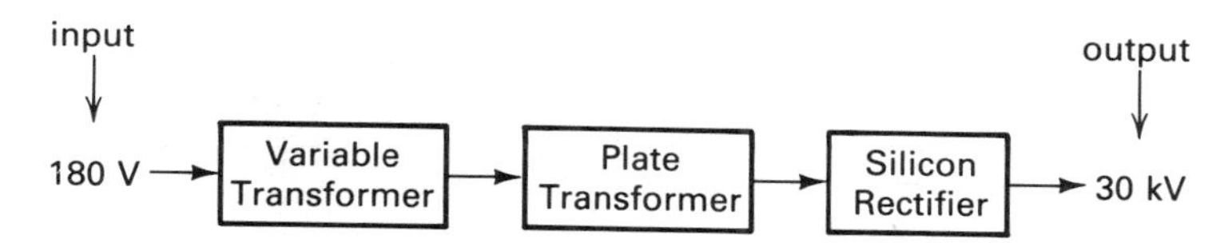

Figure 3.19 Horizontal data flow.

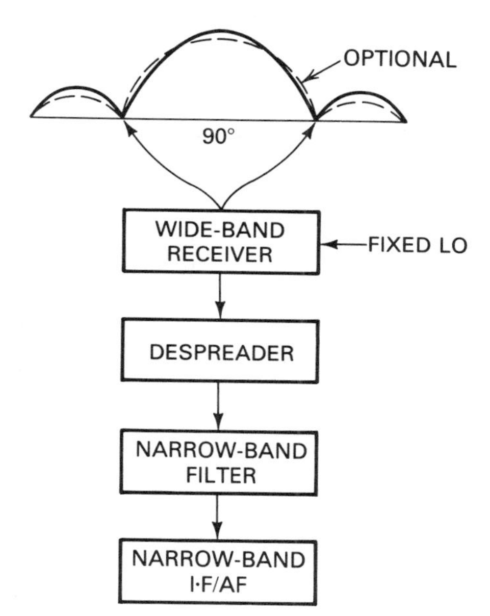

Figure 3.20 Top-to-bottom data flow. (Copyright 1983 by QST.)

Blocks and other symbols on block diagrams are *generally* drawn in a horizontal line (see Fig. 3.19) so that the data or signal flows from left to right. The flow may be vertical, however, as shown in Fig. 3.20, with data flowing from top to bottom. The flow may also be in a combined arrangement, such as in Fig. 3.21, where it is not simply from left to right.

If the diagram is complicated, there may be several horizontal layers of blocks; in such cases, the data flow will usually be left to right on the top layer, then down to the next layer, where again the data will usually flow from left to right.

3.4 INTERPRETATION

Some obvious steps must be taken in interpreting block diagrams. With the aid of Chapter 2, determine the meaning of all abbreviations. Using Fig. 3.3, make sure that you know the meaning of all the symbols. After determining what is there,

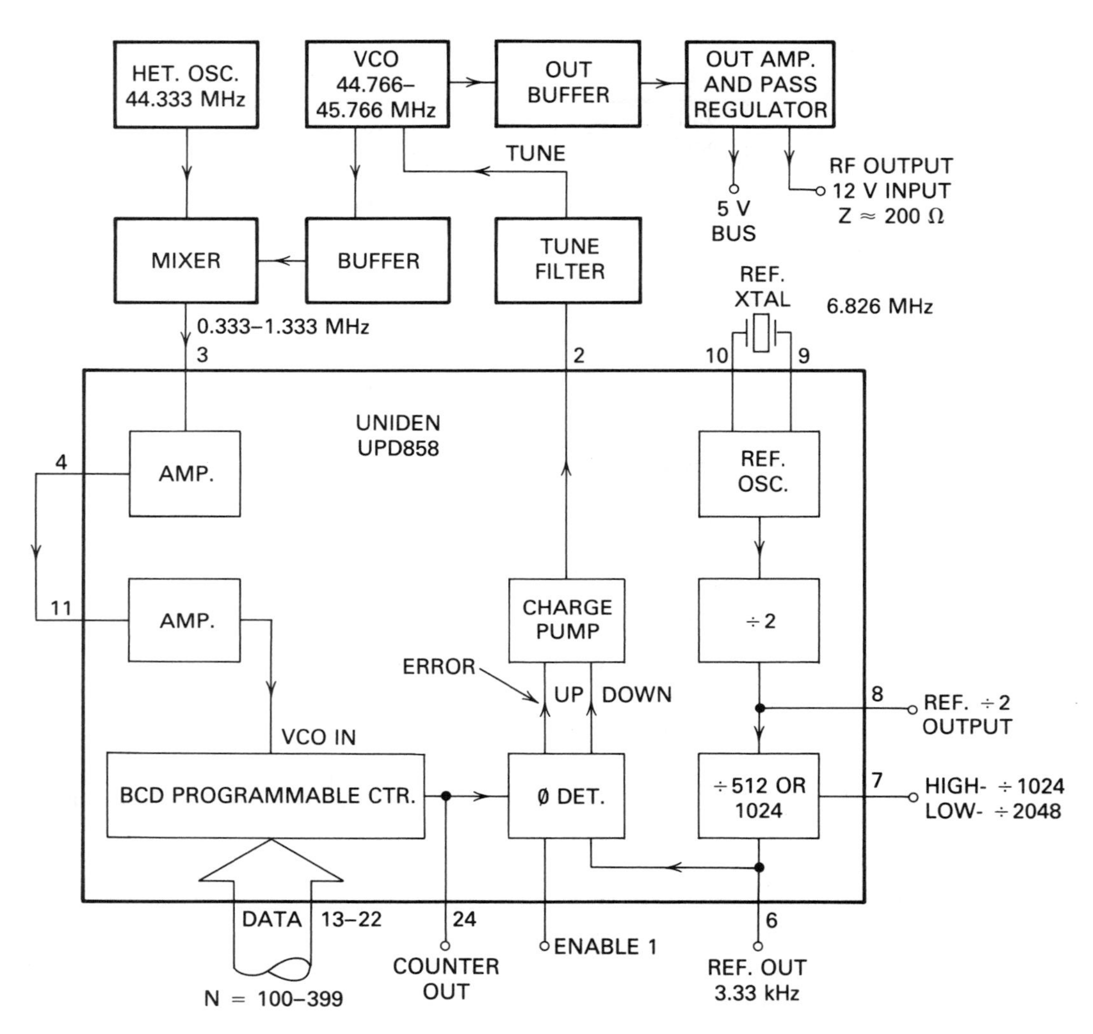

Figure 3.21 Top-to-bottom, left-to-right, and right-to-left data flow. (Copyright 1978 by QST.)

check yourself to see if you can describe the basic operation of each item; if not, refer to an electronics dictionary or textbook.

Power circuits or connections are not likely to be shown; if not, you can safely assume their existence. As you start at the input (upper left of most drawings), trace the signal flow from the input to the output (usually on the right side of the drawing), observing what each block does. If a block has larger letters than some of the other blocks, it may indicate that the stage, unit, or equipment represented by that block has greater importance or significance.

Chapter 4

Electrical and Electronic Symbols

This chapter includes most of the electrical and electronic symbols you will encounter on schematics and wiring diagrams. Logic symbols, however, are located in Chapters 7 and 8. Note that the symbols in this chapter are not listed in alphabetical order. If you think you know the name of a symbol, you can find it easier by first locating it in the index and then referring to the page indicated.

The symbols are arranged by the following types:

4.1 Straight and curved lines
4.2 Jagged lines
4.3 Coiled lines
4.4 Divided or broken lines
4.5 Squares and rectangles
4.6 Triangles
4.7 Circles
4.8 Semiconductor devices
4.9 Broken straight lines and straight lines with attached small circles and arrows
4.10 Miscellaneous shapes

For the reader who is not intimately familiar with all the symbols, this arrangement allows one to find a symbol and its word description on the basis of its shape.

Within each section, the symbols have been arranged, as much as possible, in order from simple to complex. That is, one-line symbols appear before two-line symbols. Symbols without letters within or outside the symbol come before symbols labeled with letters.

Except as noted, these symbols are ANSI/IEEE symbols as derived from American National Standard Y32.2-1975, Graphic Symbols for Electrical and Electronics Diagrams, and ANSI/IEEE Std 318A-1986, Supplement to Graphic Symbols for Electrical and Electronics Diagrams. (These standards should be consulted for further details and for symbols not included here.) Because of the desire of some publishers to have their own distinctive styles, you may encounter symbols that are not shown here. In such cases, see if the unknown symbol approximates any that are shown here, or look for a reference designation or parts list that will clarify the symbol.

Symbols labeled "SLD" will be encountered only on single-line diagrams, the diagrams used most often for microwave and coaxial circuits. They will not be encountered on most conventional diagrams.

Note that the symbols may be oriented in any direction, and that they may be drawn in various sizes. Remember that a *graphic symbol* is "a figure, mark, or character conventionally used on a diagram, document, or other display to represent an item or a concept" (ANSI/IEEE Std 991-1986).

4.1 STRAIGHT AND CURVED LINE SYMBOLS

(Transmission paths, fuses, circuit breakers, conductors)

electric or magnetic shielding;

underground cable or line

mechanical connection

associated or future

mechanical connection with fulcrum

optical shielding

wire; conductive path; conductor

x — isolated dc path in coax or waveguide

minor flow path

major flow path

bus bar with connection

F — telephone line

DIEL — dielectric path other than air for coaxial or waveguide transmission

underground cable or line

underwater line

loaded line

delay

one-way flow of power, signal, or information, to the right

one-way flow of power, signal, or information, to the right

n — one-way flow of power, signal, or information, to the right, where n is a waveform, frequency, or frequency range n one-way (to the right) direction of flow of power, signal, or information where n is a waveform, frequency, or frequency range

n one-way (to the right) direction of flow of power, signal, or information where n is a waveform, frequency, or frequency range

direction of flow of power, signal, or information, either way but not simultaneously

direction of flow of power, signal, or information, either way but not simultaneously

direction of flow of power, signal, or information, both ways, simultaneously

direction of flow of power, signal, or information, both ways, simultaneously

conductor or cable end, not connected

conductor or cable with specially insulated end

open circuit on coax or waveguide; coupling by probe to space (on VHF, UHF, SHF circuits)

short circuit on coax or waveguide circuit

movable short circuit on coax or waveguide circuit

junction of conductors

SPLICE splice

shielded single conductor

circular waveguide; overhead line

coaxial cable; ¦ indicates connection to outer shield

coaxial cable—outer shield extends only to the left

shielded single conductor

Goubau line

coaxial waveguide

rectangular waveguide

dielectric-filled metallic rectangular waveguide

solid-dielectric rectangular waveguide

gas-filled rectangular waveguide

optical fiber

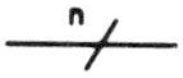

n multiple conductors where n is a number representing the number of conductors

two conductors

three conductors

two conductors

bus bar with connection

unbalanced stripline

unbalanced stripline

three conductors

ground (see also logic ground symbol in Section 4.7)

balanced stripline

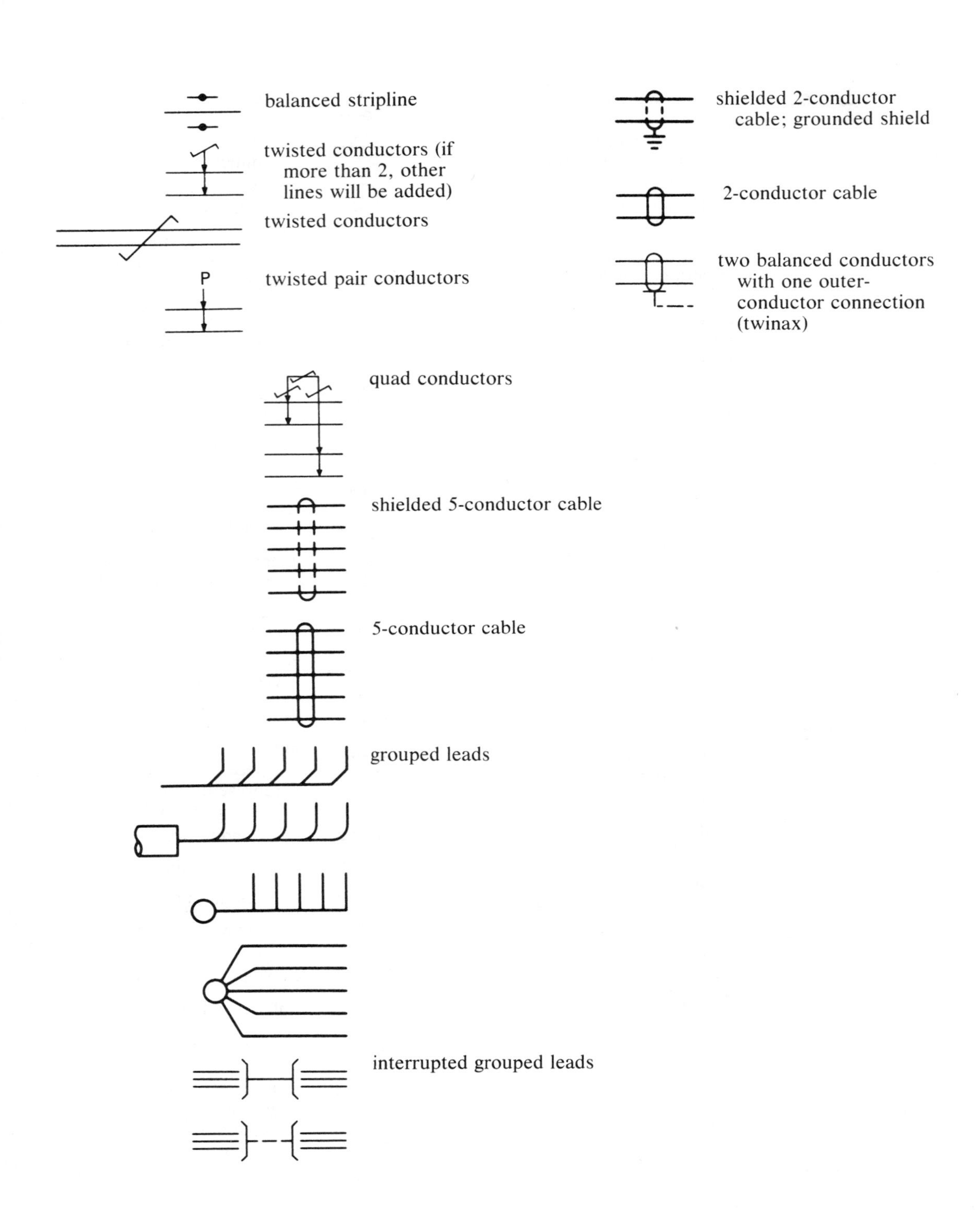
balanced stripline
twisted conductors (if more than 2, other lines will be added)
twisted conductors
P
twisted pair conductors
shielded 2-conductor cable; grounded shield
2-conductor cable
two balanced conductors with one outer-conductor connection (twinax)
quad conductors
shielded 5-conductor cable
5-conductor cable
grouped leads
interrupted grouped leads

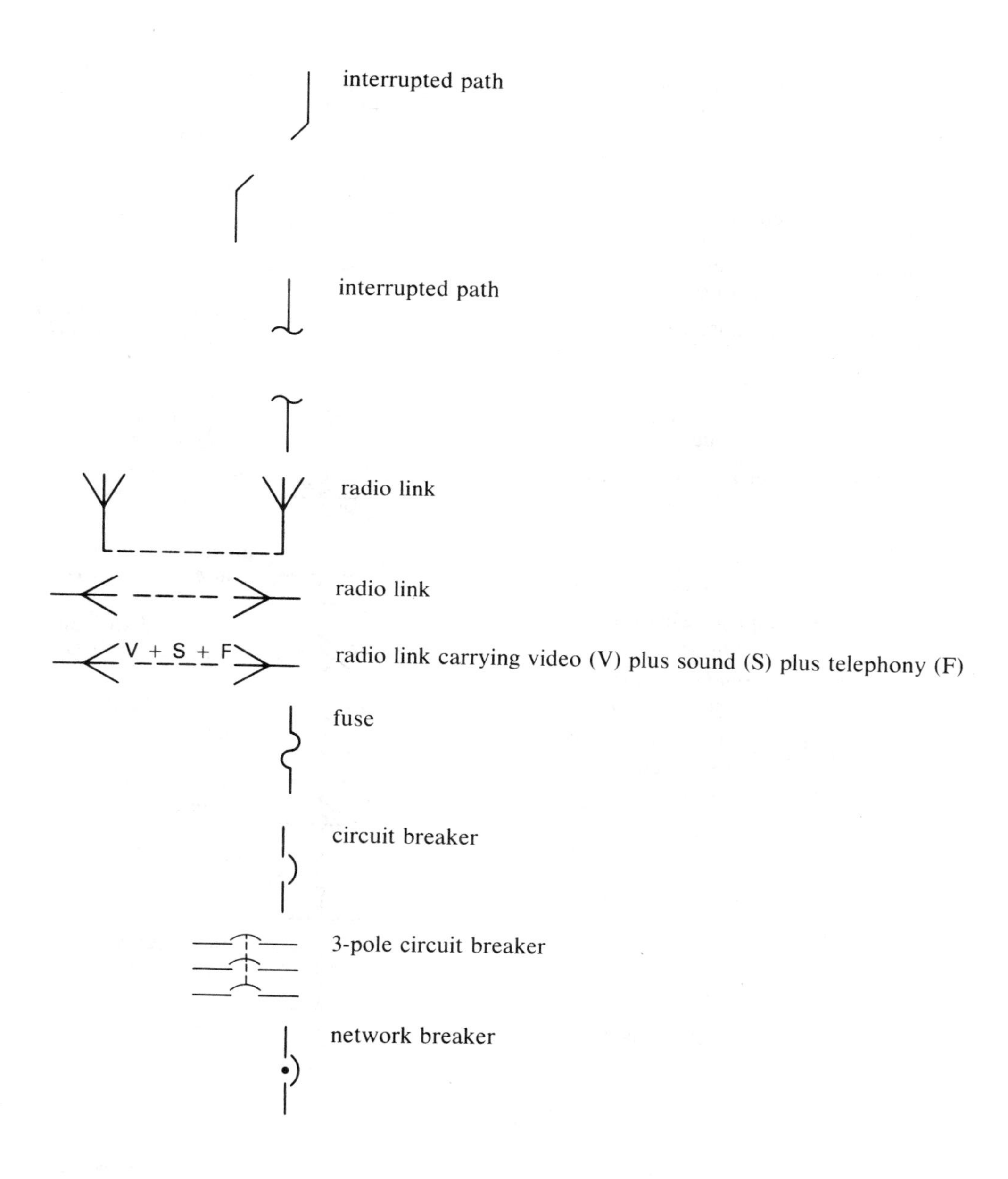

4.2 JAGGED-LINE SYMBOLS

(Resistors, resistive terminations, thermistors, photoconductive transducer, attenuators, thermopile, igniter, air or space transmission path, sawtooth voltage)

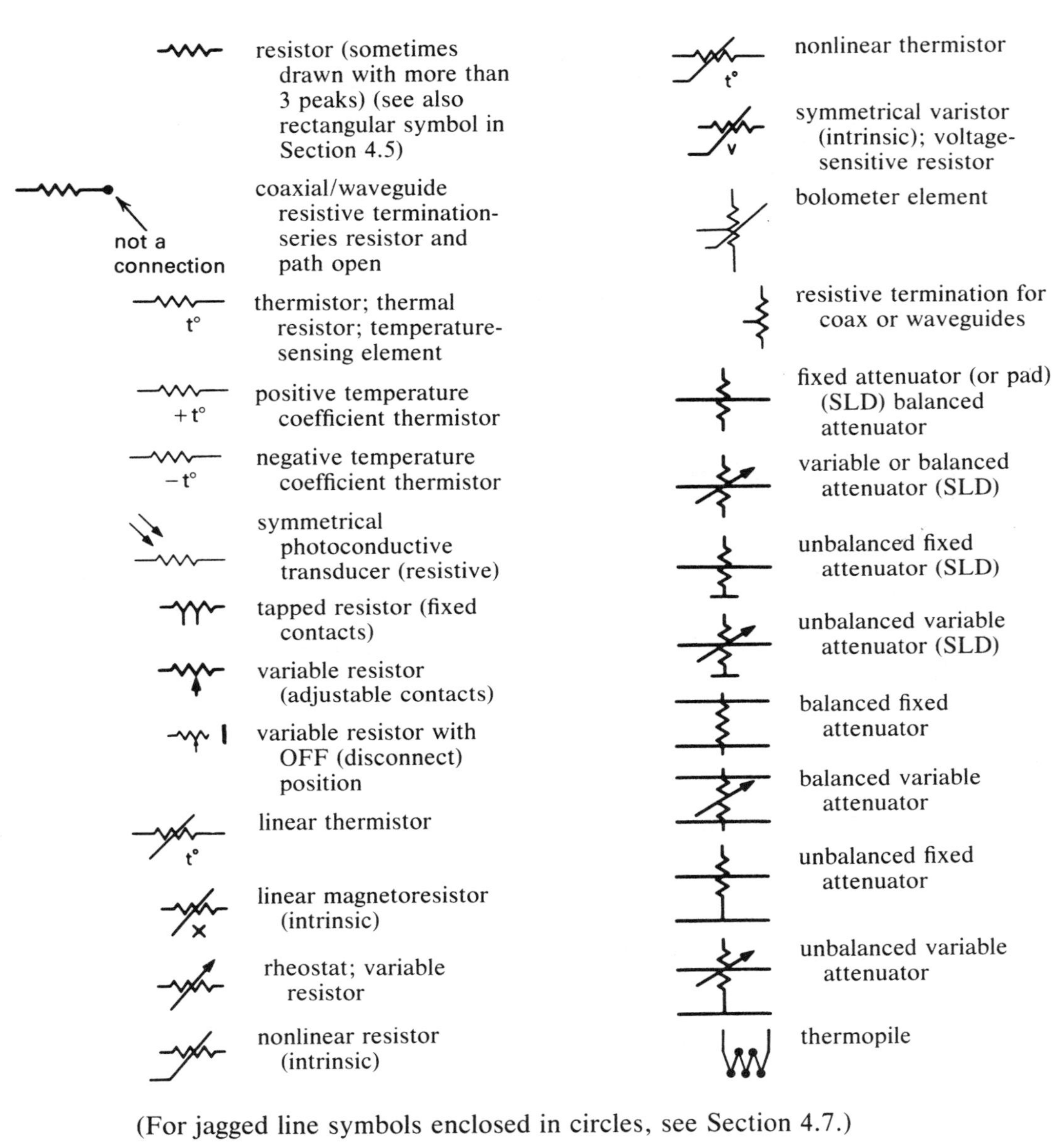

(For jagged line symbols enclosed in circles, see Section 4.7.)

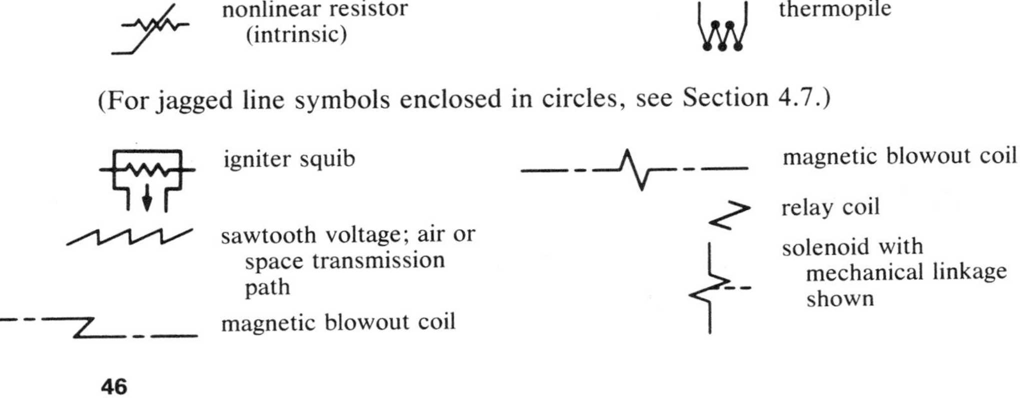

4.3 COILED LINE SYMBOLS

(Inductors, transformers, radio-frequency coils, reactors)

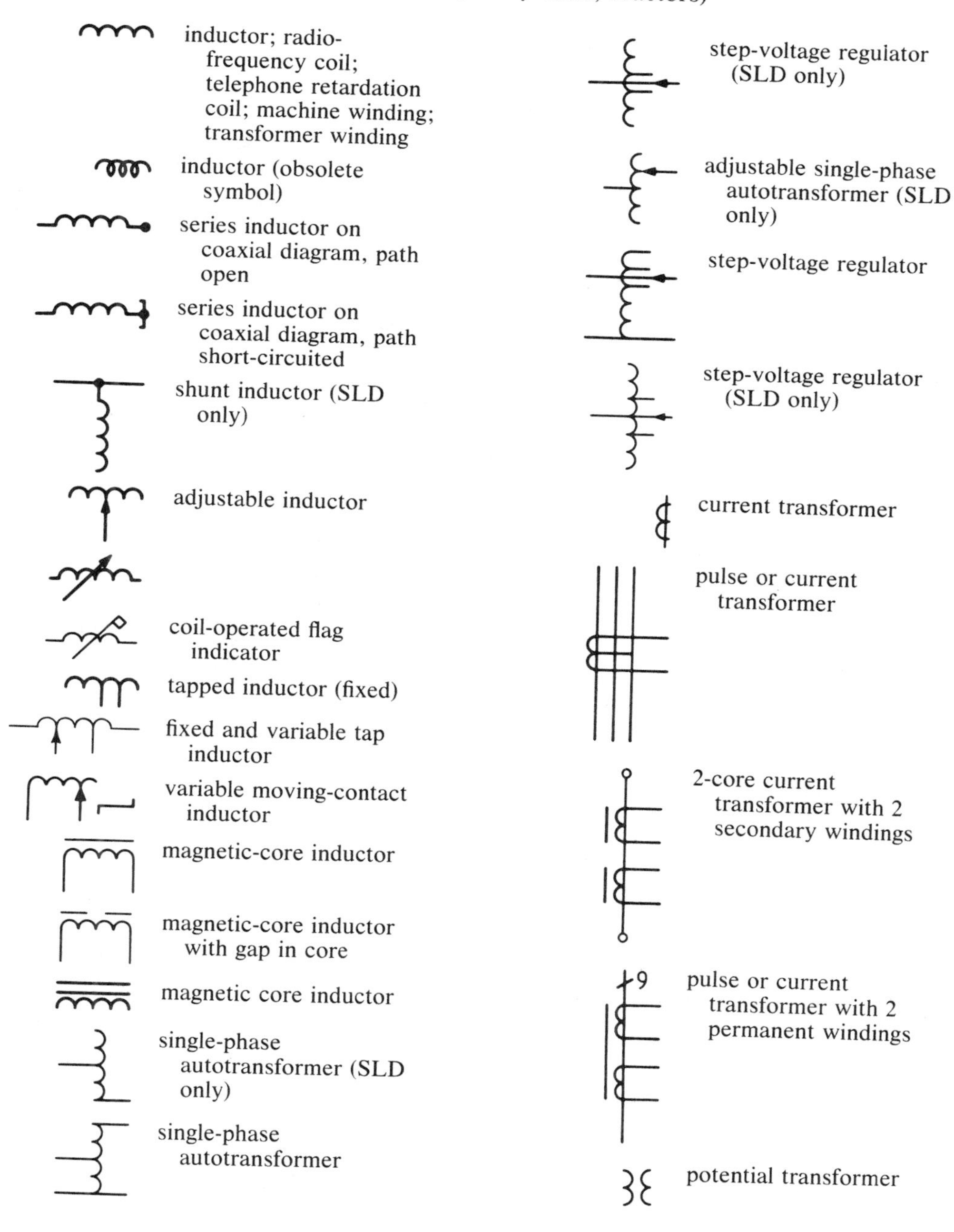

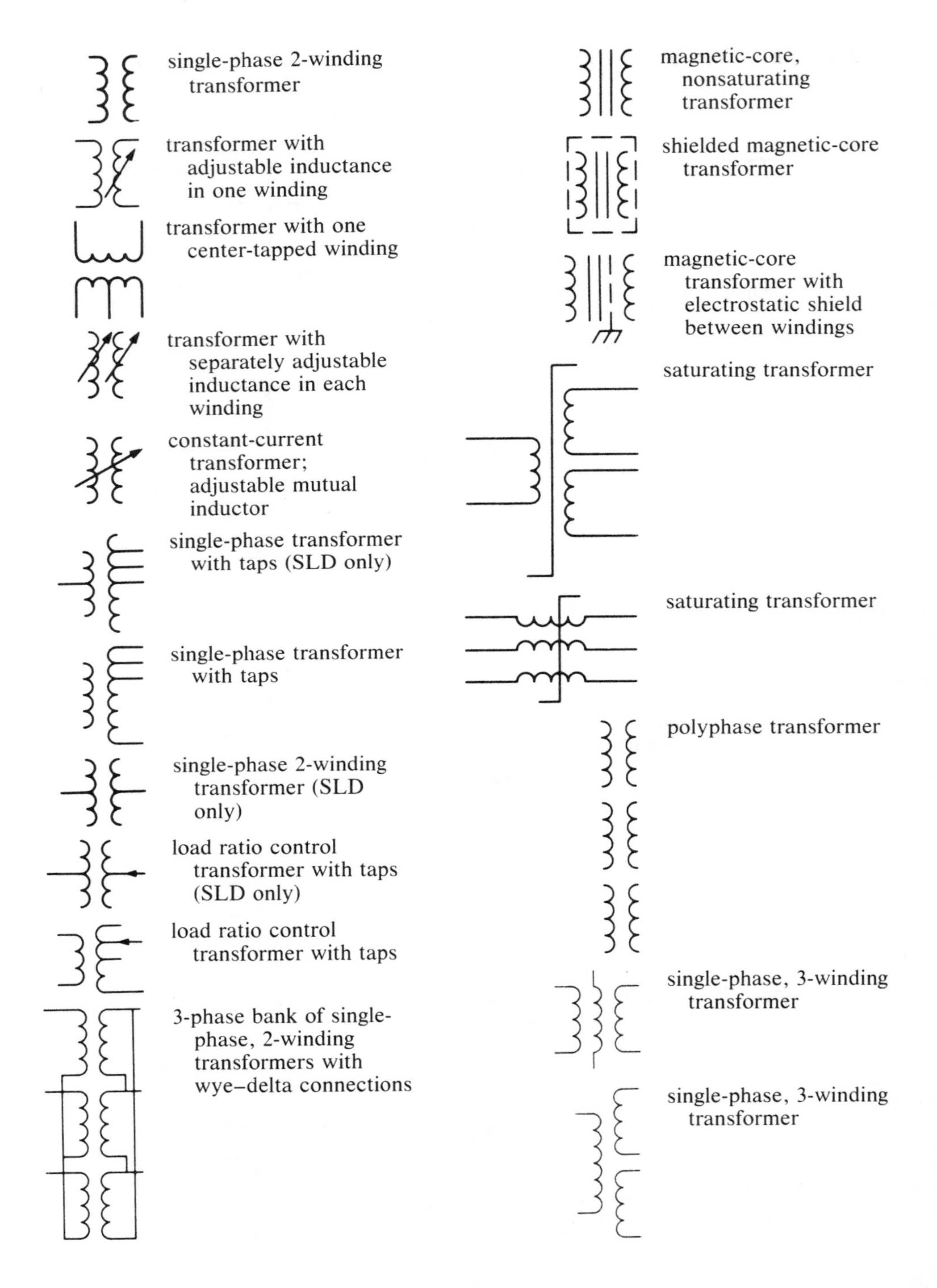
single-phase 2-winding transformer
transformer with adjustable inductance in one winding
transformer with one center-tapped winding
transformer with separately adjustable inductance in each winding
constant-current transformer; adjustable mutual inductor
single-phase transformer with taps (SLD only)
single-phase transformer with taps
single-phase 2-winding transformer (SLD only)
load ratio control transformer with taps (SLD only)
load ratio control transformer with taps
3-phase bank of single-phase, 2-winding transformers with wye–delta connections
magnetic-core, nonsaturating transformer
shielded magnetic-core transformer
magnetic-core transformer with electrostatic shield between windings
saturating transformer
saturating transformer
polyphase transformer
single-phase, 3-winding transformer
single-phase, 3-winding transformer

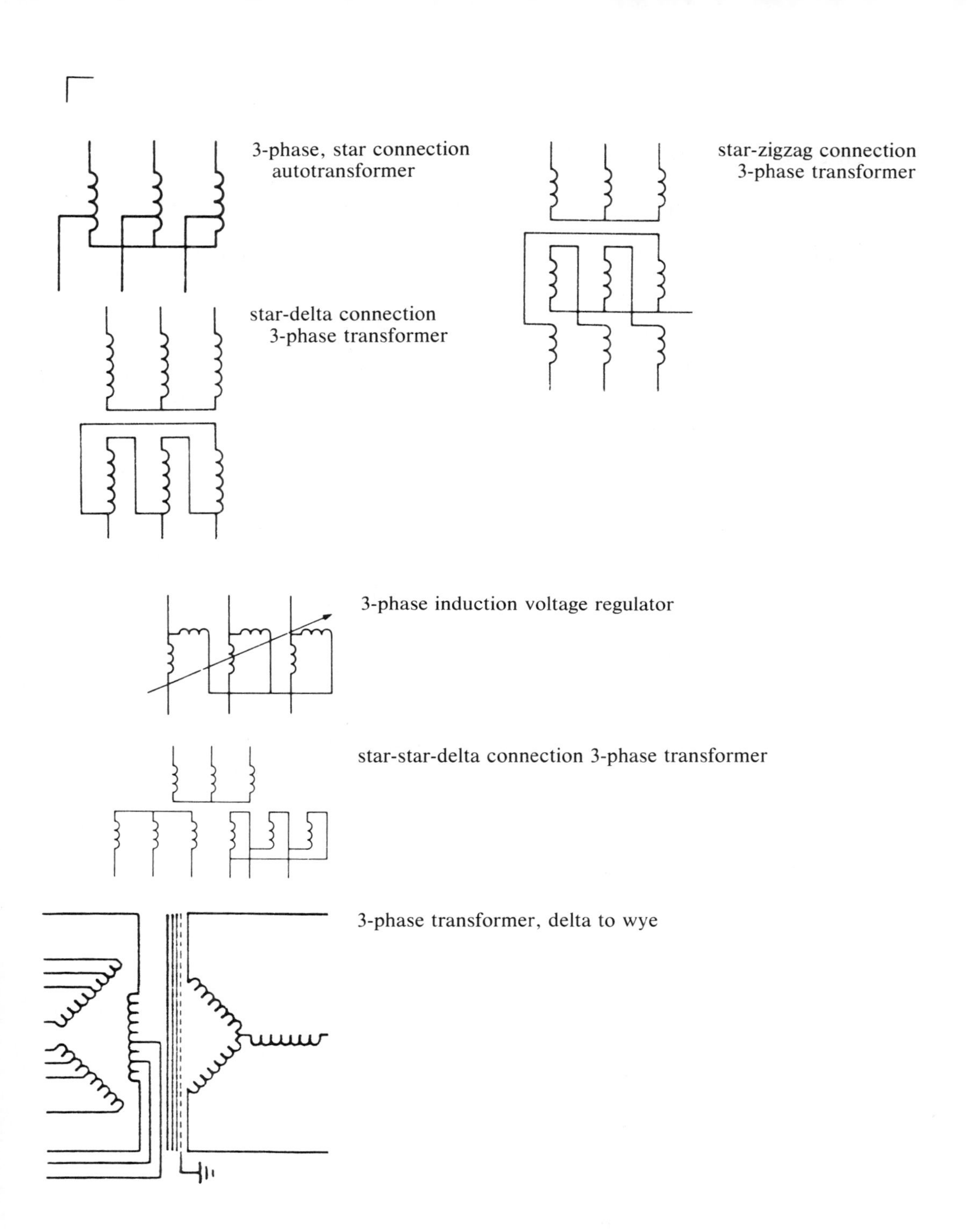

Coils plus circles: see Section 4.7, motors and generators.

4.4 DIVIDED OR BROKEN LINE SYMBOLS

(Capacitors. Refer to Section 4.9 for contactor symbols that are similar.)

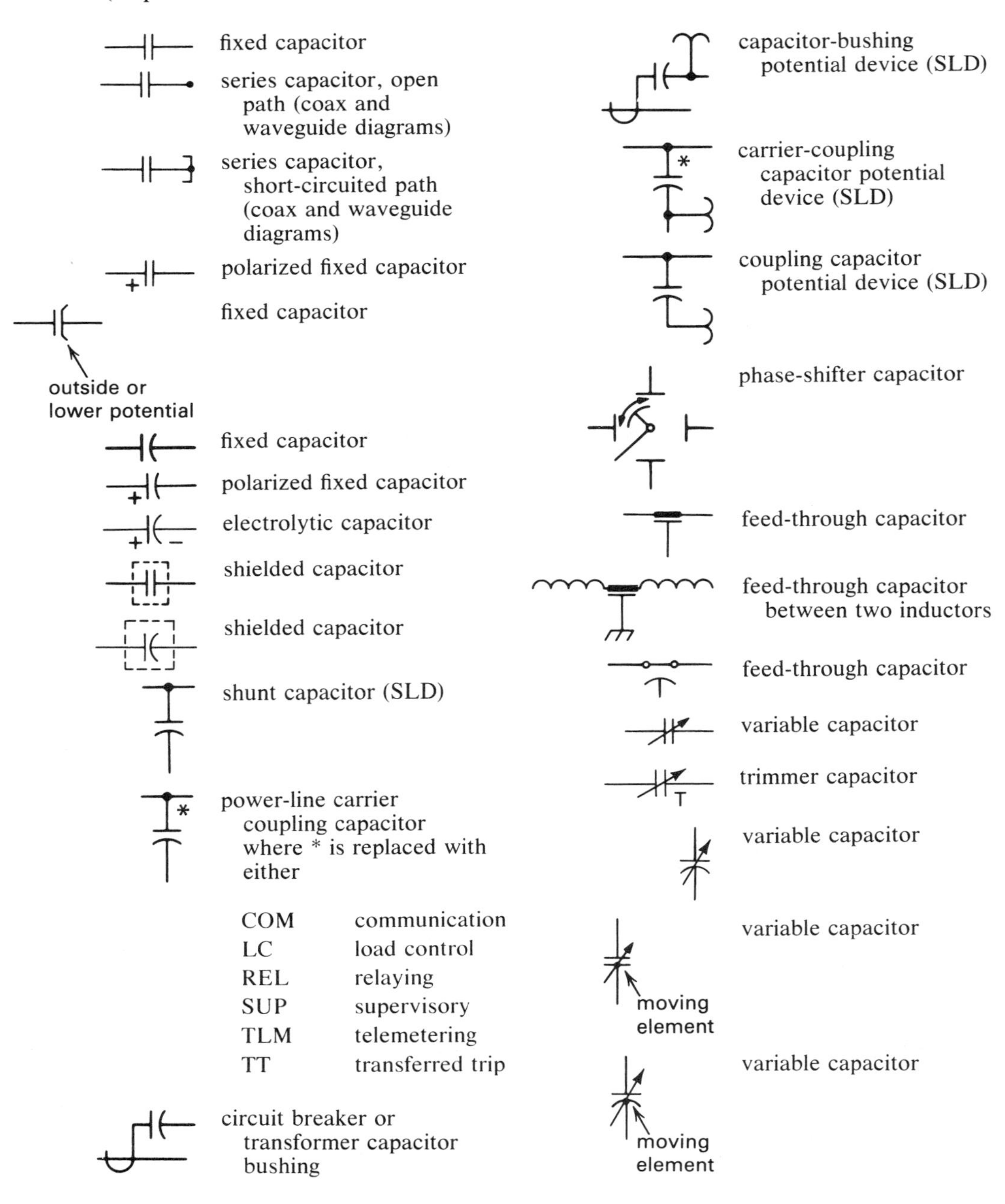

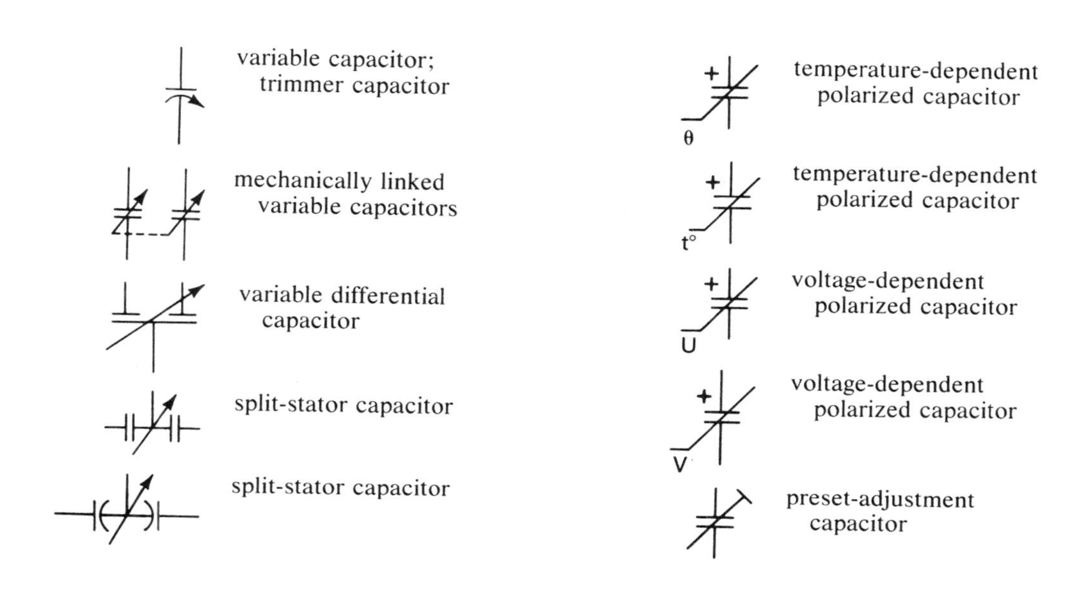

4.5 SQUARE AND RECTANGULAR SYMBOLS

The following symbols are arranged in these categories:

- small squares
- small squares—with internal upside-down triangles
- small squares divided or partitioned
- large squares
- large squares with internal switches
- large squares with arrows outside
- horizontal rectangles—small
- horizontal rectangles—medium
- horizontal rectangles—large
- vertical rectangles—small
- vertical rectangles—medium and large
- vertical rectangles—amplifiers
- vertical rectangles—function generators
- vertical rectangles with external arrows
- vertical rectangles—partitioned or divided

Small squares

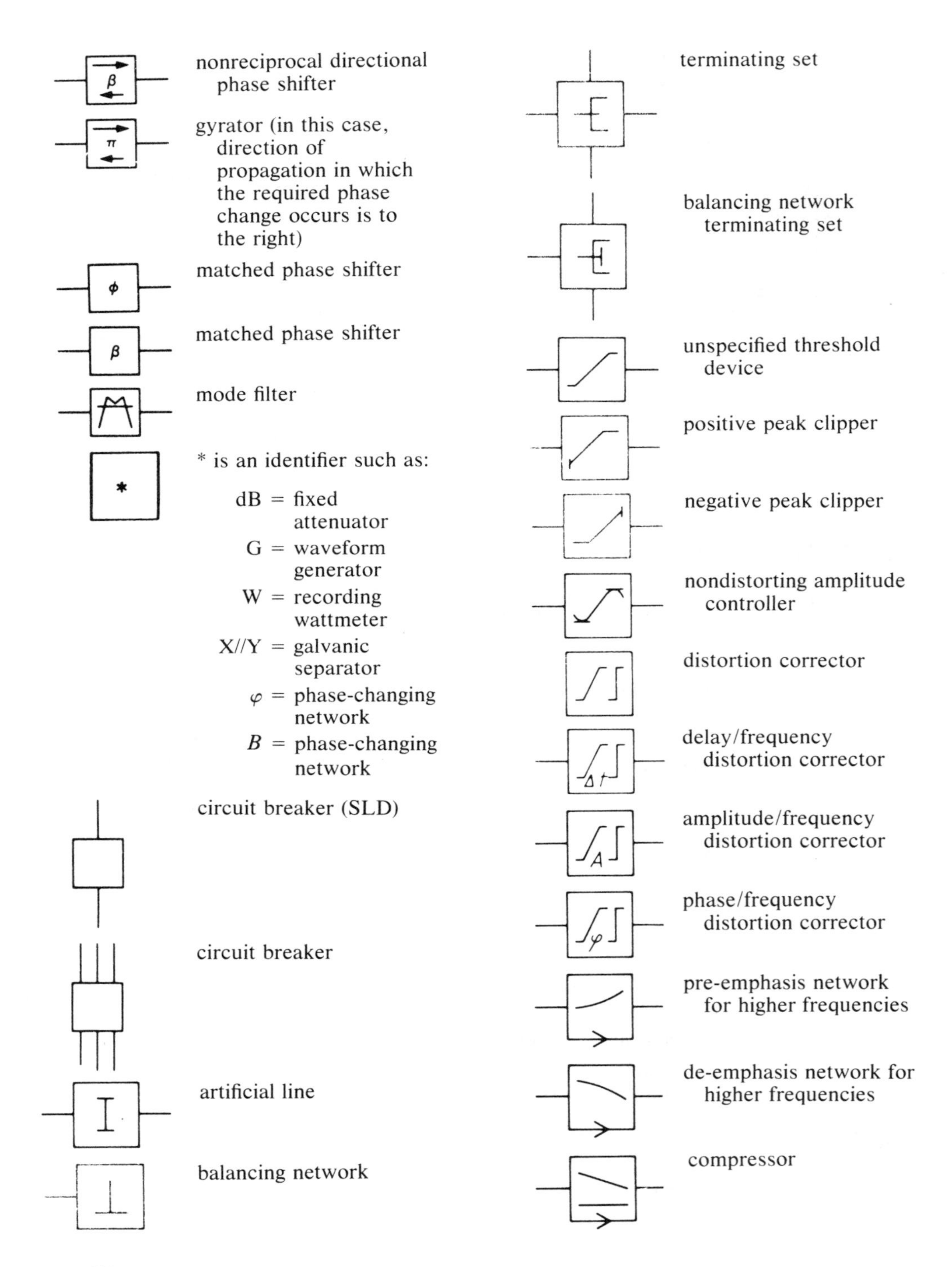
nonreciprocal directional phase shifter
gyrator (in this case, direction of propagation in which the required phase change occurs is to the right)
matched phase shifter
matched phase shifter
mode filter
* is an identifier such as:
dB = fixed attenuator
G = waveform generator
W = recording wattmeter
X//Y = galvanic separator
φ = phase-changing network
B = phase-changing network
circuit breaker (SLD)
circuit breaker
artificial line
balancing network
terminating set
balancing network terminating set
unspecified threshold device
positive peak clipper
negative peak clipper
nondistorting amplitude controller
distortion corrector
delay/frequency distortion corrector
amplitude/frequency distortion corrector
phase/frequency distortion corrector
pre-emphasis network for higher frequencies
de-emphasis network for higher frequencies
compressor

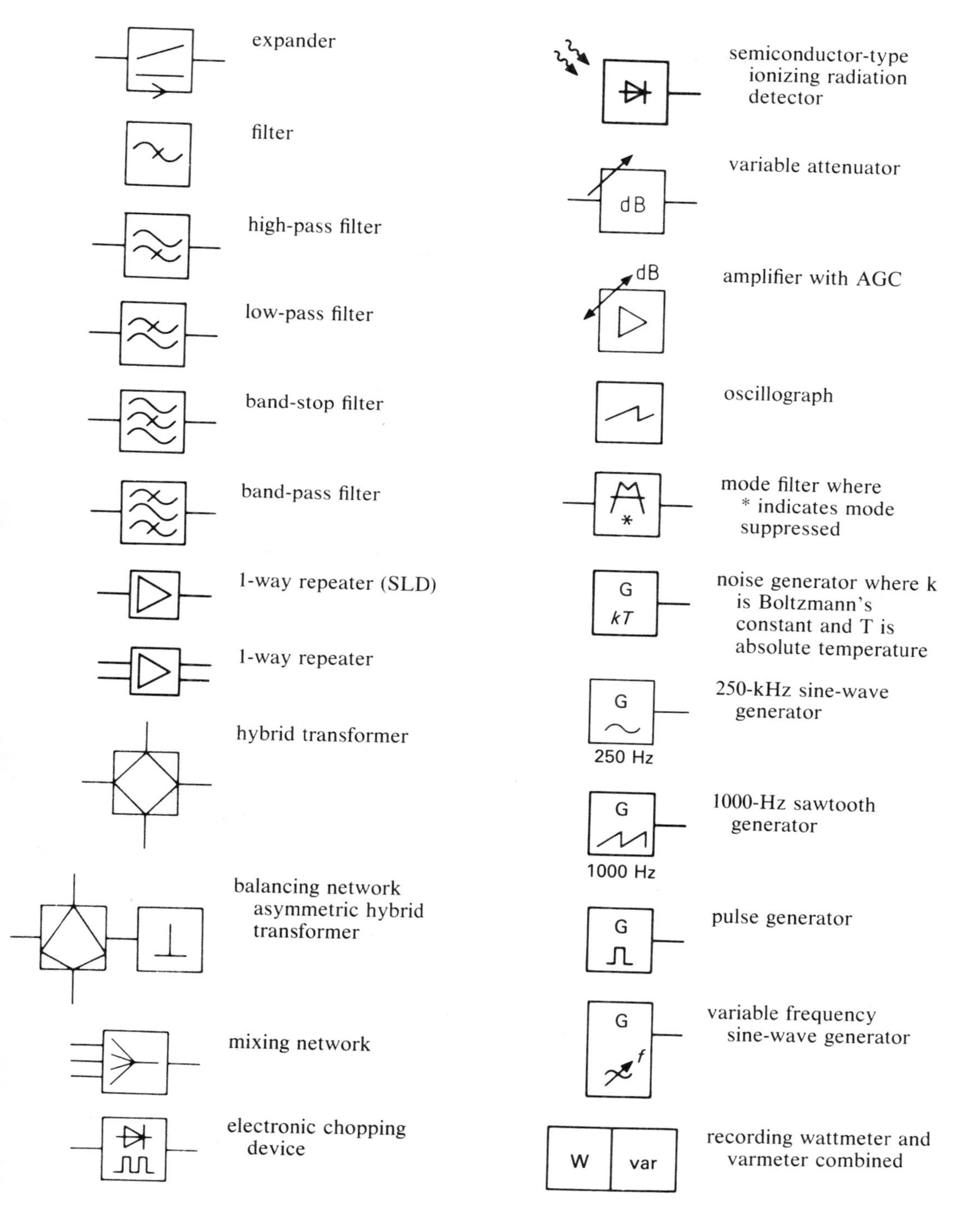

expander
filter
high-pass filter
low-pass filter
band-stop filter
band-pass filter
1-way repeater (SLD)
1-way repeater
hybrid transformer
balancing network asymmetric hybrid transformer
mixing network
electronic chopping device
semiconductor-type ionizing radiation detector
variable attenuator
dB
amplifier with AGC
dB
oscillograph
mode filter where * indicates mode suppressed
*
noise generator where k is Boltzmann's constant and T is absolute temperature
G
kT
250-kHz sine-wave generator
G
250 Hz
1000-Hz sawtooth generator
G
1000 Hz
pulse generator
G
variable frequency sine-wave generator
G
f
recording wattmeter and varmeter combined
W
var

Small squares with internal upside-down triangles

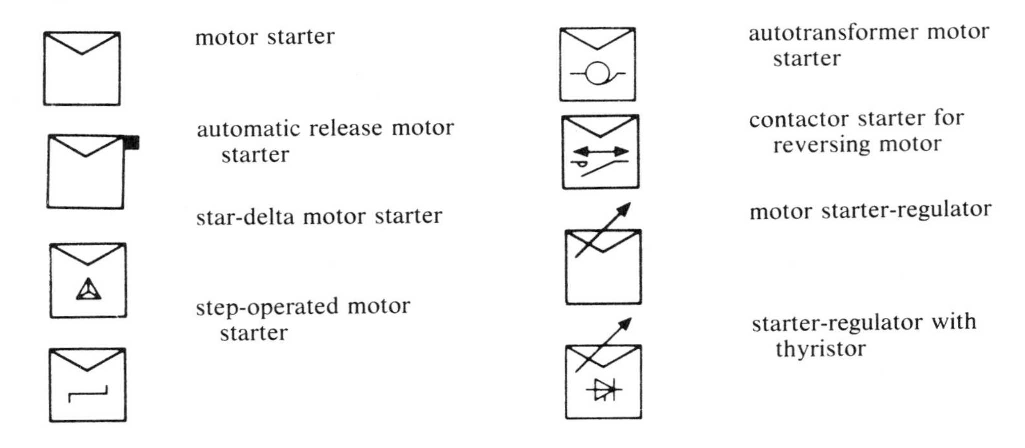

Small squares divided or partitioned

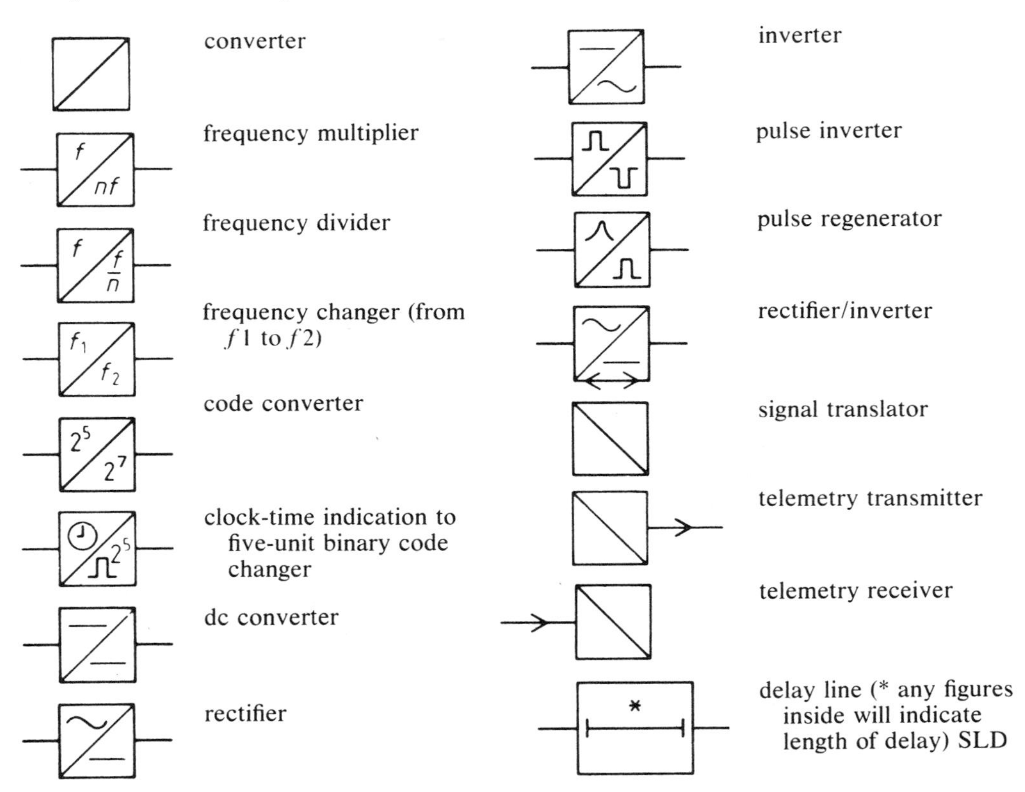

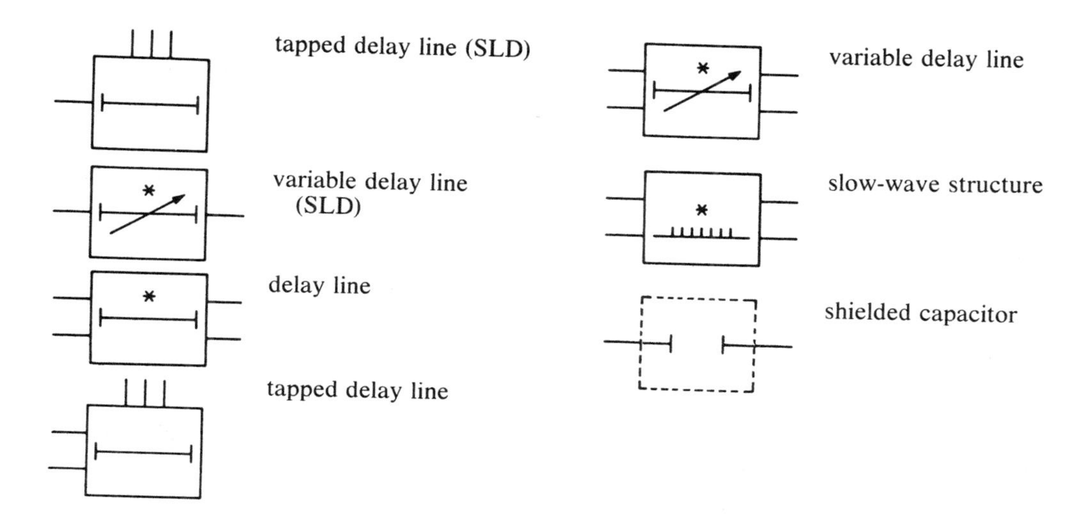

Large squares

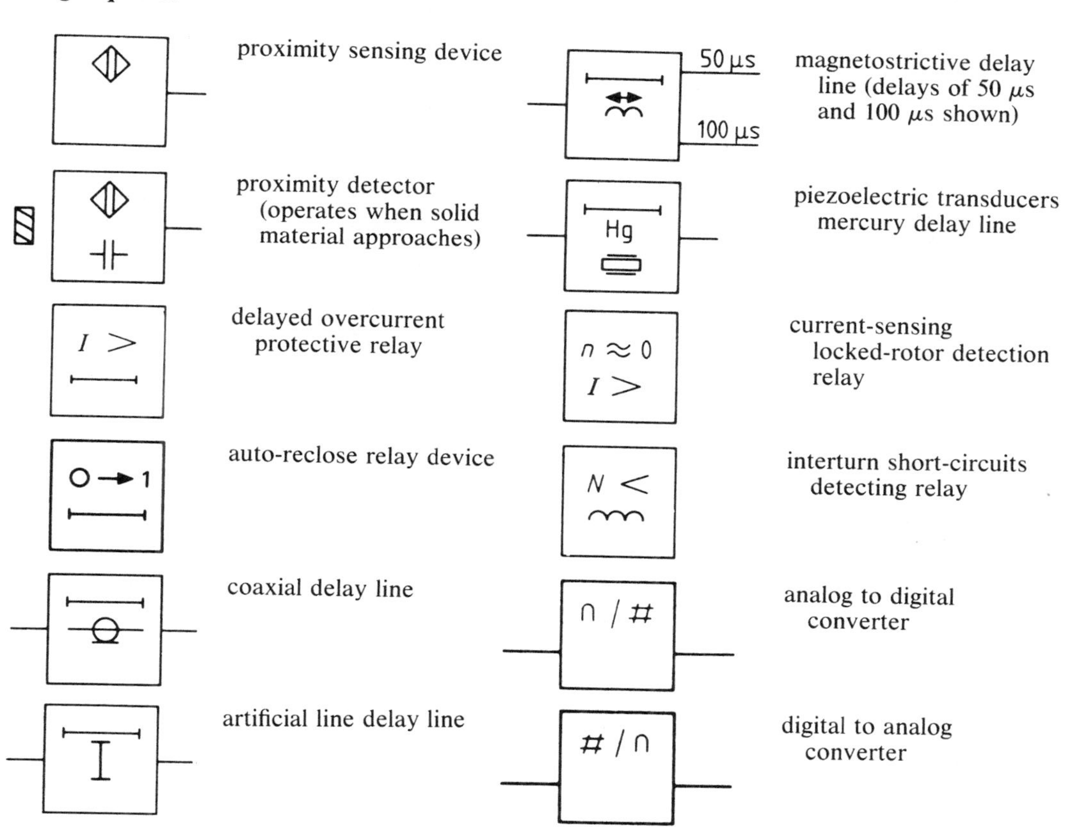

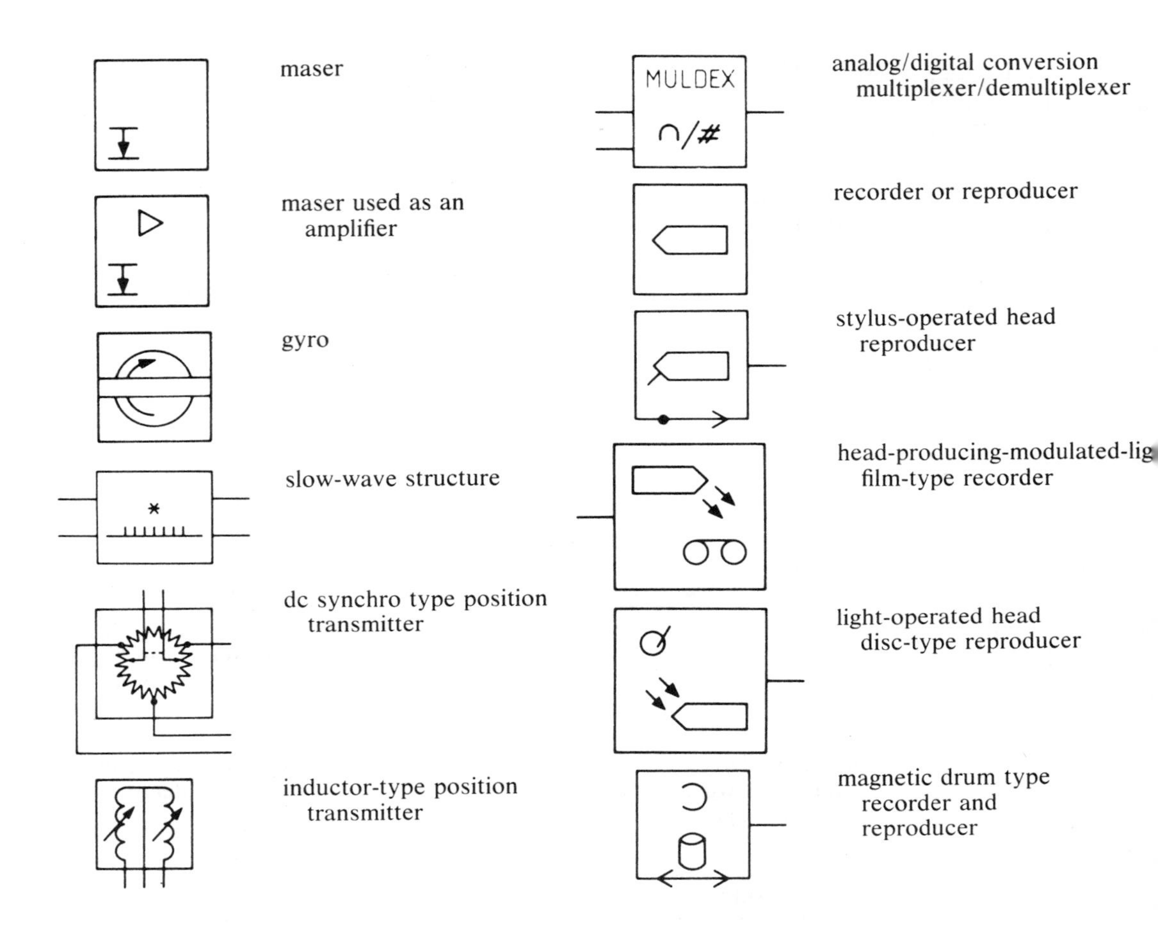

maser

maser used as an amplifier

gyro

slow-wave structure

dc synchro type position transmitter

inductor-type position transmitter

analog/digital conversion multiplexer/demultiplexer

recorder or reproducer

stylus-operated head reproducer

head-producing-modulated-lig film-type recorder

light-operated head disc-type reproducer

magnetic drum type recorder and reproducer

Large squares with internal switches

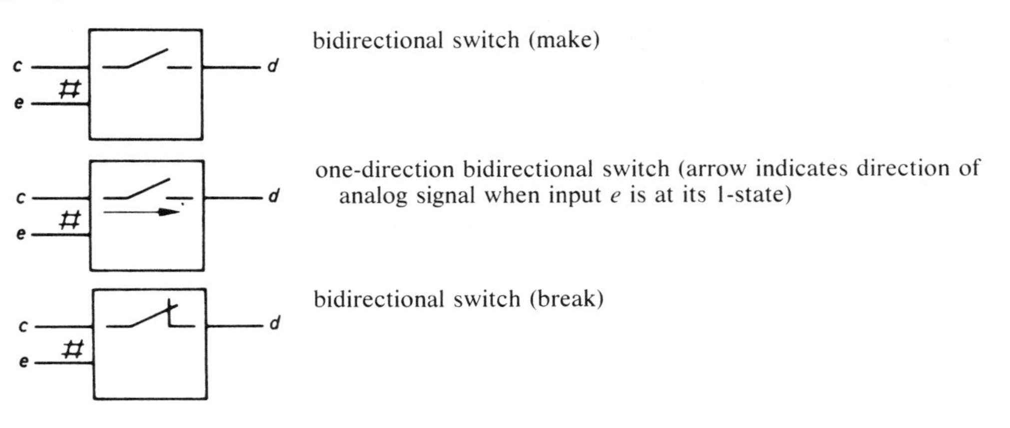

bidirectional switch (make)

one-direction bidirectional switch (arrow indicates direction of analog signal when input *e* is at its 1-state)

bidirectional switch (break)

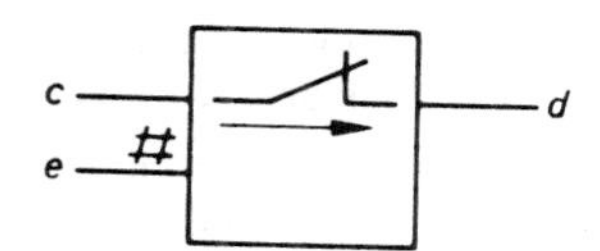

one-direction bidirectional switch (break) (arrow indicates direction of analog when input e is at its 0-state)

Large squares—arrows outside

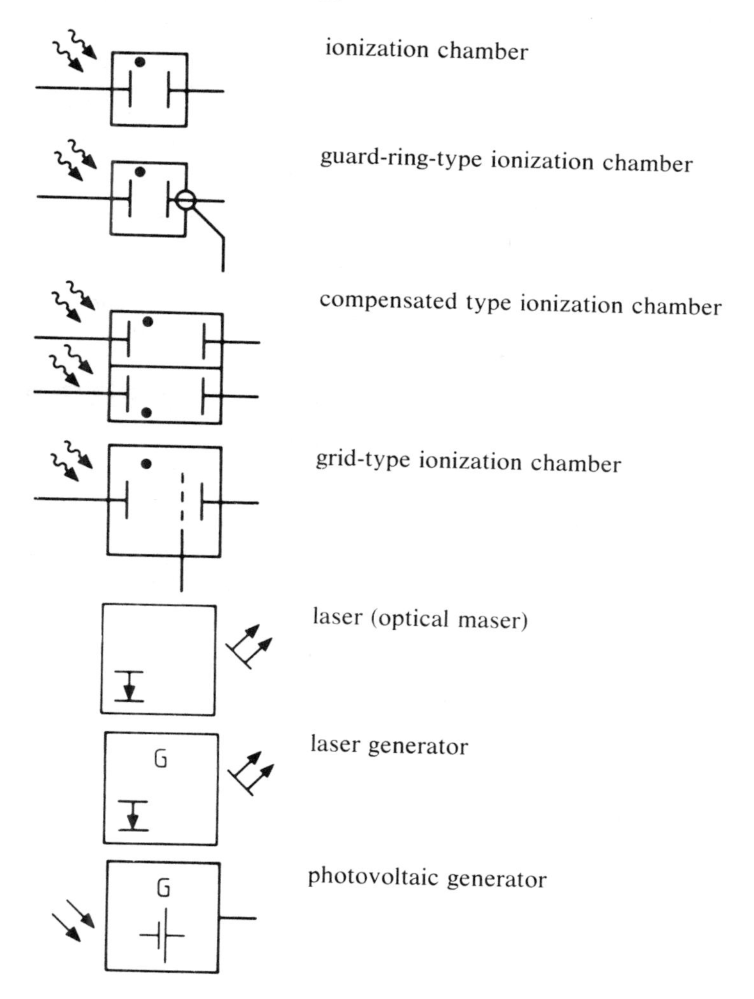

ionization chamber

guard-ring-type ionization chamber

compensated type ionization chamber

grid-type ionization chamber

laser (optical maser)

laser generator

photovoltaic generator

Horizontal rectangles—small

core of magnet

resistor (see also jagged line symbols in Section 4.2)

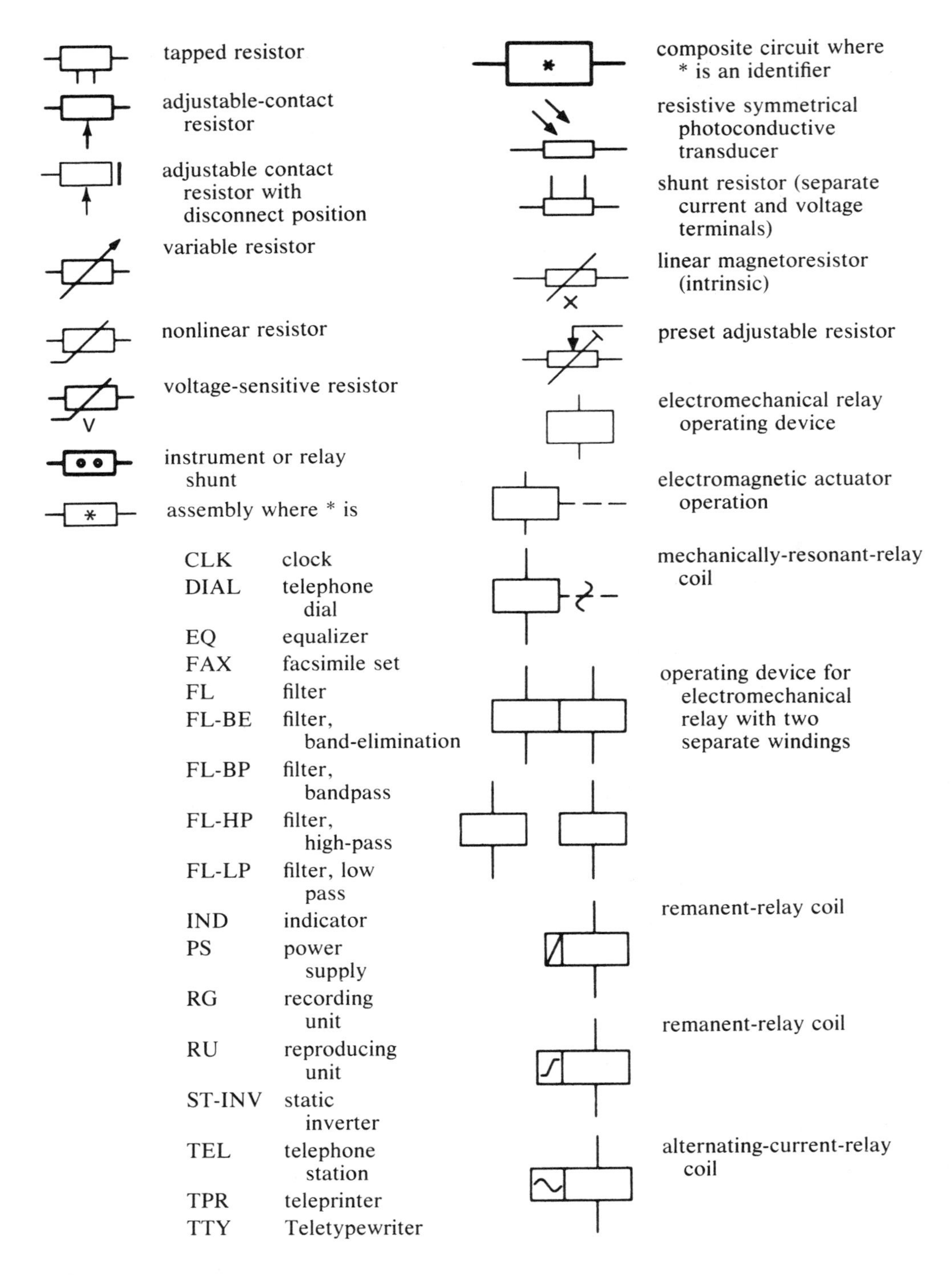

tapped resistor
adjustable-contact resistor
adjustable contact resistor with disconnect position
variable resistor
nonlinear resistor
voltage-sensitive resistor
V
instrument or relay shunt
*
assembly where * is
CLK clock
DIAL telephone dial
EQ equalizer
FAX facsimile set
FL filter
FL-BE filter, band-elimination
FL-BP filter, bandpass
FL-HP filter, high-pass
FL-LP filter, low pass
IND indicator
PS power supply
RG recording unit
RU reproducing unit
ST-INV static inverter
TEL telephone station
TPR teleprinter
TTY Teletypewriter
*
composite circuit where * is an identifier
resistive symmetrical photoconductive transducer
shunt resistor (separate current and voltage terminals)
linear magnetoresistor (intrinsic)
×
preset adjustable resistor
electromechanical relay operating device
electromagnetic actuator operation
mechanically-resonant-relay coil
operating device for electromechanical relay with two separate windings
remanent-relay coil
remanent-relay coil
alternating-current-relay coil

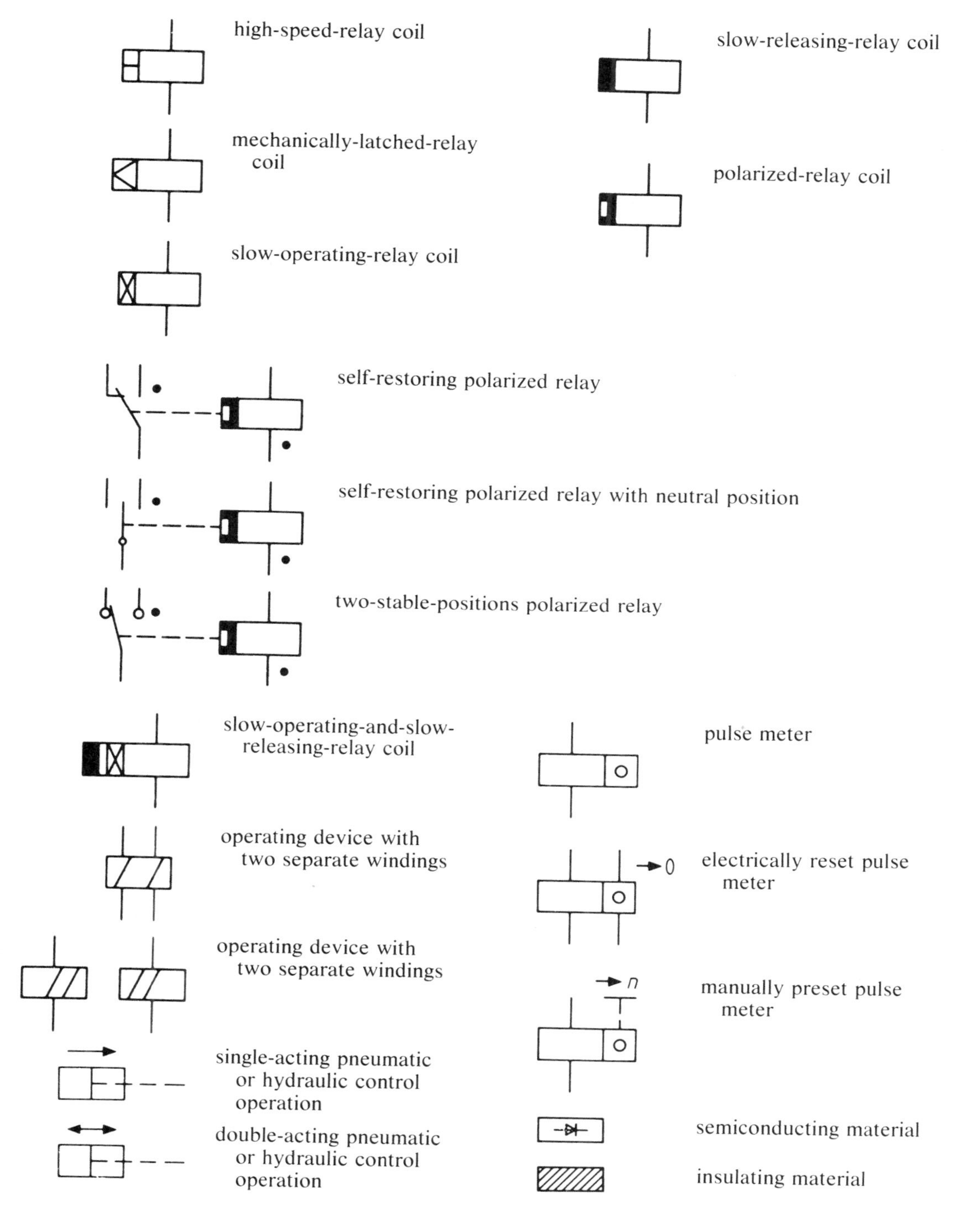
high-speed-relay coil
slow-releasing-relay coil
mechanically-latched-relay coil
polarized-relay coil
slow-operating-relay coil
self-restoring polarized relay
self-restoring polarized relay with neutral position
two-stable-positions polarized relay
slow-operating-and-slow-releasing-relay coil
pulse meter
operating device with two separate windings
electrically reset pulse meter
0
operating device with two separate windings
manually preset pulse meter
n
single-acting pneumatic or hydraulic control operation
double-acting pneumatic or hydraulic control operation
semiconducting material
insulating material

Horizontal rectangles: medium-size

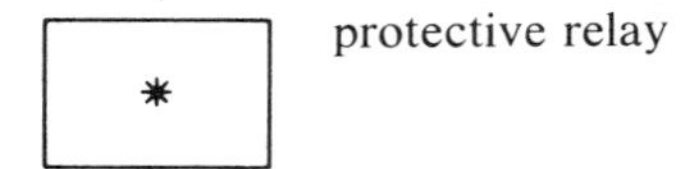

protective relay

* is replaced by one of the following qualifiers:

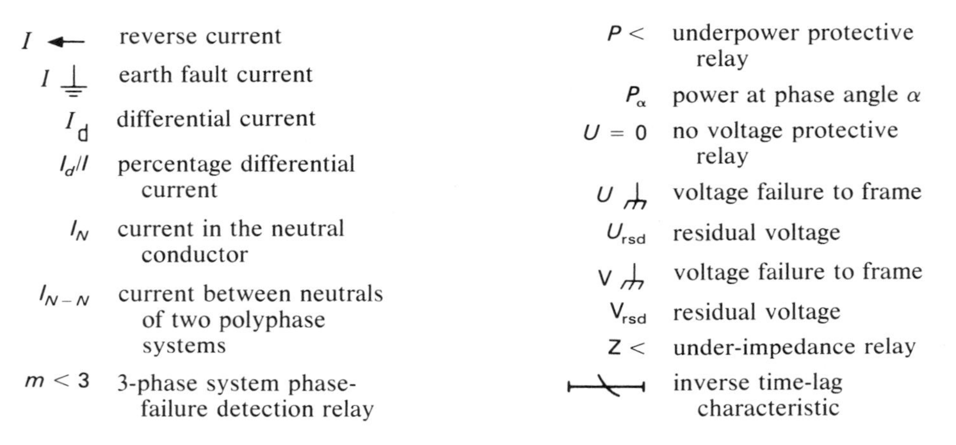

$I \leftarrow$	reverse current
I ⏚	earth fault current
I_{d}	differential current
I_d/I	percentage differential current
I_N	current in the neutral conductor
I_{N-N}	current between neutrals of two polyphase systems
$m < 3$	3-phase system phase-failure detection relay
$P <$	underpower protective relay
P_α	power at phase angle α
$U = 0$	no voltage protective relay
U (frame symbol)	voltage failure to frame
U_{rsd}	residual voltage
V (frame symbol)	voltage failure to frame
$\mathrm{V}_{\mathrm{rsd}}$	residual voltage
$Z <$	under-impedance relay
(inverse time-lag symbol)	inverse time-lag characteristic

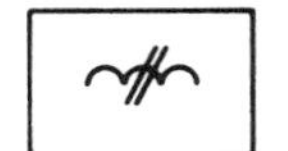

divided-conductor detection relay

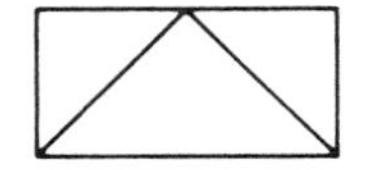

modulator, demodulator, or discriminator

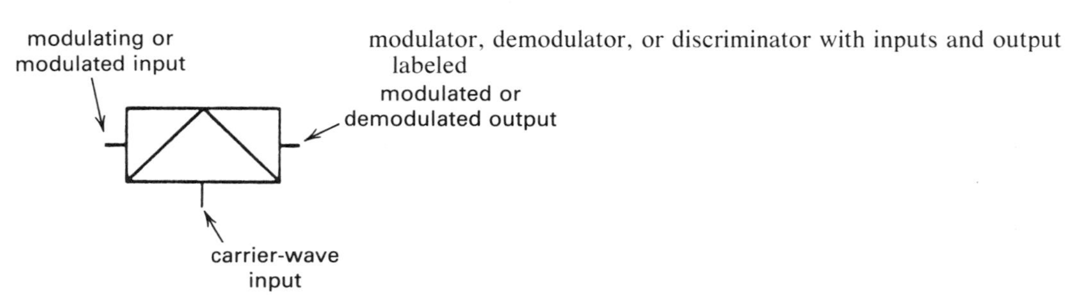

modulator, demodulator, or discriminator with inputs and output labeled

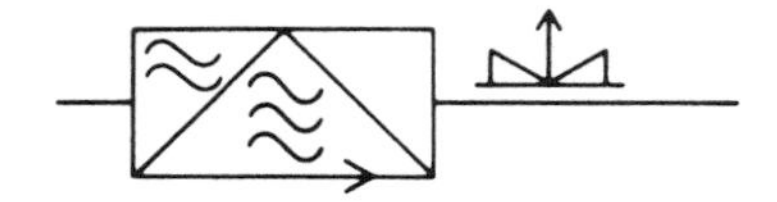

double sideband output modulator

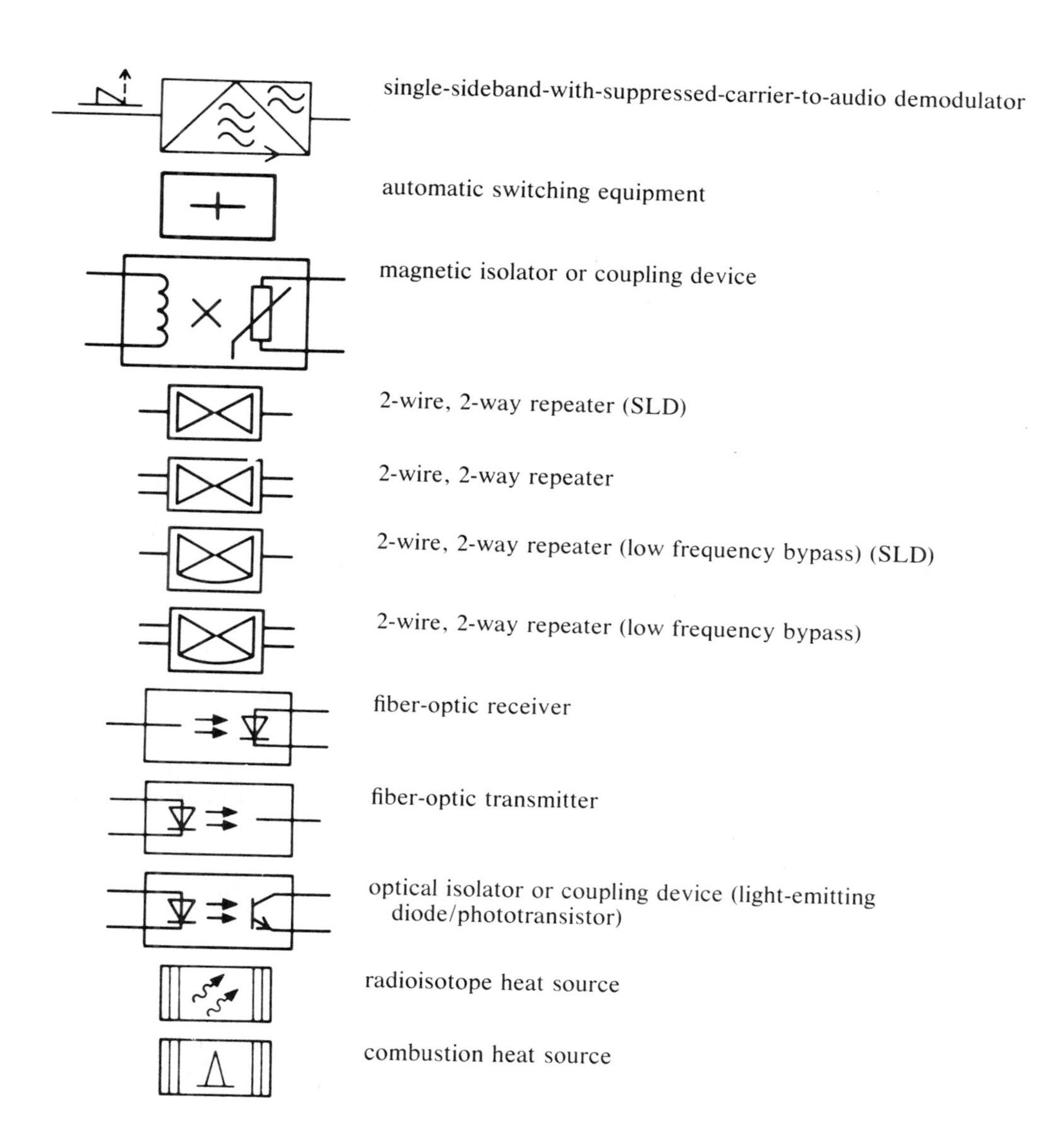

single-sideband-with-suppressed-carrier-to-audio demodulator

automatic switching equipment

magnetic isolator or coupling device

2-wire, 2-way repeater (SLD)

2-wire, 2-way repeater

2-wire, 2-way repeater (low frequency bypass) (SLD)

2-wire, 2-way repeater (low frequency bypass)

fiber-optic receiver

fiber-optic transmitter

optical isolator or coupling device (light-emitting diode/phototransistor)

radioisotope heat source

combustion heat source

Horizontal rectangles—larger size

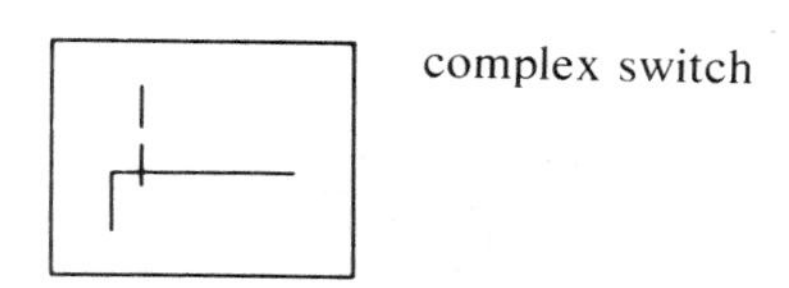

complex switch

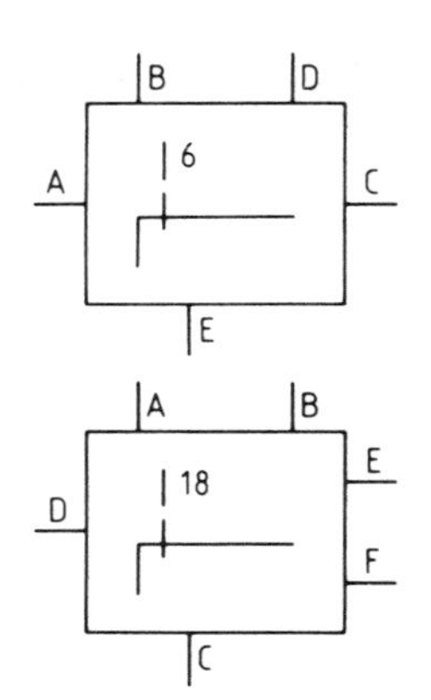

5-terminal 6-position rotary drum switch

6-terminal 18-position rotary wafer switch

Vertical rectangles—small

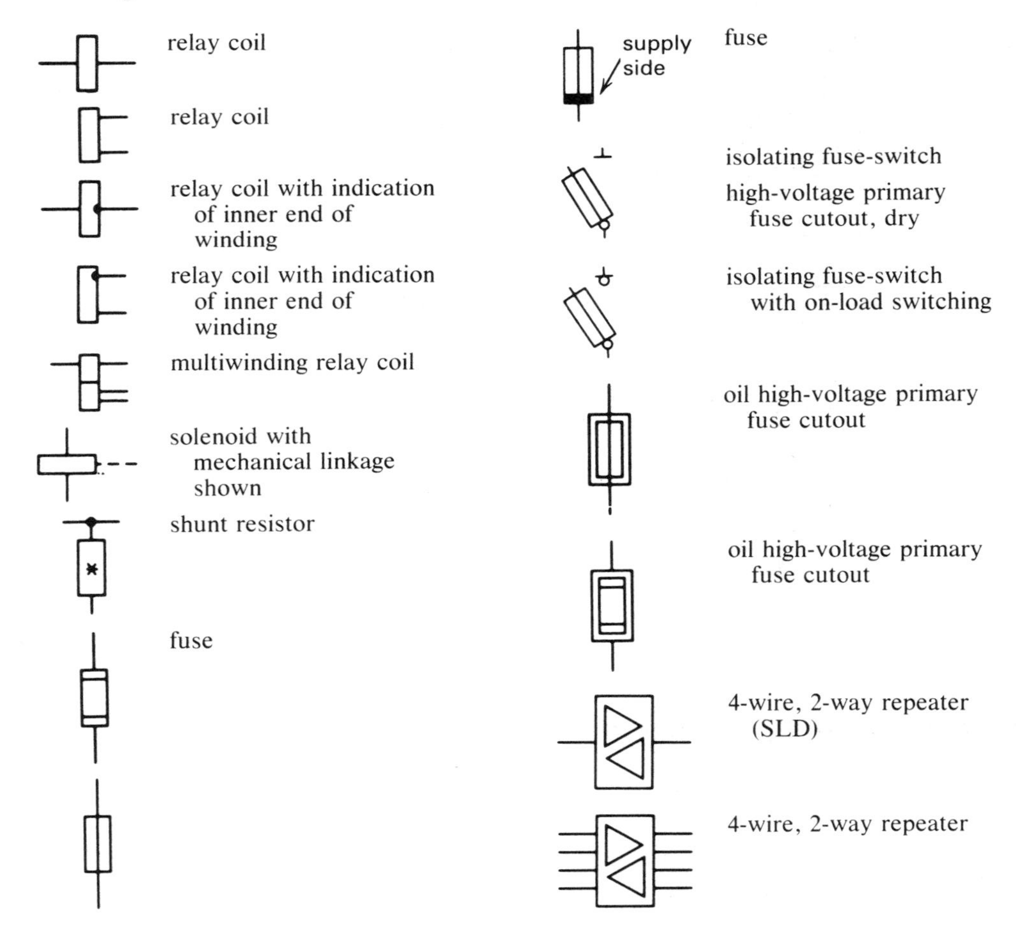

Vertical rectangles—medium and large

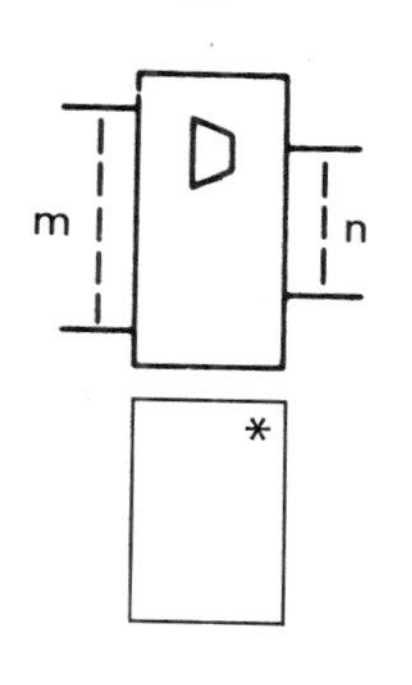

concentrator

multiplexing function where * is replaced by:

DMUX	demultiplexing function
DX	demultiplexing function
MULDEX	multiplexing/demultiplexing
MUX	multiplexing

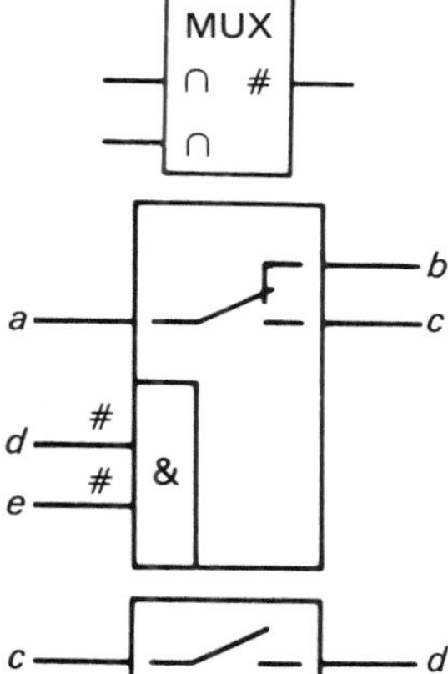

analog/digital conversion multiplexer

AND function bidirectional transfer switch (break)

common binary input independent bidirectional switches

Vertical rectangles—amplifiers

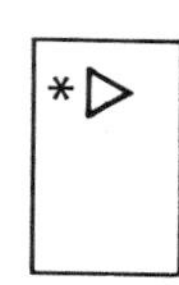

amplifier

* is replaced with one of the following qualifiers:

C control: integration occurs if in 1-state

F frequency compensation

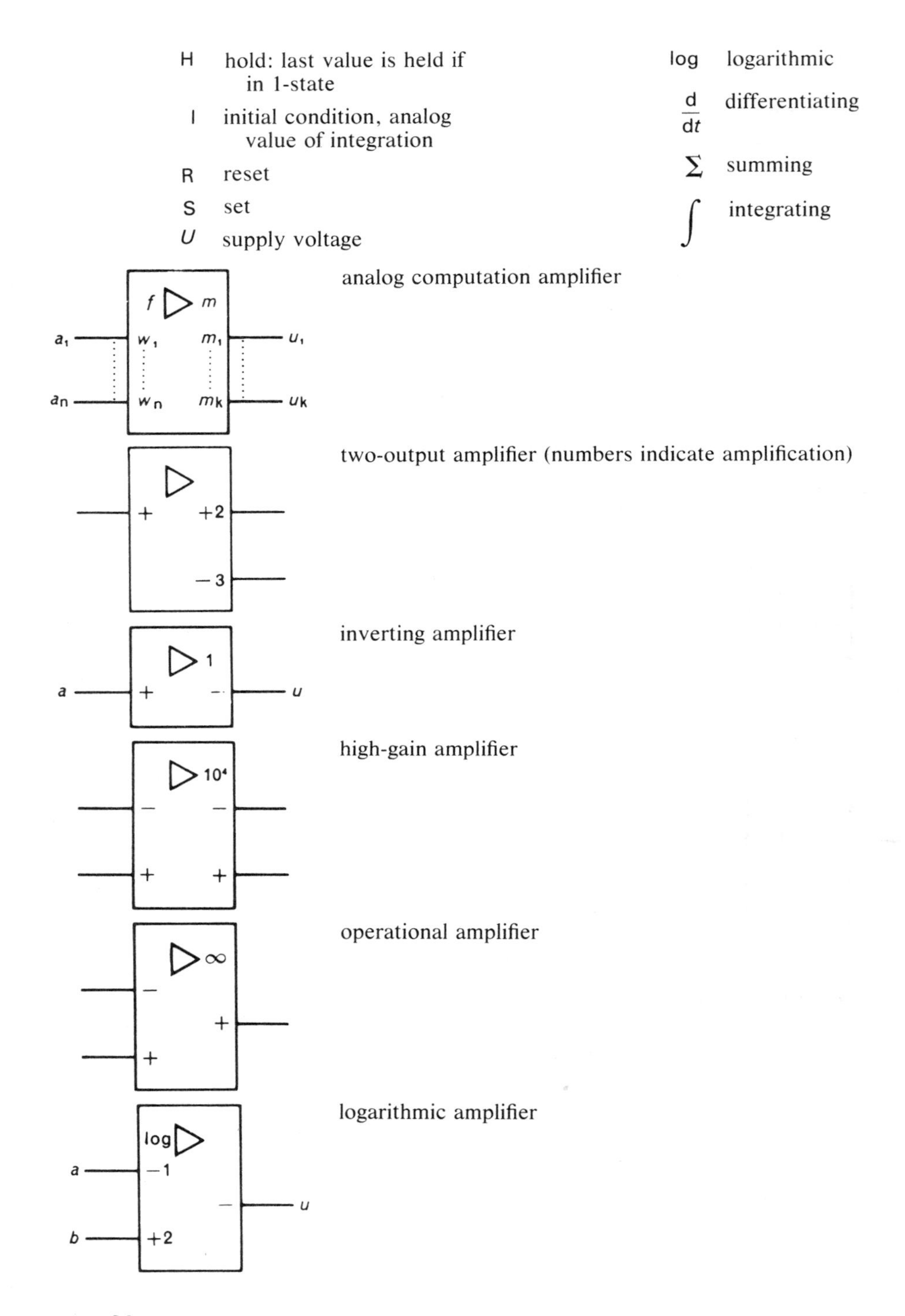
H hold: last value is held if in 1-state
I initial condition, analog value of integration
R reset
S set
U supply voltage
log logarithmic
d/dt differentiating
Σ summing
∫ integrating
analog computation amplifier
f
m
a1
an
w1
wn
m1
mk
u1
uk
two-output amplifier (numbers indicate amplification)
+
+2
−3
inverting amplifier
1
a
u
high-gain amplifier
10⁴
operational amplifier
∞
logarithmic amplifier
log
−1
+2
b

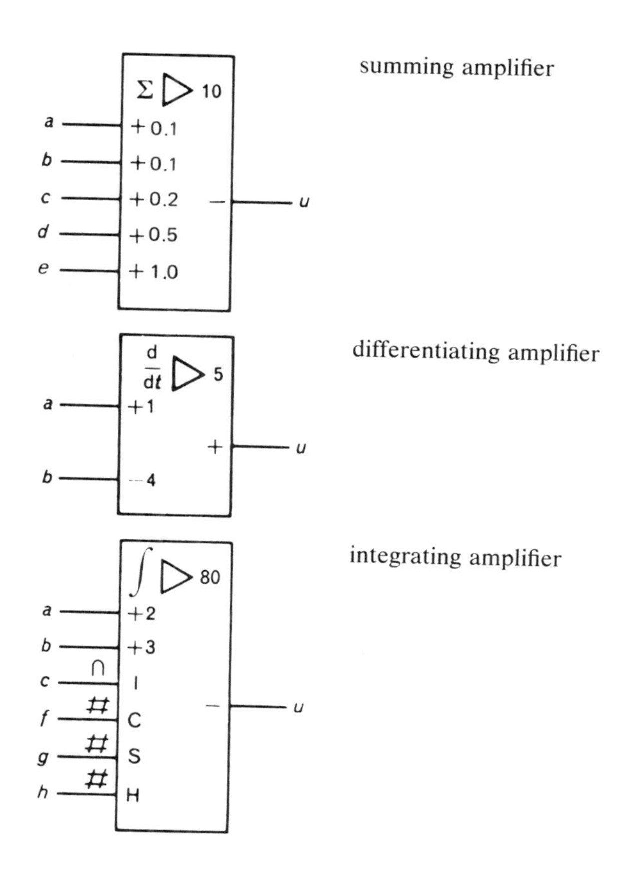

summing amplifier

differentiating amplifier

integrating amplifier

Vertical rectangles—function generators

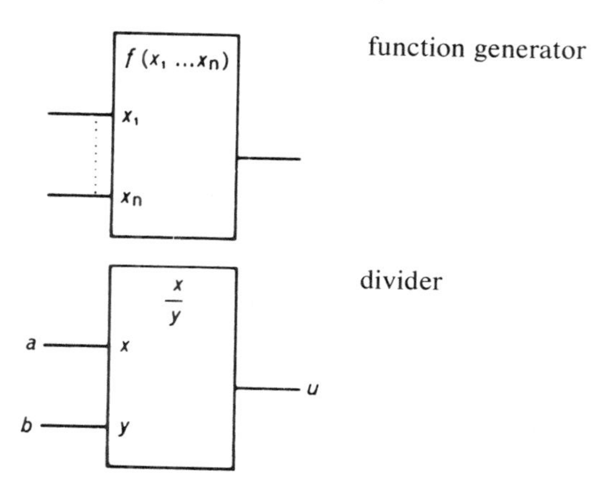

function generator

divider

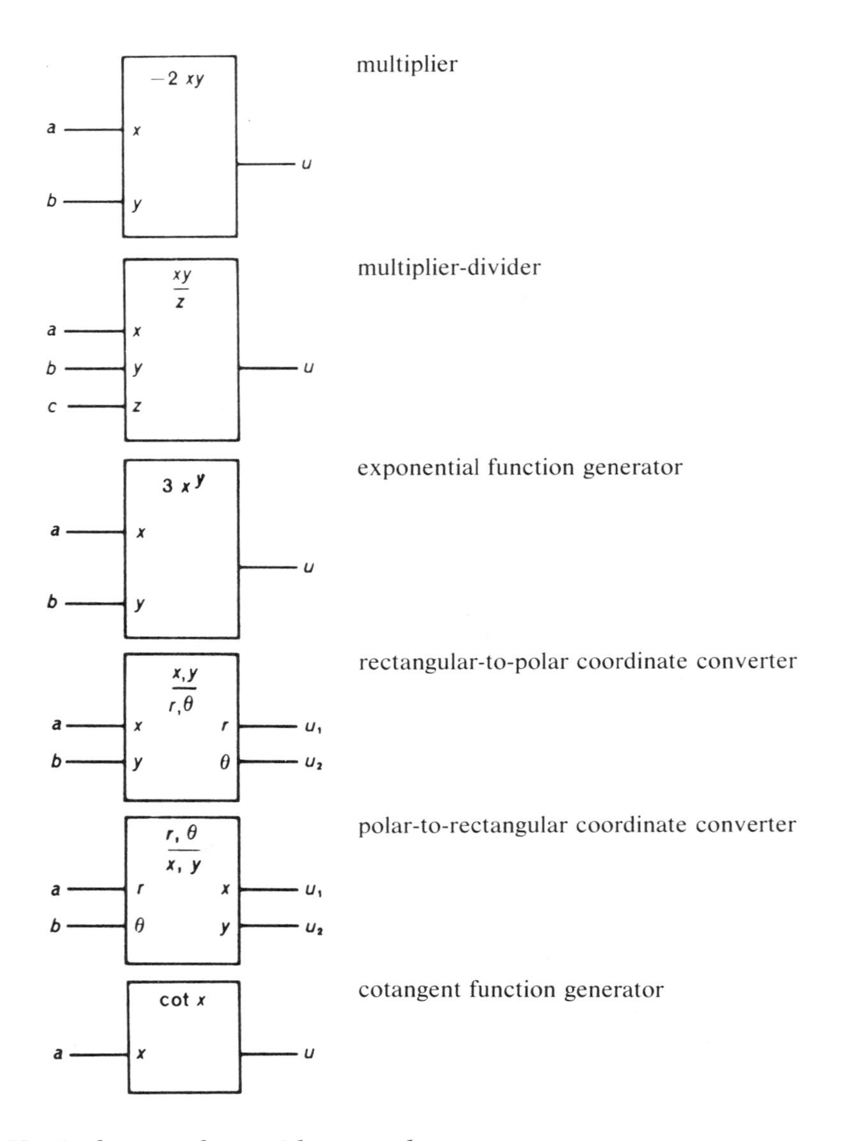

Vertical rectangles—with external arrows

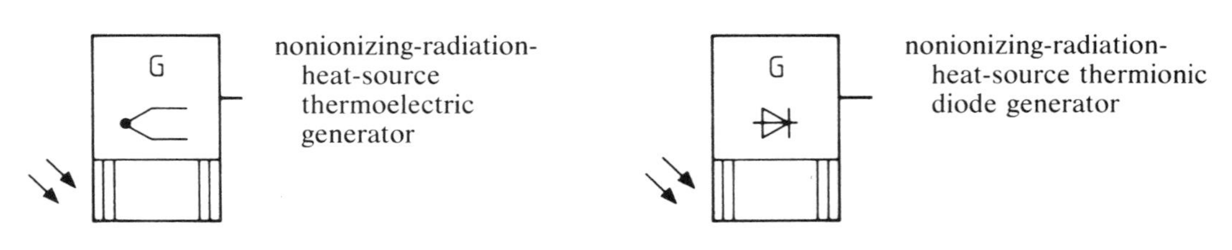

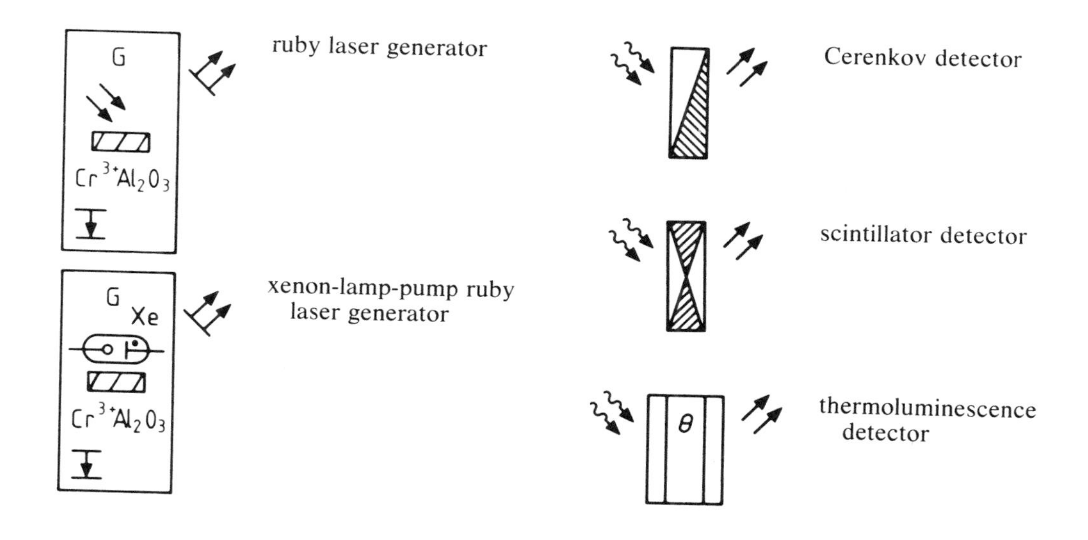

Vertical rectangles—partitioned or divided

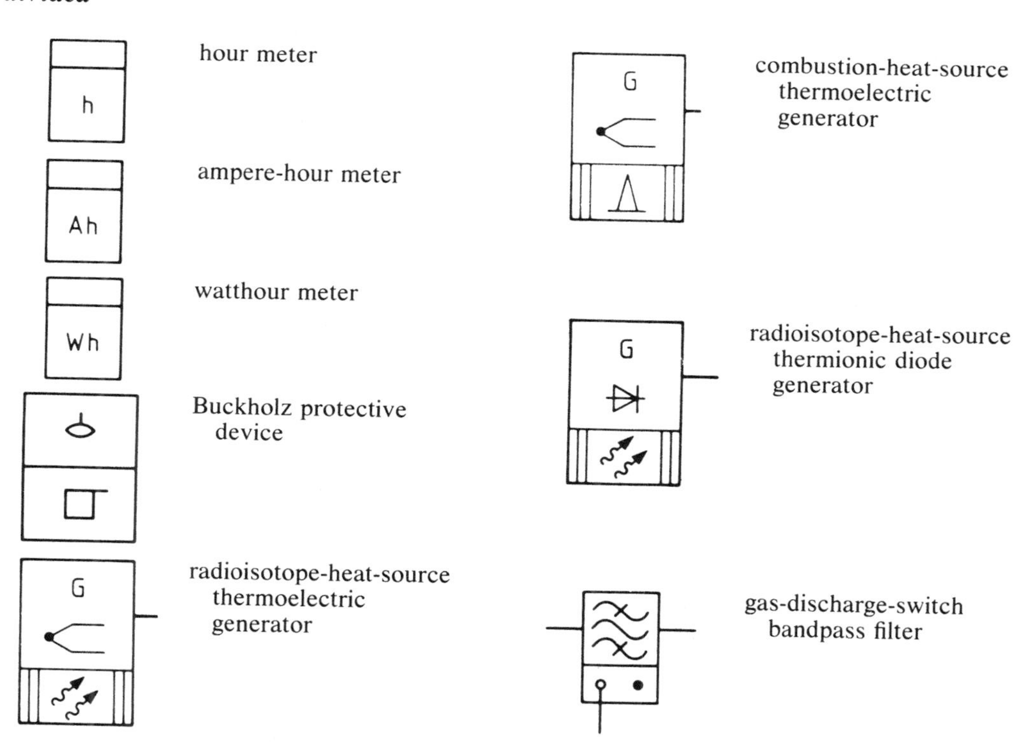

4.6 TRIANGULAR SYMBOLS AND RECTANGLES WITH TRIANGULAR ENDS

(Amplifiers, multipliers, dividers, discontinuities, phase shifters, integrators)
Note: SLD indicates for single-line drawings only.

Horizontal triangles: small open (▷) and lined (⊳)

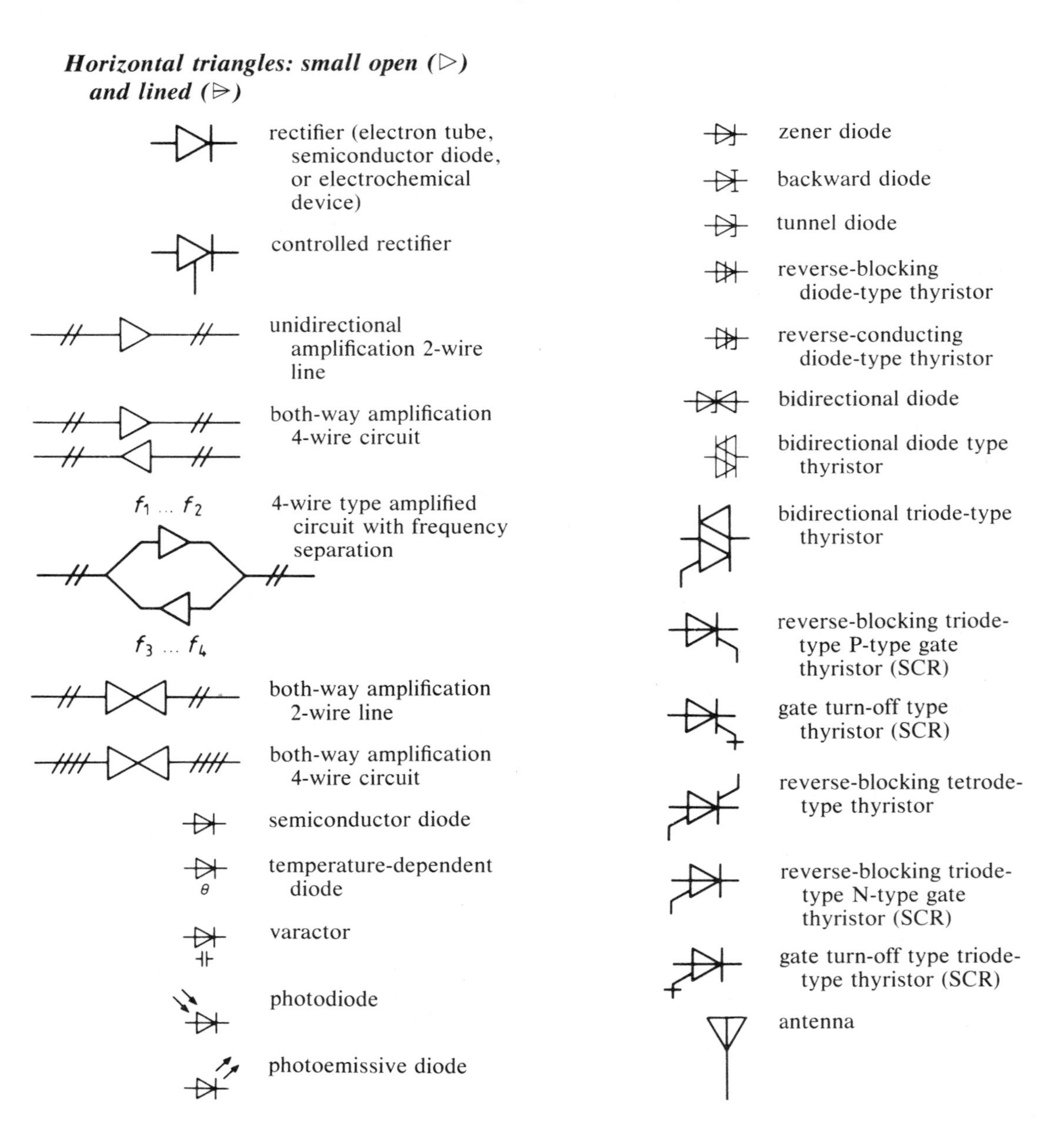

Horizontal triangles: medium

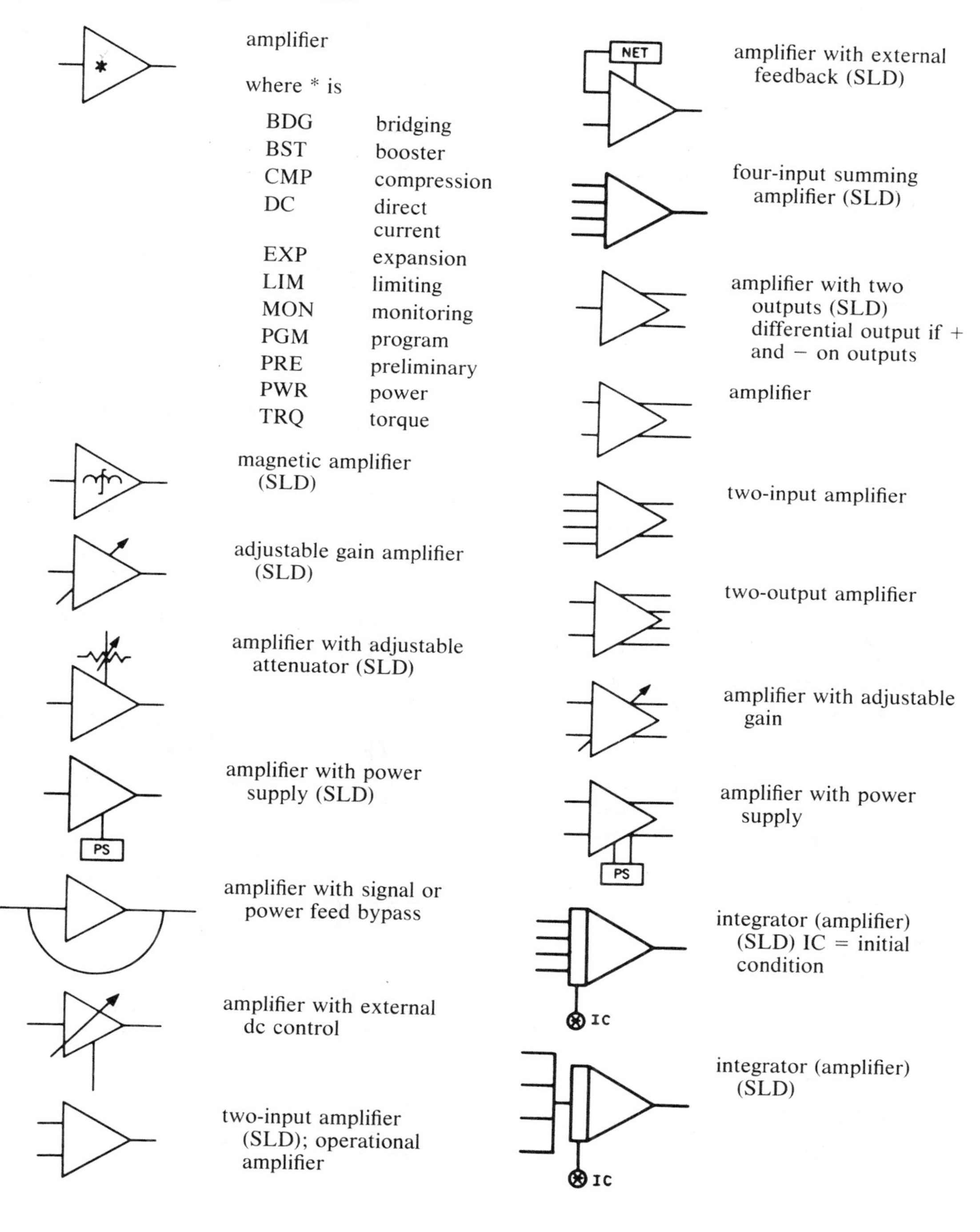

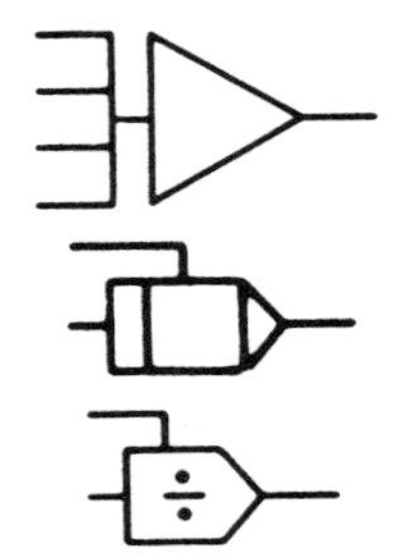

four-input summing amplifier (SLD)

generalized integrator (SLD)

electronic divider (SLD)

electronic multiplier (SLD)

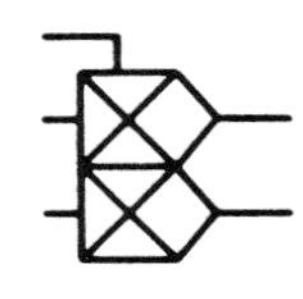

two dependent electronic multipliers (SLD)

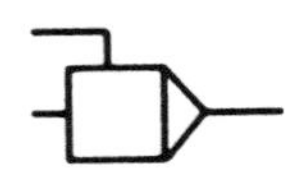

electronic function generator

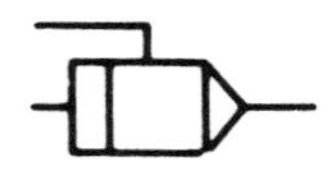

integrator

Vertical triangles

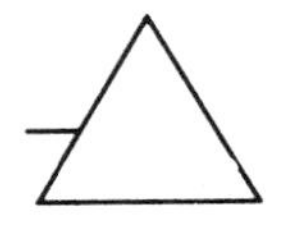

one-port terminal discontinuity

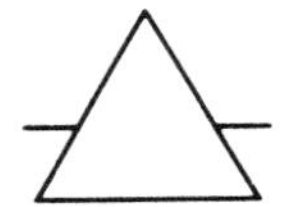

two-port discontinuity

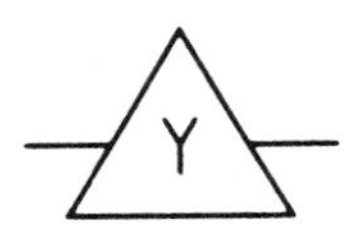

discontinuity with equivalent shunt element in parallel with guided transmission path

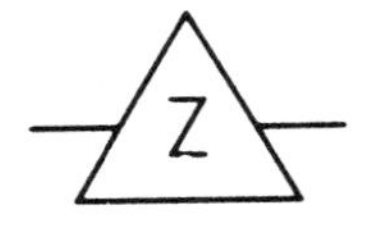

discontinuity with equivalent series element in series with guided transmission path

resistance discontinuity

conductance discontinuity

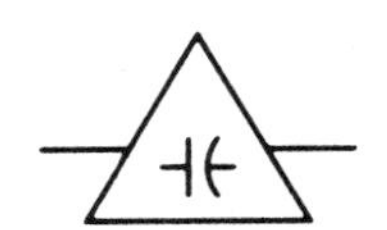

capacitive reactance discontinuity

capacitive susceptance discontinuity

inductive reactance discontinuity

inductive susceptance discontinuity

inductance-capacitance circuit discontinuity with zero reactance, infinite susceptance at resonance

inductance-capacitance circuit discontinuity with zero reactance at resonance

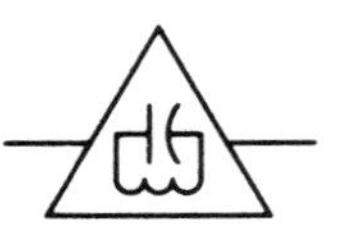

inductance-capacitance circuit discontinuity with infinite reactance at resonance

inductance-capacitance circuit discontinuity with infinite reactance, zero susceptance at resonance

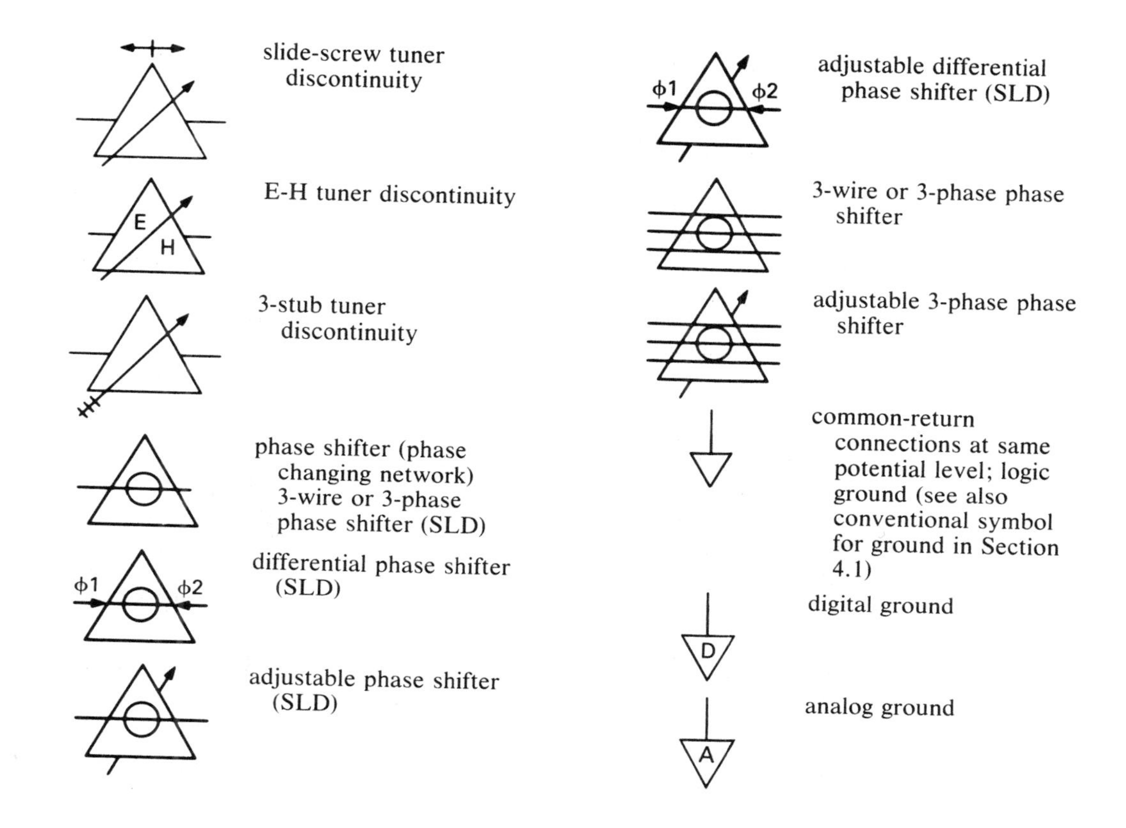

4.7 CIRCULAR-SHAPED SYMBOLS

(Meters, motors, generators, electron tubes, relays, lamps. Refer to Section 4.8 for semiconductor devices.)

basic rotating machine; relay coil

(*) where * is

A	ammeter
AH	ampere-hour meter
C	coulombmeter
CMA	contact-making (or breaking) ammeter
CMC	contact-making (or breaking) clock
CMV	contact-making (or breaking) voltmeter
CRO	oscilloscope
DB	decibel meter
DBM	decibels-referred-to-1 milliwatt meter
DM	demand meter
DTR	demand-totalizing relay
F	frequency meter

G	generator or galvanometer
$\underline{G}$	dc generator
$\underset{\sim}{G}$	ac generator
GD	ground detector
GEN	generator
GL	green signal light
GS	synchronous generator
Hz	frequency meter
I	indicating meter
INT	integrating meter
M	motor
$\underline{M}$	dc motor
$\underset{\sim}{M}$	ac motor
MA	milliammeter
MOT	motor
MS	synchronous motor
NaCl	salinity meter
NM	noise meter
OHM	ohmmeter
OP	oil pressure meter
PF	power factor meter
PH	phasemeter
PI	position indicator
RD	recording demand meter
REC	recording meter
RF	reactive factor meter
SY	synchroscope
t°	temperature meter
T	temperature meter (obsolete)
THC	thermal converter
TLM	telemeter
TT	total-time meter
UA	microammeter
V	voltmeter
VA	volt-ammeter
VAR	varmeter
VARH	varhour meter
VI	volume indicator
VU	audio-level meter
W	watthour meter
~	alternating current oscillator
μA	microammeter

$\cos\varphi$	power-factor meter
n	tachometer
A $I\sin\varphi$	reactive current ammeter
V U_d	differential voltmeter
W P_{max}	maximum demand indicator
α	coefficient scaler
θ	thermometer
φ	phase meter

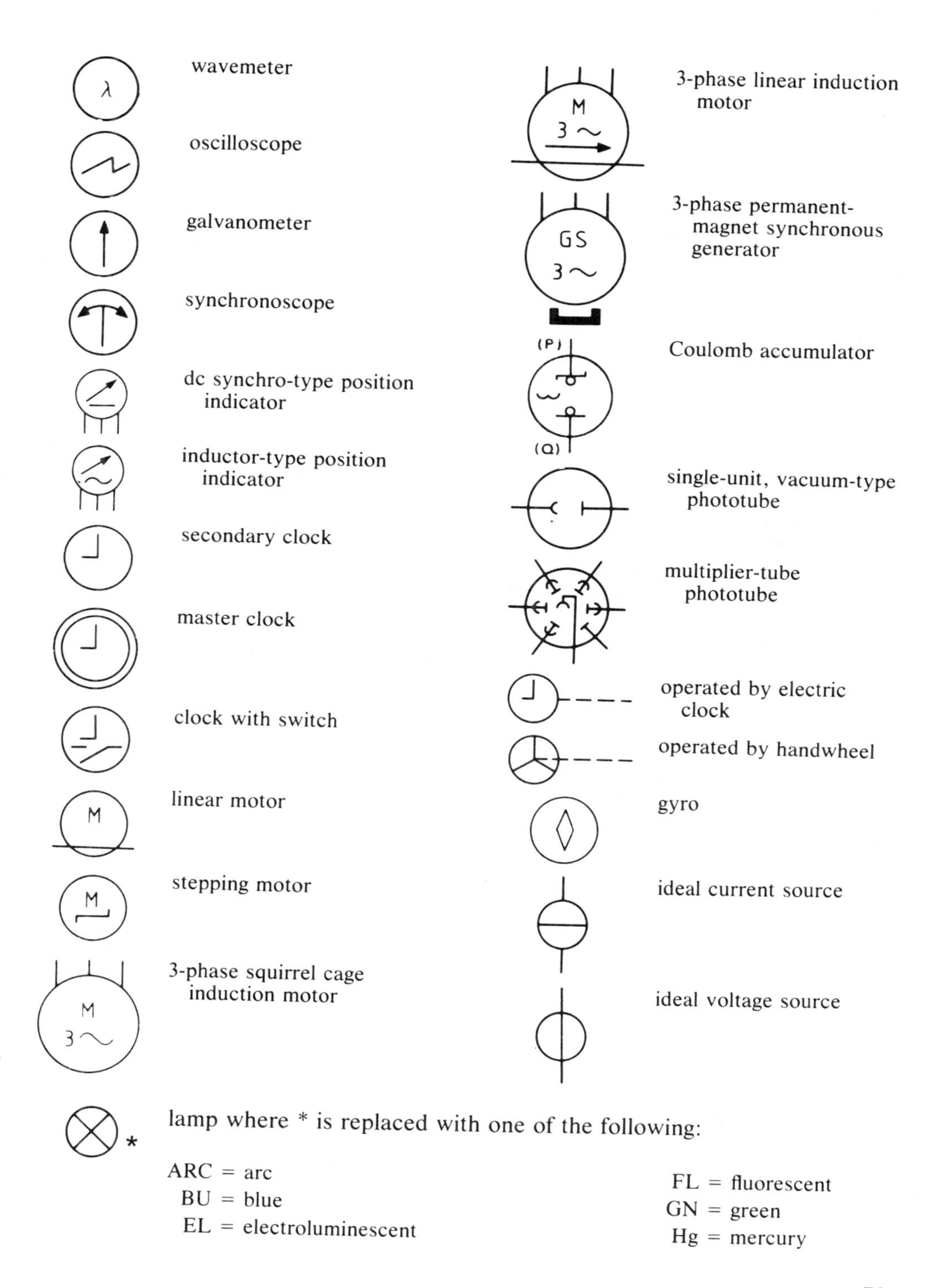

λ
wavemeter
oscilloscope
galvanometer
synchronoscope
dc synchro-type position indicator
inductor-type position indicator
secondary clock
master clock
clock with switch
M
linear motor
M
stepping motor
M
3 ∼
3-phase squirrel cage induction motor
M
3 ∼
3-phase linear induction motor
GS
3 ∼
3-phase permanent-magnet synchronous generator
(P)
(Q)
Coulomb accumulator
single-unit, vacuum-type phototube
multiplier-tube phototube
operated by electric clock
operated by handwheel
gyro
ideal current source
ideal voltage source
*
lamp where * is replaced with one of the following:
ARC = arc
BU = blue
EL = electroluminescent
FL = fluorescent
GN = green
Hg = mercury

I = iodine
IN = incandescent
IR = infrared
LED = light-emitting diode
Na = sodium vapor
Ne = neon
RD = red
UV = ultraviolet
WH = white
YE = yellow
Xe = xenon

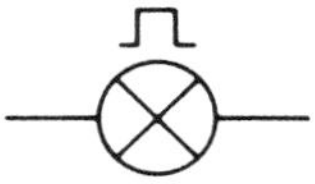

flashing signal lamp

electromechanical indicator

electromechanical position indicator

low-noise ground

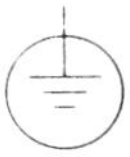

safety or protective ground

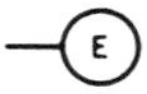

E-plane coupling by aperture to space

squirrel-cage induction motor or generator (SLD)

squirrel-cage induction motor or generator

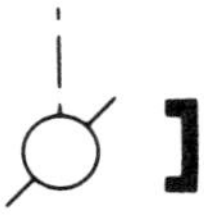

permanent-magnet-field generator or motor (SLD)

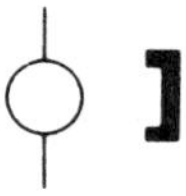

permanent-magnet-field generator or motor

microphone (SLD)

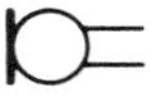

microphone

transducer or accelerometer (SLD)

electric squib (explosive)

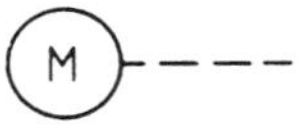

operated by electric motor

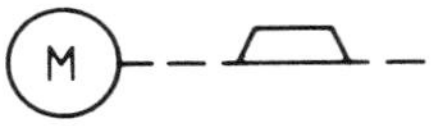

electric motor—brake applied

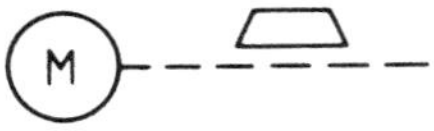

electric motor—brake released

gas-filled tube

liquid-filled tube

slip-ring and brush

coaxial waveguide

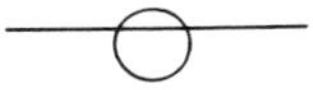

overhead line; waveguide

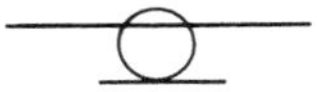

coaxial cable

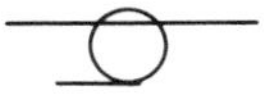

coaxial cable not maintained on the right

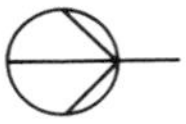

transducer or accelerometer

ballast lamp or tube

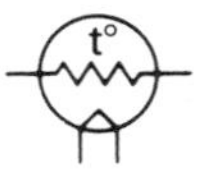

thermistor with independent integral heater

nonlinear thermistor with independent integral heater

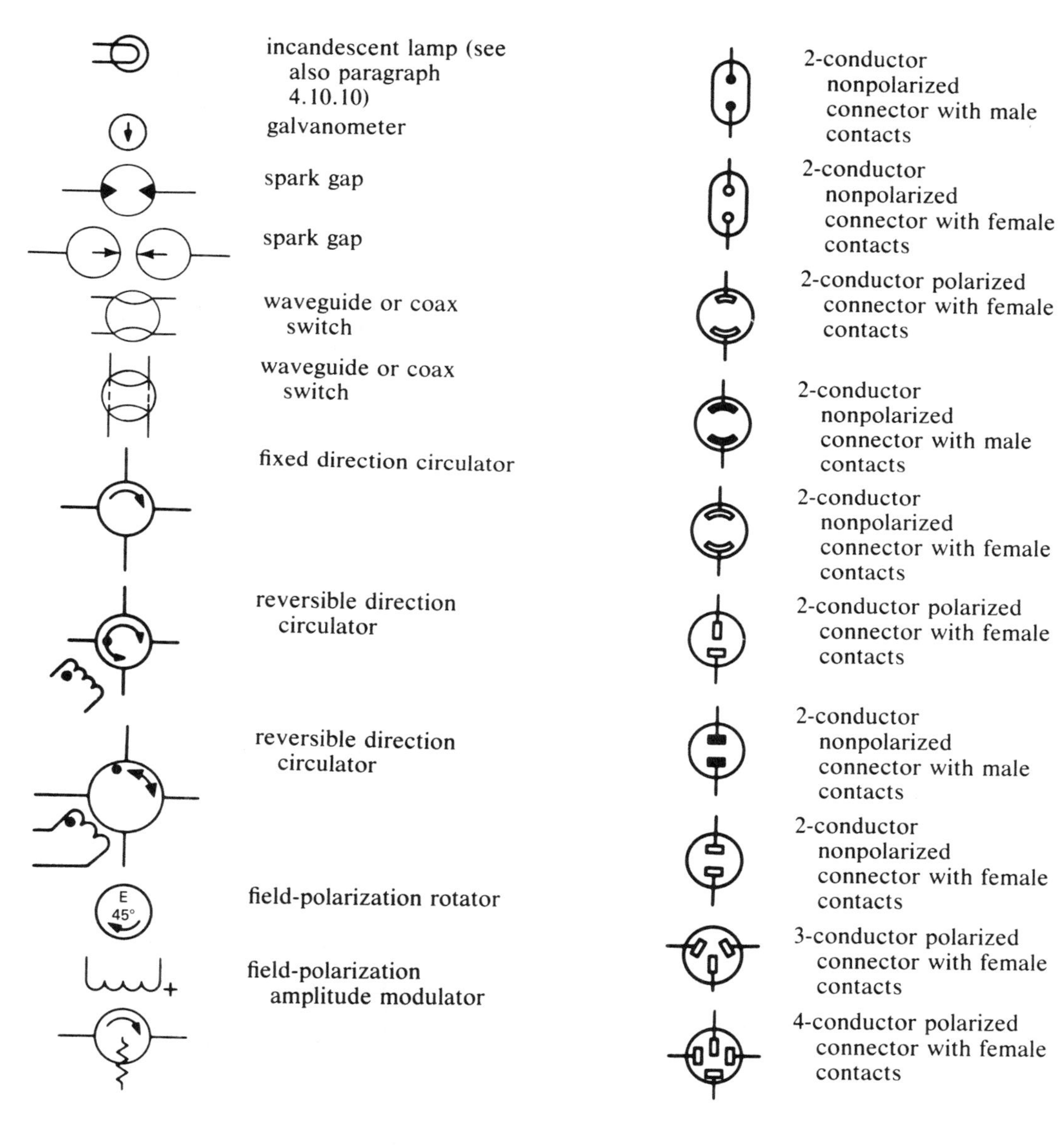

incandescent lamp (see also paragraph 4.10.10)

galvanometer

spark gap

spark gap

waveguide or coax switch

waveguide or coax switch

fixed direction circulator

reversible direction circulator

reversible direction circulator

field-polarization rotator

field-polarization amplitude modulator

2-conductor nonpolarized connector with male contacts

2-conductor nonpolarized connector with female contacts

2-conductor polarized connector with female contacts

2-conductor nonpolarized connector with male contacts

2-conductor nonpolarized connector with female contacts

2-conductor polarized connector with female contacts

2-conductor nonpolarized connector with male contacts

2-conductor nonpolarized connector with female contacts

3-conductor polarized connector with female contacts

4-conductor polarized connector with female contacts

Single circle with attachment

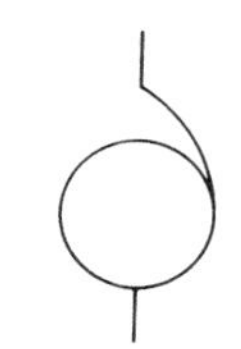

single-phase autotransformer (SLD)

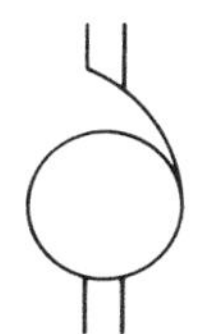

single-phase autotransformer

adjustable single-phase transformer (SLD)

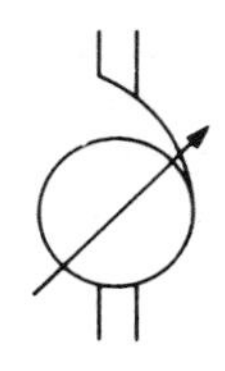

adjustable single-phase transformer

Circle within a circle

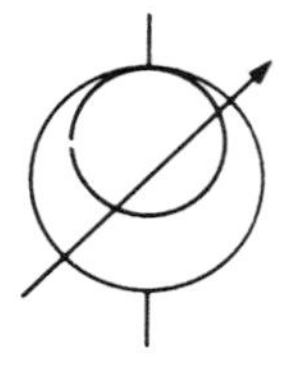

single-phase induction voltage regulator (SLD)

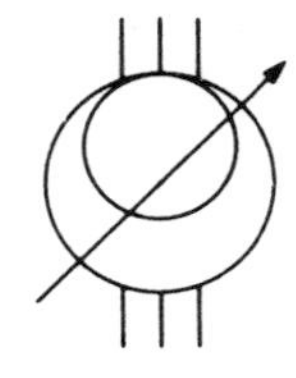

3-phase induction voltage regulator

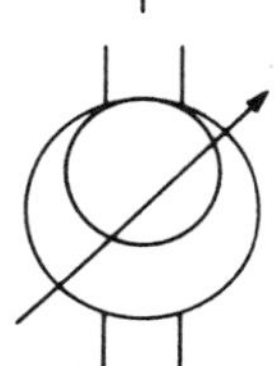

single-phase induction voltage regulator

Overlapping circles

transformer with one winding with adjustable inductance (SLD)

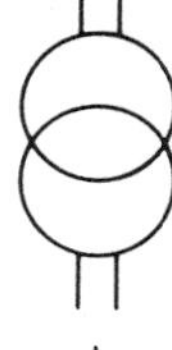

single-phase, 2-winding transformer

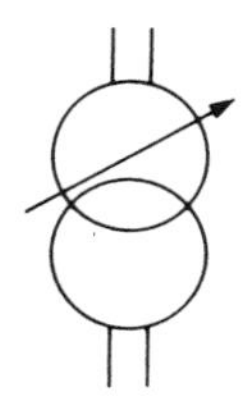

transformer with one winding with adjustable inductance (obsolete)

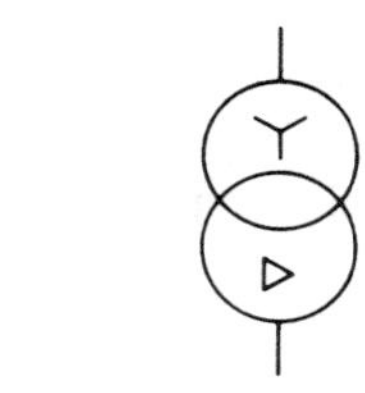

3-phase bank of single-phase, 2-winding transformers, wye (Y)–delta (▷) connections (SLD)

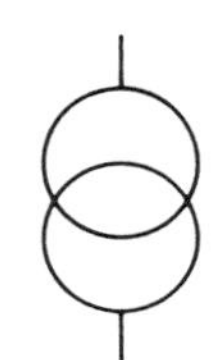

single-phase, 2-winding transformer (SLD)

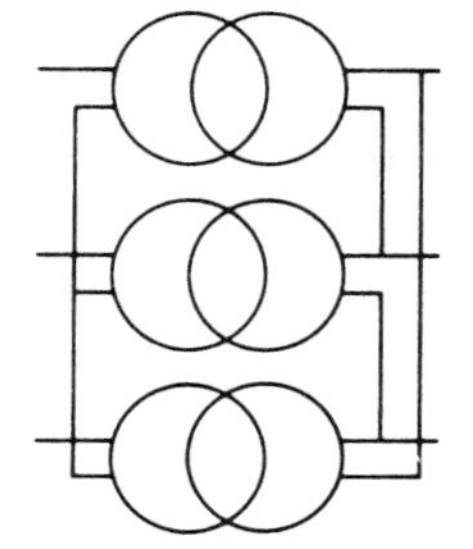

3-phase bank of single-phase, 2-winding transformers, wye (Y)–delta (▷) connections

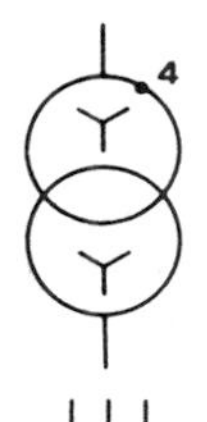

3-phase transformers with wye–wye connections (SLD)

3-phase transformer with wye–wye connections

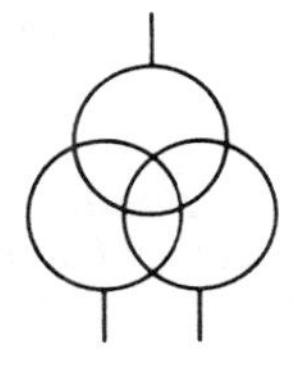

single-phase, 3-winding transformer (SLD)

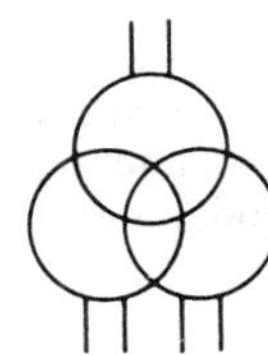

single-phase, 3-winding transformer

Circles with inner solid circle

neon lamp (ac type)

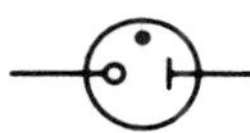

neon lamp (dc type)

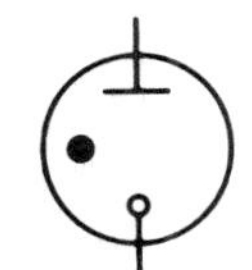

rectifier; dc voltage regulator

pool-type cathode power rectifier

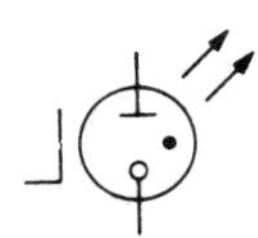

electronic flash tube

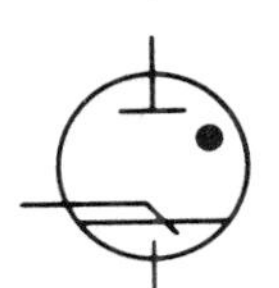

single-anode pool-type vapor rectifier with ignitor

thyratron with indirectly heated cathode

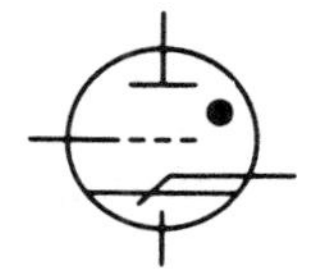

thyratron with ignitor and control grid

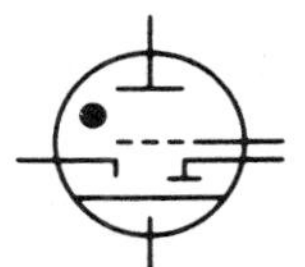

thyratron with excitor, control grid, and holding anode

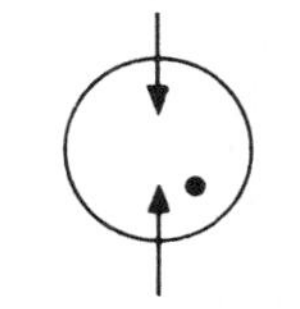

protective gas discharge tube

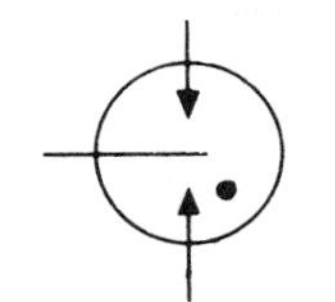

symmetric protective gas discharge tube

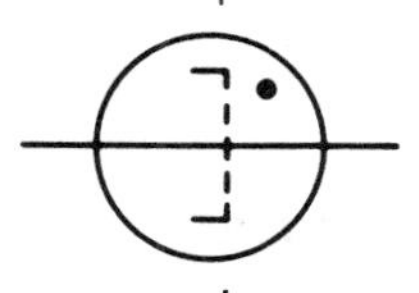

transmit-receive (T-R) tube

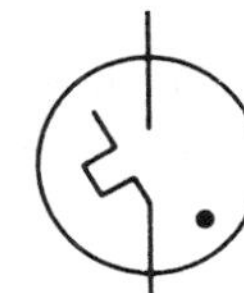

fluorescent lamp starter

Circles with exterior arrows pointing away

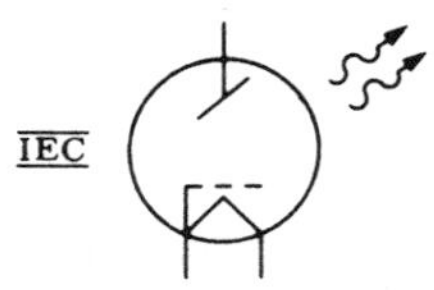

X-ray tube with filamentary cathode and focusing grid

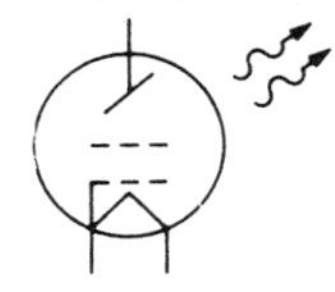

X-ray tube with control grid, filamentary cathode, and focusing cup

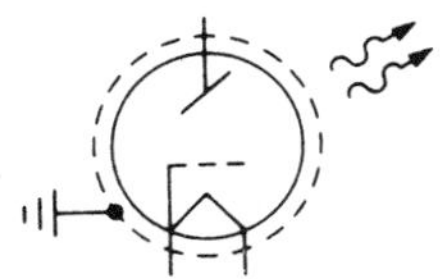

X-ray tube with grounded electrostatic shield

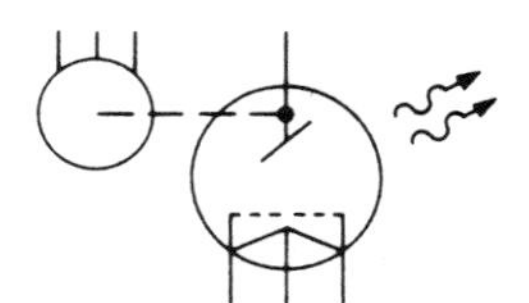

X-ray tube with double focus with rotating anode

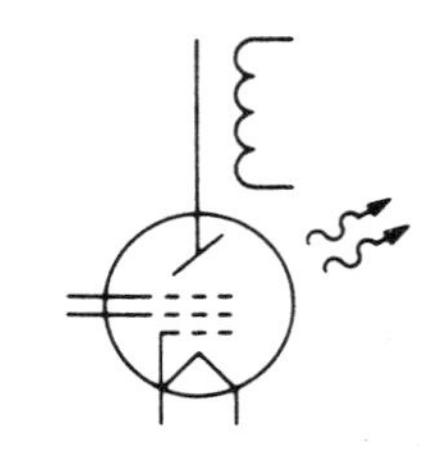

X-ray tube with multiple accelerating electrode electrostatically and electromagnetically focused

Circles with exterior arrows pointing toward circles

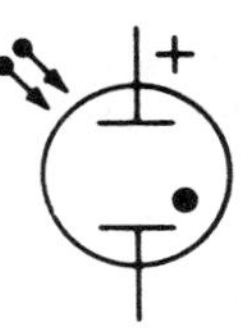

ionization chamber

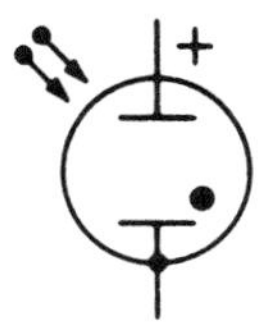

metal enclosed ionization chamber (one collector connected to enclosure)

Faraday cup

counter tube

counter tube with guard ring

Circles with internal broken lines

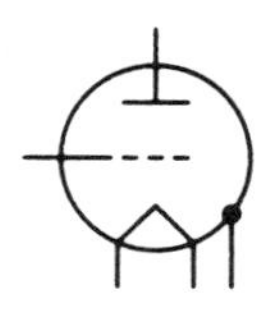

triode with directly heated cathode and envelope connected to base terminal

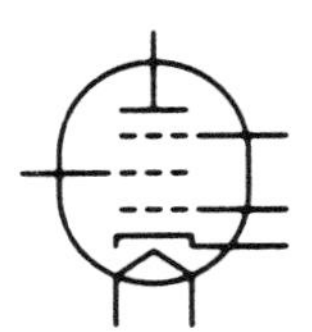

equipotential-cathode pentode

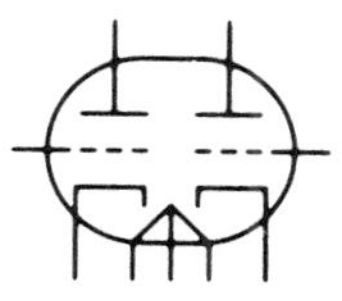

equipotential-cathode twin triode

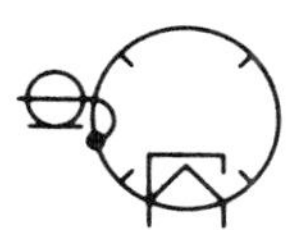

resonant type magnetron with coaxial output

transit-time split-plate type magnetron with stabilizing deflecting electrodes and internal circuit

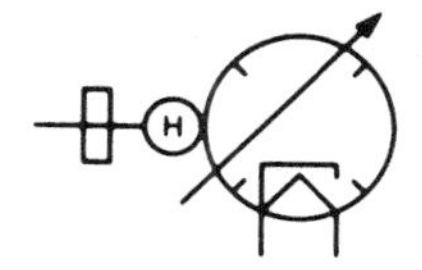

tunable magnetron, aperture coupled

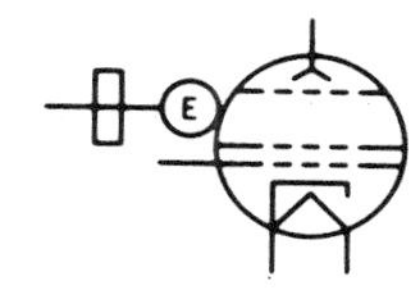

reflex klystron, integral cavity, aperture coupled

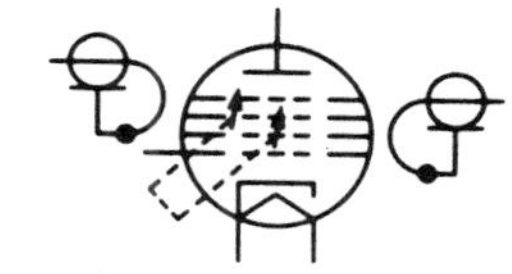

double-cavity klystron, integral cavity

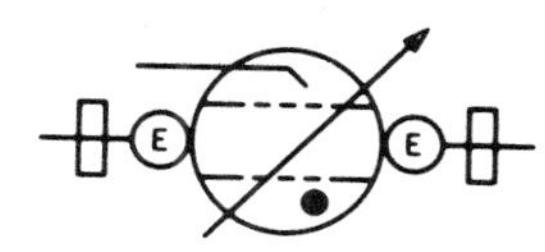

gas-filled transmit-receive (TR) tube with tunable integral cavity, aperture coupled

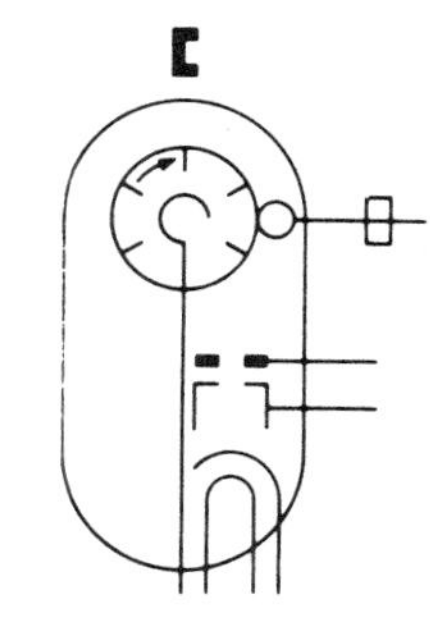

backward-wave oscillator tube with window-coupler to rectangular waveguide

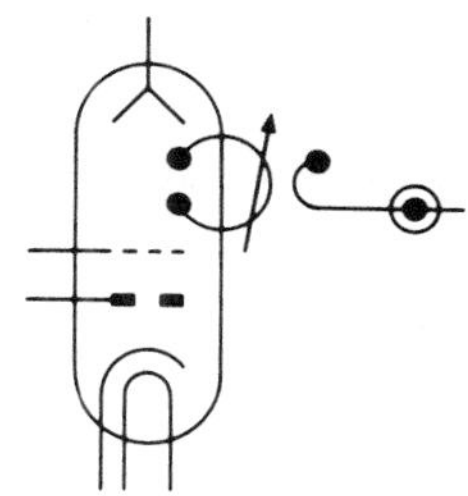

reflex klystron with loop coupler to coaxial output

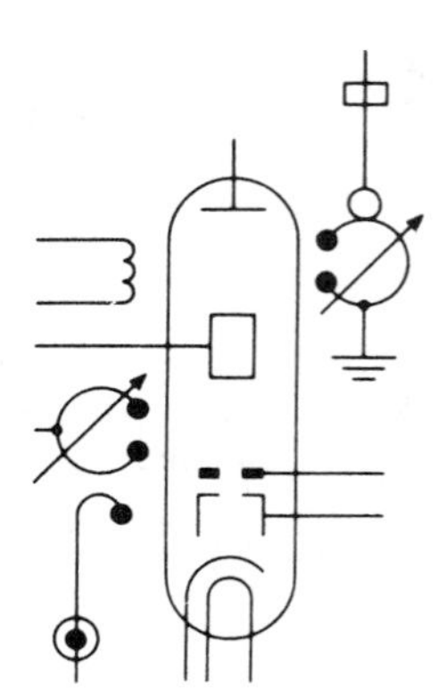

klystron with output window-coupler to rectangular waveguide

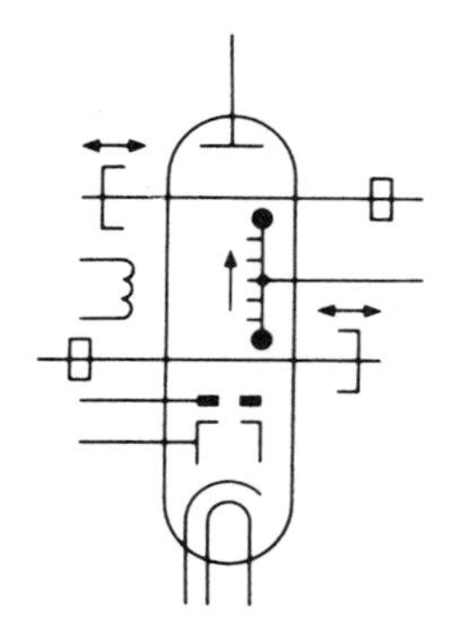

O-type forward traveling wave amplifier tube

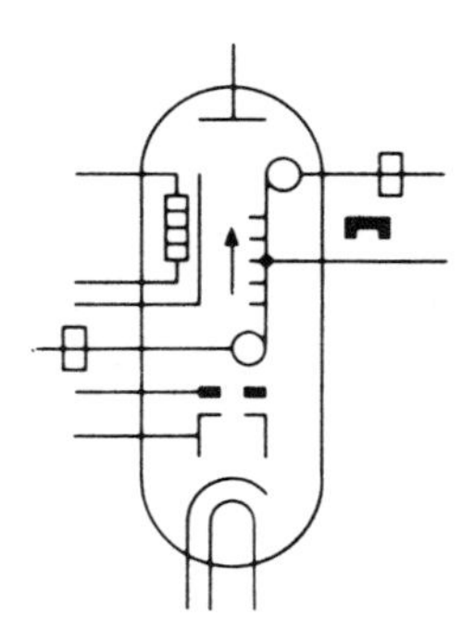

M-type forward traveling wave amplifier tube with window-couplers to rectangular waveguides

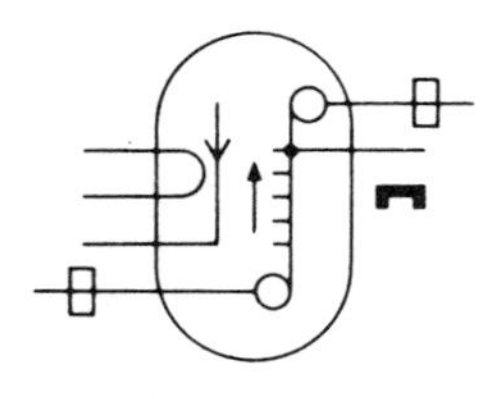

M-type backward wave amplifier tube with window-couplers to rectangular waveguides

Stretched circles and related cylindrical shapes

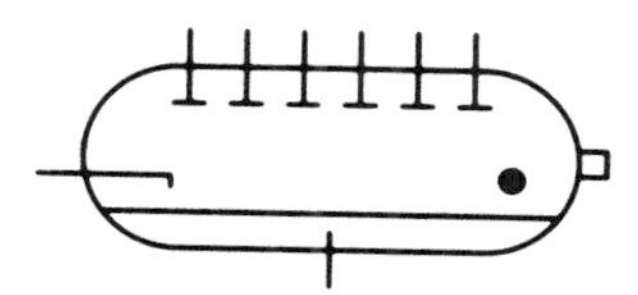

6-anode metallic-tank pool-type vapor rectifier

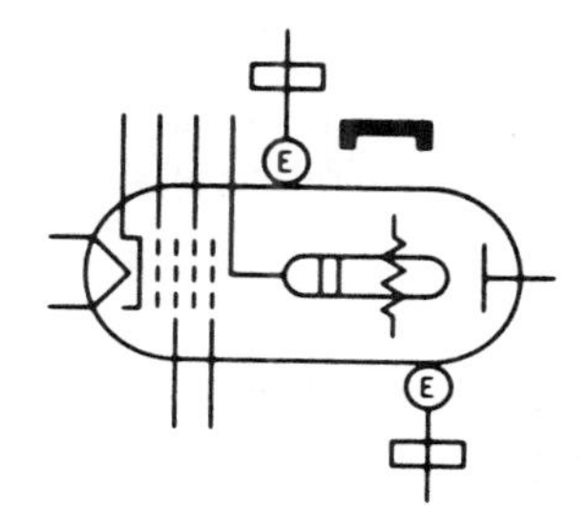

forward-wave traveling-wave-tube amplifier

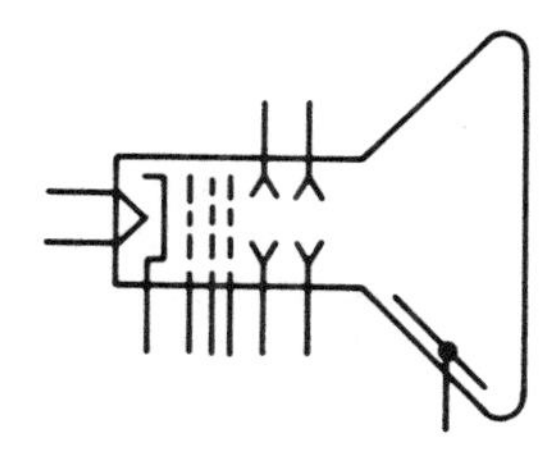

cathode-ray tube with electric field deflection

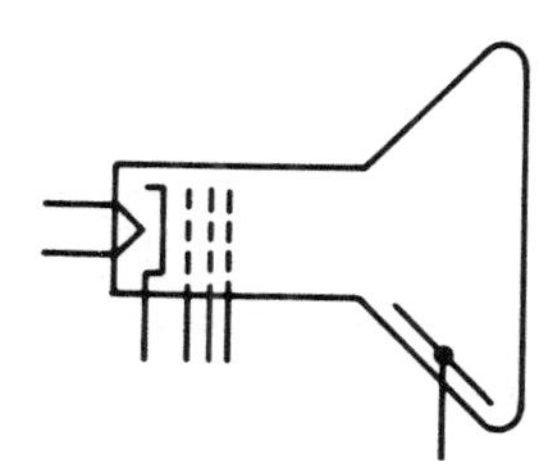

single-gun cathode-ray tube with electromagnetic deflection

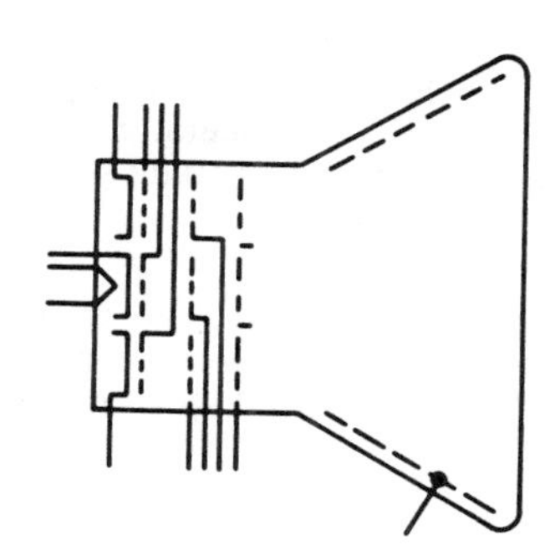

three-gun cathode-ray tube with electromagnetic deflection

Single circle, single winding

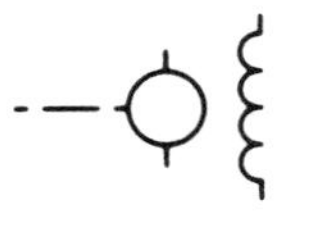

dc generator or motor, separately excited (SLD only)

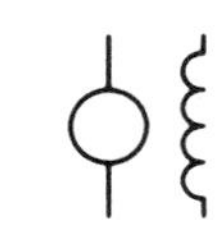

dc generator or motor, separately excited

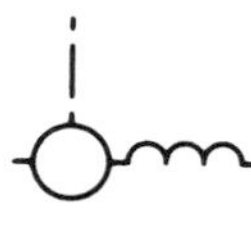

dc or ac series motor or 2-wire generator (SLD only)

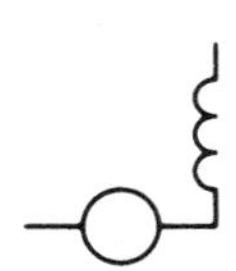

dc or ac series motor or 2-wire generator

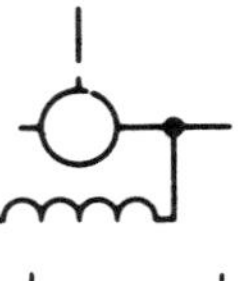

dc shunt motor or 2-wire generator (SLD only)

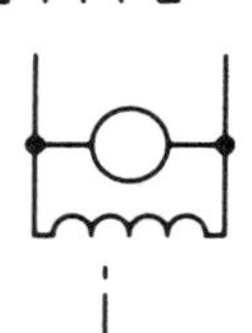

dc shunt motor or 2-wire generator

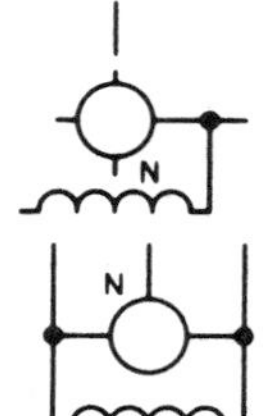

dc 3-wire shunt generator (SLD only)

dc 3-wire shunt generator

Single circle plus 2 windings

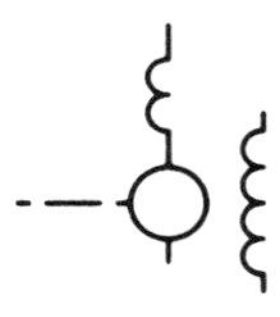

dc generator or motor, separately excited, with commutating or compensating field winding (SLD only)

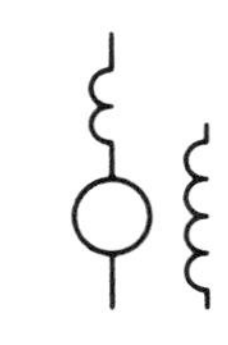

dc generator or motor, separately excited, with commutating or compensating field winding

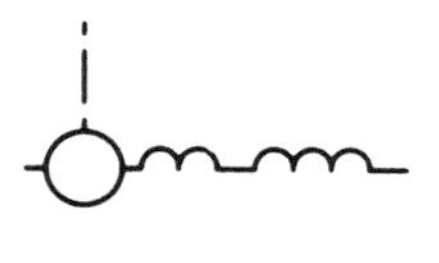

dc series motor or 2-wire generator with commutating or compensating field winding (SLD only)

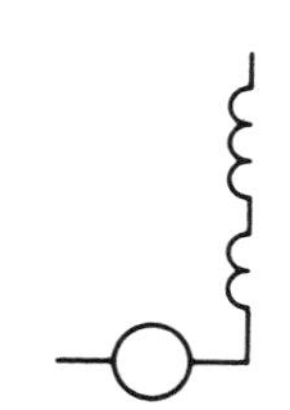

dc series motor or 2-wire generator with commutating or compensating field winding

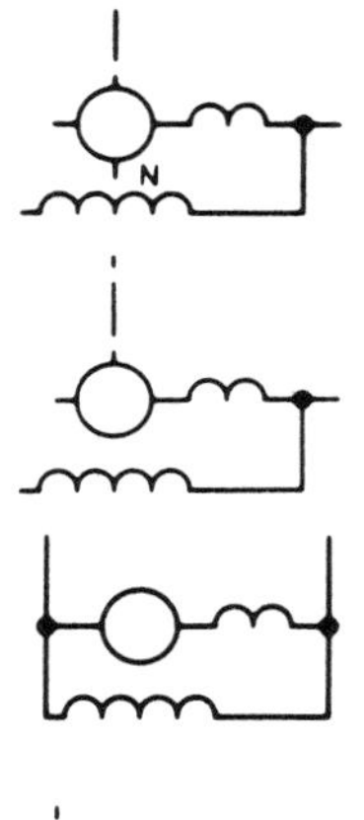

dc 3-wire shunt generator, with commutating or compensating field winding (SLD only)

dc shunt motor or 2-wire generator, with commutating or compensating field winding (SLD)

dc shunt motor or 2-wire generator, with commutating or compensating field winding

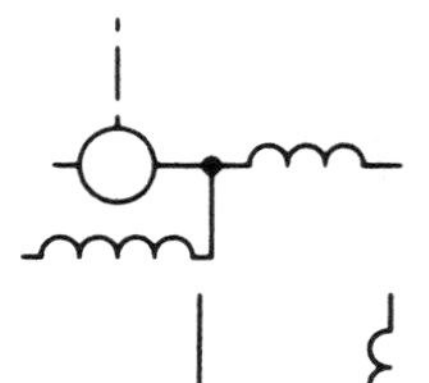

dc compound motor or 2-wire generator or stabilized shunt motor (SLD)

dc compound motor or 2-wire generator or stabilized shunt motor

Single circle plus 3 or more windings

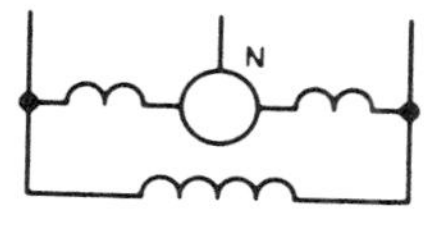

dc 3-wire shunt generator, with commutating or compensating field winding

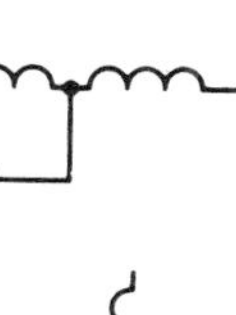

dc compound motor or 2-wire generator or stabilized shunt motor, with commutating or compensating field winding (SLD)

dc compound motor or 2-wire generator or stabilized shunt motor, with commutating or compensating field winding

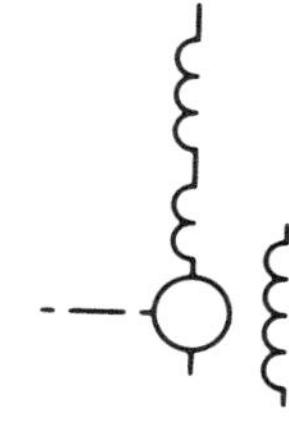

dc motor or generator, compositely excited, with commutating or compensating field winding (SLD)

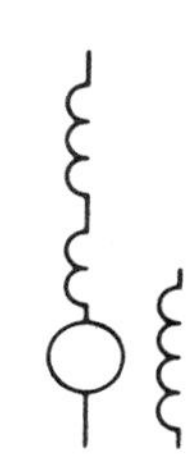

dc motor or generator, compositely excited, with commutating or compensating field winding

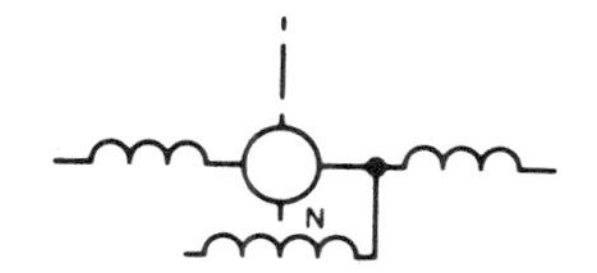

dc 3-wire compound compound generator (SLD)

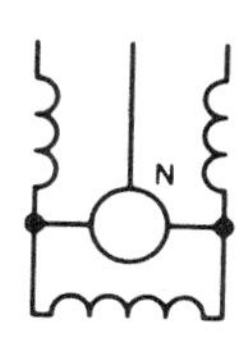

dc 3-wire compound generator

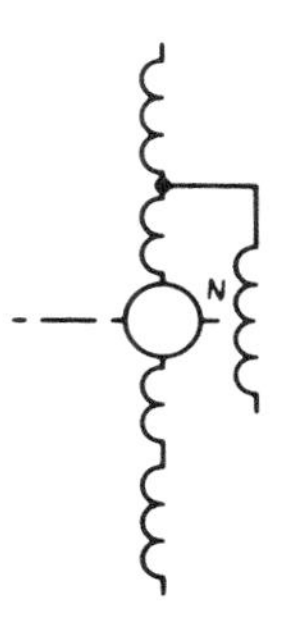

dc 3-wire compound generator with commutating or compensating field winding (SLD)

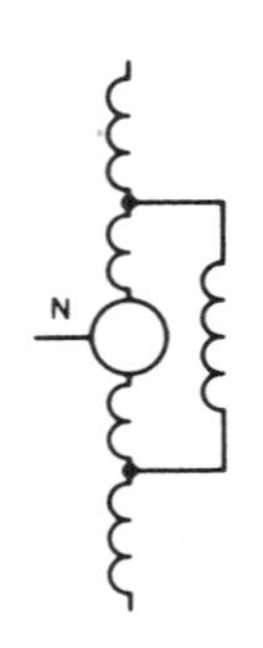

dc 3-wire compound generator, with commutating or compensating field winding

Half circle

microphone (SLD)

microphone

electrical bell (SLD)

electrical bell

single-stroke electrical bell (SLD)

single-stroke electrical bell

buzzer (SLD)

buzzer

Half circle plus whole circle

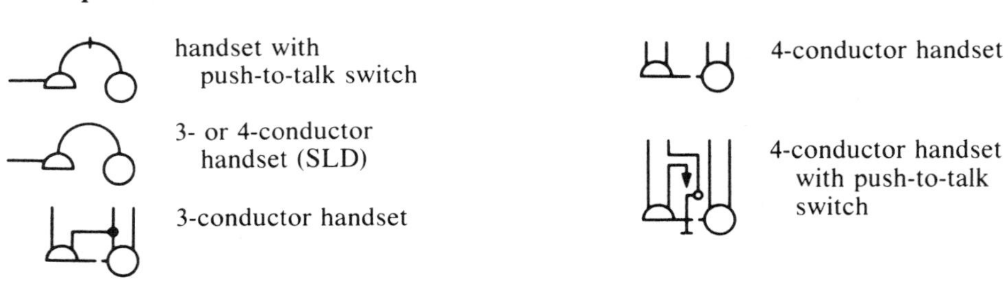

handset with push-to-talk switch

3- or 4-conductor handset (SLD)

3-conductor handset

4-conductor handset

4-conductor handset with push-to-talk switch

Single doughnut type circle

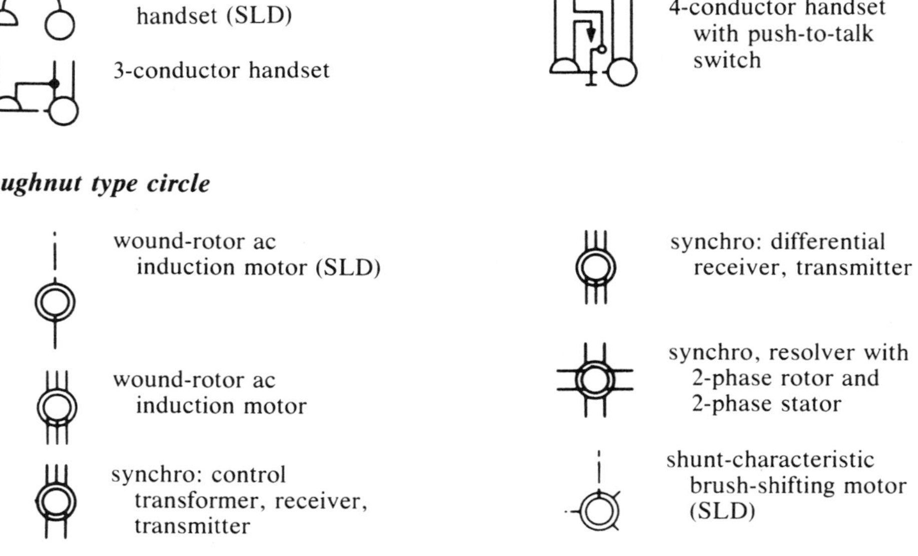

wound-rotor ac induction motor (SLD)

wound-rotor ac induction motor

synchro: control transformer, receiver, transmitter

synchro: differential receiver, transmitter

synchro, resolver with 2-phase rotor and 2-phase stator

shunt-characteristic brush-shifting motor (SLD)

shunt-characteristic brush-shifting motor

series-characteristic brush-shifting motor with 3-phase rotor (SLD)

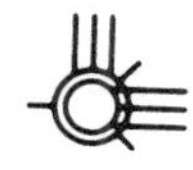

series-characteristic brush-shifting motor with 3-phase rotor

series-characteristic brush-shifting motor with 6- or 8-phase rotor (SLD)

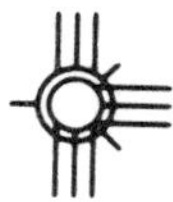

series-characteristic brush-shifting motor with 6- or 8-phase rotor

ohmic-drop exciter with 3- or 6-phase input (SLD)

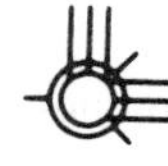

ohmic-drop exciter with 3- or 6-phase input

4.8 SEMICONDUCTOR SYMBOLS (CIRCLES ARE OPTIONAL) (See also page 68.)

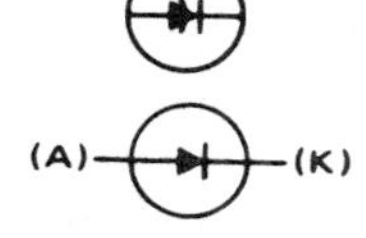

capacitive diode (varactor) (obsolete)

semiconductor diode; semiconductor rectifier; metallic rectifier; A = anode, K = cathode

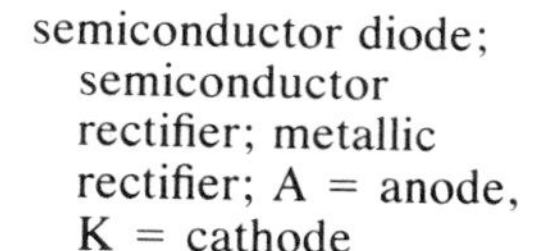

temperature-dependent diode

storage diode

unidirectional diode; zener diode

tunnel diode

backward diode; tunnel rectifier

reverse-blocking diode type thyristor

PIN-type diode

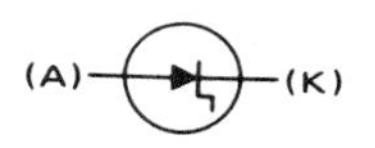

step recovery diode

tunnel diode

zener diode; voltage regulator

zener diode; voltage regulator

tunnel diode

backward diode, tunnel rectifier

backward diode, tunnel rectifier

varactor

varactor

bidirectional diode

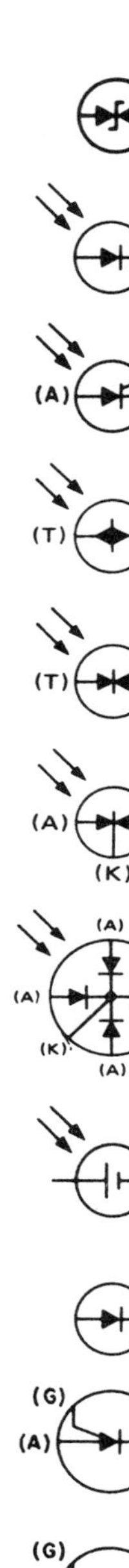

bidirectional diode

photosensitive
photodiode

light-activated thyristor

NPN photosensitive
photo-duo-diode

PNP photosensitive
photo-duo-diode

photosensitive
photodiode;
2-segment, with
common cathode lead

photosensitive
photodiode; 4
quadrant, with
common cathode lead

photovoltaic transducer;
solar cell

light-emitting diode

N-type gate
reverse-blocking
triode-type thyristor
(SCR)

gate turn-off type
triode-type thryistor
(SCR)

P-type gate,
reverse-blocking,
triode-type thyristor
(SCR)

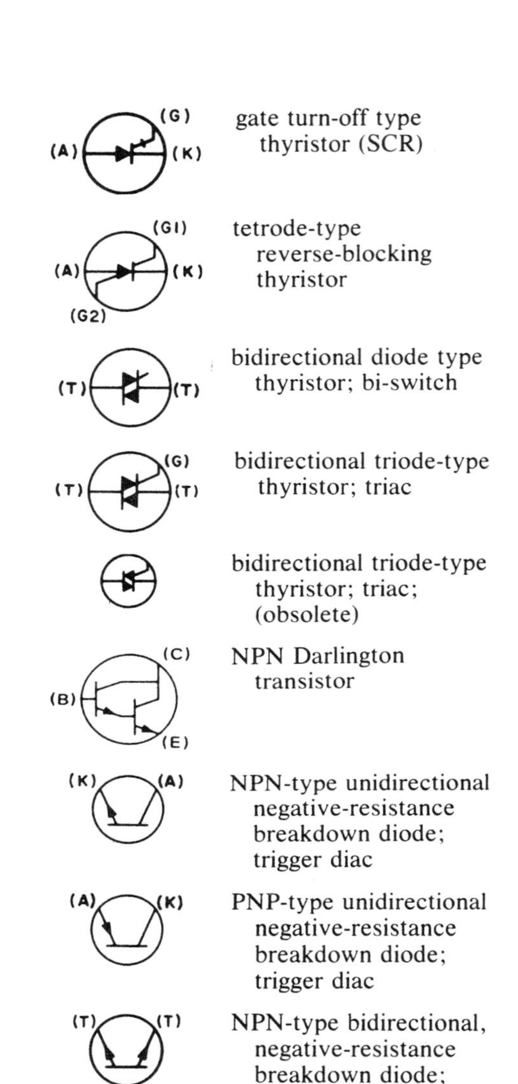

gate turn-off type
thyristor (SCR)

tetrode-type
reverse-blocking
thyristor

bidirectional diode type
thyristor; bi-switch

bidirectional triode-type
thyristor; triac

bidirectional triode-type
thyristor; triac;
(obsolete)

NPN Darlington
transistor

NPN-type unidirectional
negative-resistance
breakdown diode;
trigger diac

PNP-type unidirectional
negative-resistance
breakdown diode;
trigger diac

NPN-type bidirectional,
negative-resistance
breakdown diode;
trigger diac

PNP-type bidirectional
negative-resistance
breakdown diode;
trigger diac

reverse-blocking
diode-type thyristor

reverse-blocking
diode-type thyristor

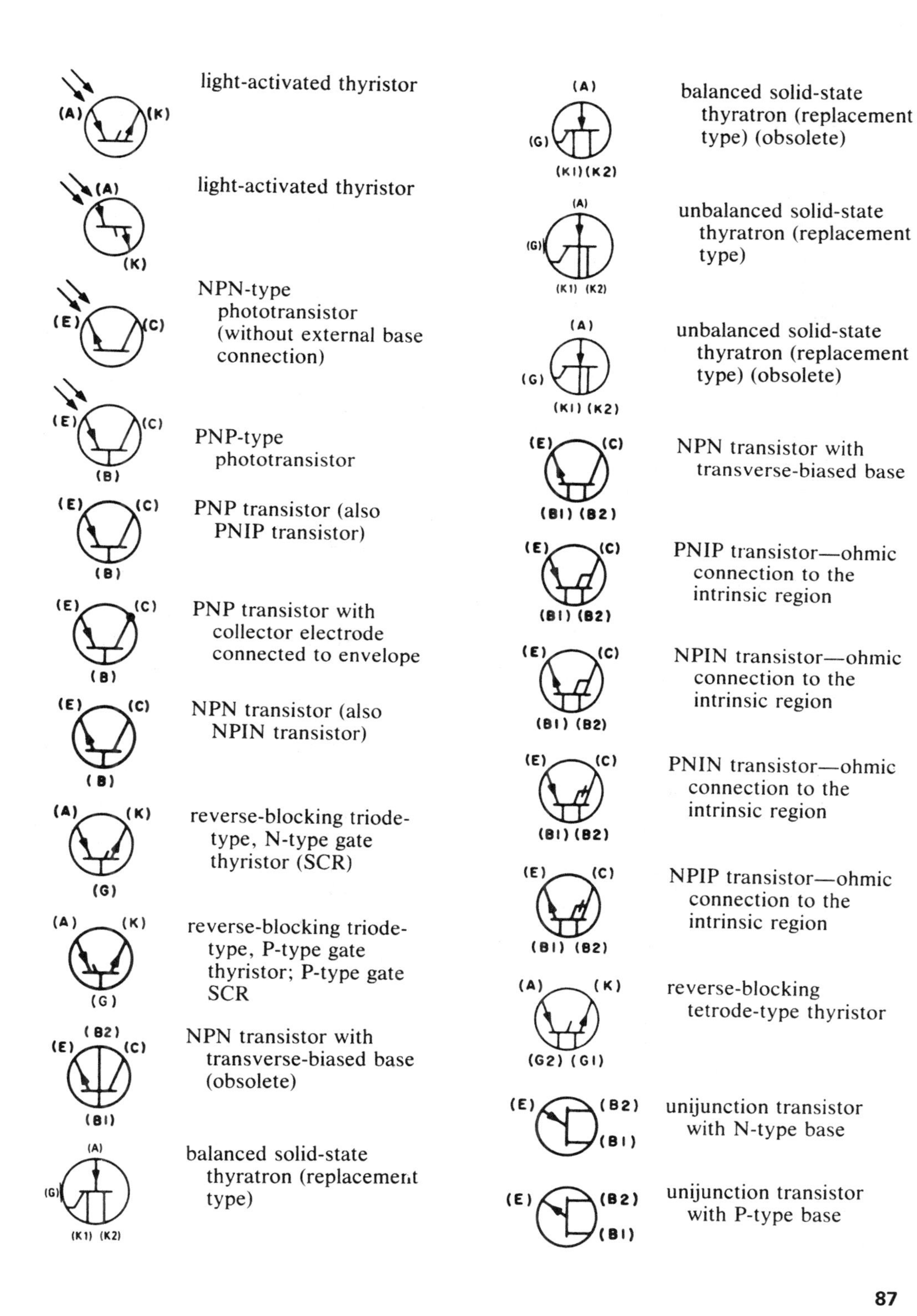
(A)
(K)
light-activated thyristor
(A)
(K)
light-activated thyristor
(E)
(C)
NPN-type phototransistor (without external base connection)
(E)
(C)
(B)
PNP-type phototransistor
(E)
(C)
(B)
PNP transistor (also PNIP transistor)
(E)
(C)
(B)
PNP transistor with collector electrode connected to envelope
(E)
(C)
(B)
NPN transistor (also NPIN transistor)
(A)
(K)
(G)
reverse-blocking triode-type, N-type gate thyristor (SCR)
(A)
(K)
(G)
reverse-blocking triode-type, P-type gate thyristor; P-type gate SCR
(B2)
(E)
(C)
(B1)
NPN transistor with transverse-biased base (obsolete)
(A)
(G)
(K1) (K2)
balanced solid-state thyratron (replacement type)
(A)
(G)
(K1)(K2)
balanced solid-state thyratron (replacement type) (obsolete)
(A)
(G)
(K1) (K2)
unbalanced solid-state thyratron (replacement type)
(A)
(G)
(K1) (K2)
unbalanced solid-state thyratron (replacement type) (obsolete)
(E)
(C)
(B1) (B2)
NPN transistor with transverse-biased base
(E)
(C)
(B1) (B2)
PNIP transistor—ohmic connection to the intrinsic region
(E)
(C)
(B1) (B2)
NPIN transistor—ohmic connection to the intrinsic region
(E)
(C)
(B1) (B2)
PNIN transistor—ohmic connection to the intrinsic region
(E)
(C)
(B1) (B2)
NPIP transistor—ohmic connection to the intrinsic region
(A)
(K)
(G2) (G1)
reverse-blocking tetrode-type thyristor
(E)
(B2)
(B1)
unijunction transistor with N-type base
(E)
(B2)
(B1)
unijunction transistor with P-type base

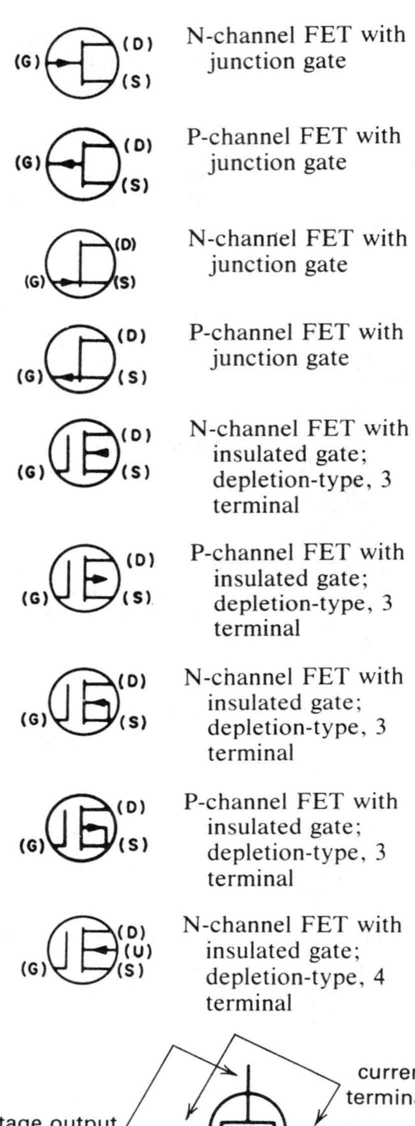

N-channel FET with junction gate

P-channel FET with junction gate

N-channel FET with junction gate

P-channel FET with junction gate

N-channel FET with insulated gate; depletion-type, 3 terminal

P-channel FET with insulated gate; depletion-type, 3 terminal

N-channel FET with insulated gate; depletion-type, 3 terminal

P-channel FET with insulated gate; depletion-type, 3 terminal

N-channel FET with insulated gate; depletion-type, 4 terminal

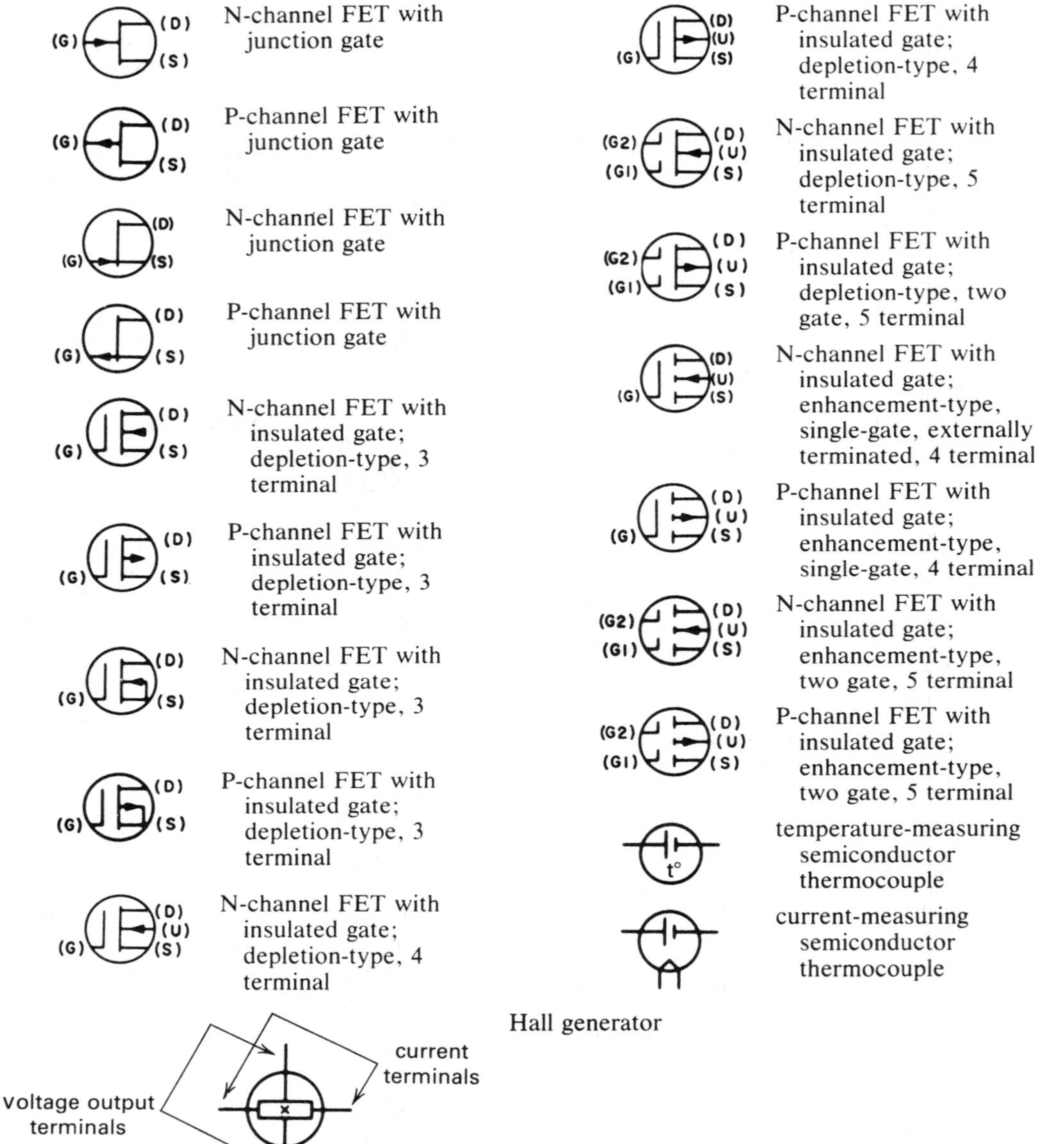

P-channel FET with insulated gate; depletion-type, 4 terminal

N-channel FET with insulated gate; depletion-type, 5 terminal

P-channel FET with insulated gate; depletion-type, two gate, 5 terminal

N-channel FET with insulated gate; enhancement-type, single-gate, externally terminated, 4 terminal

P-channel FET with insulated gate; enhancement-type, single-gate, 4 terminal

N-channel FET with insulated gate; enhancement-type, two gate, 5 terminal

P-channel FET with insulated gate; enhancement-type, two gate, 5 terminal

temperature-measuring semiconductor thermocouple

current-measuring semiconductor thermocouple

Hall generator

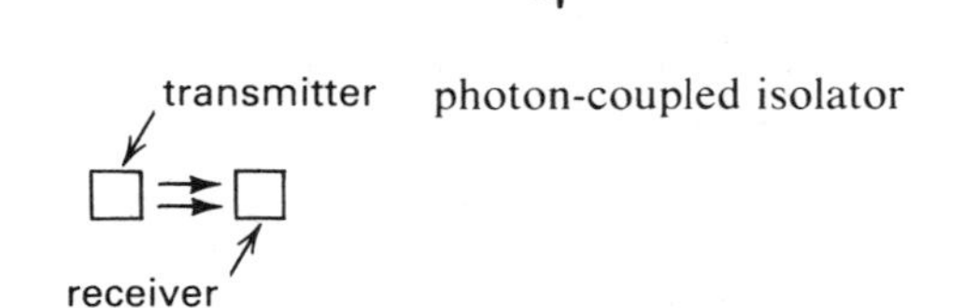

photon-coupled isolator

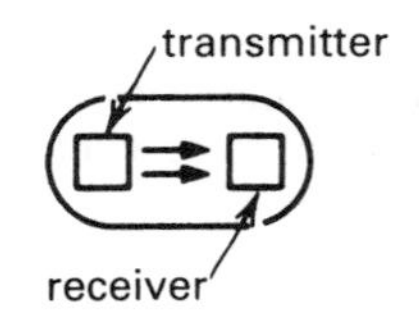

photon-coupled isolator (single-package type)

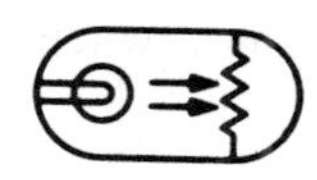

incandescent lamp and symmetrical photoconductive transducer type photon-coupled isolator

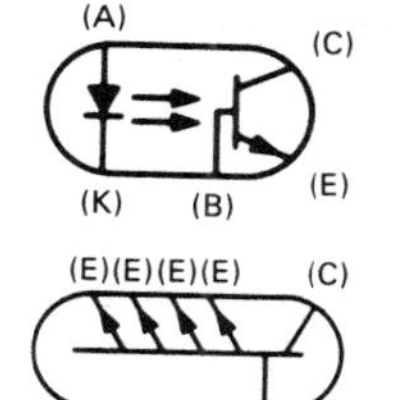

photoemissive diode and phototransistor type photon-coupled isolator

NPN transistor with multiple emitters (with four emitters shown)

4.9 BROKEN STRAIGHT LINES AND STRAIGHT LINES WITH ATTACHED SMALL CIRCLES AND ARROWS

(Switches, relays) (See Section 4.10.6 for parallel lines.)

Switches

single-throw switch

thermostat—closes on rising temperature

knife switch

thermostat with contact-motion direction clarified

horn-gap switch

double-throw switch

Switches—frequently in vertical position

switch or make contact

passing make contact

circuit breaker

isolator

thermostat—closes on rising temperature

make contact of a multiple contact assembly

changeover make-before-break contact

changeover break-before-make contact

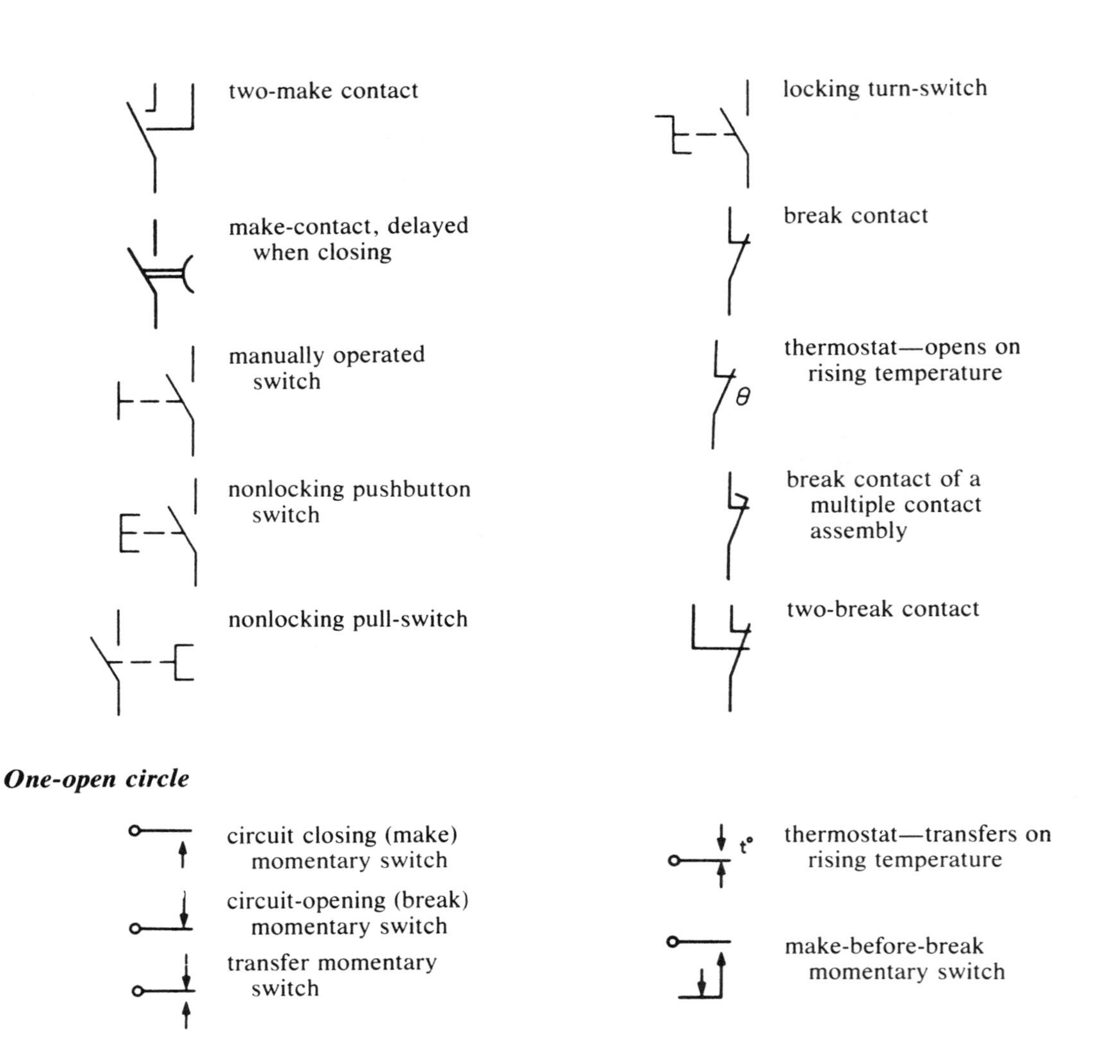

One-open circle—switch in vertical position

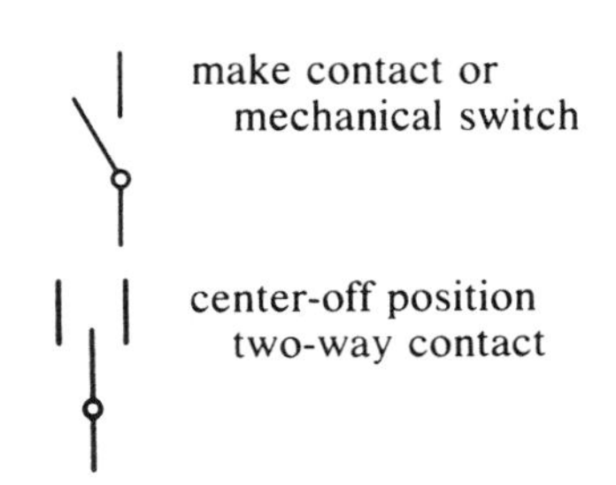

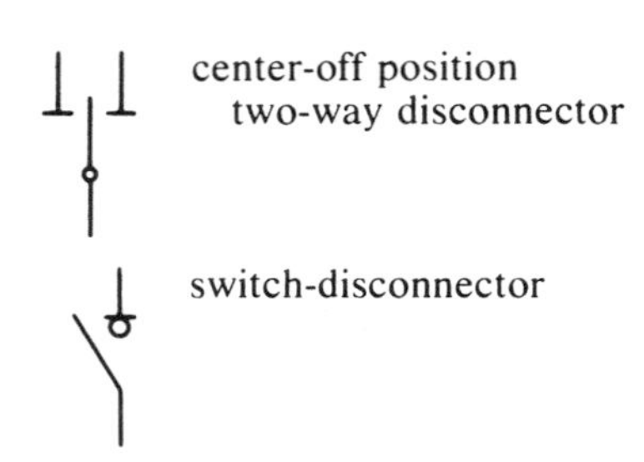

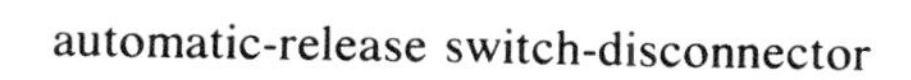

automatic-release switch-disconnector

Two-open circles

time-delay closing open switch

circuit-closing toggle switch

time-delay opening open switch

circuit-opening toggle switch

time-delay closing closed switch

time-delay opening closed switch

thermostat—opens on rising temperature

time-delay closing open switch

time-delay opening closed switch

time-delay opening open switch

time-delay closing closed switch

circuit closing pushbutton

circuit opening pushbutton

circuit opening interlock

circuit closing interlock

foot-operated switch (foot pressure opens)

foot-operated switch (foot pressure closes)

flow-actuated switch, closes on increase in flow

flow-actuated switch, opens on increase in flow

pressure-actuated switch, closes on rising pressure

pressure-actuated switch, opens on rising pressure

liquid-level-actuated switch, closes on rising level

liquid-level-actuated switch, opens on rising level

temperature-actuated switch, closes on rising temperature

temperature-actuated switch, opens on rising temperature

normally open, spring-returned limit switch

normally open held-closed spring-returned limit switch

normally closed spring-returned limit switch

normally closed held-open spring-returned limit switch

Three open circles

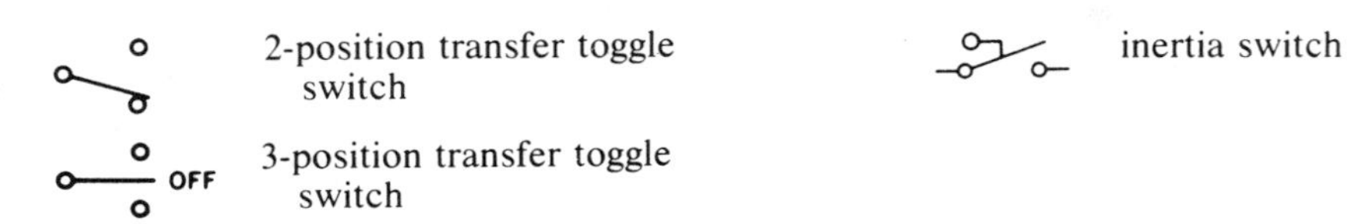

Four open circles

Multiple open circles

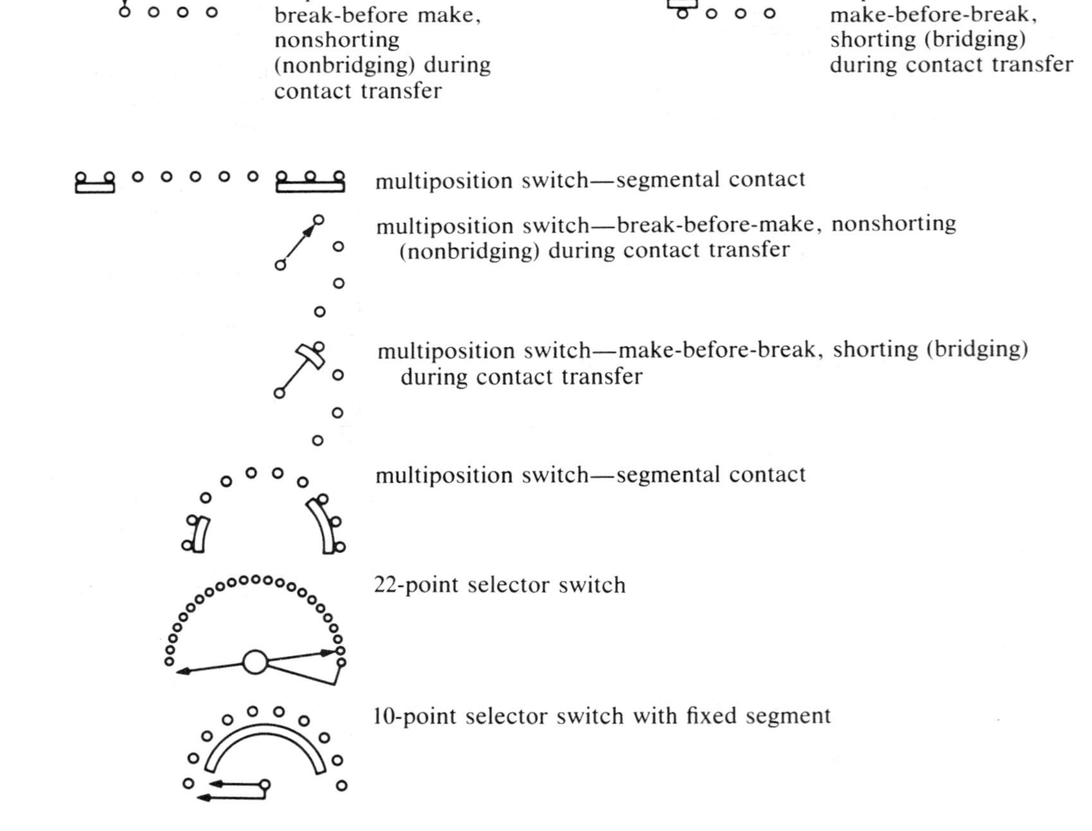

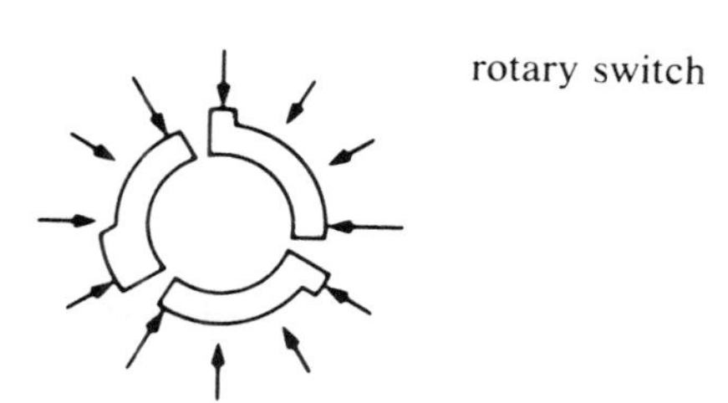

rotary switch

Open circles with jag at end of line

(key-type switch)

2-position with locking transfer and break contacts

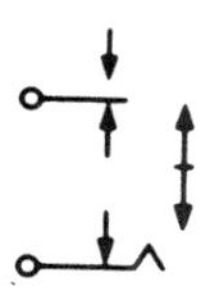

3-position with nonlocking transfer and locking break contacts

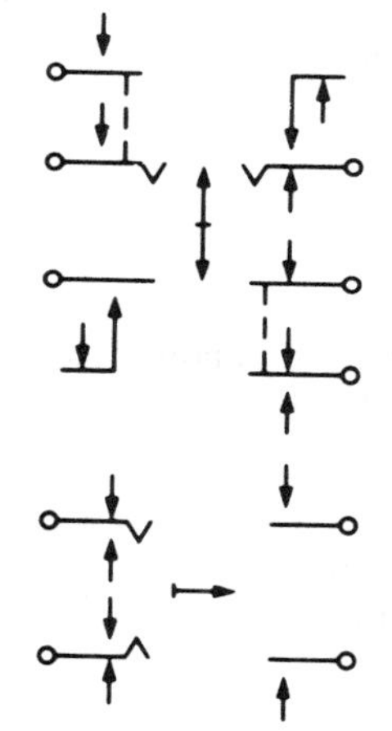

3-position, multicontact combination

2-position, half of key switch normally operated, multicontact combination

Open circles with offset arrows

circuit-closing toggle switch

circuit opening toggle switch

transfer toggle switch

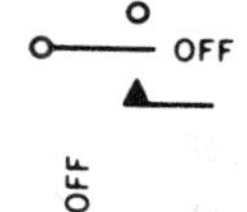

3-position 1-pole toggle switch

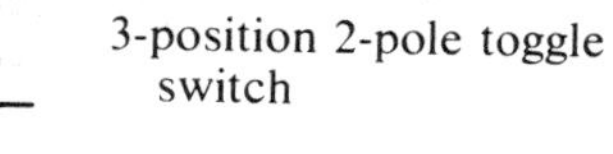

3-position 2-pole toggle switch

One open circle with jag at end of line

(locking switches used in key switches and jacks)

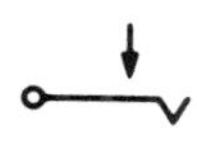

circuit closing (make)

circuit opening (break)

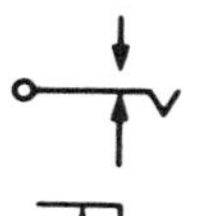

transfer, 2-position

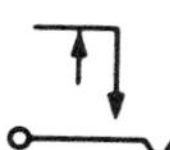

make-before-break

One or more open circles with jag at end of line

(plugs and jacks)

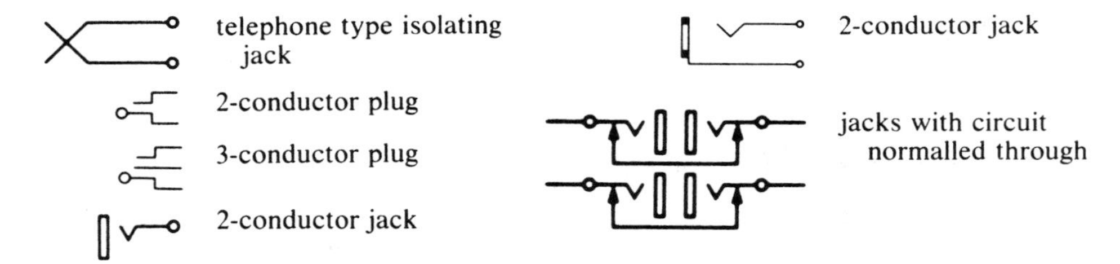

Half-circle or triangle

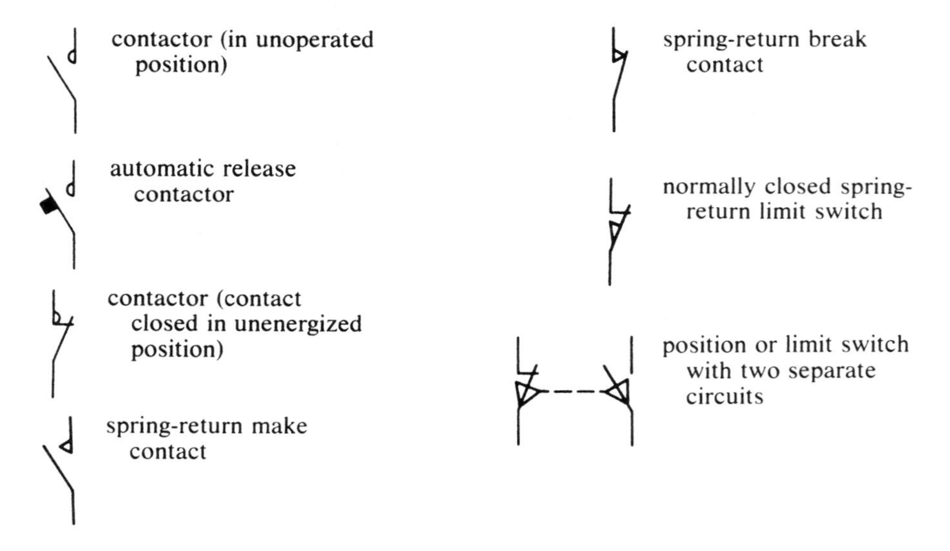

Parallel lines

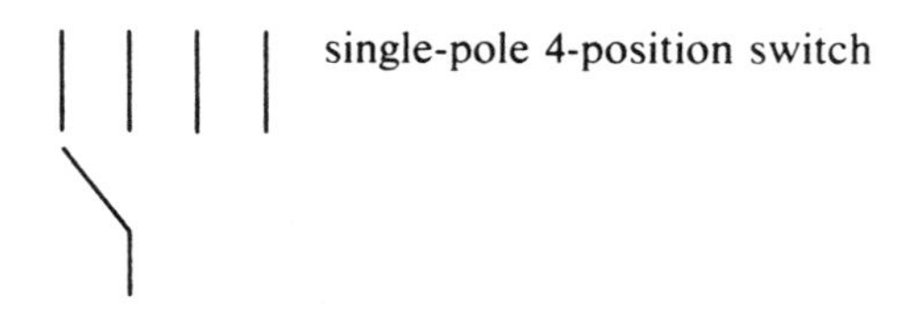

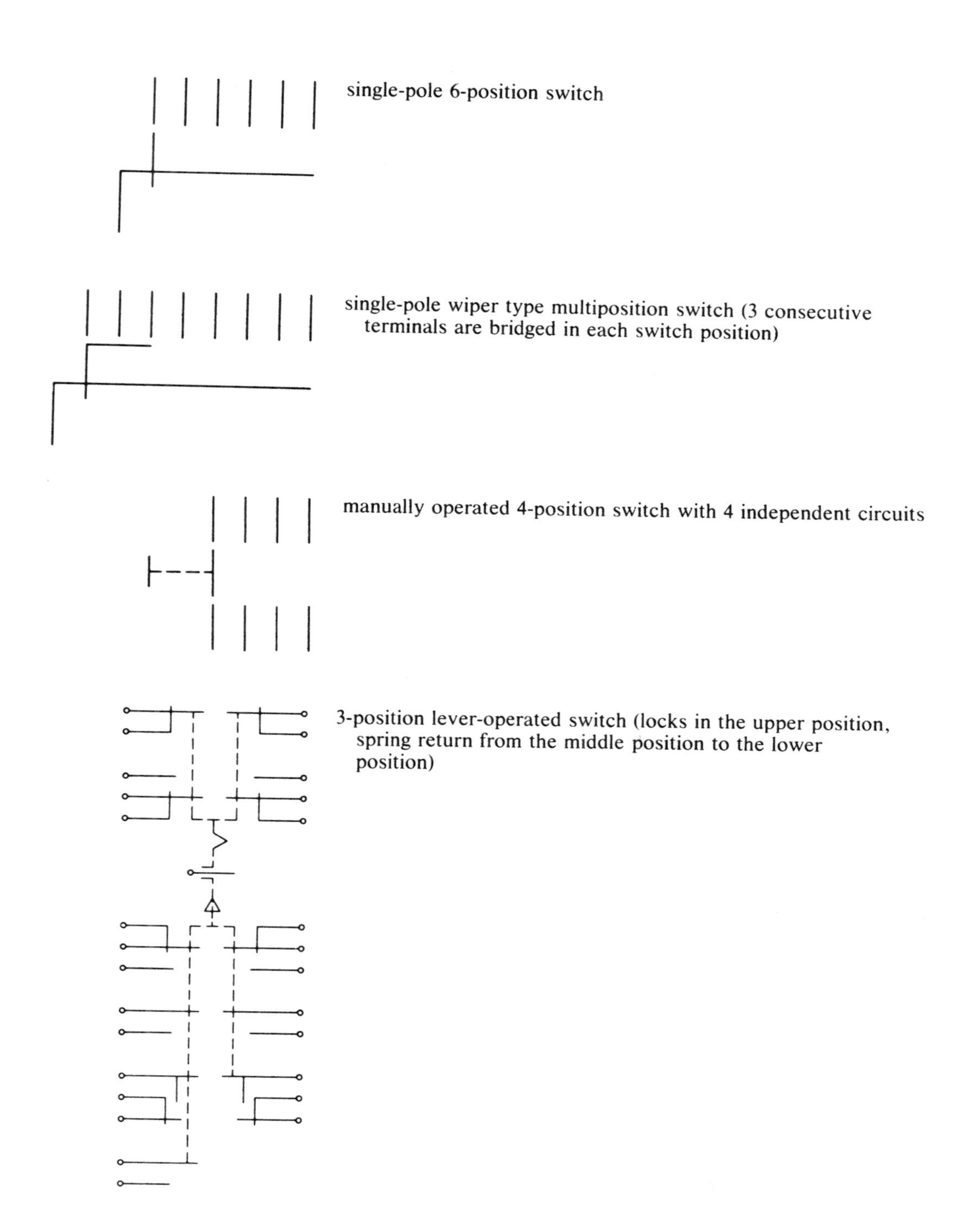

single-pole 6-position switch

single-pole wiper type multiposition switch (3 consecutive terminals are bridged in each switch position)

manually operated 4-position switch with 4 independent circuits

3-position lever-operated switch (locks in the upper position, spring return from the middle position to the lower position)

Miscellaneous

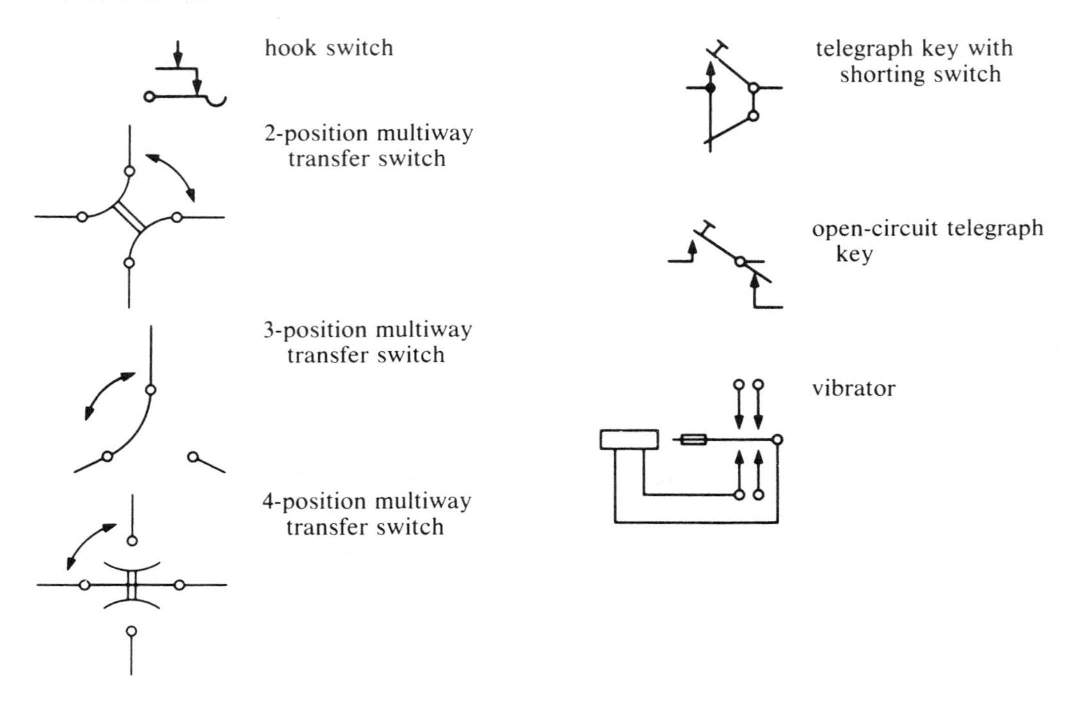

Common switches (copyright 1984 by QST)

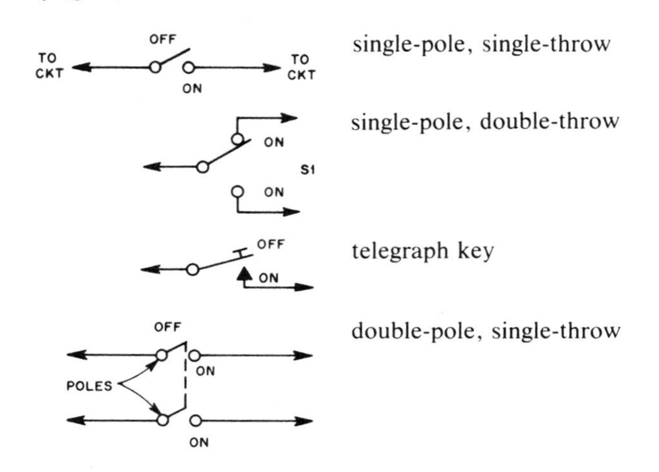

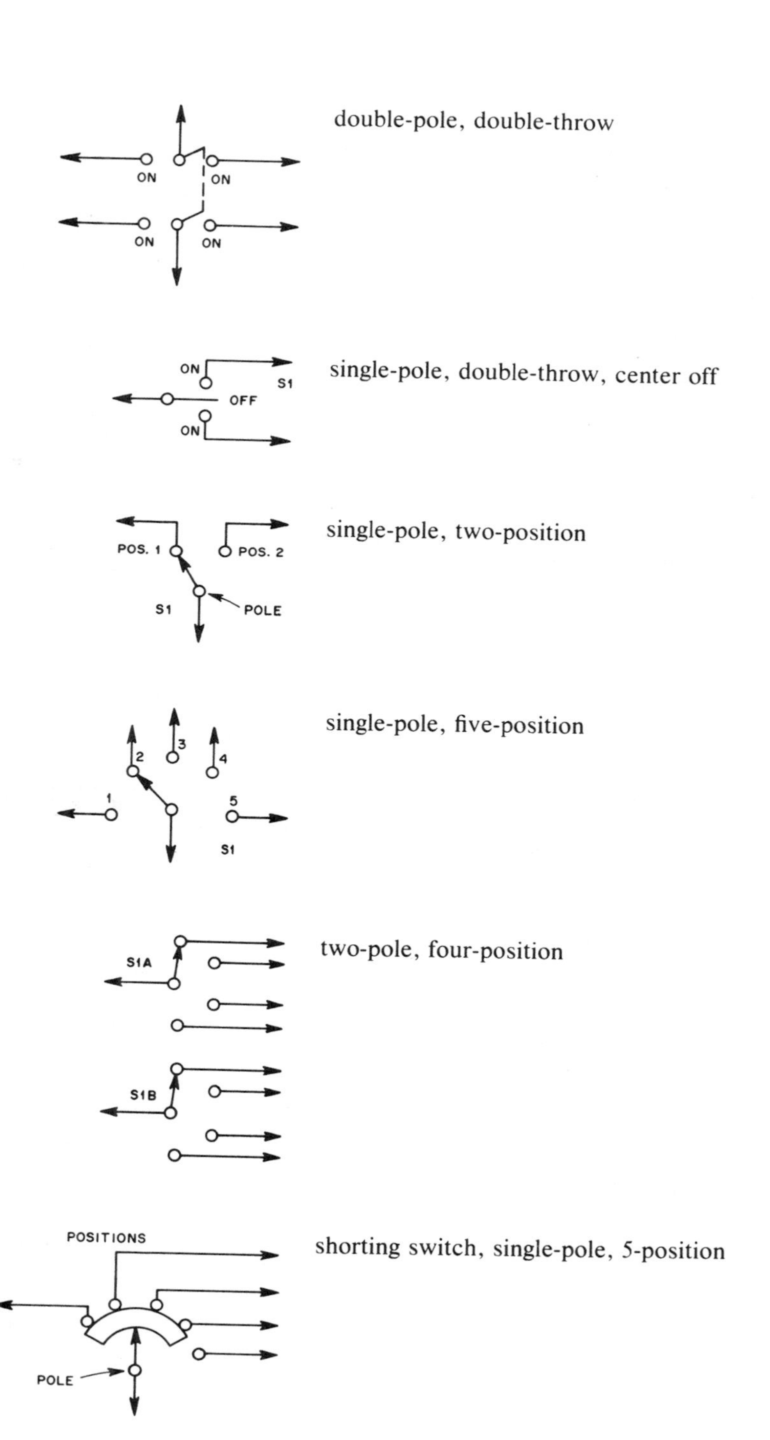
double-pole, double-throw
ON
ON
ON
ON
single-pole, double-throw, center off
ON
S1
OFF
ON
single-pole, two-position
POS. 1
POS. 2
S1
POLE
single-pole, five-position
1
2
3
4
5
S1
two-pole, four-position
S1A
S1B
shorting switch, single-pole, 5-position
POSITIONS
POLE

Relay Contact Forms (Copyright 1986 Hearst Business Communications, Inc.)

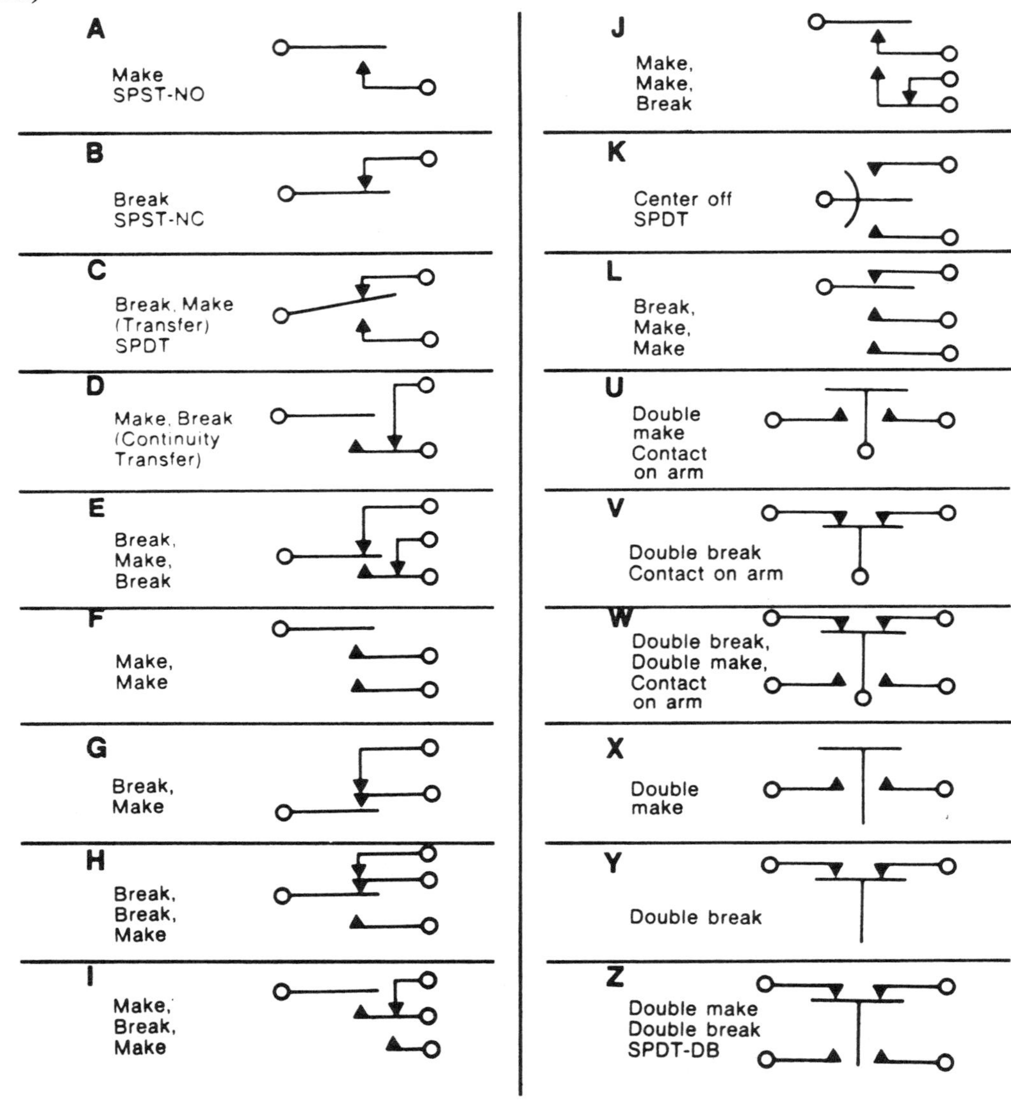

4.10 MISCELLANEOUS SHAPES

The symbols in this section are divided into the following groups:

4.10.1 Irregular-Shaped Blocks

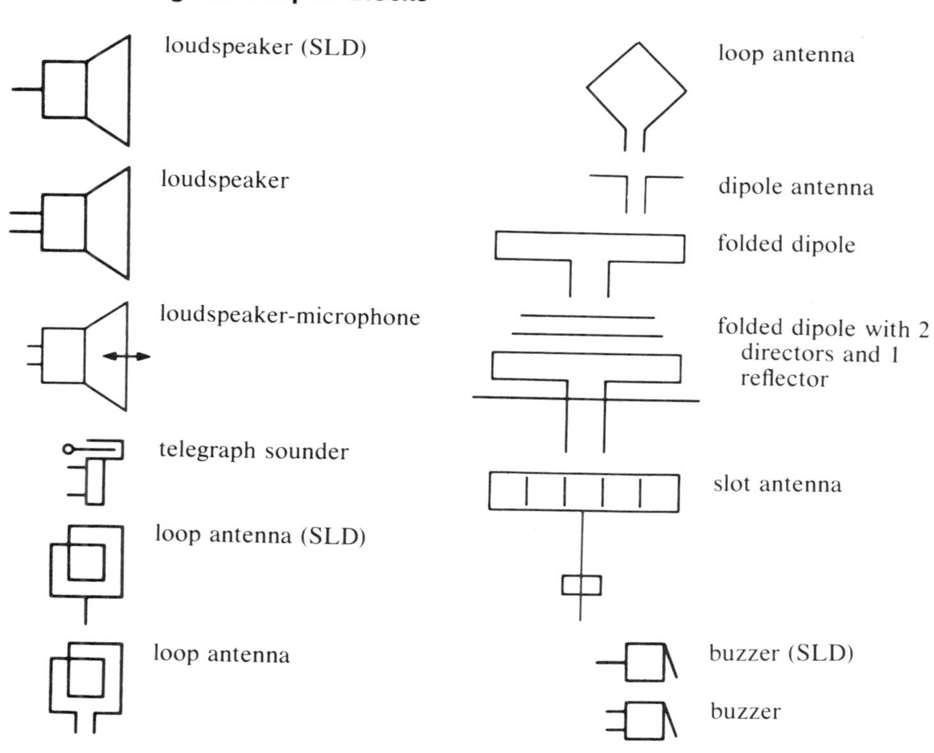

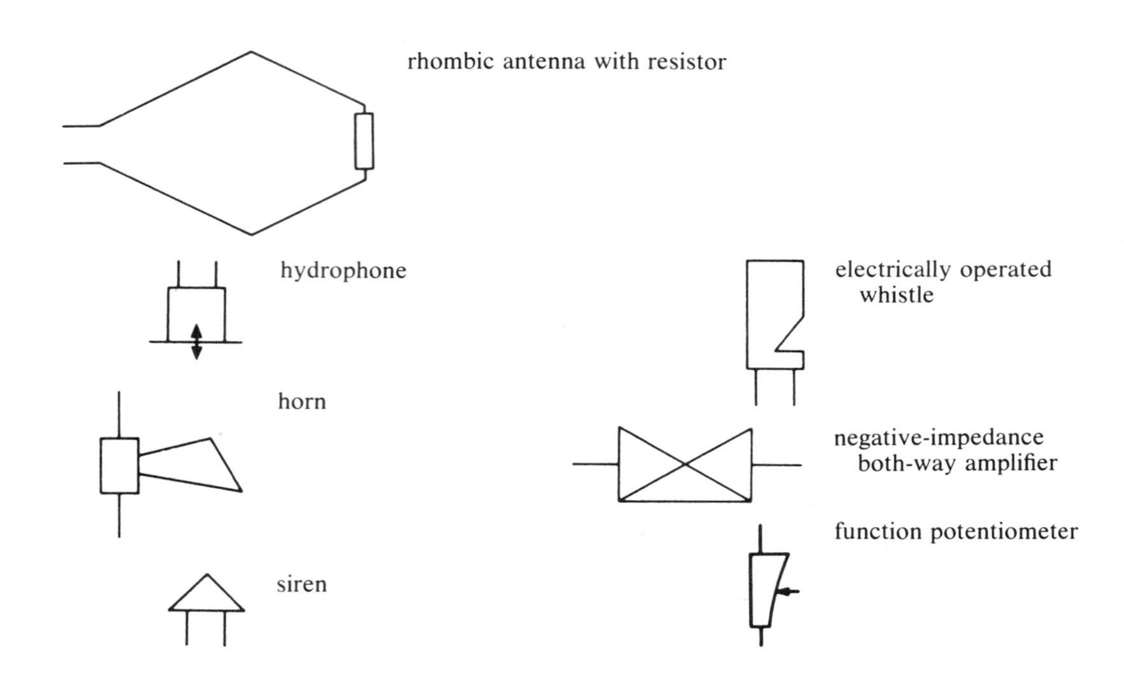

4.10.2 Rectangle With Triangular Side (Boat)

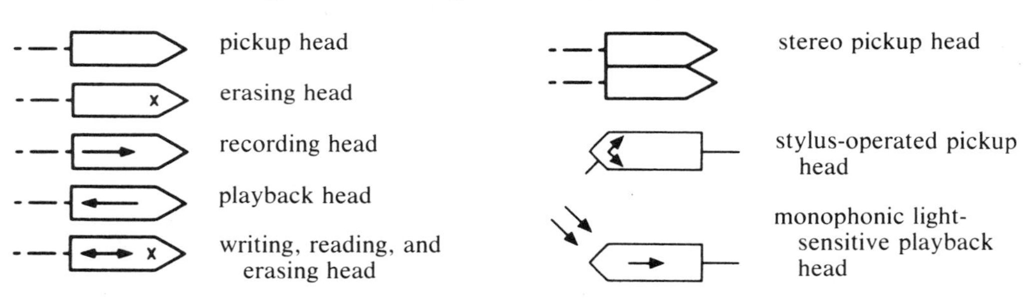

4.10.3 Rectangle With Two Curved Sides

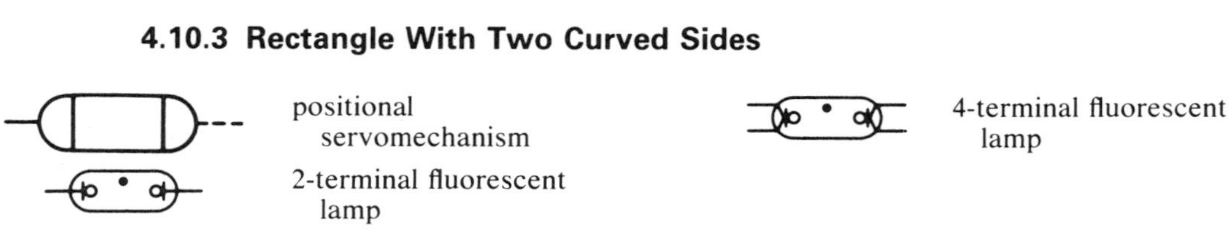

4.10.4 Y Shape

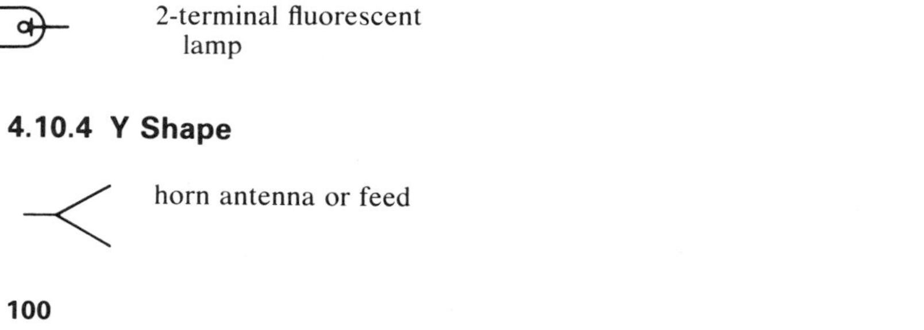

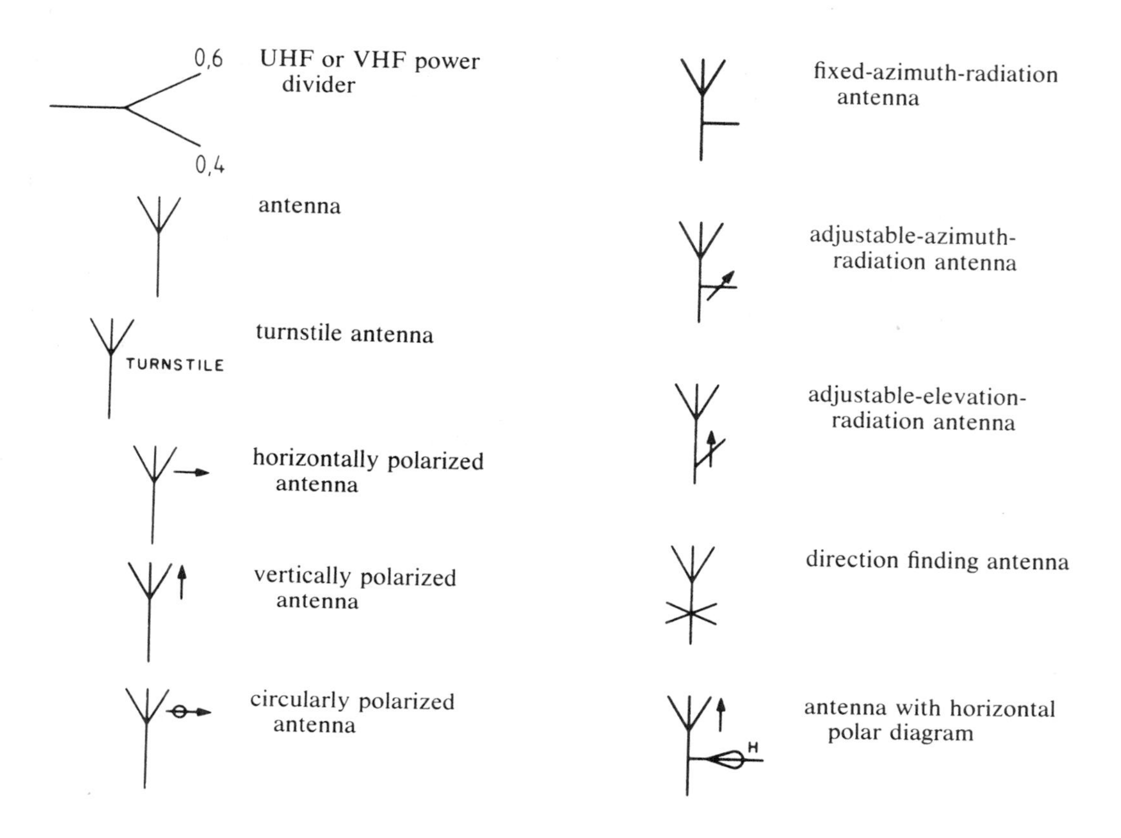

4.10.5 Cone Shape

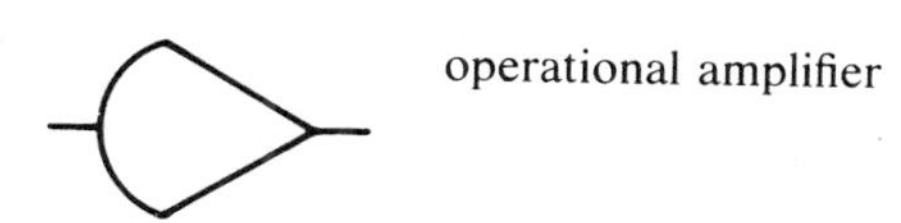

4.10.6 Parallel Lines—Horizontal & Vertical

(antenna counterpoise, batteries, lightning or surge arresters, switching systems)

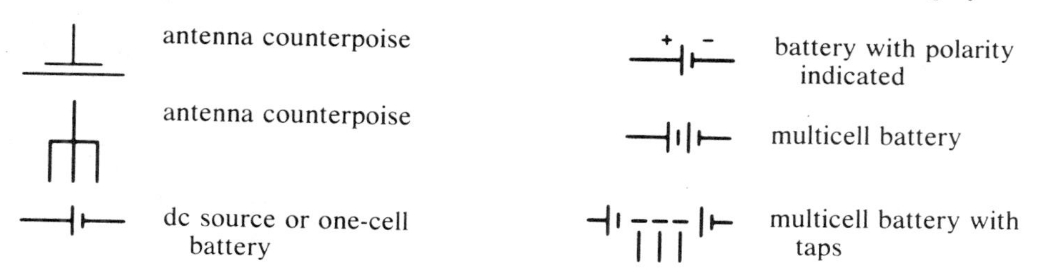

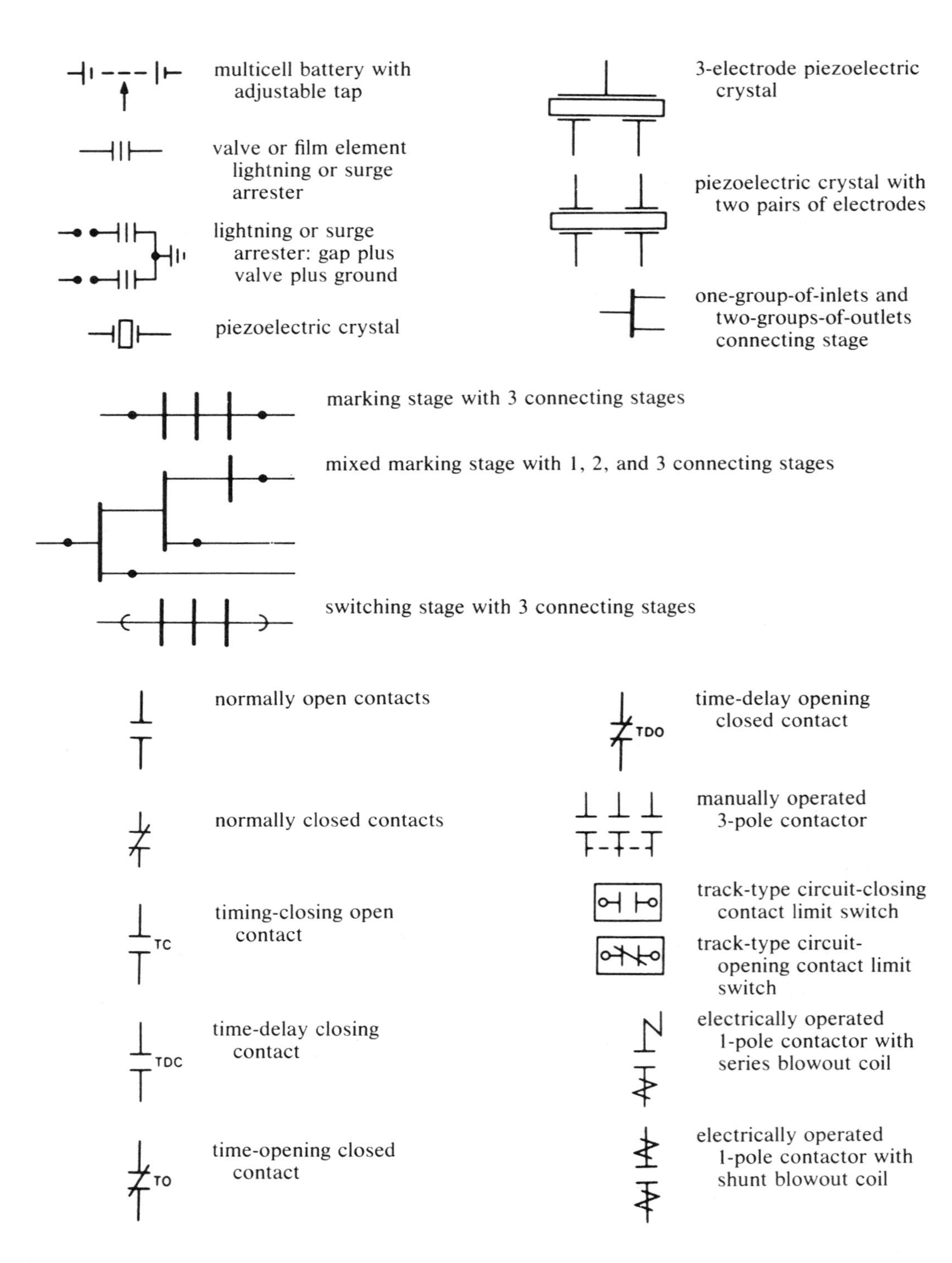

multicell battery with adjustable tap
valve or film element lightning or surge arrester
lightning or surge arrester: gap plus valve plus ground
piezoelectric crystal
3-electrode piezoelectric crystal
piezoelectric crystal with two pairs of electrodes
one-group-of-inlets and two-groups-of-outlets connecting stage
marking stage with 3 connecting stages
mixed marking stage with 1, 2, and 3 connecting stages
switching stage with 3 connecting stages
normally open contacts
normally closed contacts
TC
timing-closing open contact
TDC
time-delay closing contact
TO
time-opening closed contact
TDO
time-delay opening closed contact
manually operated 3-pole contactor
track-type circuit-closing contact limit switch
track-type circuit-opening contact limit switch
electrically operated 1-pole contactor with series blowout coil
electrically operated 1-pole contactor with shunt blowout coil

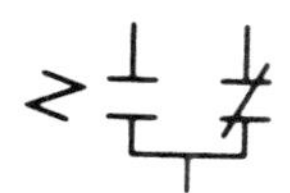

relay with transfer contacts

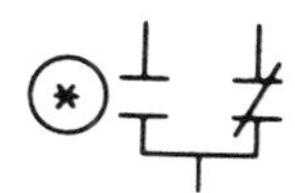

relay with transfer contacts

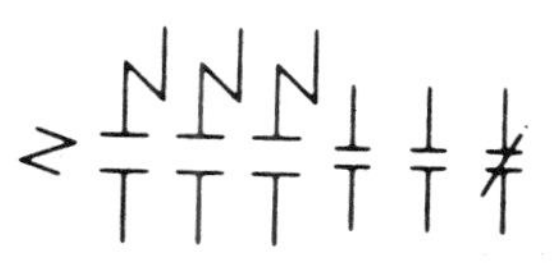

electrically operated 3-pole contactor with series blowout coils; 2 open and 1 closed auxiliary contacts

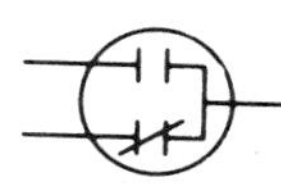

1-pole double throw switch in evacuated envelope

4.10.7 Signal Identifiers and Waveforms

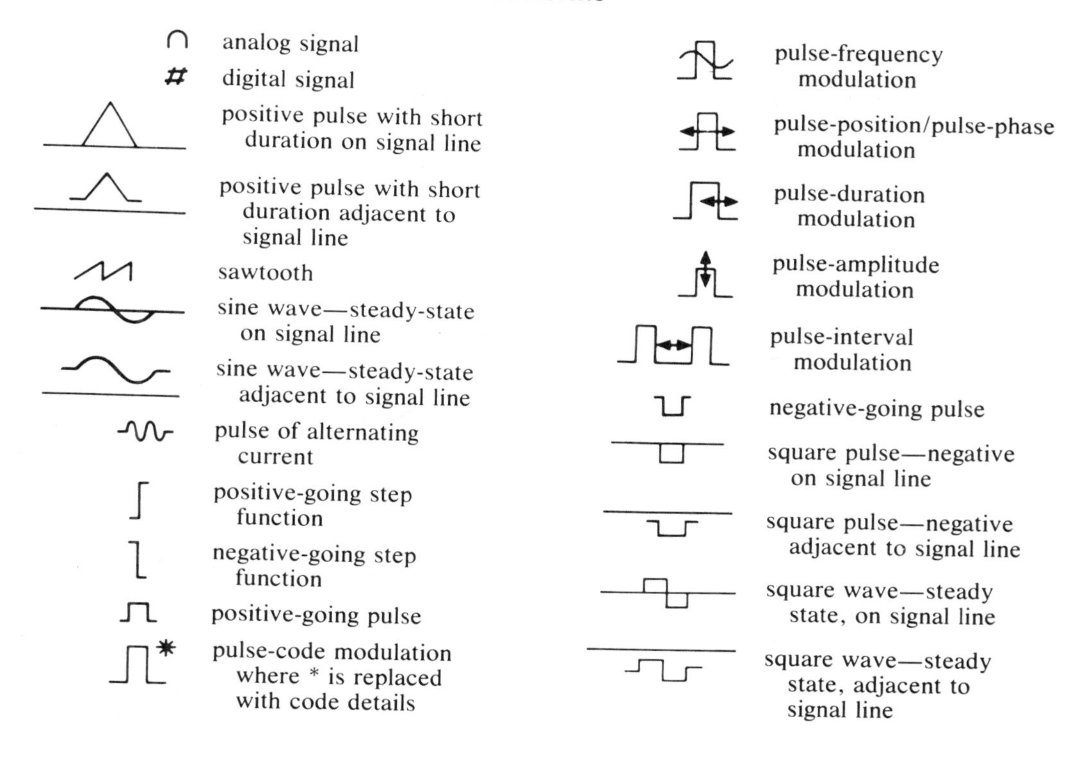

4.10.8 Diamond Shapes

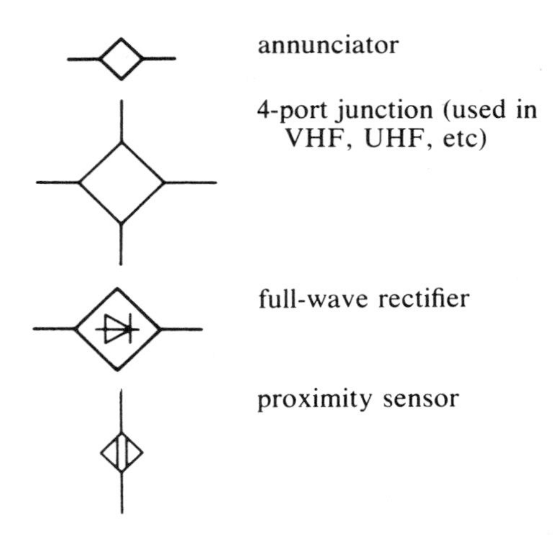

annunciator

4-port junction (used in VHF, UHF, etc)

full-wave rectifier

proximity sensor

touch sensor

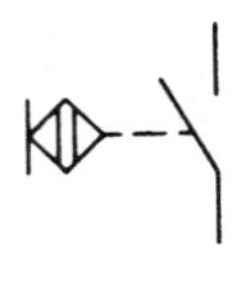

make-contact touch-sensitive switch

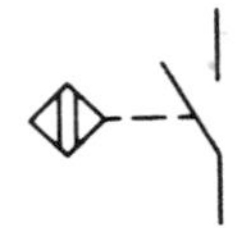

make-contact proximity switch

4.10.9 Vertical Arrows With Related Symbols

(symbol elements for frequency spectrum diagrams)

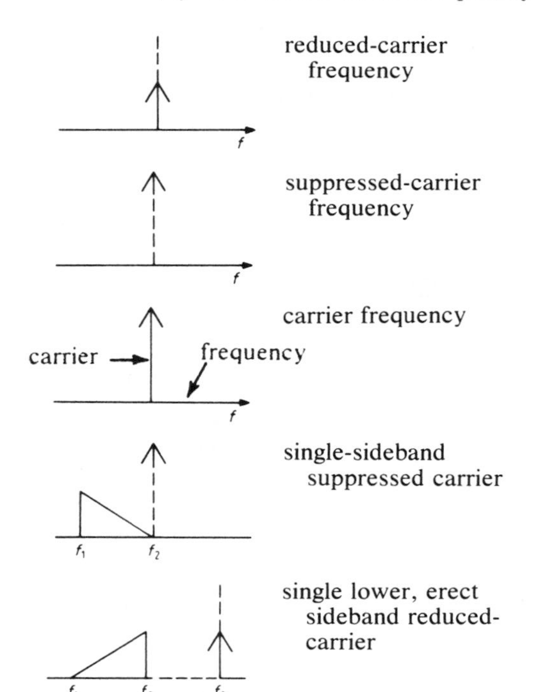

reduced-carrier frequency

suppressed-carrier frequency

carrier frequency

single-sideband suppressed carrier

single lower, erect sideband reduced-carrier

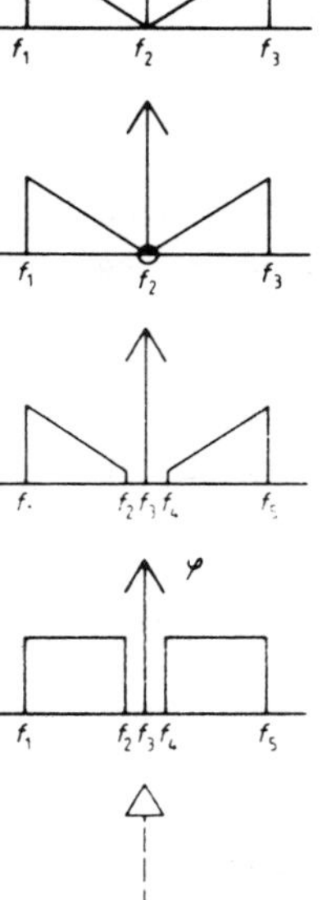

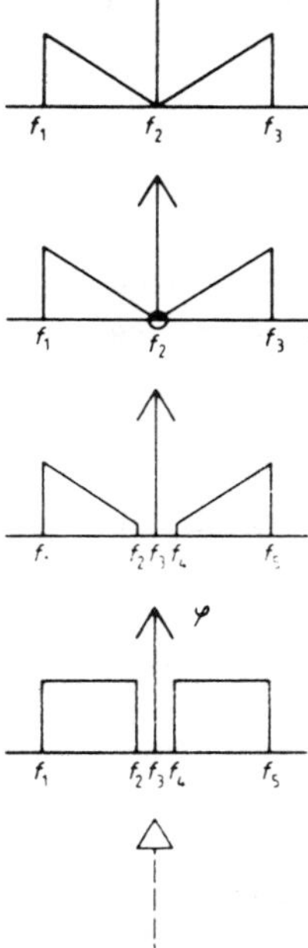

double-sideband amplitude-modulated carrier

double-sideband amplitude-modulated carrier (modulating frequencies down to zero are transmitted)

double-sideband amplitude-modulated carrier (lower mod. frequencies are not transmitted)

double-sideband phase modulated carrier (if φ is replaced with f, frequency modulation is indicated)

suppressed pilot frequency

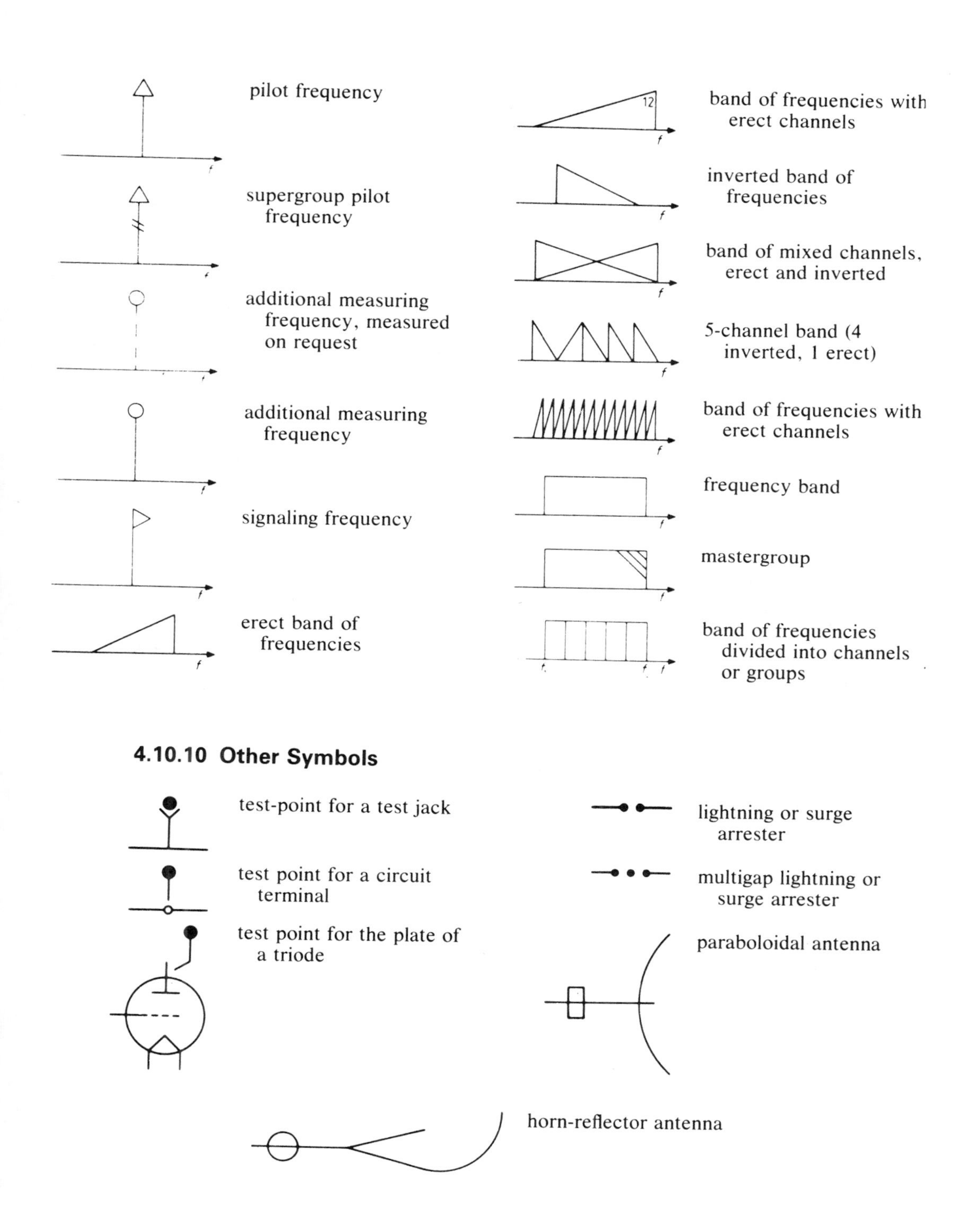
pilot frequency
band of frequencies with erect channels
supergroup pilot frequency
inverted band of frequencies
band of mixed channels, erect and inverted
additional measuring frequency, measured on request
5-channel band (4 inverted, 1 erect)
additional measuring frequency
band of frequencies with erect channels
frequency band
signaling frequency
mastergroup
erect band of frequencies
band of frequencies divided into channels or groups
4.10.10 Other Symbols
test-point for a test jack
lightning or surge arrester
test point for a circuit terminal
multigap lightning or surge arrester
test point for the plate of a triode
paraboloidal antenna
horn-reflector antenna

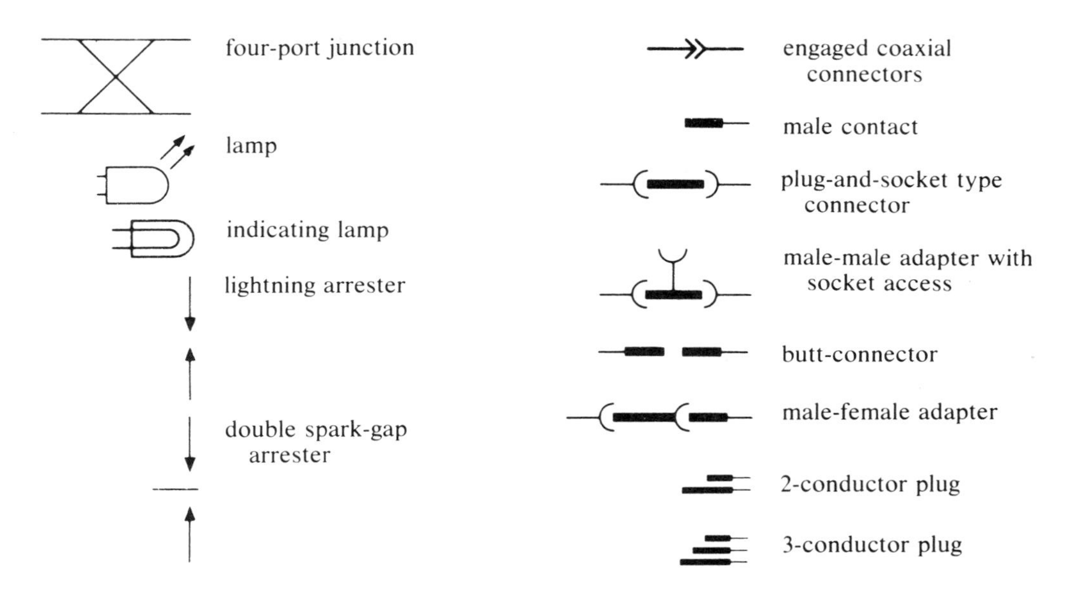

4.10.11 Schematic Diagram Notes (© reproduced with permission of Hewlett-Packard Co.)

Chapter 5

Interpreting Conventional (Nonlogic) Schematics

A schematic diagram such as Fig. 5.1 is a drawing that shows the functional relationship of components in electronic equipment by means of graphic symbols and interconnecting lines. Without a schematic, it would be impossible to understand and maintain most electronic equipment. With a schematic, however, a technician or engineer has a practical way to troubleshoot and align such equipment.

A schematic may cover an entire system, subsystem, assembly, or subassembly. It can be a partial drawing, or it can be simplified or detailed. In either case it serves as a road map showing the flow of power and signals.

The symbols show the function of the components rather than their physical appearance. Most of the graphic symbols used in electronics were shown in Chapter 4.

Although a schematic can be used as a guide for wiring of electronic equipment, it is not, strictly speaking, a wiring diagram. A schematic, for example, does not necessarily show the components in their desired location as a wiring diagram would. Two components could be shown side by side on a schematic diagram but could be placed several inches apart physically.

As an aid for the user, the schematic may include pertinent voltage and resistance values at crucial points in the circuit. Waveforms may also be shown. Component values (such as ohms for resistors) will be placed near the component or in a parts list on the schematic.

At times a schematic may be combined with parts of wiring diagrams, logic diagrams, and pictorial drawings. Because the schematic is a symbolic drawing, it is not drawn to any scale.

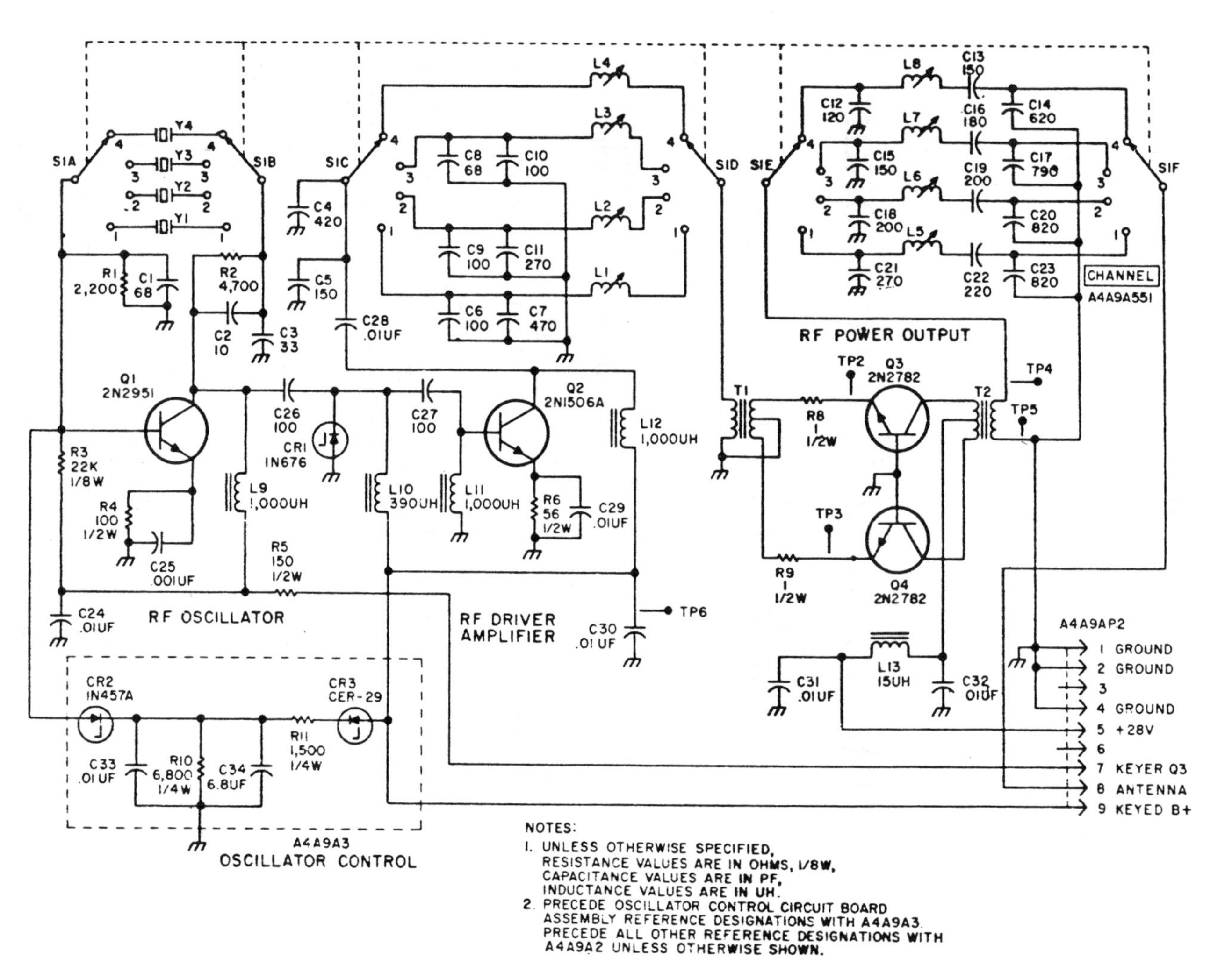

Figure 5.1 Conventional schematic example. (From MIL-HDBK-6303101.)

5.1 IDENTIFYING COMPONENTS

A first step in interpreting schematics is identifying all components. Some components will be obvious from their symbols, which were identified in Chapter 4. As explained in Chapter 2, reference designators can also be used to identify components or to confirm the symbol interpretation.

5.1.1 Position of Components on Schematics

As long as the electrical connections of a component are not changed, a component's symbol may be positioned in several ways without changing the operation of the component or the complete circuit. This is illustrated in Fig. 5.2.

Components such as terminal boards, switches, and relays may be shown physically separated on subdivided on a schematic, with one part or half in one area of the schematic and the other part in still another area. Such separations do not exist in fact but make the schematic easier to read.

When multielement parts such as dual capacitors are separated on a schematic diagram, the parts are identified by suffix letters on the reference designators. For example, if capacitor C1 has two electrically separate sections, they will be identified as C1A and C1B. This method of identifying subdivisions of a complete part is also used when the individual parts are shown enclosed in one unit, as in Fig. 5.3.

Suffix letters (A, B, C, etc.) used to identify rotary switch parts or sections are assigned sequentially from the actuator end, with A being the section closest to the actuator. The designator may be modified further if both sides of a rotary switch section are used to perform separate switching functions, as shown in Fig. 5.4.

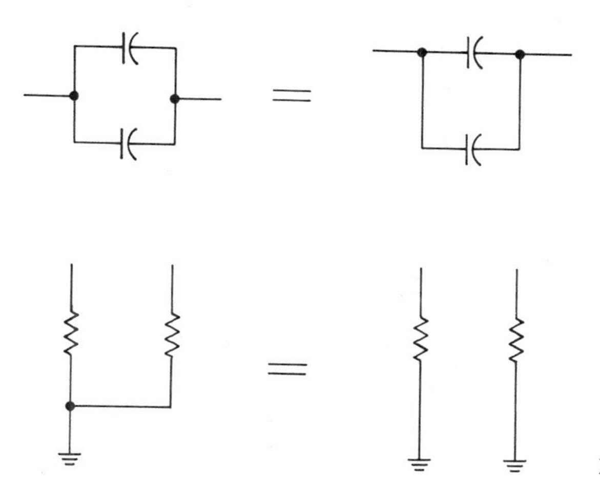

Figure 5.2 Component positioning.

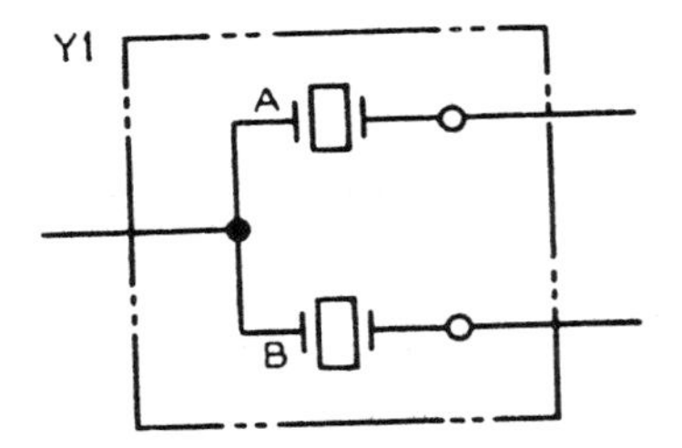

Figure 5.3 Identification of parts by suffix letters. (From USAS Y14.15-1966.)

Functionally separated connectors or terminal boards on a diagram may be identified by jagged lines (⊐) or by the words "PART OF," as shown in Fig. 5.5. Sometimes, however, they may be separated without either designation. Note, too, as mentioned in Chapter 4, that the orientation of a graphic symbol does not change its function; that is, the symbol can be rotated, turned upside down, or formed as a mirror image without affecting its interpretation.

5.1.2 Part Values

Whereas it is essential to know the numerical value of a part when you go to replace it, it is not usually necessary to know these values in order to understand most circuits. However, you may need to pay particular attention to part values

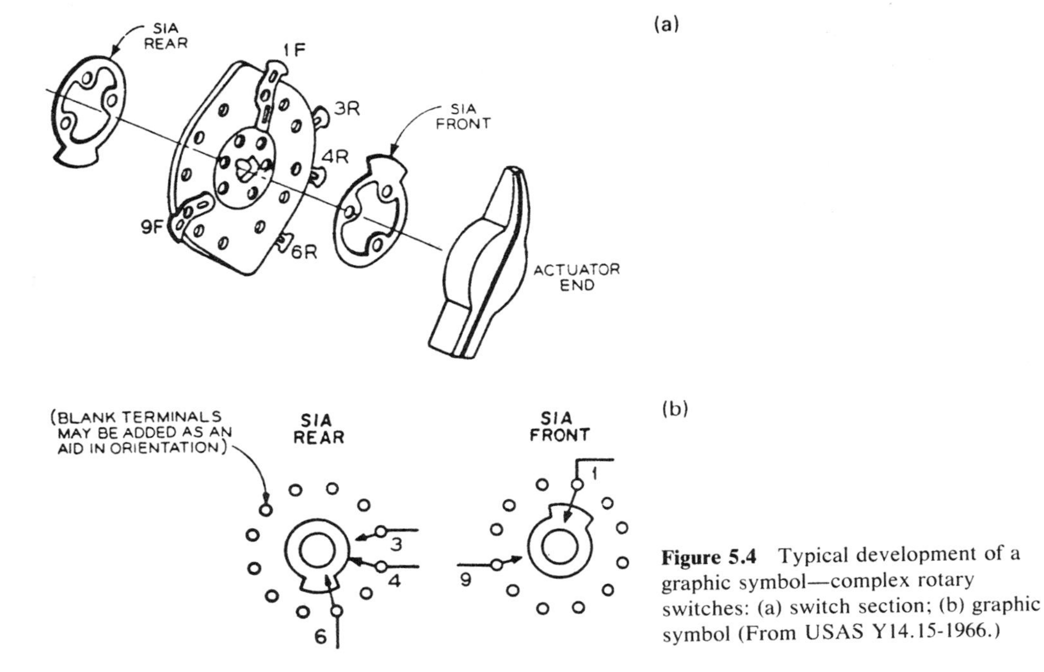

Figure 5.4 Typical development of a graphic symbol—complex rotary switches: (a) switch section; (b) graphic symbol (From USAS Y14.15-1966.)

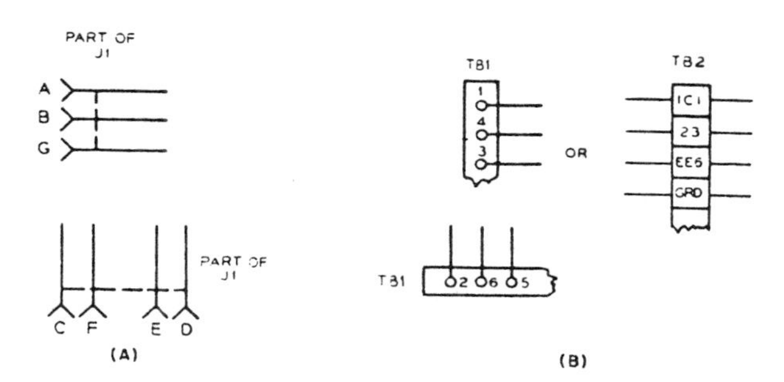

Figure 5.5 Identification of portions of items. (From USAS Y14.15-1966.)

in delay circuits and resonant circuits. For example, the resonant frequency of a circuit can be determined by the values of capacitance and inductance in the circuit. When a component value is pertinent to your understanding of a circuit, look for the value next to the component, as shown in Fig. 5.6. To interpret these values, look for qualifying notes as shown in Fig. 5.7 and Table 5.1. If the value is not shown next to the component, look in the parts list (Fig. 5.8).

5.1.3 Parts Lists

For some components, particularly resistors and capacitors, the value or electrical specification (10 K, 0.01 μF, etc.) may be printed on the schematic near the

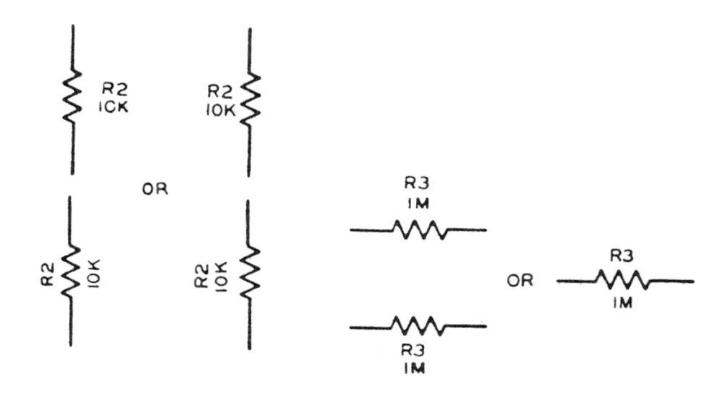

Figure 5.6 Methods of numerical value and reference designation placement. (From USAS Y14.15-1966.)

UNLESS OTHERWISE SPECIFIED: RESISTANCE VALUES ARE IN OHMS. CAPACITANCE VALUES ARE IN MICROFARADS.

or

CAPACITANCE VALUES ARE IN PICOFARADS.

An alternative note for specifying capacitance values is:

CAPACITANCE VALUES SHOWN AS NUMBERS EQUAL TO OR GREATER THAN UNITY ARE IN pF AND NUMBERS LESS THAN UNITY ARE IN μF.

Figure 5.7 Qualifying notes for part values. (From USAS Y14.15-1966.)

TABLE 5.1 MULTIPLIERS

Multiplier	Prefix	Symbol Method 1	Symbol Method 2
10^{12}	tera	T	T
10^{9}	giga	G	G
10^{6} (1,000,000)	mega	M	M
10^{3} (1000)	kilo	k	K
10^{-3} (.001)	milli	m	MILLI
10^{-6} (.000001)	micro	μ	U
10^{-9}	nano	n	N
10^{-12}	pico	p	P
10^{-15}	femto	f	F
10^{-18}	atto	a	A

Source: USAS Y14.15-1966.

component. However, when the parts values become too lengthy to be placed directly on the schematic, they are generally placed in a parts list on the schematic.

As shown in Fig. 5.8, a parts list will include the manufacturer's part number for unique or special components. Information on hardware such as nuts and bolts may be included but can be safely ignored in reading the schematic.

C1—Trimmer, approx. 3-35 pF, Arco 403 or equiv.
C2, C3—Four Arco 469 trimmer capacitors in parallel (680-3120 pF).
J1, J2—Builder's option.
L1—40 turns no. 24 on Amidon T-50-6 core (12 μH).
L2, L3—6.5-10.6 μH, Cambion 2060-3 or equiv.
L4—1-mH molded choke, Nytronics WEE-1000 or equiv.
L5, L7—100-μH molded choke, Nytronics WEE-100 or equiv.
L6—330-μH molded choke, Nytronics WEE-330 or equiv.
L8—25 turns no. 24 on Amidon T-50-6 core (8μH).
L9—11 turns no. 20, ¾ inch (19 mm) dia.
L10—18 turns no. 20 on amidon T-68-2 core (2 μH).
M1—Ammeter, full scale 2 to 3A, General Electric 50-250100LJLJ or equiv. Any meter of smaller scale may be shunted.
Q1-Q3—J308 or MPF102.
Q5—2N3866 or equiv. (see table).
R1—10-kΩ 10-turn potentiometer, Bourns 3509S or equiv.
R2, R3—Band-limit resistors (see text).
T1—Primary, 7 turns no. 22; secondary, 1 turn no. 18 on Amidon T-50-6 core.
U1—Dual CMOS flip-flop, CD4027, Radio Shack 276-2427 or equiv.

Figure 5.8 Typical parts list. (Copyright 1979 by QST.)

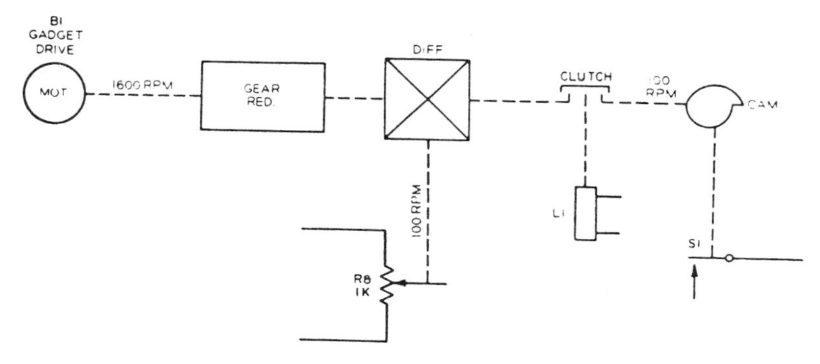

Figure 5.9 Typical schematic diagram showing mechanical linkages (From USAS Y14.15-1966.)

5.1.4 Mechanical Linkages

A dashed (---) line between components as shown in Fig. 5.9 indicates that the components are mechanically connected. If one of the components is turned or moved, the other component will also move. The linkage can be between a mechanical component and an electrical or electronic component or between sections of a single component.

A stepping motor, for example, may be linked to a rheostat to change the resistance in steps. In a multisection rotary switch, the sections will be connected by a common shaft so that as the shaft is rotated, connections in all sections will change. As mechanical linkage lines do not usually have arrows, you will have to determine which component is the controlling or initiating component.

5.1.5 Diagram Notes

Notes on schematics may provide additional circuit information, such as shown in Table 5.2.

TABLE 5.2 DIAGRAM NOTES

Dc resistance of windings and coils (if more than 1 ohm)
Critical input or output impedance values
Wiring requirements for critical ground points, shielding, pairing, etc.
Power or voltage ratings of parts
Indication of operational controls or circuit functions
Caution notation for electrical hazards at maintenance points
Circuit voltage values at significant points
Significant circuit resistance values at designated reference points

Source: USAS Y14.15-1966.

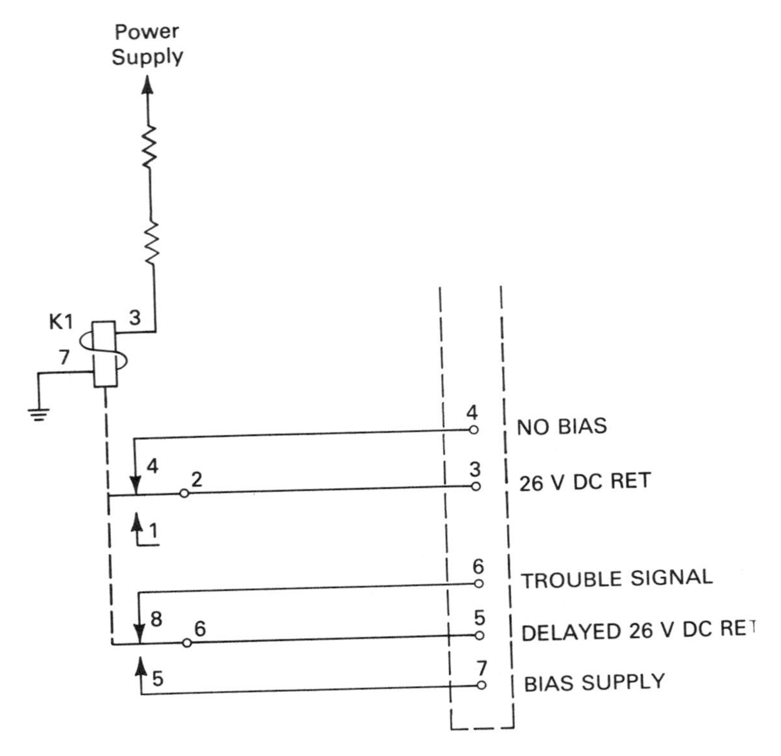

Figure 5.10 Relay shown in deenergized position.

5.2 STATE OF A MULTIPOSITION COMPONENT

Note that relays on a schematic are shown in their deenergized position (see Fig. 5.10). When energized they will connect the signal or power to the other set of contacts. Multiposition switches can be shown in any position.

5.3 TERMINALS

Terminal circles are used occasionally to prevent misinterpretation of a schematic. Figure 5.11(a) shows the terminals for a complex rotary switch (S1B, S1C). Figure 5.11(b) shows terminal markings for shielded parts; notice that the markings are outside the enclosure. Refer to Fig. 5.12 for terminals for relays (K1), key switches (S1), and jacks (J1).

Terminals may be indicated by letters or numbers on a diagram even if there are no terminal markings on the part. To determine their meaning, check the

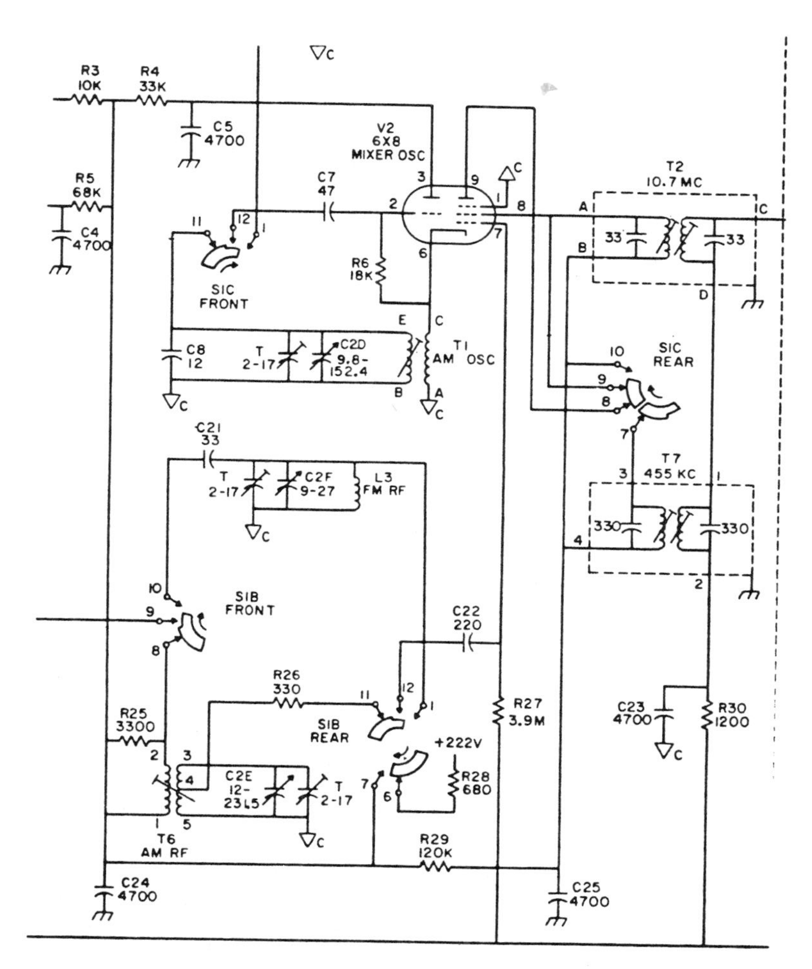

Figure 5.11 (a) Schematic showing terminals for complex switch. (From USAS Y14.15-1966.)

schematic for explanatory notes (see Fig. 5.12) or terminal orientation diagrams as illustrated in Figs. 5.11(b), 5.12, 5.13, and 5.14.

Rotary-type adjustable resistors can be rotated in a clockwise (CW) or counterclockwise (CCW) direction. This direction may be indicated on diagrams as in Fig. 5.15. In the illustration, the letters CW indicate the terminal closest to the movable contact (←) when the contact is turned to its clockwise limit. The motion is specified as seen from the knob end of the control. Notice that the terminal numbers are shown in Fig. 5.15(b), and fixed taps (4, 5) are shown in Fig. 5.15(c).

As different positions of a switch will cause different circuit functions to occur, a schematic may show position-to-function relations near the switch sym-

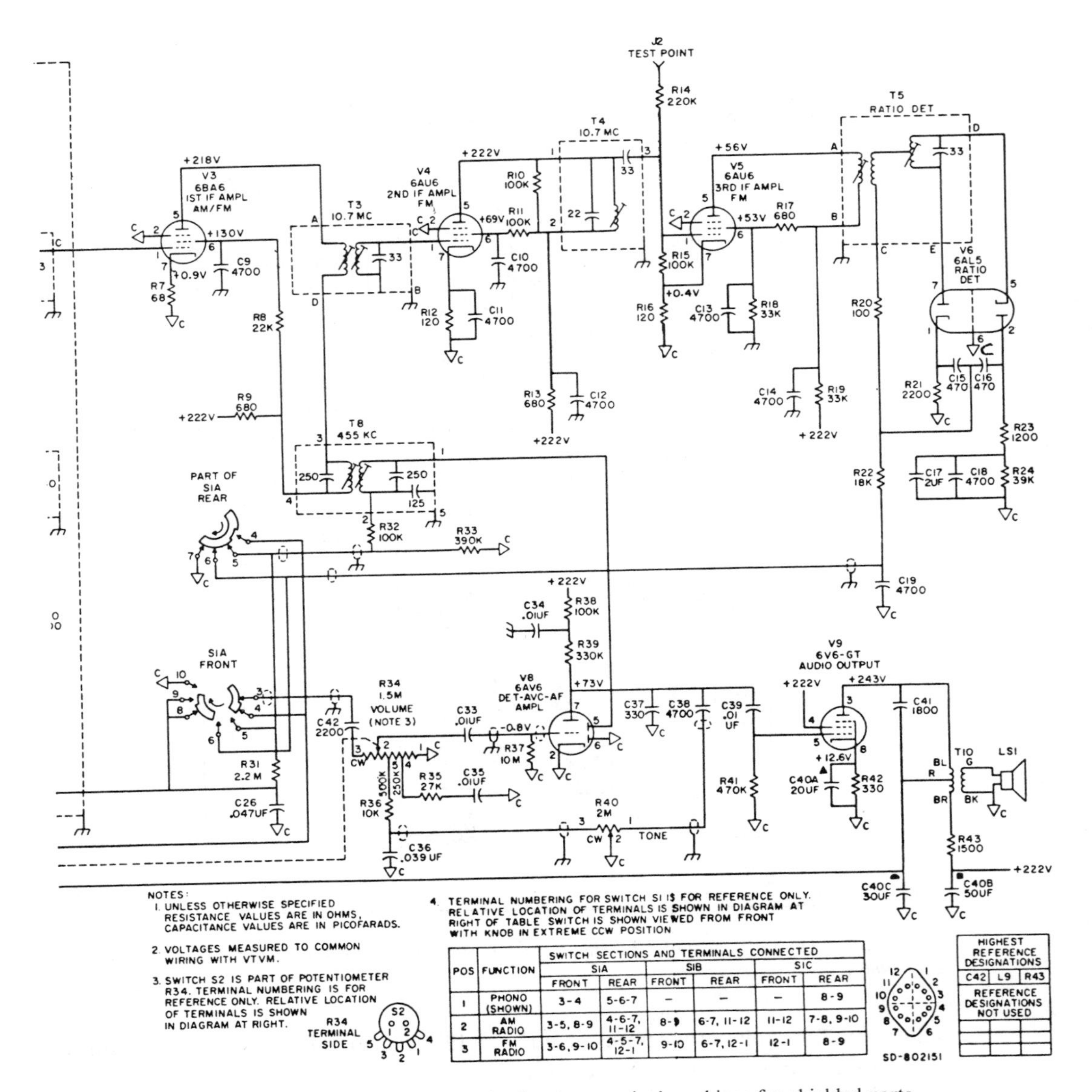

POS	FUNCTION	SWITCH SECTIONS AND TERMINALS CONNECTED					
		S1A		S1B		S1C	
		FRONT	REAR	FRONT	REAR	FRONT	REAR
1	PHONO (SHOWN)	3-4	5-6-7	—	—	—	8-9
2	AM RADIO	3-5, 8-9	4-6-7, 11-12	8-9	6-7, 11-12	11-12	7-8, 9-10
3	FM RADIO	3-6, 9-10	4-5-7, 12-1	9-10	6-7, 12-1	12-1	8-9

Figure 5.11 (b) Schematic showing terminal markings for shielded parts.

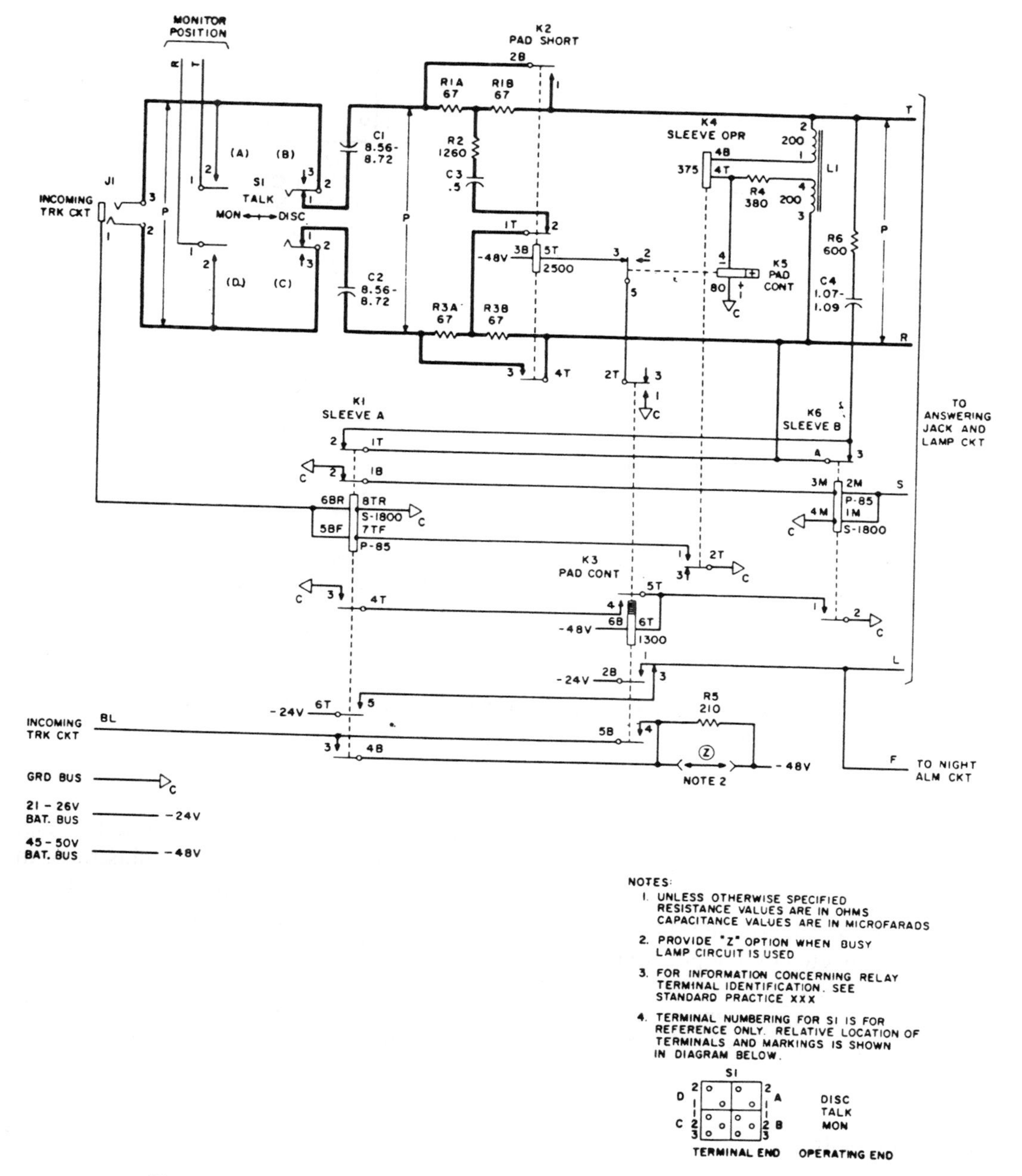

Figure 5.12 Schematic showing preferred terminal representation of spring contacts for relays, jacks, and key switches. (From USAS Y14.15-1966.)

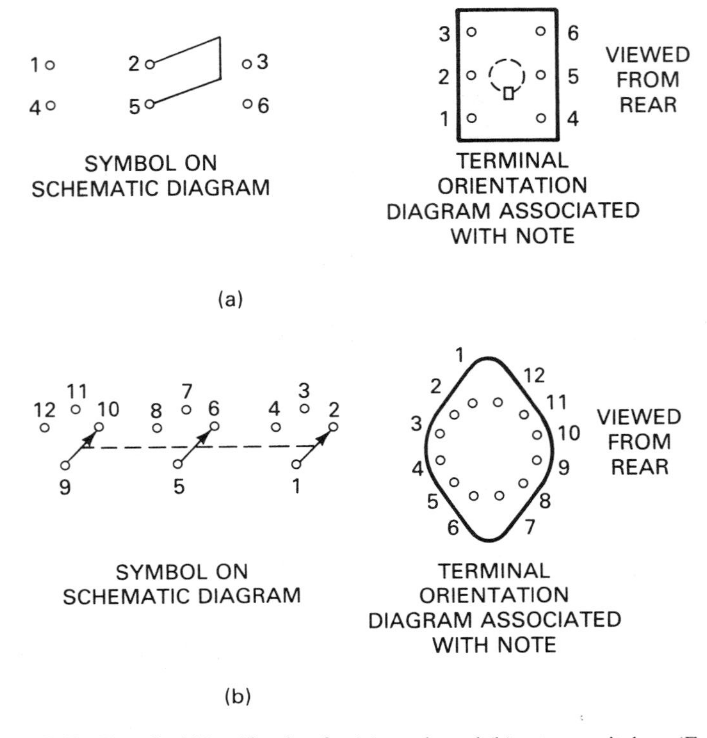

Figure 5.13 Terminal identification for (a) toggle and (b) rotary switches. (From USAS Y14.15-1966.)

bol or elsewhere in a table on the drawing. This technique is illustrated in Fig. 5.16 for a multipurpose rotary switch. For complicated switches, as illustrated in Fig. 5.17, the supplementary information will be presented only in tabular form. The dashes in the TERM column of Fig. 5.17 indicate that the terminals are connected in that position. For instance, if S1 is in position 3 (OPERATE), terminal 1 is connected to terminal 4, terminal 5 to 8, and 9 to 12.

Coaxial connector contacts may be drawn as in Fig. 5.18 (for single-line diagrams) or as in Fig. 5.19 (for complete diagram representation).

5.4 TEST POINTS

Test points are identified on diagrams by the letters "TP" with a number suffix or they may be identified with a special distinguishing symbol such as ☆.

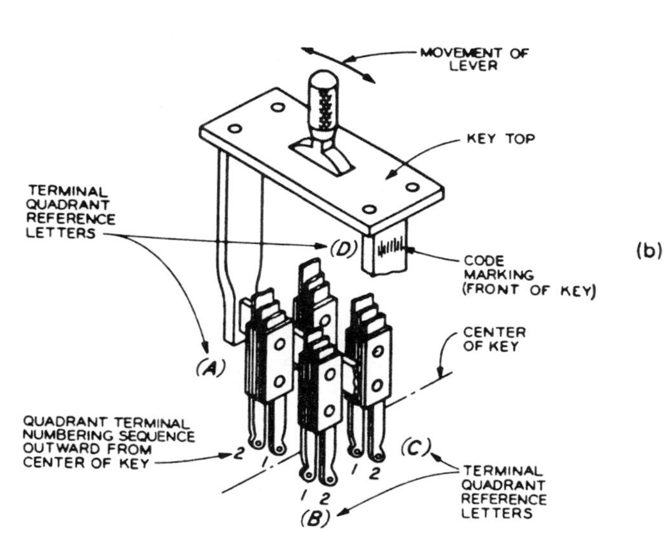

Figure 5.14 Typical lever switch: (a) terminal identification and orientaton; (b) relationship of keytop front and spring terminal quadrants. (From USAS Y14.15-1966.)

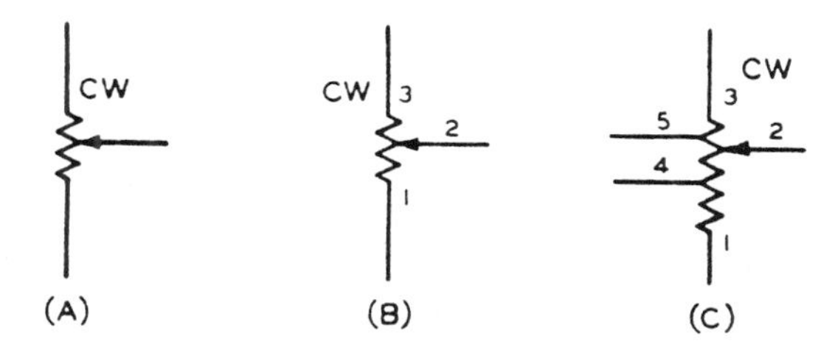

Figure 5.15 Terminal identification for adjustable resistors. (From USAS Y14.15-1966.)

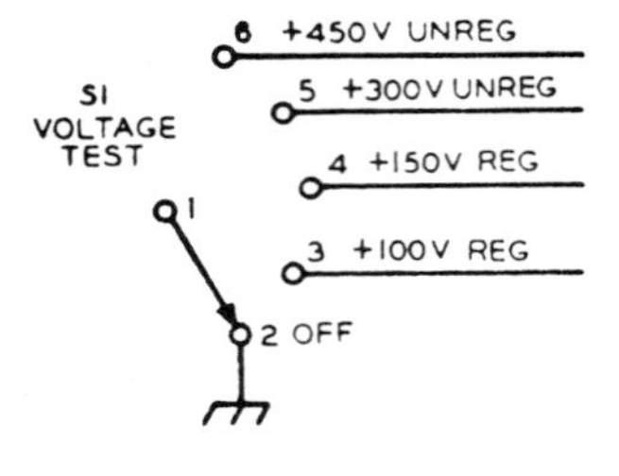

FUNCTIONS SHOWN AT SYMBOL

S1 VOLTAGE TEST	
FUNCTION	TERM.
OFF	1-2
+ 100V REG	1-3
+ 150V REG	1-4
+300V UNREG	1-5
+450V UNREG	1-6

FUNCTIONS SHOWN IN TABULAR FORM

Figure 5.16 Position-function relationship for rotary switches (optional methods). (From USAS Y14.15-1966.)

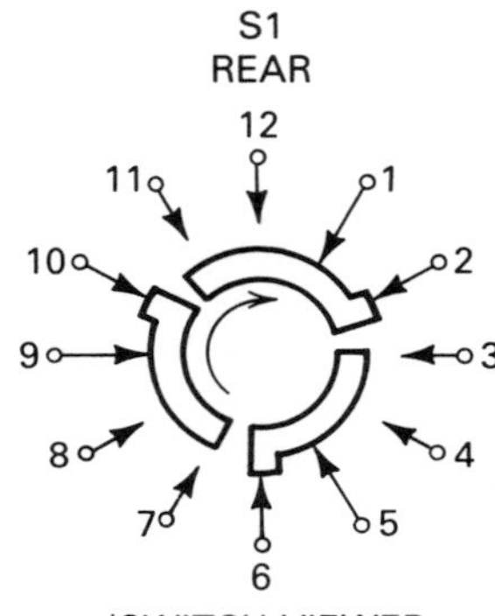

(SWITCH VIEWED FROM FRONT)

SYMBOL ON SCHEMATIC DIAGRAM

S1 (REAR)		
POS	FUNCTION	TERM.
1	OFF (SHOWN)	1–2, 5–6, 9–10
2	STANDBY	1–3, 5–7, 9–11
3	OPERATE	1–4, 5–8, 9–12

FUNCTIONS SHOWN IN TABULAR FORM

Figure 5.17 Position-function relationships for rotary switches (tabular method only). (From USAS Y14.15-1966.)

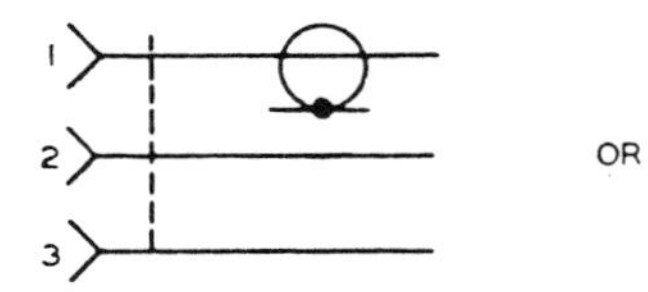

OR

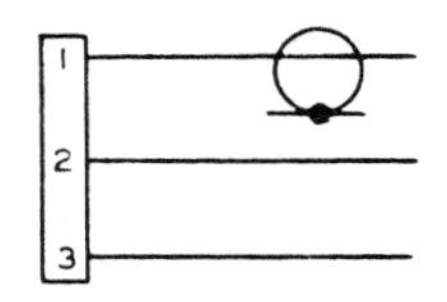

Figure 5.18 Single-line diagram representation of coaxial connector. (From USAS Y14.15-1966.)

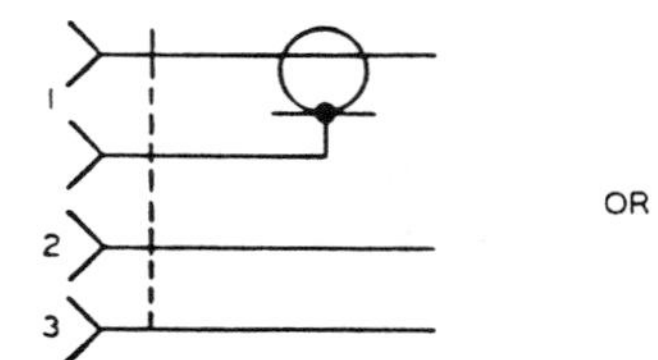

OR

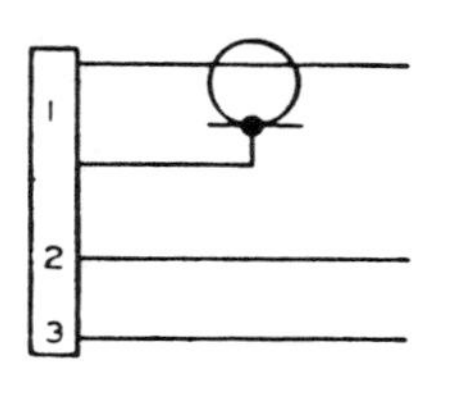

Figure 5.19 Complete diagram representation of coaxial connector. Notice shield continuity. (From USAS Y14.15-1966.)

5.5 TRANSMISSION PATHS (Connecting Lines)

5.5.1 Junctions and Crossovers

Junctions and crossovers of connecting lines are shown in several ways on schematics, as illustrated in Fig. 5.20. Junctions, of course, indicate an electrical connection, while crossovers indicate that there is no connection. If interpretation of a schematic becomes difficult or impossible, consider the possibility that some crossover point could actually be a junction where the dot was accidentally left off.

5.5.2 Elimination of Connecting Lines

While schematics show all transmission paths, they do not always show these with connecting lines. To show all connecting lines could make the schematic unreasonably complex. Lines can be, and usually are, eliminated or shortened for power lines and ground connections. Thus the circuit in Fig. 5.21(a) is more likely to be seen as that in Fig. 5.21(b). It is left to the reader to mentally draw the missing links in the connection. Note that Fig. 5.21 shows only the power circuit line elimination; the same technique is also used on other circuits. As the ground

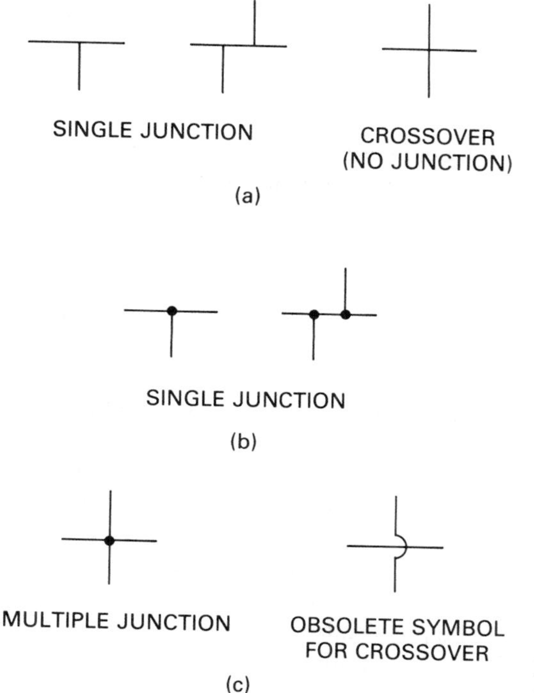

Figure 5.20 Junctions and crossovers. (From USAS Y14.15-1966.)

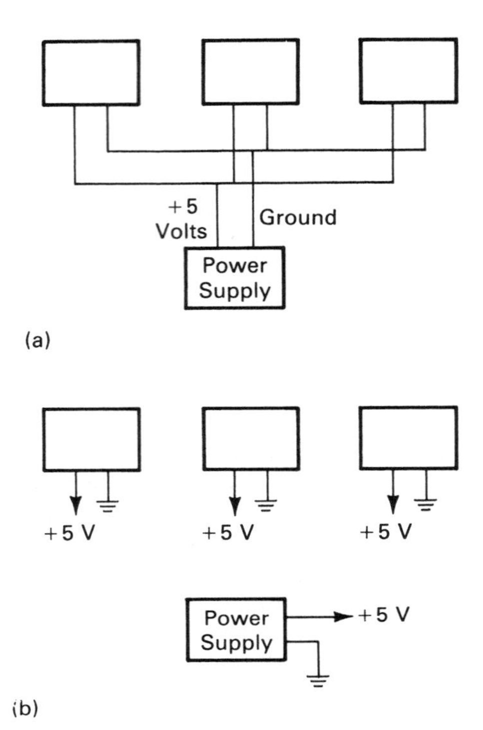

Figure 5.21 Power circuit connection line elimination: (a) diagram with cluttered connecting lines; (b) diagram with connecting lines eliminated.

or chassis is a common return, the ground symbol is used to eliminate numerous connecting lines. In general, a connecting line can be eliminated by terminating it at its beginning with the appropriate designation (+5 V in Fig. 5.21).

5.5.3 Signal Paths

To understand any schematic, one must be able to trace the signal path or flow. Signal connections can be represented by three methods: point-to-point, highway, and interrupted flow.

In the *point-to-point* method (see Fig. 5.22) each signal is represented separately with a continuous line.

In the *highway* method (see Fig. 5.23) two or more signals are blended together in a single line. With this technique the flow of a group of related signals can readily be illustrated. Note that any signal that has been blended out of a line is no longer present on that line. Each signal blended in or blended out of a line will have an identification.

A connection path may be physically broken or *interrupted* as shown by the broken connecting line in Fig. 5.24. In this discussion, however, we are looking at interruption of lines on a drawing or drawings, not physical paths.

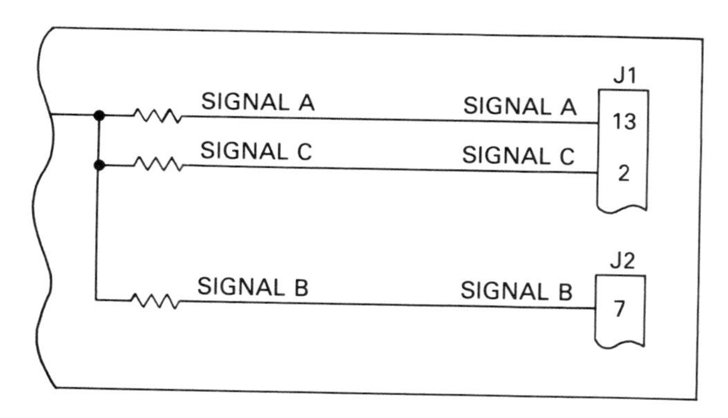

Figure 5.22 Point-to-point method for signal flow. (From MIL-HDBK-63038-1.)

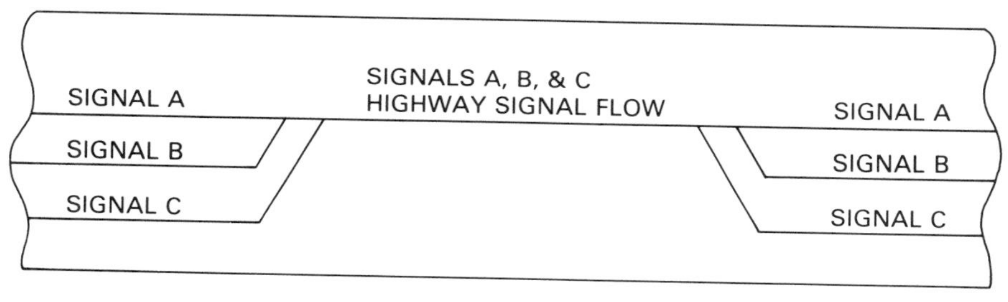

Figure 5.23 Highway method for signal flow. (From MIL-HDBK-63038-1.)

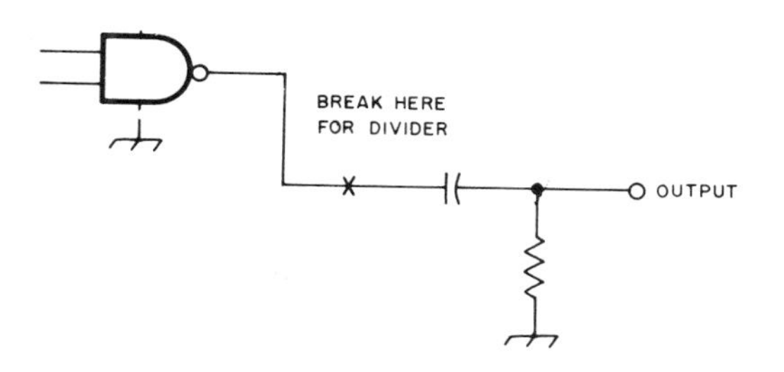

Figure 5.24 Temporary line breaks.

Two basic methods are used for showing interrupted paths: U.S. Standard USAS Y14.15-1966, shown in Figs. 5.25, 5.26, and 5.27, and military practice, shown in Figs. 5.28 through 5.35.

When connecting lines are interrupted, letters, numbers, or abbreviations at the point of interruption indicate the destination, as shown in Fig. 5.25. When groups of connecting lines are interrupted, they may be bracketed as shown in Figs. 5.26 and 5.27. Notice that the dashed line in Fig. 5.27 is not a continuation of one of the bracketed lines.

In the interrupted flow method, special symbols are used to show interruption of signal flow. Within a single sheet of a diagram, oval connectors, signal returns, and breakoff symbols may be used to show interrupted flow.

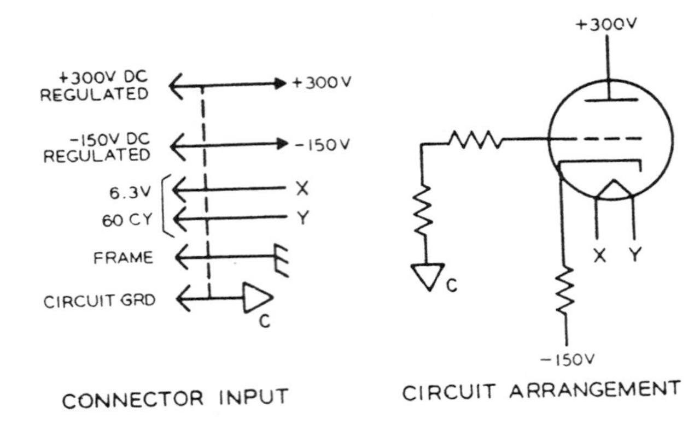

Figure 5.25 Identification of interrupted lines. (From USAS Y14.15-1966.)

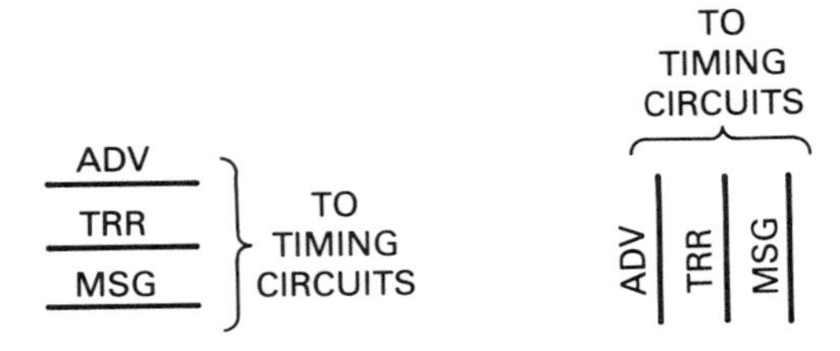

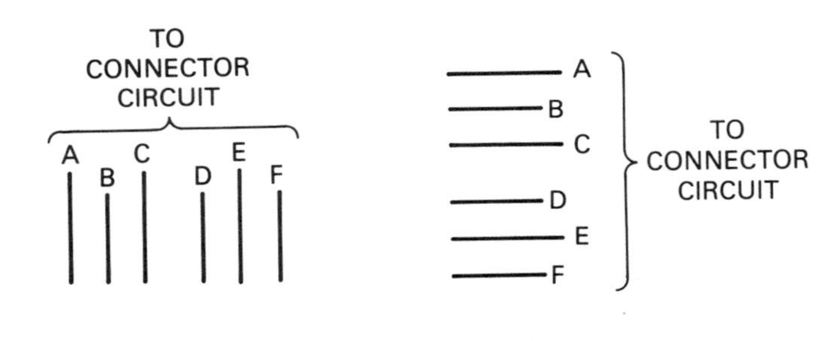

Figure 5.26 Typical arrangement of line identifications and destinations. (From USAS Y14.15-1966.)

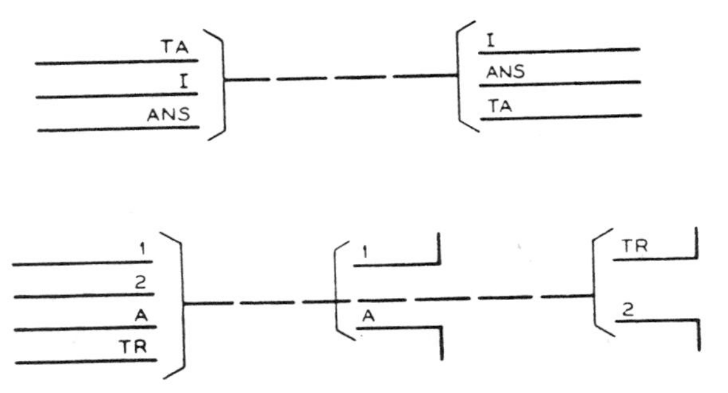

Figure 5.27 Typical interrupted lines interconnected by dash lines. (From USAS Y14.15-1966.)

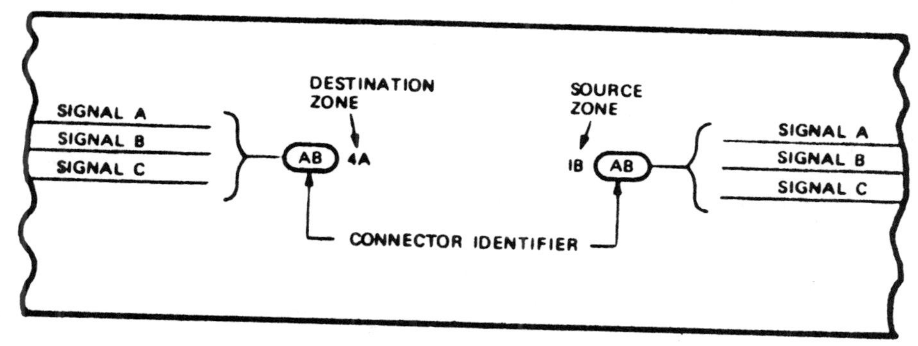

Figure 5.28 Oval connectors used to continue signals. (From MIL-HDBK-63038-1.)

Oval connectors (see Figs. 5.28 and 5.29) are used to continue signals from one area of a sheet to another area. With oval connectors, any number of signals may be bracketed together. Each signal is identified at its source bracket and destination bracket. An oval connector has a unique letter identifier inside the oval. The position of the source and destination connectors is identified by zone numbers.

Signal returns (see Fig. 5.30) are used to continue signal returns within a single sheet of a diagram. Signal returns have unique number identifiers inside the network. Each return is labeled the first time it appears on the diagram (usually on the left edge of the diagram).

Breakoff symbols (ʃ) (see Fig. 5.31) are used to continue power forms, clock pulses, or other multiuse minor signals. Each signal is identified adjacent to its

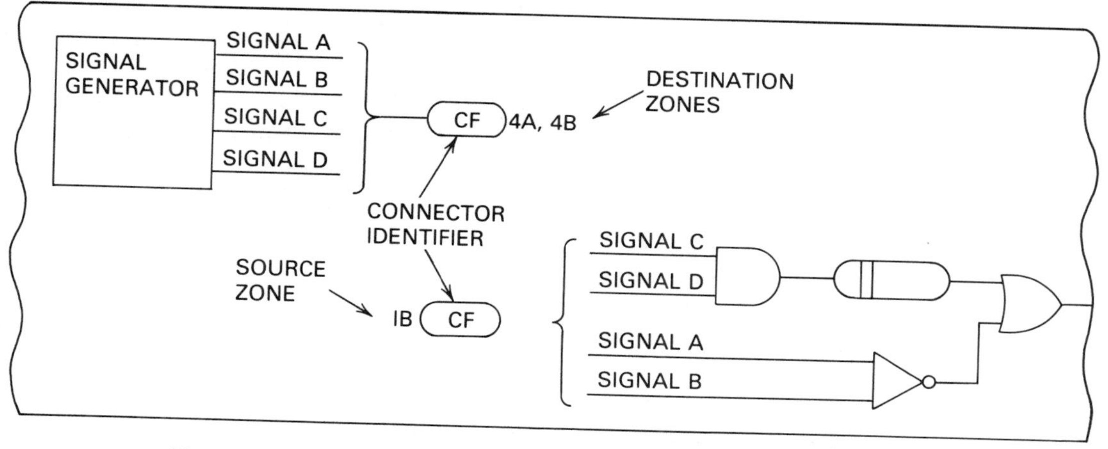

Figure 5.29 Oval connectors used to continue signals to more than one destination. (From MIL-HDBK-63038-1.)

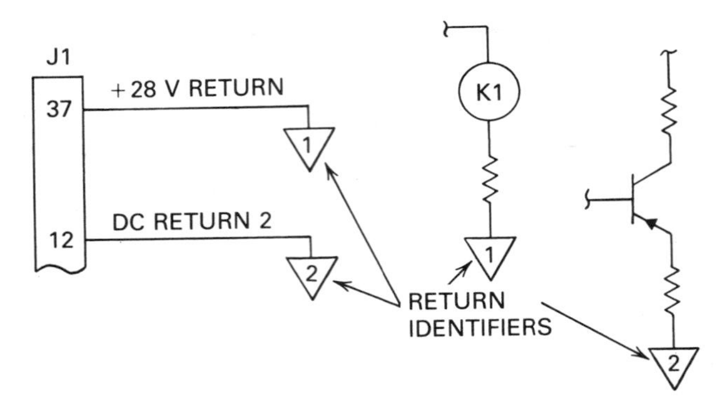

Figure 5.30 Signal returns used to continue signals. (From MIL-HDBK-63038-1.)

breakoff symbol. The source of signals is generally shown at the left edge of the diagram. Only power forms, clock pulses, and other multiuse minor signals use the breakoff symbol technique.

Between sheets of a diagram, boat symbols or oval connectors can be used. Boat symbols (see Fig. 5.32) are used to continue signals from the right edge of one sheet to the left edge of the following sheet within a multisheet diagram. Boat symbols are used for single signals only. Note that they have a unique letter inside the boat.

Oval connectors are used to continue signals from one sheet of a diagram to another just as within a single sheet of a diagram.

In multisheet diagrams, vertical zones are identified by numbers, horizontal

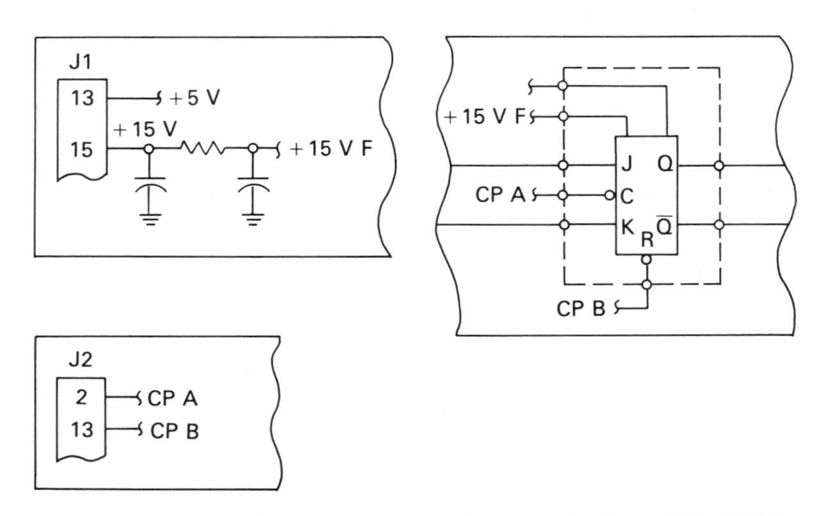

Figure 5.31 Breakoff symbols used to continue signals. (From MIL-HDBK-63038-1.)

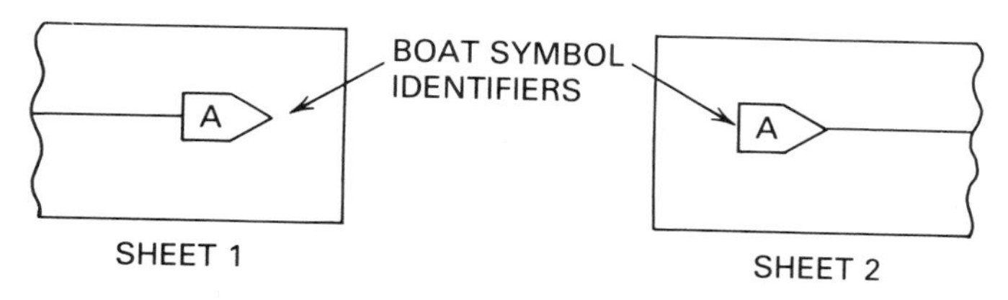

Figure 5.32 Boat symbols used to continue signals between adjacent sheets of a diagram. (From MIL-HDBK-63038-1.)

zones by letters. Horizontal zones are limited to 10 and are numbered as below, even if all zones are not used on any sheet:

Sheet 1	Starts with zone 1
Sheet 2	Starts with zone 11
Sheet 3	Starts with zone 21, etc.

Between diagrams either the block, oval connector, or pyramid diagram technique may be used. As shown in Fig. 5.33, the block technique for continuing signals between diagrams includes figure name and number, connector and pin numbers, and zone numbers.

Oval connectors (see Fig. 5.34) are used to insert source and destination figure numbers before zone references. The pyramid diagram technique is shown in Fig. 5.35.

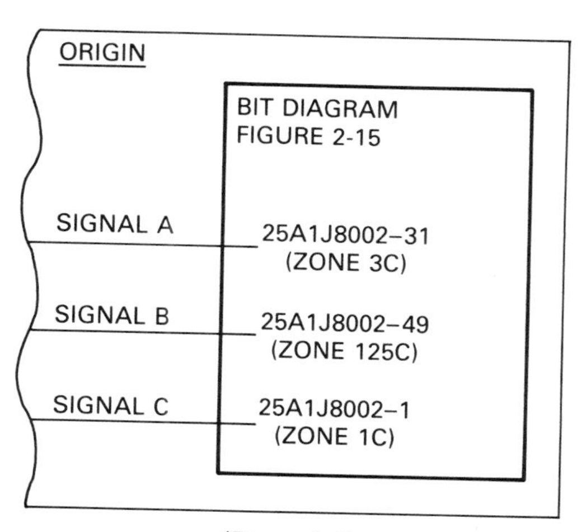

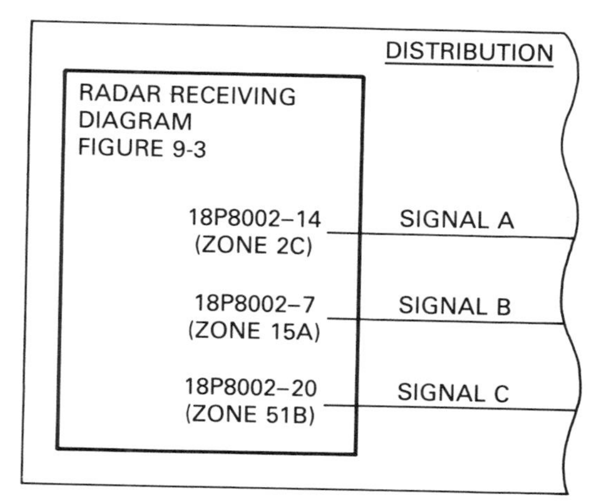

Figure 5.33 Block technique for continuing signals between diagrams. (From MIL-HDBK-63038-1.)

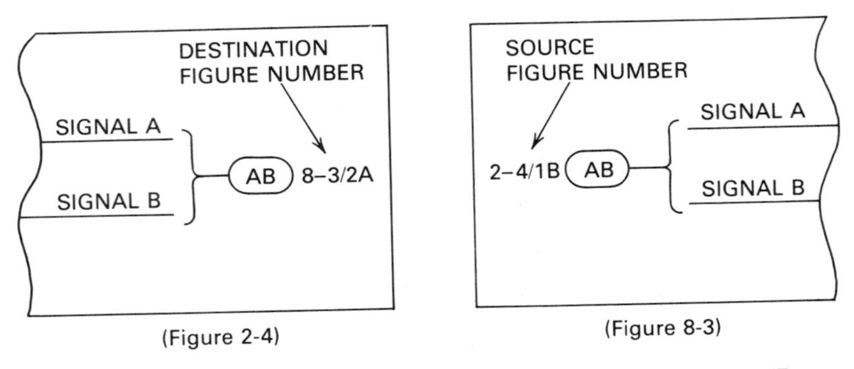

Figure 5.34 Oval connectors used to continue signals between diagrams. (From MIL-HDBK-63038-1.)

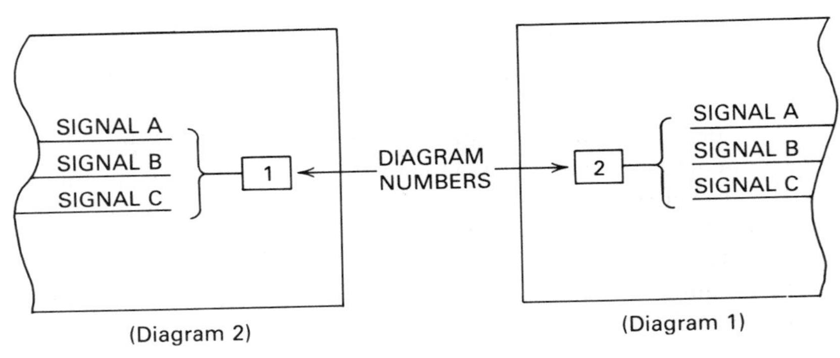

Figure 5.35 Pyramid diagrams interdiagram signal continuation. (From MIL-HDBK-63038-1.)

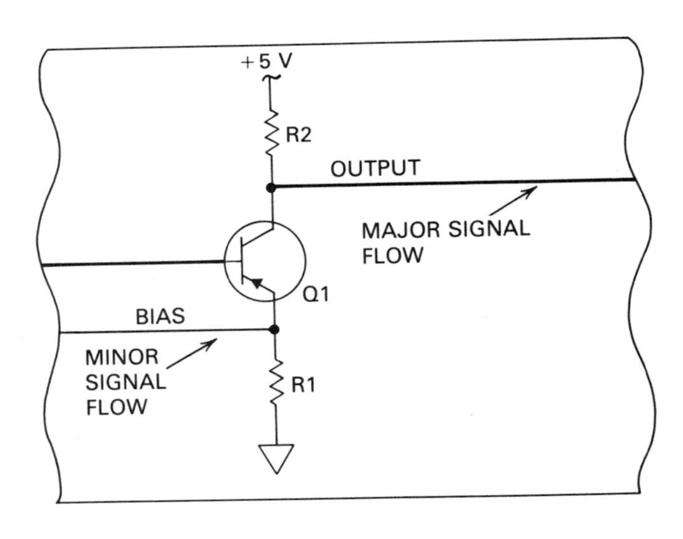

Figure 5.36 Line width code used to indicate signal importance. (From MIL-HDBK-63038-1.)

5.5.4 Signal Importance and Type

Various codes of line width and arrowheads are used to indicate signal importance and type. Wide lines are used to represent major signals as shown in Fig. 5.36. Special arrowheads (see Fig. 5.37) are used to indicate signal types.

The relative importance of signals can also be indicated by the way signal junctions are represented, as indicated in Figs. 5.38 and 5.39.

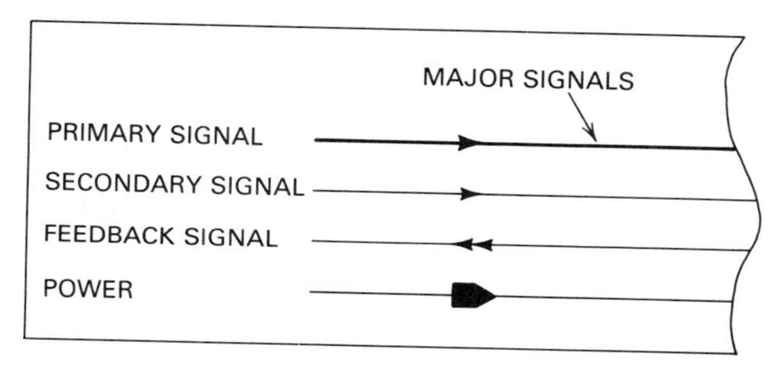

Figure 5.37 Arrowheads used to indicate signal types. (From MIL-HDBK-63038-1.)

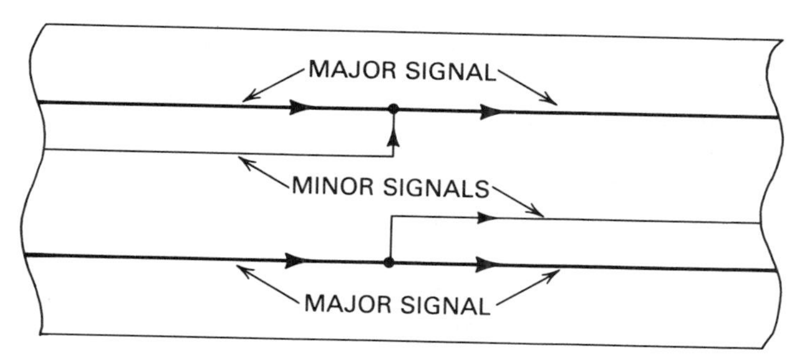

Figure 5.38 Subordinate junctions used to indicate differences in signal importance. (From MIL-HDBK-63038-1.)

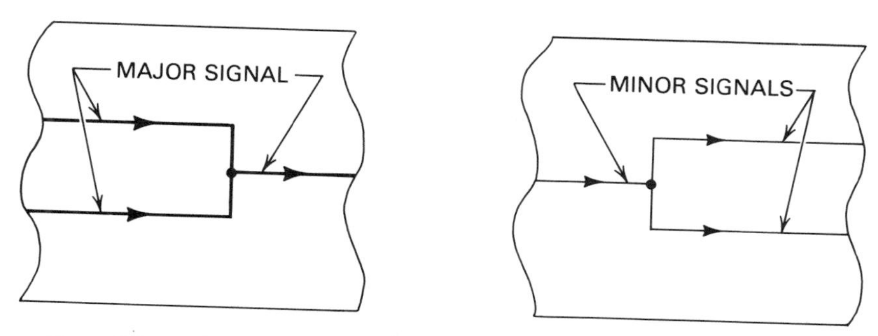

Figure 5.39 Coordinate junctions used to indicate equality in signal importance. (From MIL-HDBK-63038-1.)

5.6 INTERPRETATION

5.6.1 Layout

Whether a schematic is simple or complex, it consists of several sections, stages, or assemblies. These subdivisions may be identified by words such as OSCILLATOR or TIMING next to a component or group of components. If they are not identified in this manner, look for groupings that can be determined by *dashed lines,* which may indicate shielding of a stage, etc.; *boundary or phantom lines,* which enclose groups of parts; or a noticeable amount of *white space* between parts. The center of each stage will be an electron tube, transistor, or an integrated circuit. Look for coupling components such as resistors, capacitors, and transformers.

Ideally, a schematic will be arranged so that the signal flow can easily be traced from input to output or from source to load. With this pattern, which is the same as for block diagrams, the drawing can be "read" from left to right for simpler one-layer drawings. For complex diagrams with more than one layer, the pattern will start at the upper left corner of the diagram, go from the left to the right edge of the drawing, then down to the next layer, where it will go from left to right again, as shown in Fig. 5.40. In some cases external connections will be shown on the outer edges of the diagram.

Unfortunately, this type of layout is not always followed. But the first step in interpreting the schematic is to determine if this practice has been followed.

It is vital to find the inputs and outputs. Look for inputs at the left edge or top of the diagram; look for outputs on the right edge or bottom.

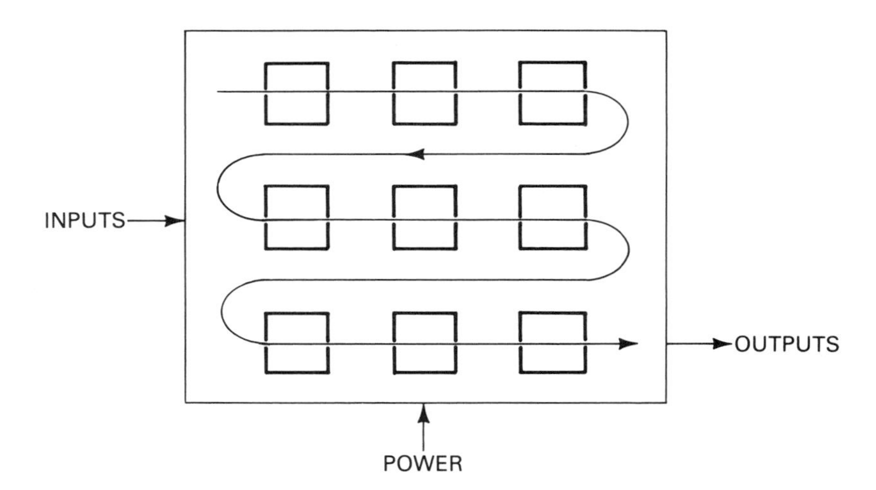

Figure 5.40 Ideal schematic layout showing signal flow.

5.6.2 Waveforms

Waveforms showing voltage or current levels are given at crucial points on some schematics as an aid to understanding and troubleshooting the circuit. Figure 5.41 shows some of the more common waveforms. Variations of these forms and exponential, trapezoidal, and triangular forms may also be shown. Notice that these waveforms are simplified representations; actual waveforms, as seen on an oscilloscope, will not have vertical lines or square corners.

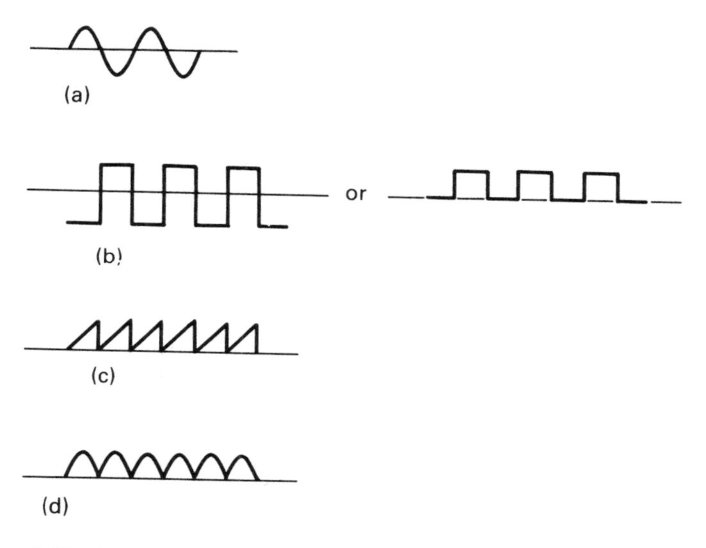

Figure 5.41 Representative ideal waveforms: (a) sine wave; (b) square wave; (c) sawtooth wave; (d) rectified sine wave.

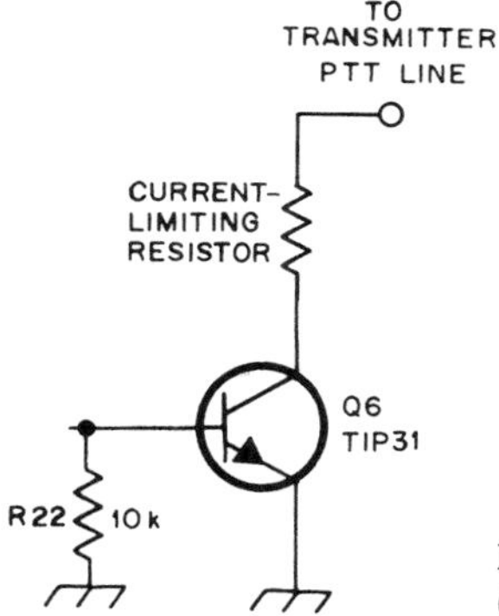

Figure 5.42 Resistor used to limit current. (Copyright 1983 by QST.)

5.6.3 Common Circuits

As a reminder, resistors are used to limit current (Fig. 5.42) and as voltage dividers (Fig. 5.43). Capacitors and inductances are used in filters, as shown in Fig. 5.44. Diodes have a variety of uses as shown in Figs. 5.45 through 5.50. Some

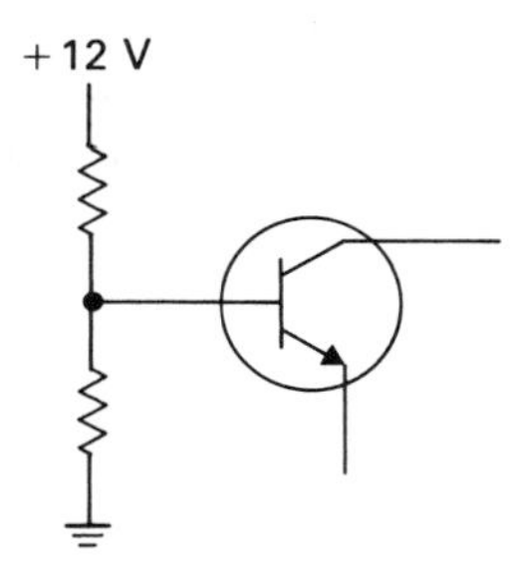

Figure 5.43 Resistors used as voltage dividers.

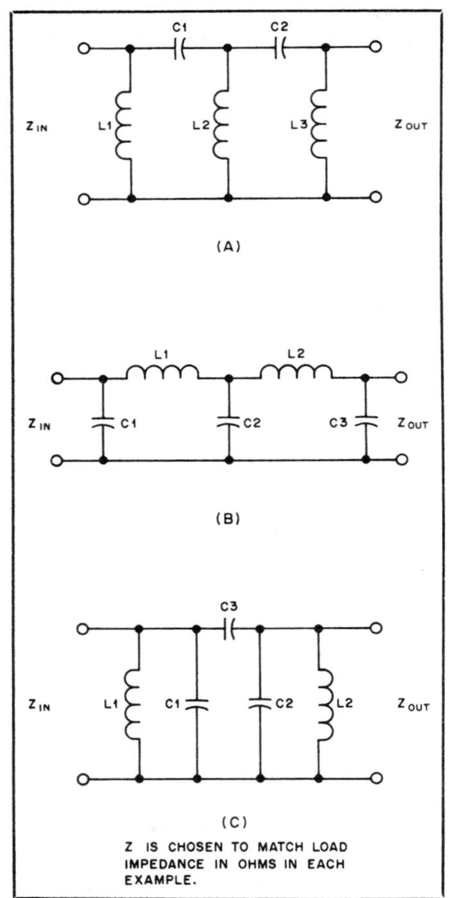

Figure 5.44 Capacitors and inductances used in filters: (a) high pass; (b) low pass; (c) bandpass. (Copyright 1984 by QST.)

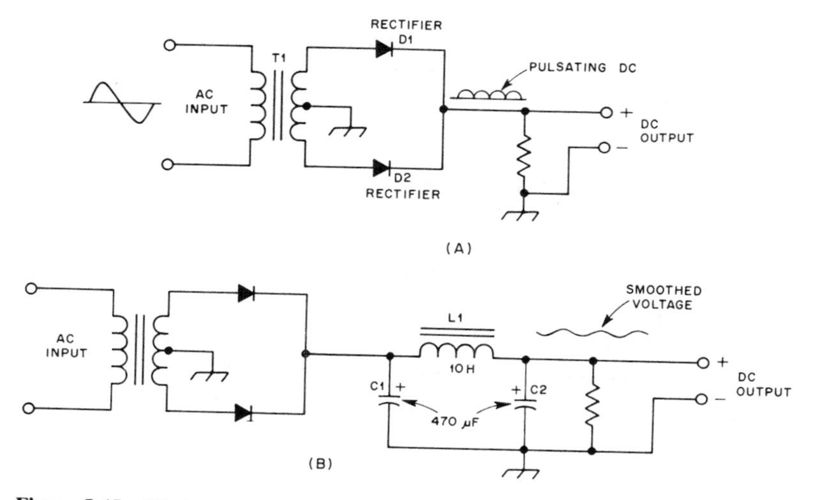

Figure 5.45 Diodes used as power-supply rectifiers. (Copyright 1984 by QST.)

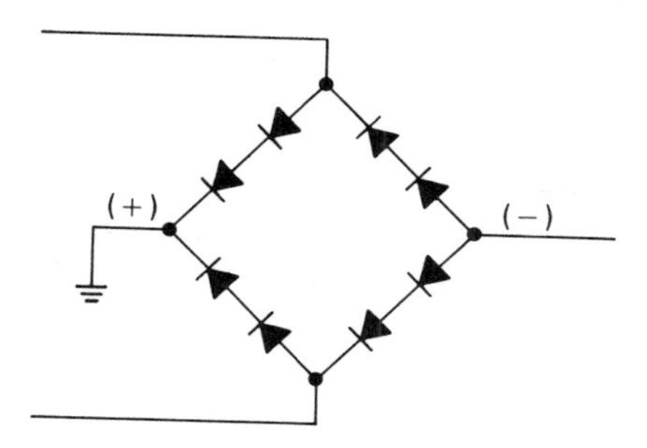

Figure 5.46 Diodes used in series to avoid voltage breakdown.

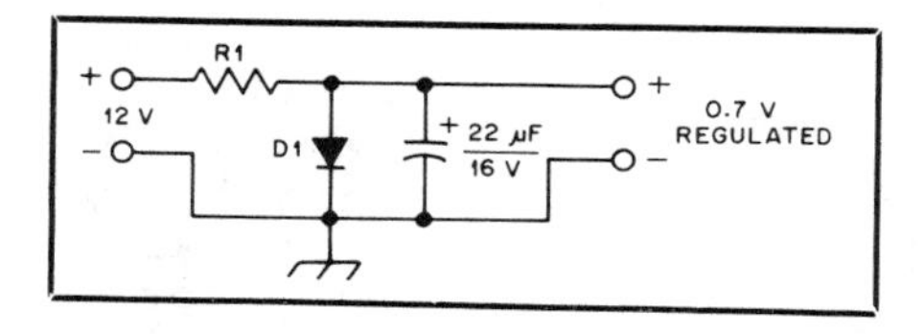

Figure 5.47 Diode used to provide voltage regulation. (Copyright 1984 by QST.)

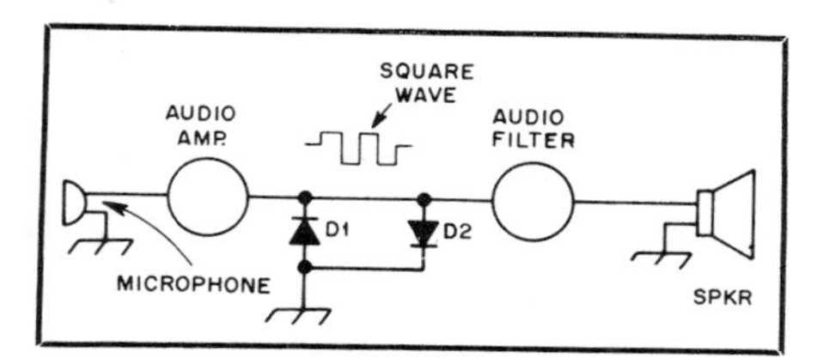

Figure 5.48 Diodes used to limit voltage peaks. (Copyright 1984 by QST.)

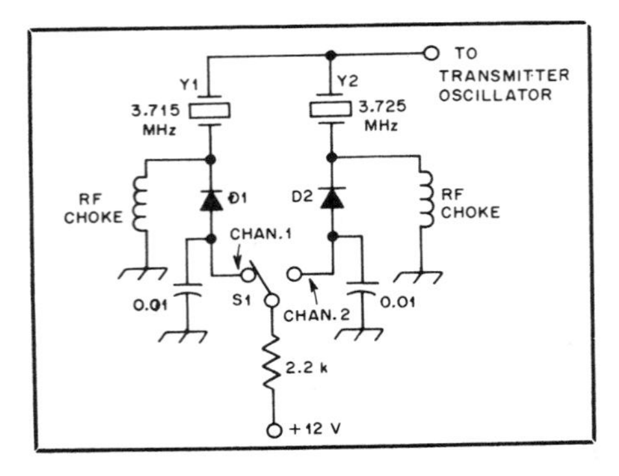

Figure 5.49 Diodes used as electronic switches. (Copyright 1984 by QST.)

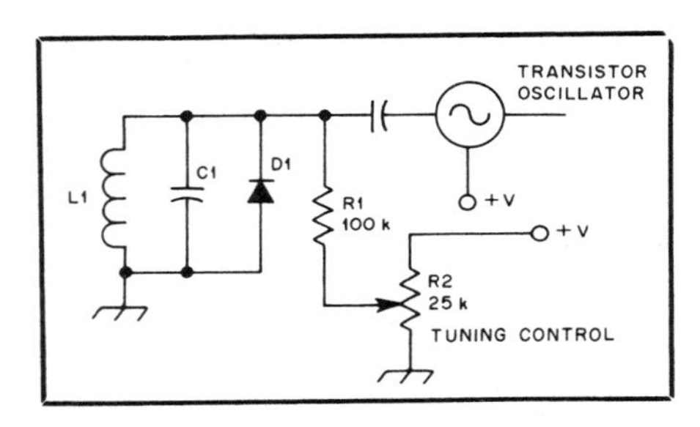

Figure 5.50 Diode used as a variable capacitor. (Copyright 1984 by QST.)

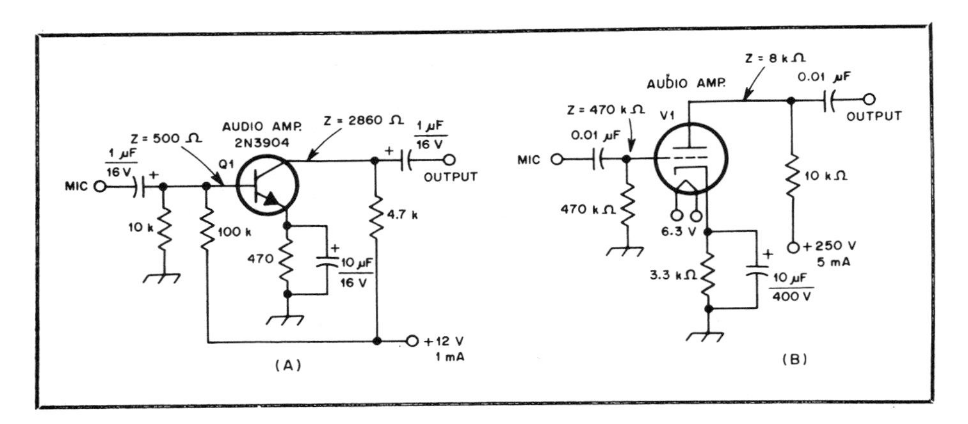

Figure 5.51 Comparison of transistor audio amplifier with tube amplifier. (Copyright 1984 by QST.)

common transistor circuits are shown in Figs. 5.51, 5.52, and 5.53. Figure 5.54 shows the same circuit drawn in two noticeably different ways. Figure 5.55 shows some of the ways an operational amplifier can be used.

Obviously, there are many more types of circuits than shown here. If interpretation of a particular part of a schematic seems impossible, do not overlook the fact that there could be an error in the drawing.

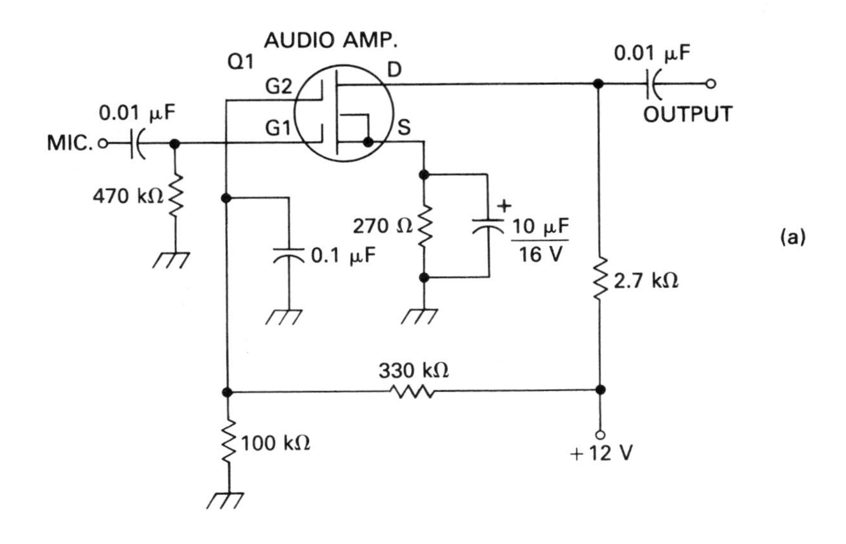

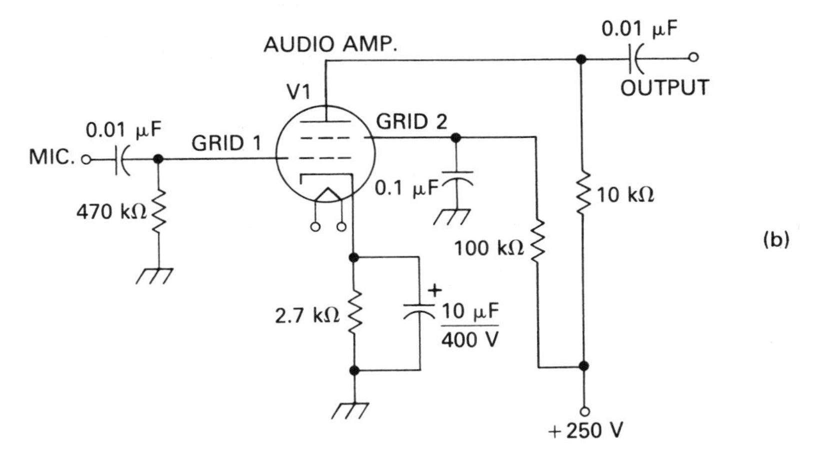

Figure 5.52 Audio amplifier with (a) dual-gate MOSFET and (b) tetrode tube. (Copyright 1984 by QST.)

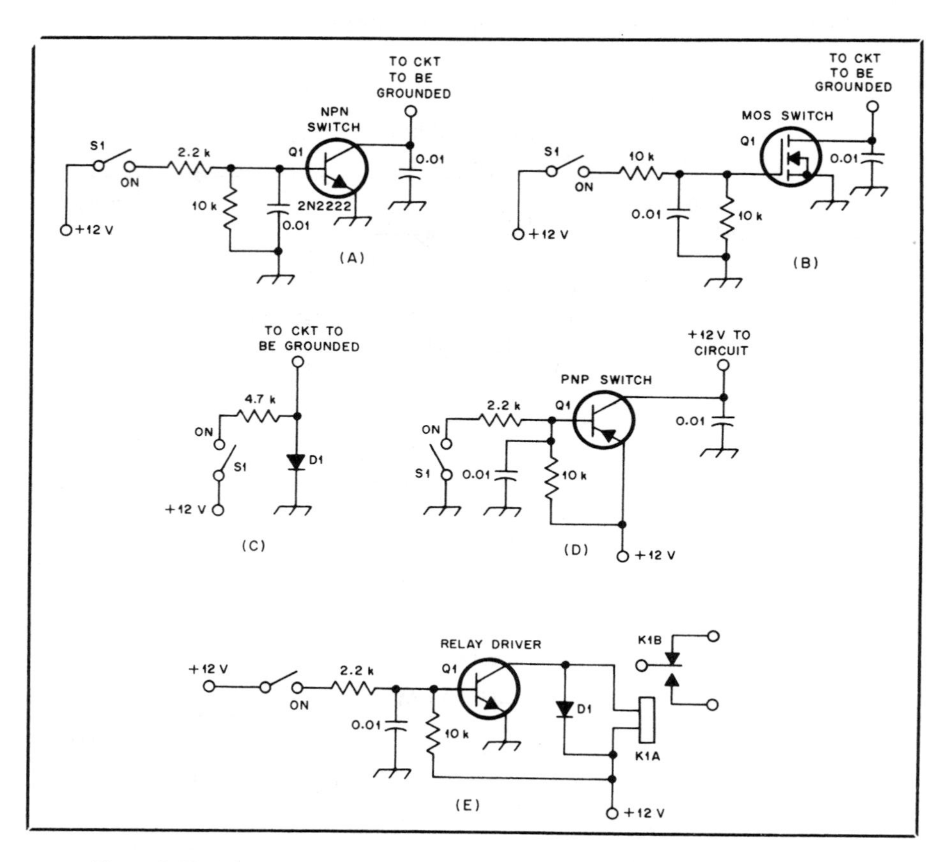

Figure 5.53 Diode and transistors used as switches: (a) NPN transistor switch; (b) power FET switch; (c) simple diode switch; (d) PNP transistor switch; (e) relay driver. (Copyright 1984 by QST.)

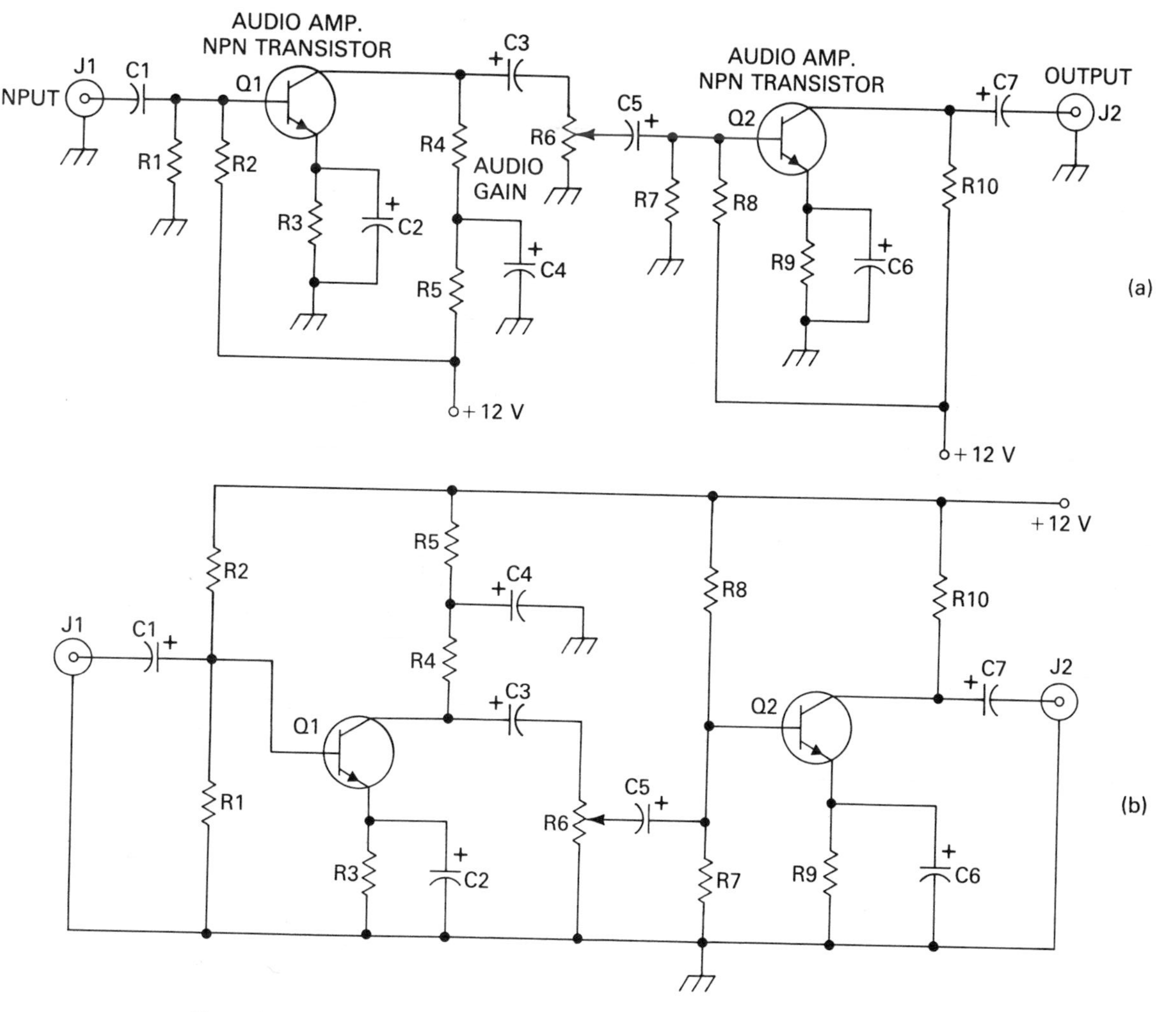

Figure 5.54 Identical circuit drawn in different ways. (Copyright 1984 by QST.)

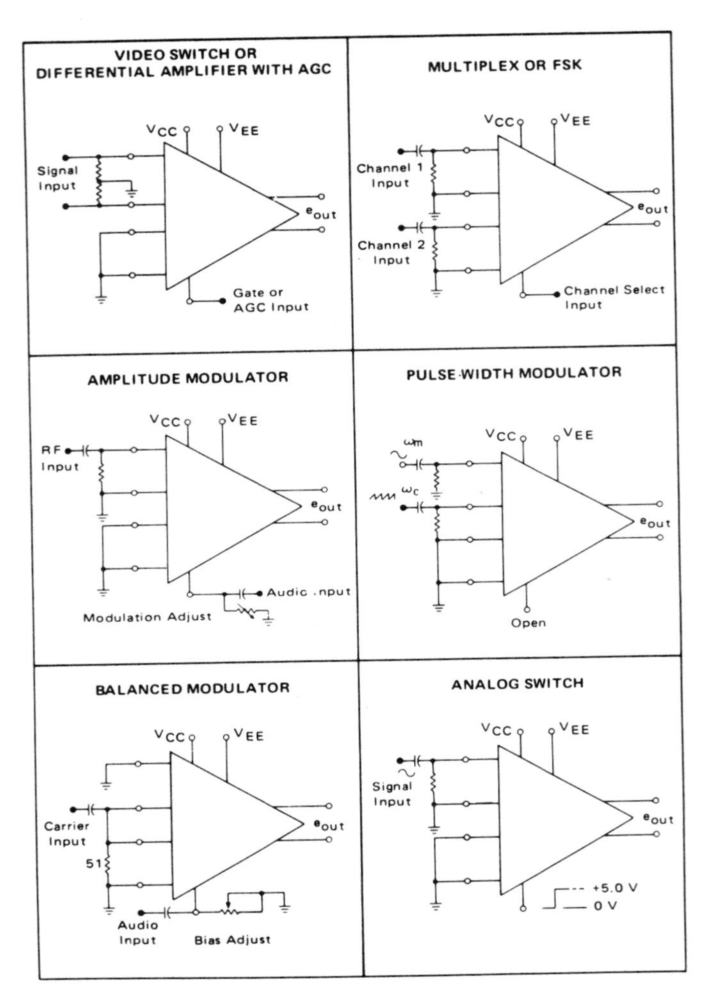

Figure 5.55 Operational amplifier configurations. (Copyright by Motorola, Inc. Used by permission.)

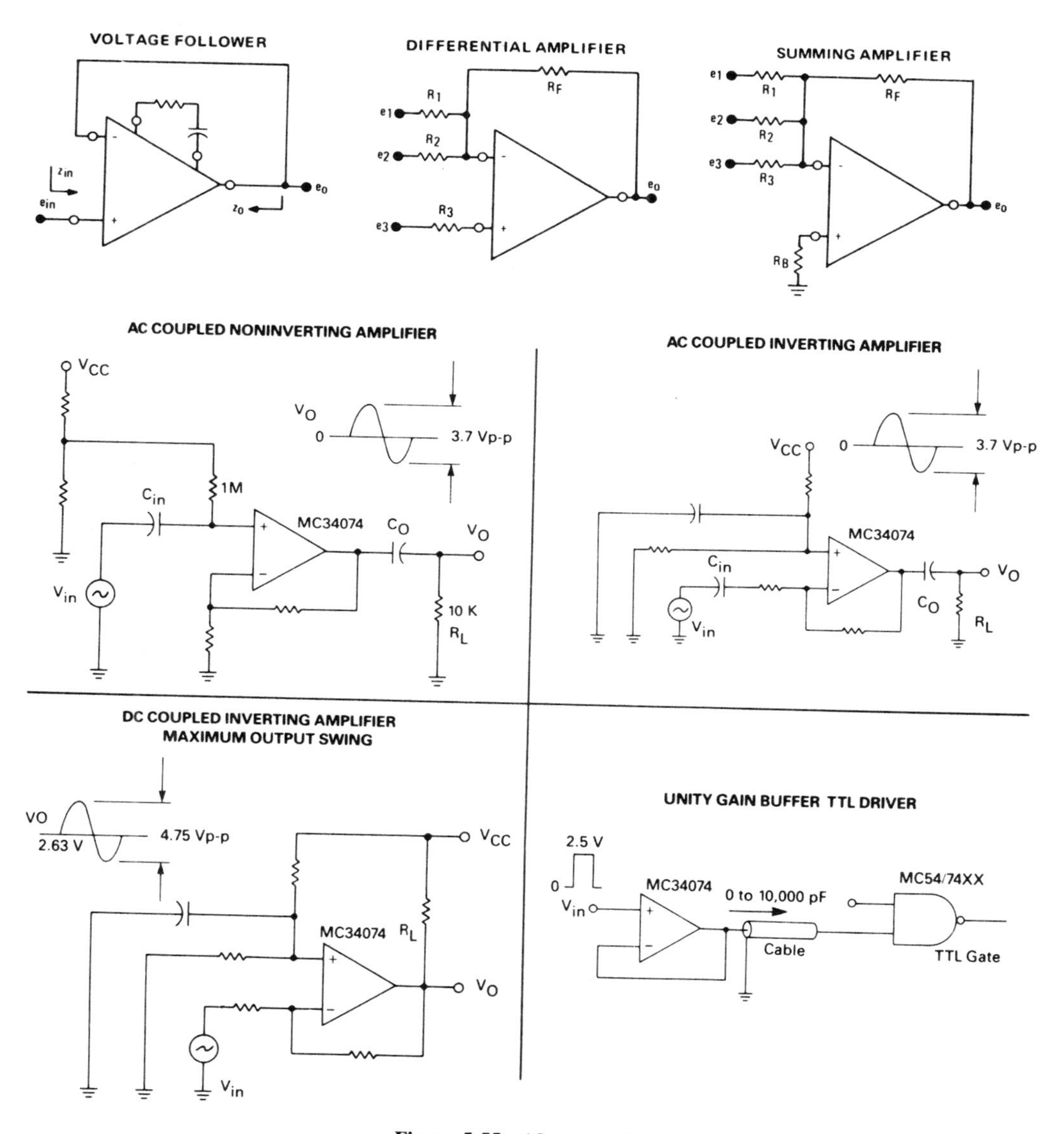

Figure 5.55 *(Continued)*

HIGH SLEW-RATE INVERTER

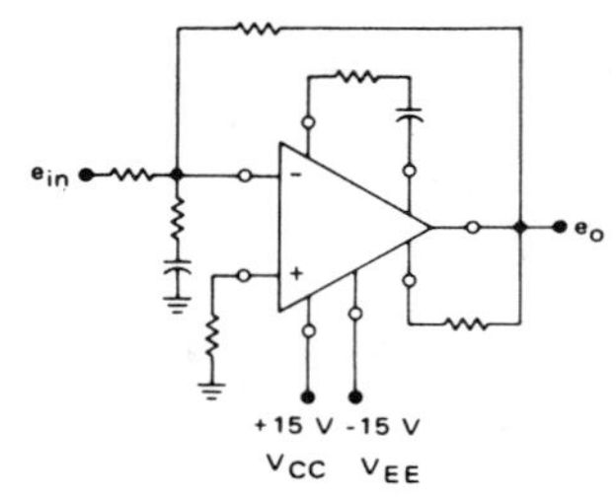

OUTPUT NULLING CIRCUIT

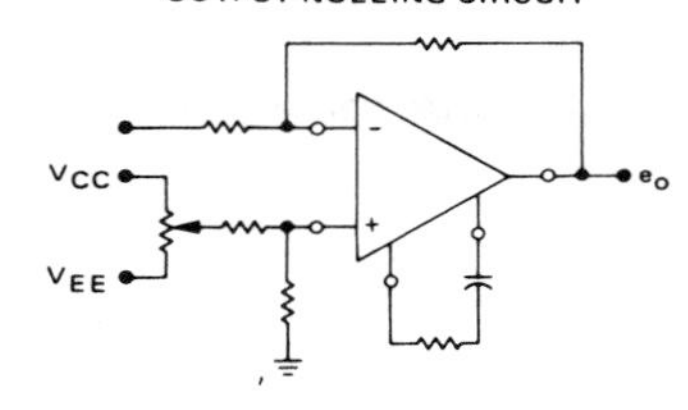

DIFFERENTIAL AMPLIFIER WITH ± 20 V COMMON-MODE INPUT VOLTAGE RANGE

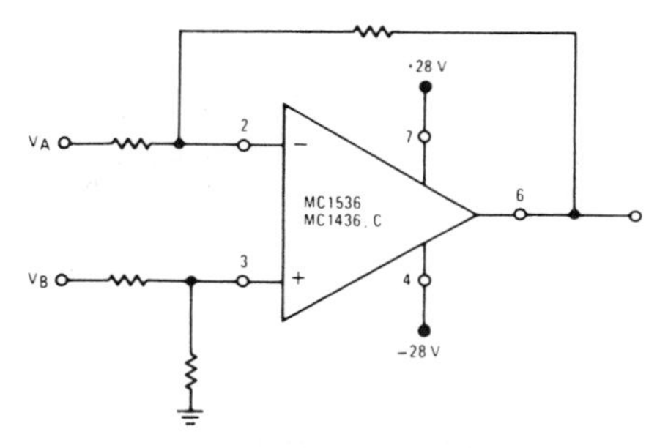

OUTPUT LIMITING CIRCUIT

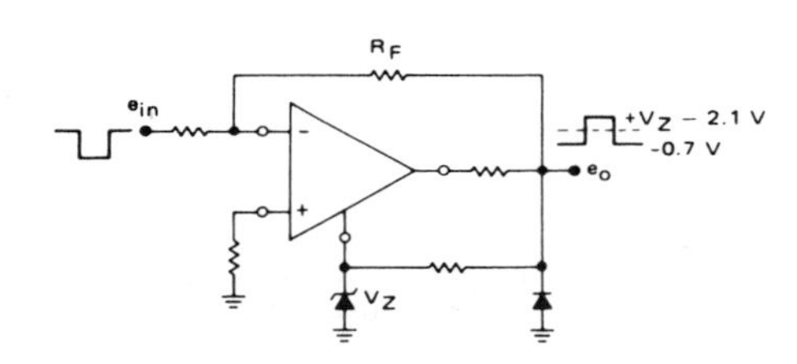

TYPICAL NONINVERTING X10 VOLTAGE AMPLIFIER

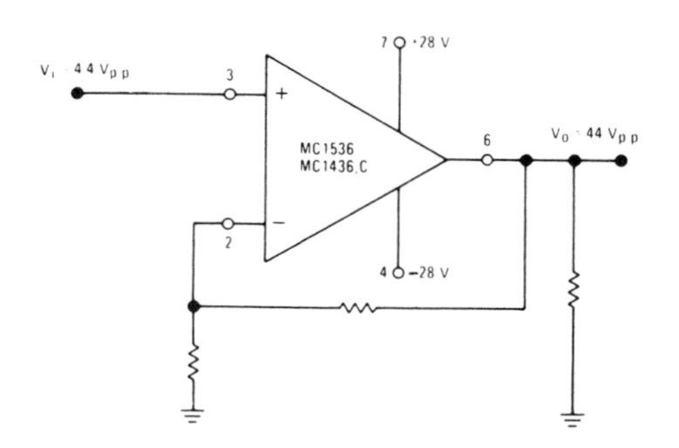

LOW-DRIFT SAMPLE AND HOLD

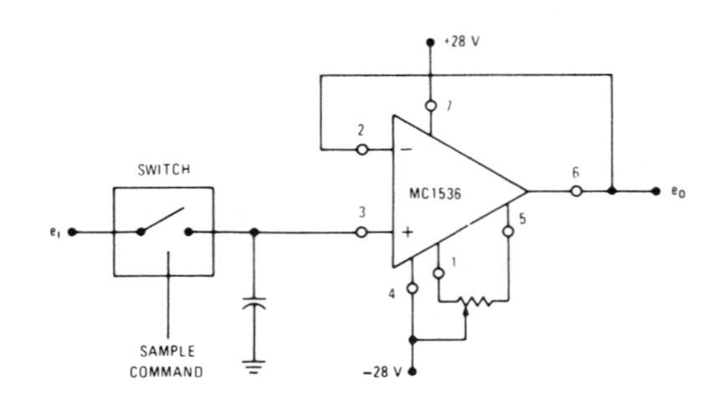

Figure 5.55 *(Continued)*

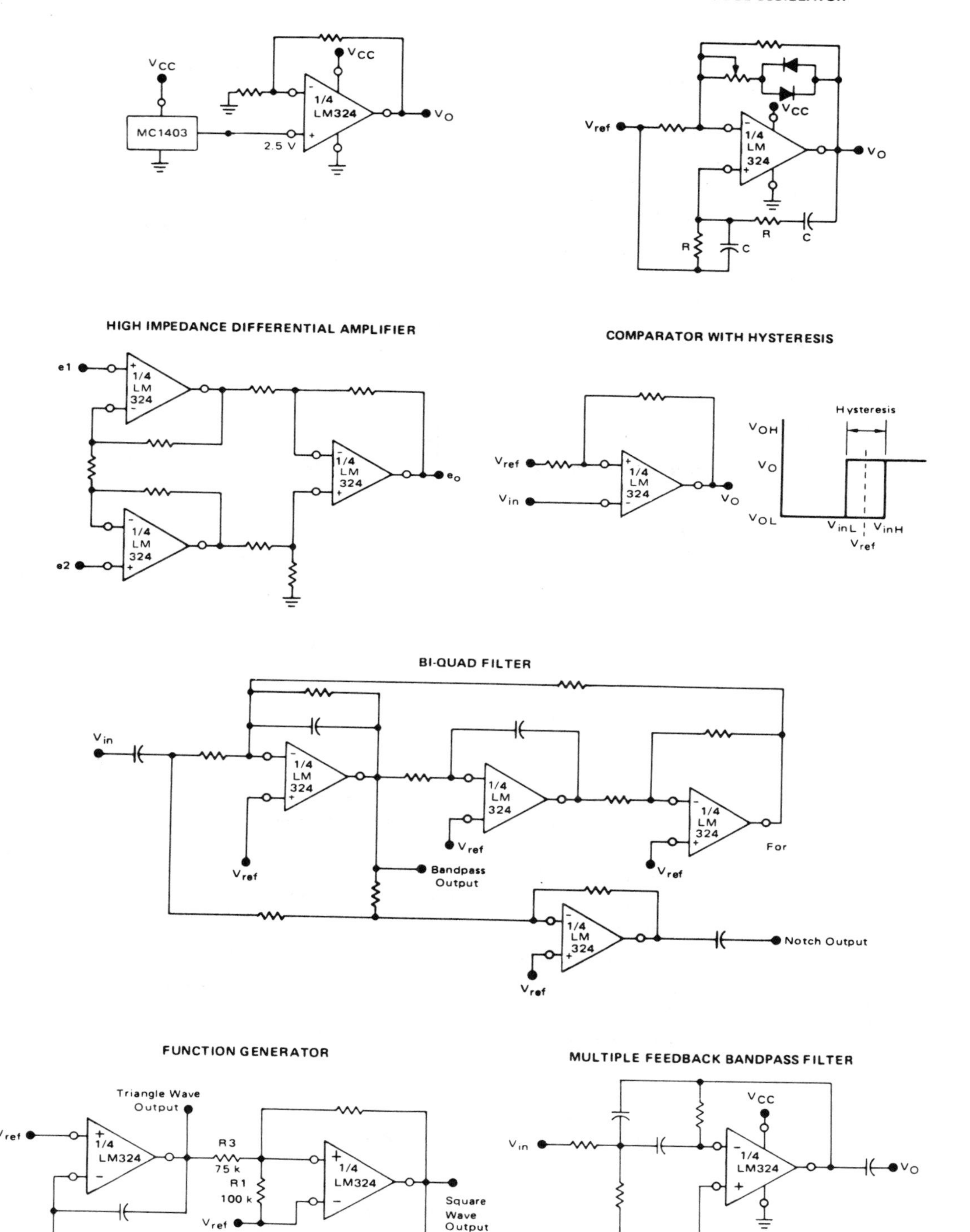

Figure 5.55 *(Continued)*

FAST SETTLING INVERTER

BASIC INVERTING AMPLIFIER

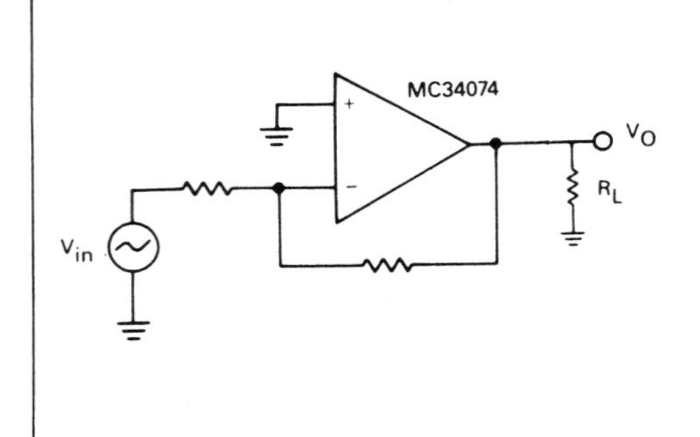

BASIC NON-INVERTING AMPLIFIER

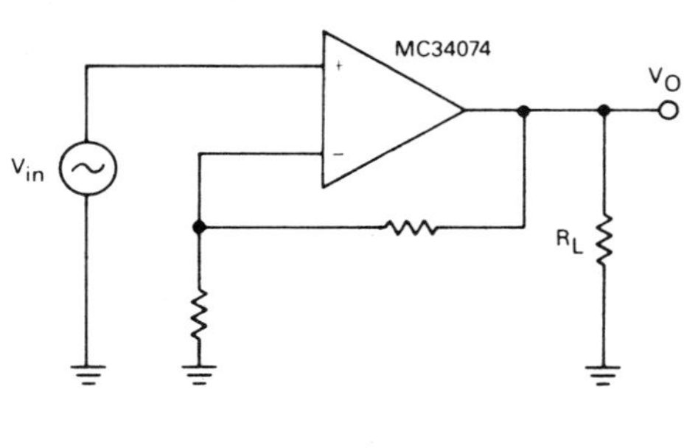

UNITY GAIN BUFFER (A_V = +1)

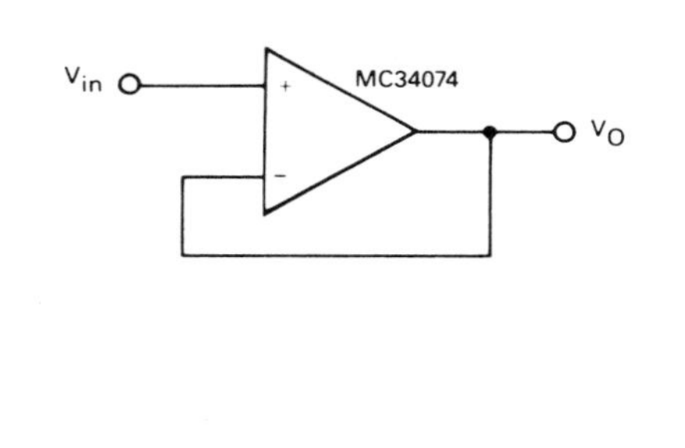

HIGH IMPEDANCE DIFFERENTIAL AMPLIFIER

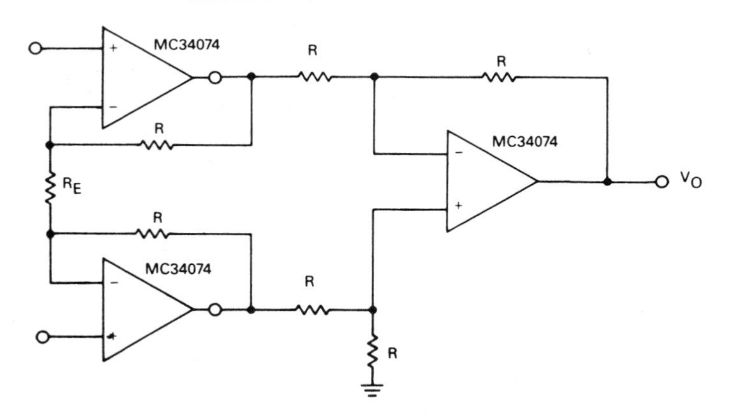

Figure 5.55 *(Continued)*

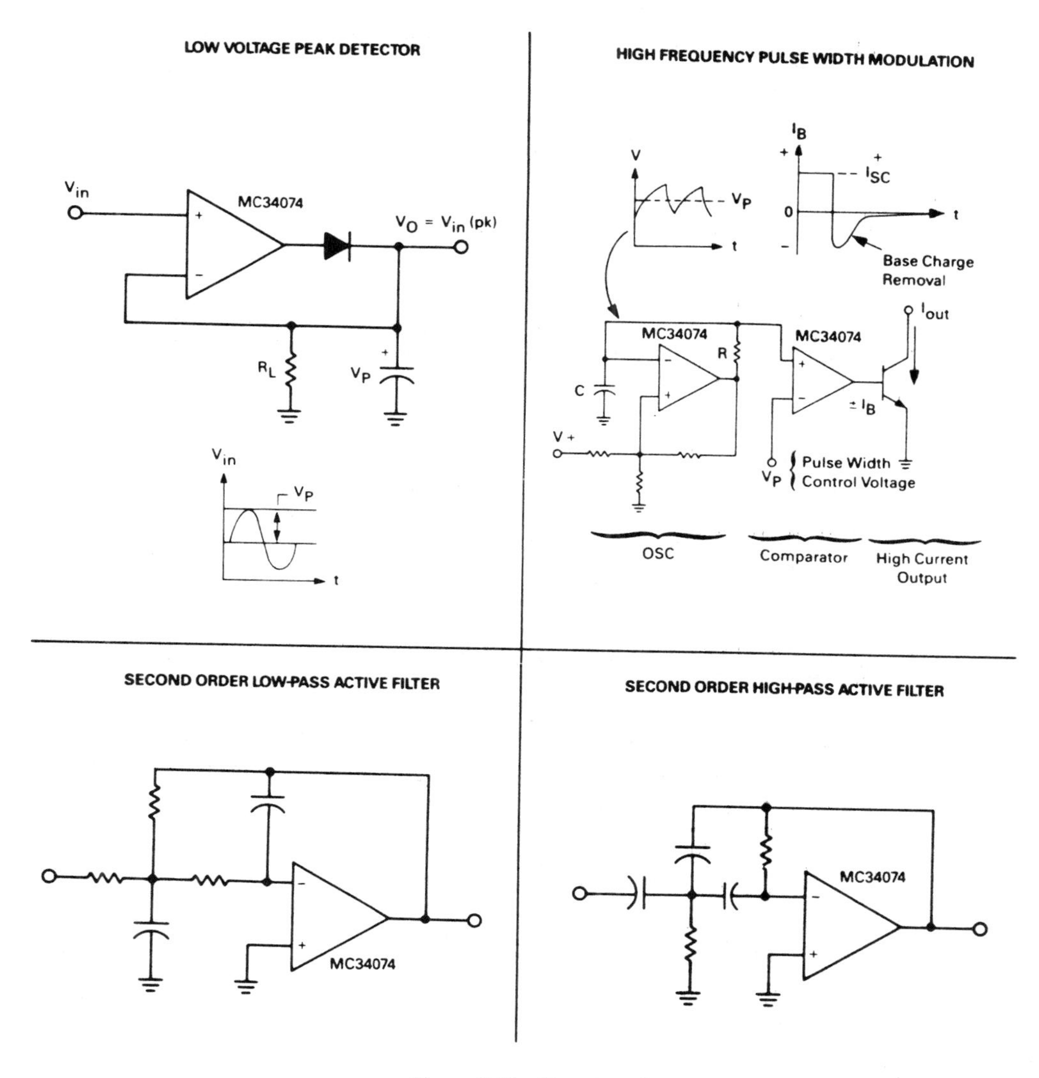

Figure 5.55 *(Continued)*

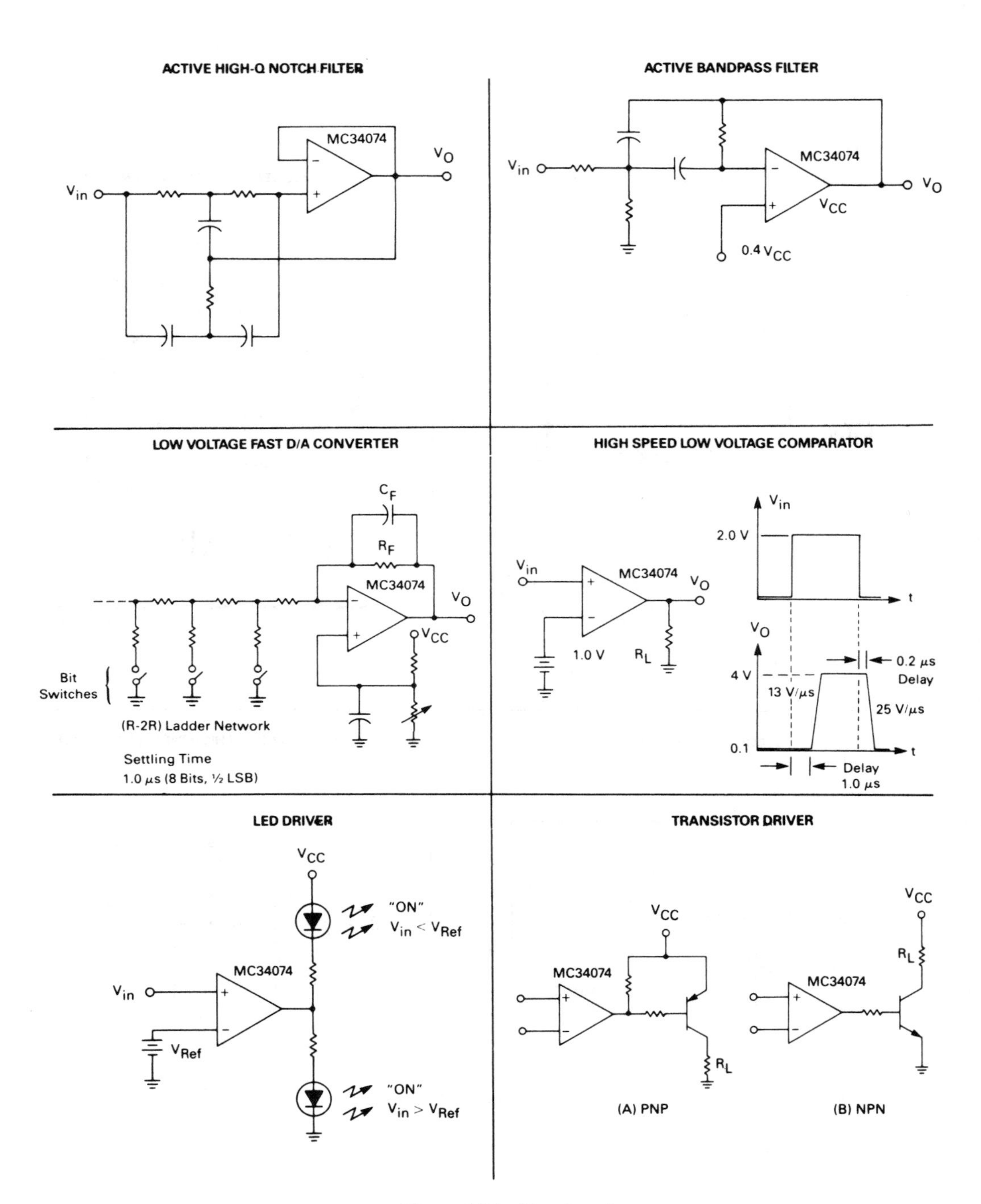

Figure 5.55 *(Continued)*

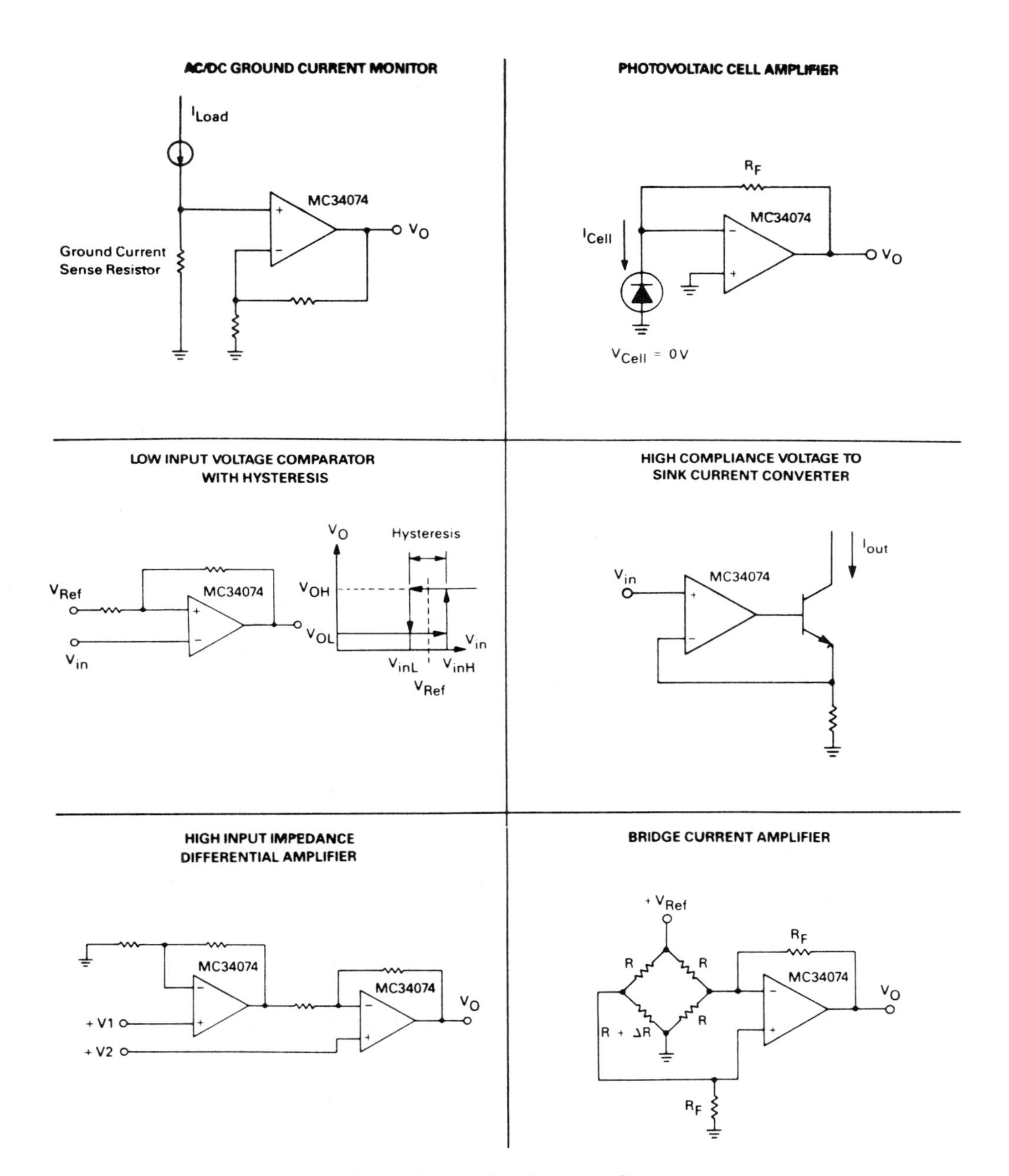

Figure 5.55 *(Continued)*

Chapter 6

Connection Diagrams*

Whereas a wiring or connection diagram may be combined with a schematic diagram, for more complex equipment the wiring diagram will be separate. Connection diagrams show the electrical connections or wiring between component devices or parts; they supplement schematic diagrams by relating circuit information with the actual wiring and location of component devices or parts.

Some of the common wiring symbols used on connection diagrams are shown in Fig. 6.1. Connection information is available as (1) wiring (point-to-point) diagrams, (2) ladder diagrams, (3) cable or highway diagrams, (4) wire lists, and (5) interconnection diagrams.

The *point-to-point diagram* is useful mainly for the representation of wiring information for less complex equipment that may consist of relatively few items. In these diagrams (Fig. 6.2) separate connecting lines represent the actual terminal-to-terminal connections that are to be provided on the equipment. They may also show terminal, connector, and wire identifiers that appear on the hardware.

Grouped lines may indicate related circuits. Heavier lines may indicate ground returns or power circuits. Parts are likely to be represented by squares or rectangles rather than by pictorial symbols or functional symbols used on schematics. Terminal strips may be drawn in any desired order, contrary to the physical layout, to eliminate undesired line crossings. Wires used to connect terminals on the same component are called *jumpers* or *straps* (Fig. 6.3).

* Except for the material on ladder diagrams, some of the information in this chapter has been extracted from USA Standard Drafting Practices, Electrical and Electronics Diagrams, USAS Y14.15-1966, published by the American Society of Mechanical Engineers. Information on ladder diagrams was extracted from James D. Bethune, *Basic Electronic and Electrical Drafting*, 2nd ed., Prentice-Hall, Inc., Englewood Cliffs, N.J., 1985.

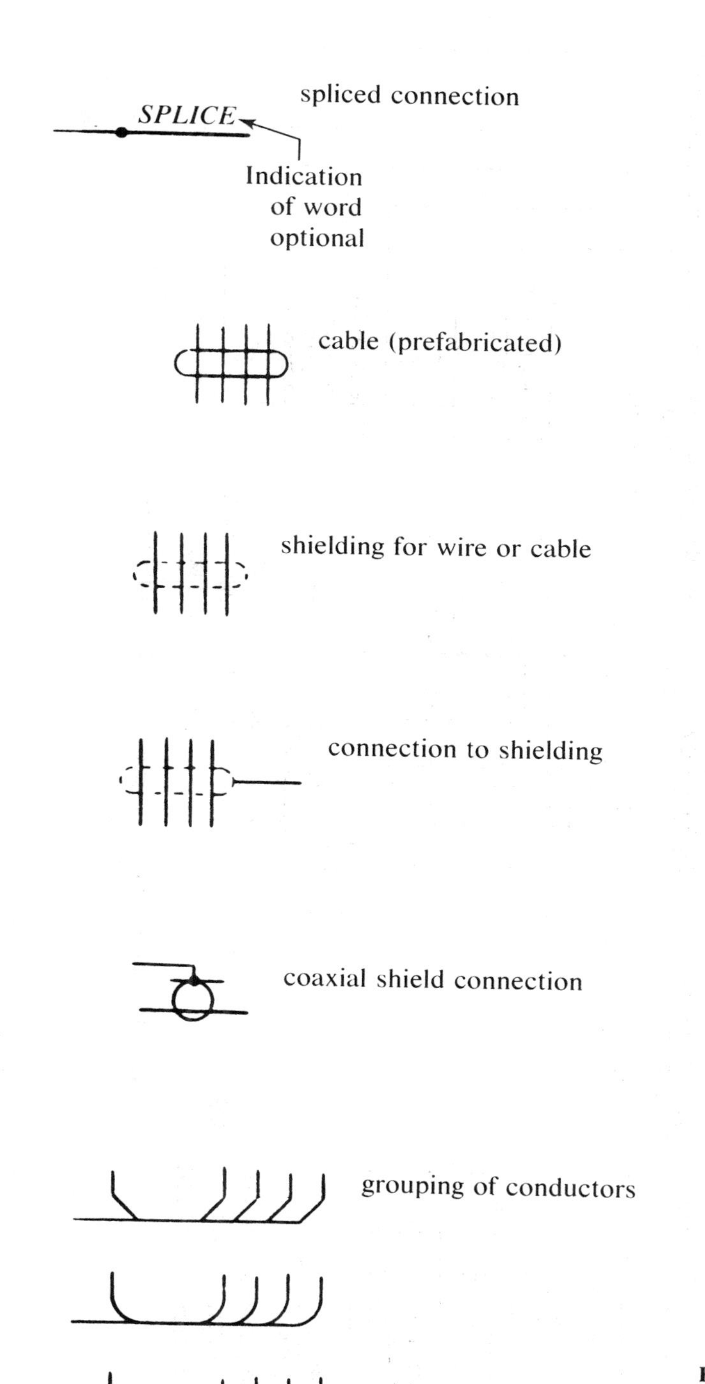

Figure 6.1 Graphic symbols commonly used on connection diagrams. (From USAS Y14.15-1966.)

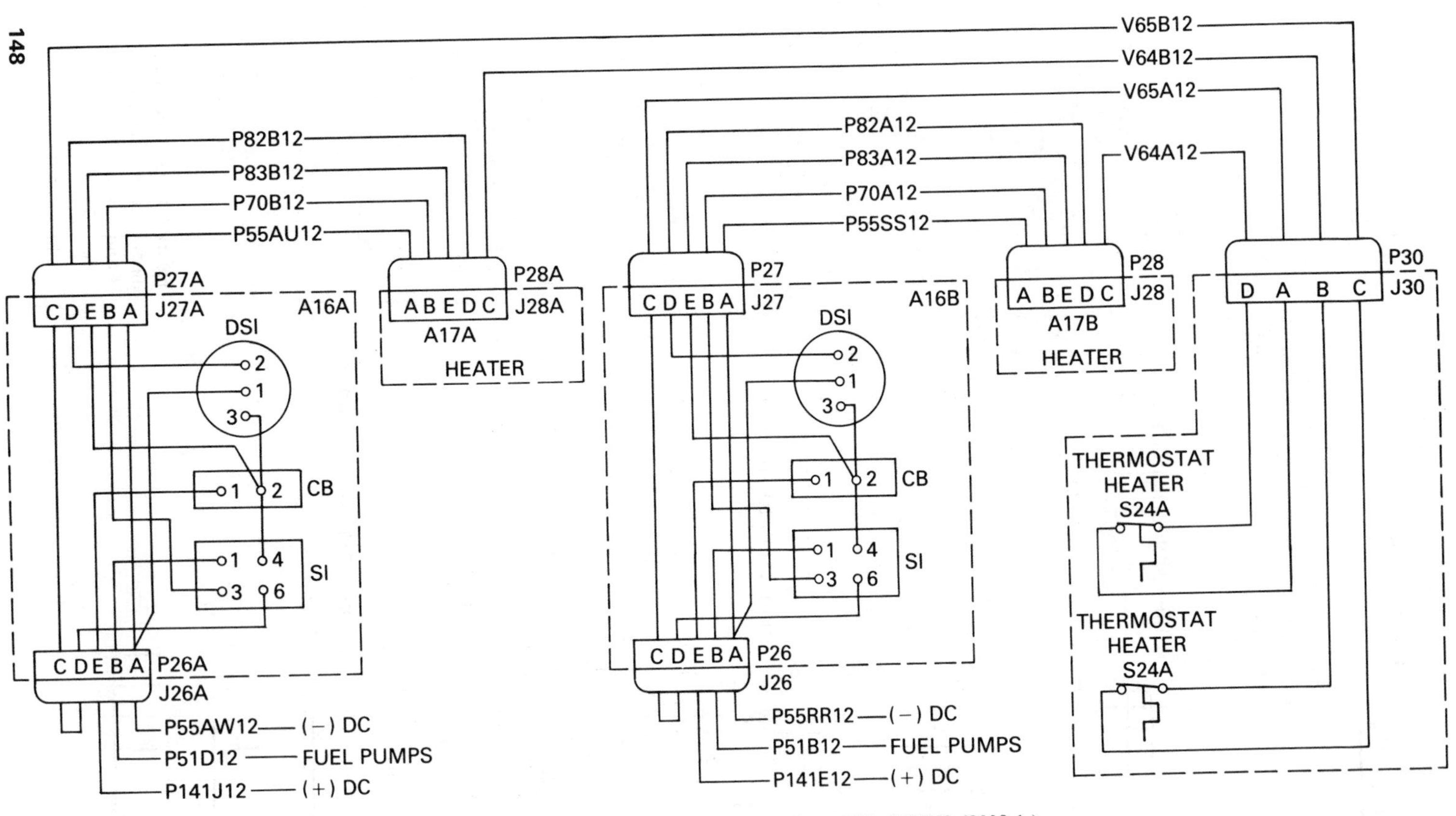

Figure 6.2 Example of wiring diagram. (From MIL-HDBK-63038-1.)

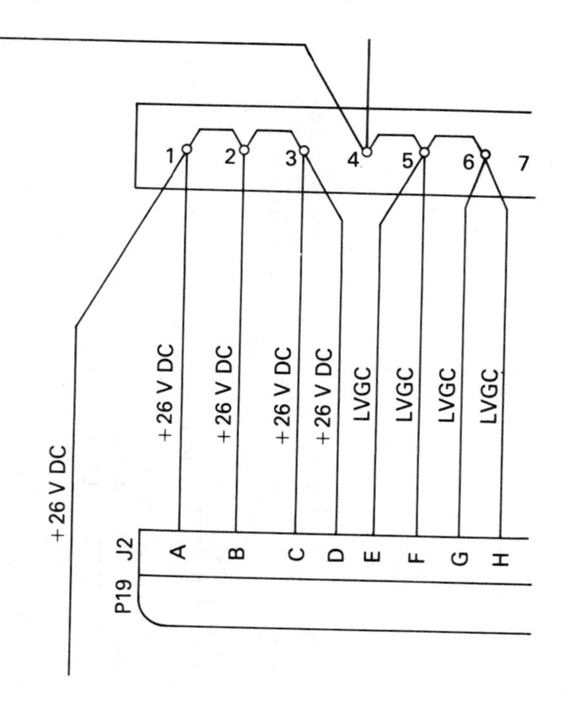

Figure 6.3 Example of straps or jumpers.

Ladder diagrams are wiring diagrams that show the components used to operate and control electrical motors. Although not needed for consumer electronics, they are essential for circuits controlling motors for radar and telemetry antennas. As shown in Fig. 6.4, ladder diagrams have two parts: a horizontal upper portion and a vertical/horizontal ladder portion. The upper portion defines the power applications, and the lower portion defines the control functions.

To read a ladder diagram, follow the upper portion from left to right across the diagram, and then go down the left side of the lower portion and across each individual horizontal line from left to right.

Figure 6.5 shows the symbols and abbreviations used on ladder diagrams. As ladder diagrams include many switches and coils whose symbols appear the same, it is essential to notice the labels so as to tell which switch is related to which coil. For example, in Fig. 6.4 coil (2CR) activates contacts 2CR.

Switches and contacts are shown in their normal position before being activated by current. For example, a normally closed switch is one that is closed unless it is opened by a current; therefore, this switch is drawn in the closed position.

Figure 6.6 shows how to read ladder diagrams. It is a control portion showing the circuitry needed to start and stop an electric motor. Following electric circuit practice, we consider current to flow from positive to negative rather than

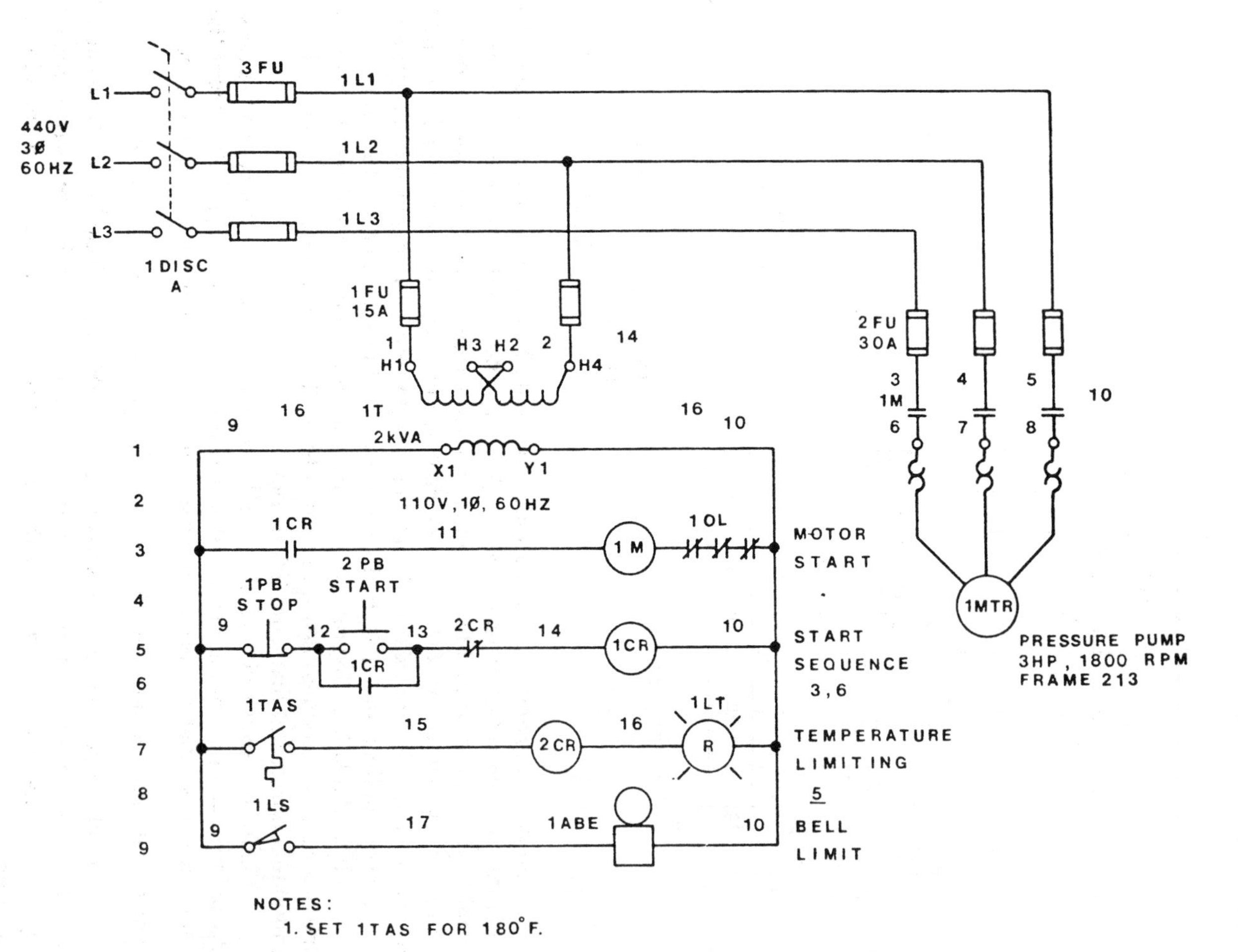

Figure 6.4 Example of a ladder diagram. (From James D. Bethune, *Basic Electronic and Electrical Drafting*, 2nd ed., © 1985. Reprinted by permission of Prentice-Hall, Inc., Englewood Cliffs, New Jersey.)

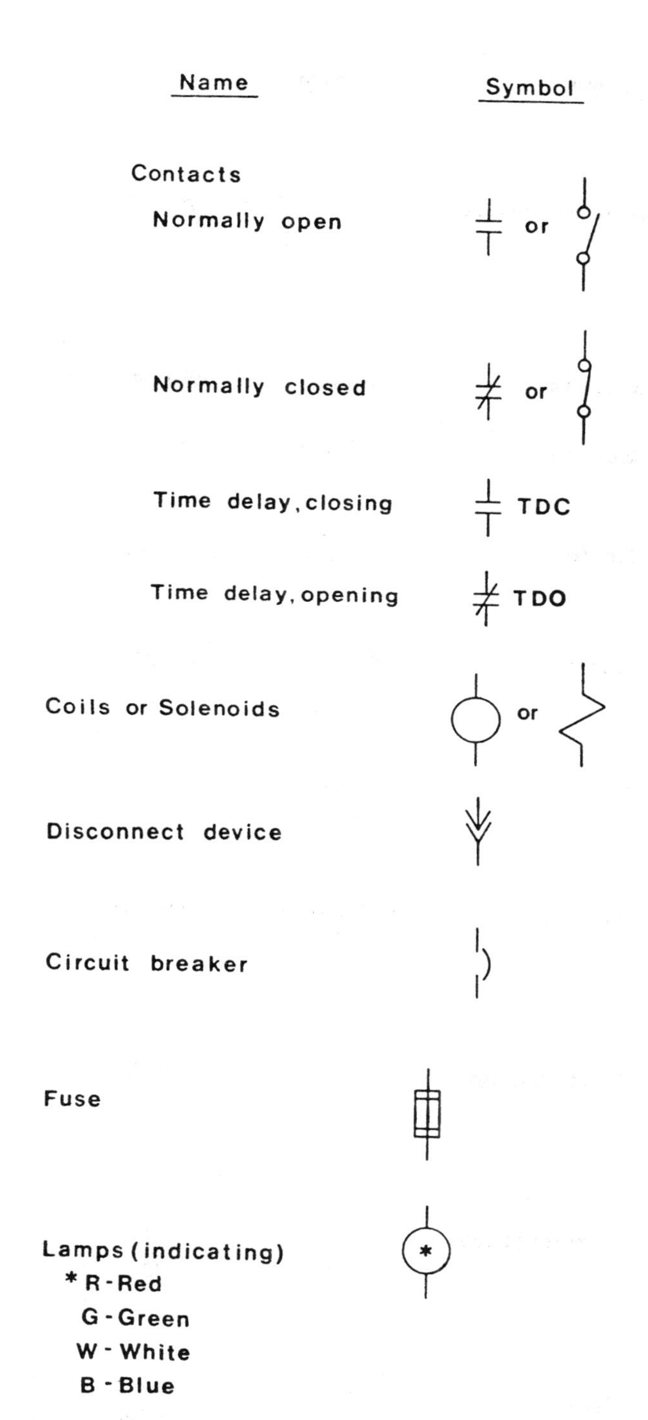

Figure 6.5 Symbols for ladder diagrams. (From James D. Bethune, *Basic Electronic and Electrical Drafting*, 2nd ed., © 1985. Reprinted by permission of Prentice-Hall, Inc.)

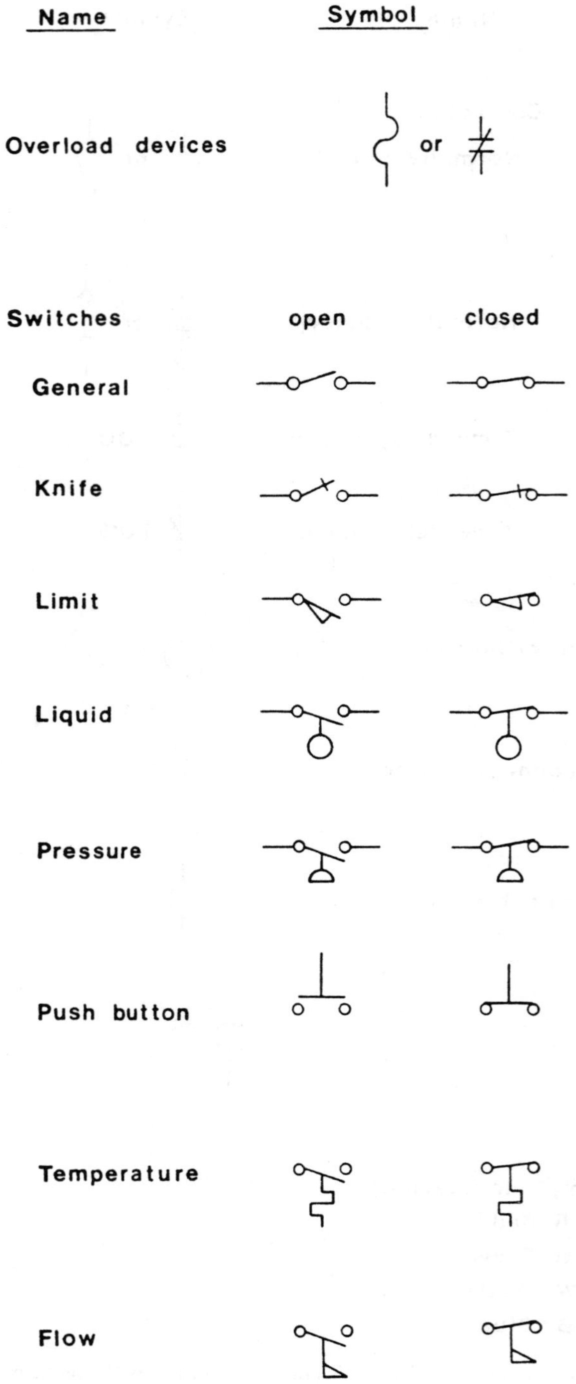

Name	Symbol	
Overload devices	or	
Switches	open	closed
General		
Knife		
Limit		
Liquid		
Pressure		
Push button		
Temperature		
Flow		

Figure 6.5 (*Continued*)

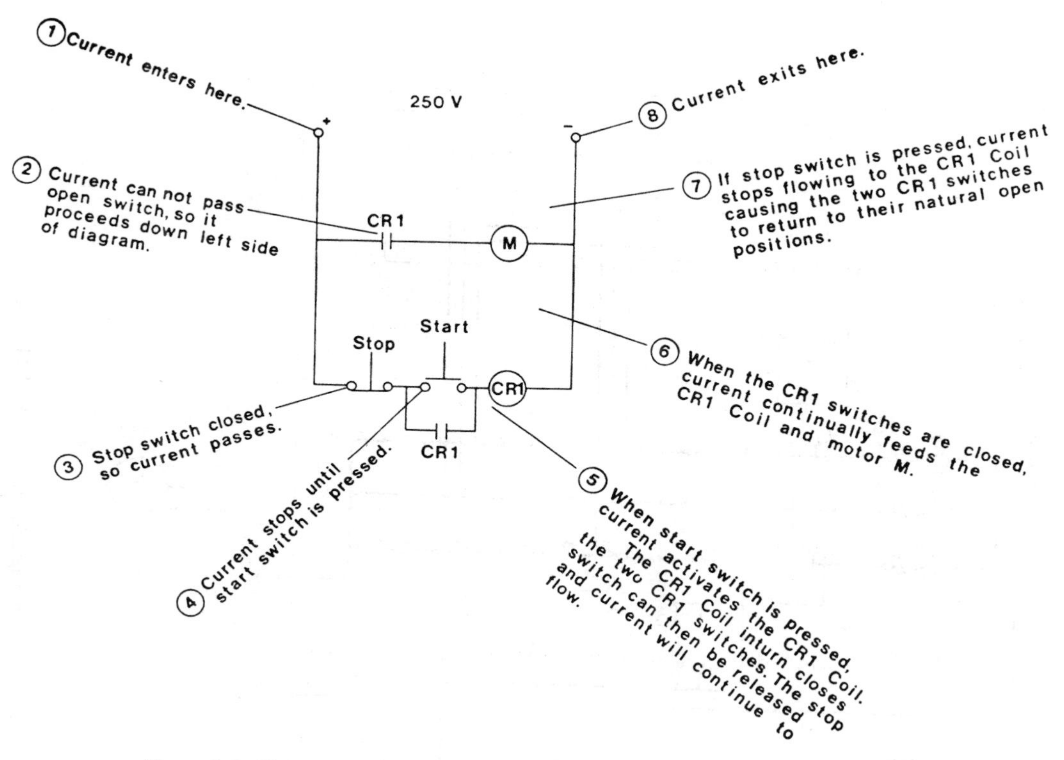

Figure 6.6 How to read a ladder diagram. (From James D. Bethune, *Basic Electronic and Electrical Drafting*, 2nd ed., © 1985. Reprinted by permission of Prentice-Hall, Inc.)

from negative to positive as in electron flow. Note in the diagram that there is a completed path only when the start switch is pressed.

The *highway* or *cable connection diagram* (Fig. 6.7) is basically similar to the point-to-point diagram with the exception that groups of connecting lines (conductors) are merged into lines called highways (cables or trunklines) instead of being shown in their entire run as individual lines. The highway diagram is used when a point-to-point diagram would be hopelessly complex because of so many individual lines.

Wire (or wiring) lists provide wiring data in a tabular format for cables and complex wire runs. In some cases there may be one list in hardware numerical order. The list shows wire origin (component and terminal) to wire destination and the terminal, connector, and wire identifiers that appear on the hardware. The table may also indicate wire color and size (see Fig. 6.8). Wire lists are sometimes called *to-from diagrams*.

An *interconnection diagram* shows the connections between major assemblies, units, and equipment as shown in Fig. 6.9. The internal connections within the major items are usually omitted.

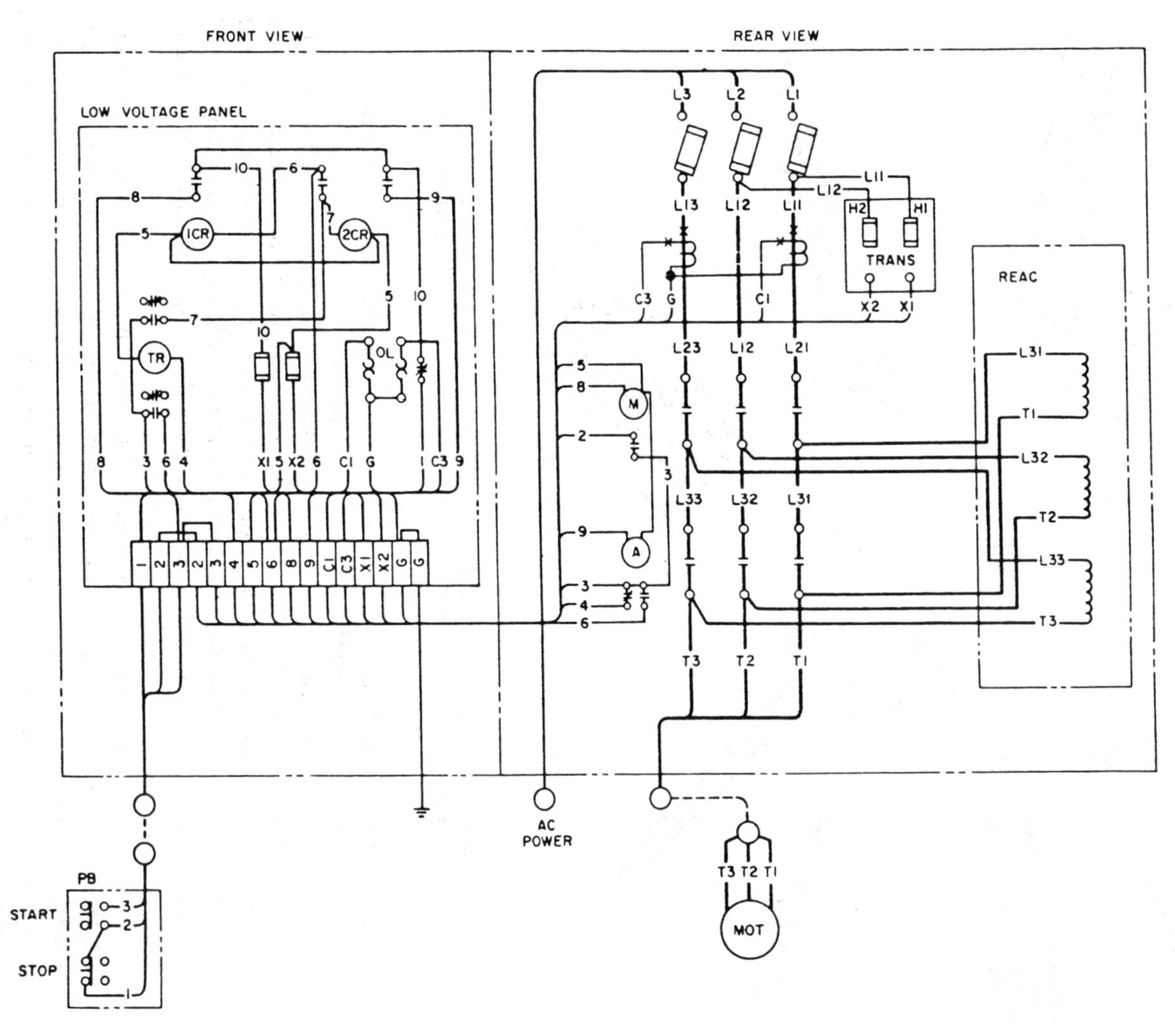

Figure 6.7 Typical highway or cable connection diagram. (From USAS Y14.15-1966.)

FROM	SHD GRP	WIRE ID	TO	SHD GRP	ROUTE CODE
A17 (Cont)					
J2–A		381	A17 J1–A		
J2–B		382	A17 J1–B		
J2–C		383	A17 J1–C		
J2–D		384	A17 J1–D		
S1	FS	385	A39	FT	
S1–1		351	A17 TB1–1		
S1–2		BLK 385	A39		
S1–3		353	A17 TB1–3		
S1–4		352	A17 TB1–2		
S1–5		RED 385	A39		
S1–6		354	A17 TB1–4		
S2–1		368	A17 S3–1		
S2–1		376	A17 TB2–1		
S2–2		355	A17 TB1–5		
S3–1		368	A17 S2–1		
S3–1		369	A17 S4–1		
S3–2		356	A17 TB1–6		
S4–1		369	A17 S3–1		
S4–1		370	A17 S5–1		
S4–2		357	A17 TB1–7		
S5–1		370	A17 S4–1		
S5–1		371	A17 S6–1		
S5–2		358	A17 TB1–8		
S6–1		371	A17 S5–1		
S6–1		372	A17 S7–1		
S6–2		359	A17 TB1–9		
S7–1		372	A17 S6–1		
S7–1		373	A17 S8–1		
S7–2		360	A17 TB1–10		
S8–1		373	A17 S7–1		
S8–1		374	A17 S9–4		
S8–2		361	A17 TB2–4		

FROM	SHD GRP	WIRE ID	TO	SHD GRP	ROUTE CODE
A17 (Cont)					
TB1–8		386H	TB12–8		A–D
TB1–9		359	A17 S6–2		
TB1–9		386J	TB12–9		A–D
TB1–10		360	A17 S7–2		
TB1–10		386K	TB12–10		A–D
TB2		387	TB13		A–D
TB2–1		376	A17 S2–1		
TB2–1		387A	TB13–2		A–D
TB2–2		388	A20 1		
TB2–2		387B	TB13–4		A–D
TB2–3		389	A20 2		
TB2–3		387C	TB13–6		A–D
TB2–4		361	A17 S8–2		
TB2–4		387D	TB12–11		A–D
TB2–6		362	A17 S9–1		
TB2–5		387E	TB12–12		A–D
TB2–6		363	A17 S9–2		
TB2–6		387F	TB12–13		A–D
TB2–7		364	A17 S9–3		
TB2–7		387G	TB12–14		A–D
TB2–8		365	A17 S10–1		
TB2–8		387H	TB13–12		A–D
TB2–9		366	A17 S10–2		
TB2–9		387J	TB13–13		A–D
TB2–10		367	A17 S10–3		
TB2–10		387K	TB13–14		A–D
A18					
E1		705	A18 STP	FU	
STP	FU	705	A18 E1		
J1–A		416	P10–A		
J1–A		420	A18 J2–A		
J1–B		417	P10–B		

Figure 6.8 Example of a wire list. (From MIL-HDBK-63038-1.)

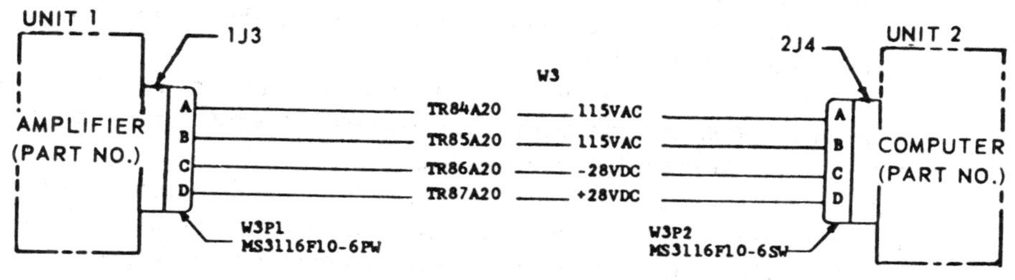

Figure 6.9 Simple point-to-point interconnection diagram. (From ANSI Y14.15a-1971.)

Chapter 7

Conventional Logic Symbols

For years the standard for logic symbols has been either Military Standard MIL-STD-806B or American National Standard ANSI Y32.14-1973. The distinctive and rectangular shaped symbols in these standards are considered conventional logic symbols. Newer logic symbols are covered in Chapter 8.

A *logic symbol* is a figure used on a diagram to represent a logic function such as an electronic gate or flip-flop or combinations thereof. The symbol includes an IC number (a reference designation) such as U4, and may also include a type or part number. It does not show the power supply and ground connections. A logic symbol is the predominant symbol in a logic diagram, but it may also be used on traditional schematics as necessary to simplify such diagrams.

Without logic symbols, diagrams of digital circuits would be hopelessly complicated. Figure 7.1, for example, shows a two-input AND gate in its logic symbol form compared to its equivalent schematic. Even for simple schematics, this logic symbol is necessary to make the diagram a manageable size. Obviously, more complicated logic elements, many of which have thousands of gates, could never be replaced with their equivalent schematics, as the resulting drawing would be several square yards in area.

Note, too, that most users will treat a logic element as a ''black box'' and are therefore concerned only with what goes in and what comes out. How the black box behaves internally is of no concern.

As shown in Fig. 7.2, the orientation of a logic symbol does not change its meaning. While symbols are frequently drawn horizontally, from left to right, they have the same effect on the circuit no matter how they are turned. The size of a logic symbol is not important. The element will have the same effect whether it is drawn large or small, as indicated in Fig. 7.3.

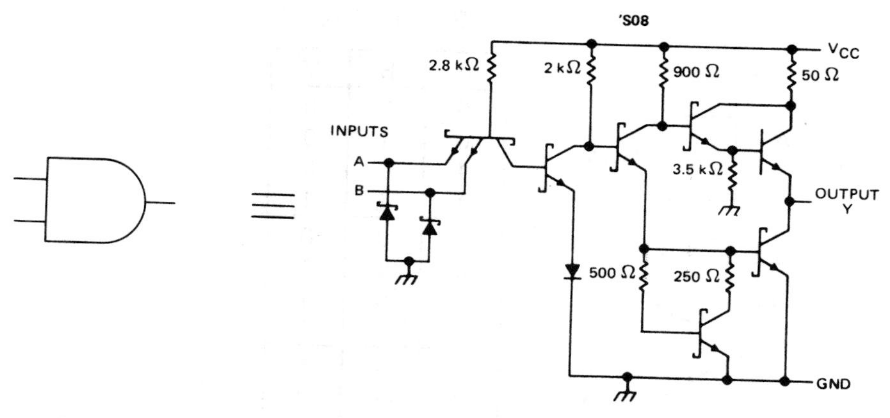

Figure 7.1 Logic symbol for equivalent schematic. (Courtesy of Texas Instruments Incorporated.)

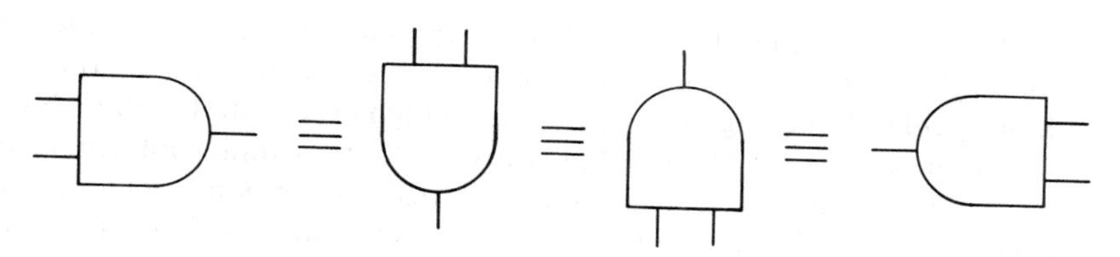

Figure 7.2 Orientation of logic symbols does not affect their meaning.

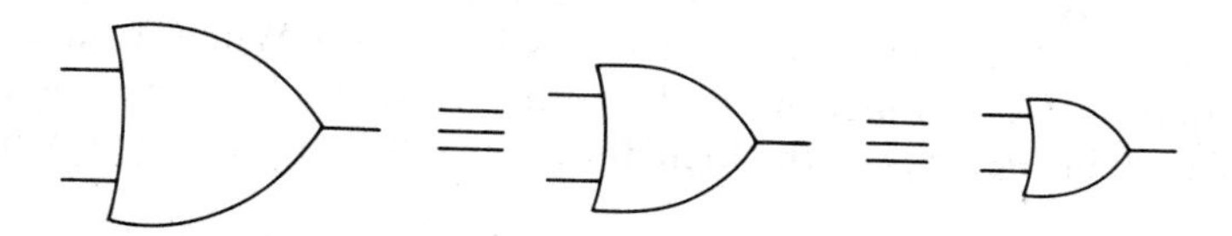

Figure 7.3 The size of logic symbols has no meaning.

The basic logic symbols or functions are the *AND* gate [Fig. 7.4(a)], the *OR* gate [Fig. 7.4(b)], and the *inverter* or *NOT* gate [Fig. 7.4(c)]. As shown in the illustration, A and B are input signals, and Y is the output signal. A, B, and Y are random letters; other letters could also have been used in the illustration. In a logic diagram, of course, A, B, and Y would be replaced with signal mnemonics.

The function of the AND gate can be expressed by the Boolean equation, $Y = A \cdot B$, which means that the output (Y) is true when and only when each input (A, B) is true. For the OR gate, the equation is $Y = A + B$, meaning that the output will be true if either input is true. The output of the inverter is expressed by $Y = \overline{A}$; that is, the output of the inverter will be the inverse (NOT) of the input.

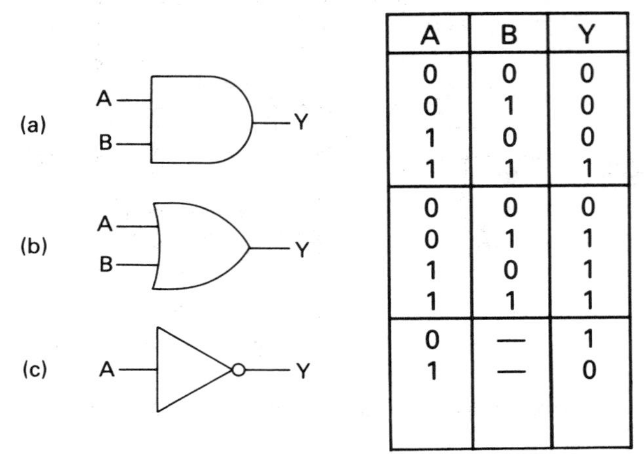

TRUTH TABLE

A	B	Y
0	0	0
0	1	0
1	0	0
1	1	1
0	0	0
0	1	1
1	0	1
1	1	1
0	—	1
1	—	0

Figure 7.4 Basic logic symbols.

The operation of these gates is explained by the truth table on Fig. 7.4. A 0 indicates a false (generally a low-level voltage) signal and a 1 represents a true (or high-level voltage) signal. Thus Y has a high (1) output for the AND gate when the two inputs are high (1), but has a low (0) output under all other conditions.

These gates are considered basic because they are building blocks for more complicated logic functions and because they are used so often by themselves.

The inverter symbol may be replaced with a small circle or bubble, as shown in Fig. 7.5, placed next to the appropriate symbol. The bubble indicates that the logic level will be negated or reversed. (A negation bar over a signal name also indicates that the signal is active when low.) When a bubble is present on the input of a logic function, it also indicates that a low-level signal will be the active signal; in such cases, the logic function will be activated only by a low-level signal.

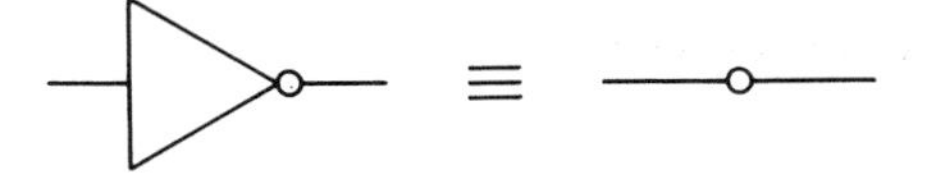

Thus

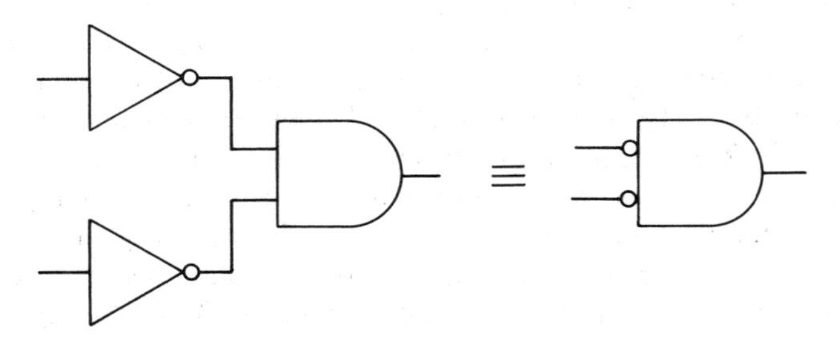

Figure 7.5 Use of bubble in logic symbols.

◣ positive polarity (obsolete)

◺ negative polarity

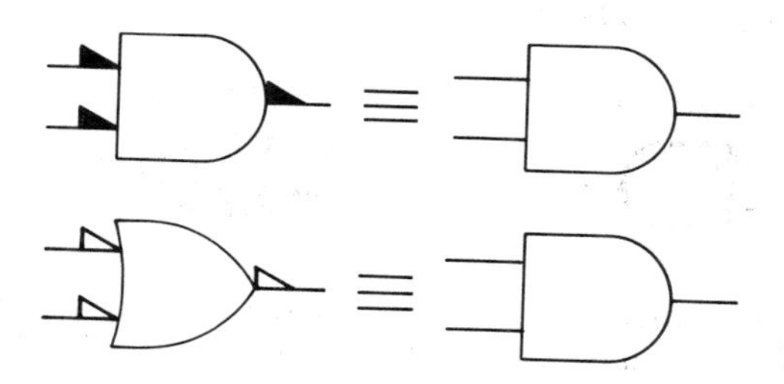

Figure 7.6 Polarity indicators.

In Fig. 7.5, for example, the AND gate at the lower right will act as an AND gate only if both inputs are low-level signals. If either or both of the inputs are high-level signals, the output of the AND gate will remain at 0.

If the negation bubble is not present on an input, the input will always be "looking for" an active-high signal; if the bubble is present, the input will look for an active-low signal.

When the negation bubble is on the output of a logic function, the signal will be inverted. If the function has, for example, an output of A with no bubble shown, it will have an output of $\overline{A}$ when the bubble is present.

Figure 7.6 shows positive and negative polarity indicators.

> In describing the operation of electronic logic devices, the symbol H is used to represent a "high level," which is a voltage within the more-positive (less-negative) of the two ranges of voltages used to represent the binary variables. L is used to represent a "low level," which is a voltage within the less-positive (more-negative) range.
>
> A function table for a device shows (implicitly or explicitly) all the combinations of input conditions and the resulting output conditions.
>
> In graphic symbols, inputs or outputs that are active when at the high level are shown without polarity indication. The polarity indicator ◺ denotes that the active (1) state of an input or output with respect to the symbol to which it is attached is the low level.
>
> Assume two devices having the following function tables:

DEVICE 1
FUNCTION TABLE

A	B	Y
H	H	H
H	L	L
L	H	L
L	L	L

DEVICE 2
FUNCTION TABLE

A	B	Y
H	H	H
H	L	H
L	H	H
L	L	L

Positive logic. By assigning the relationships H = 1, L = 0 at both input and output, device 1 can perform the AND function and device 2 can perform the OR function. Such a consistent assignment is referred to as positive logic. The corresponding logic symbols would be

Negative logic. Alternatively, by assigning the relationship H = 0, L = 1 at both input and output, device 1 can perform the OR function and device 2 can perform the AND function. Such a consistent assignment is referred to as negative logic. The corresponding logic symbols would be

Mixed logic. The use of the polarity indicator symbol on some inputs (⊾) automatically invokes a mixed-logic convention. That is, positive logic is used at the inputs and outputs that do not have polarity indicators; negative logic is used at the inputs and outputs that have polarity indicators.

FUNCTION TABLE

A	B	Z
H	H	L
H	L	H
L	H	H
L	L	H

This may be shown either of two ways:

FUNCTION TABLE

A	B	Z
H	H	L
H	L	L
L	H	L
L	L	H

This may be shown either of two ways:

Note that one can easily convert from the symbology of positive logic merely by substituting a polarity indicator (⊾) for each negation indicator (○) while leaving the distinctive shapes alone. To convert from the symbology of negative logic, a polarity indicator is substituted for each negation indicator and the OR shape is substituted for the AND shape, or vice versa. (© Reproduced with permission of Hewlett-Packard Co.)

Figure 7.7 shows most of the basic logic symbols, both distinctive shapes and rectangular shapes.

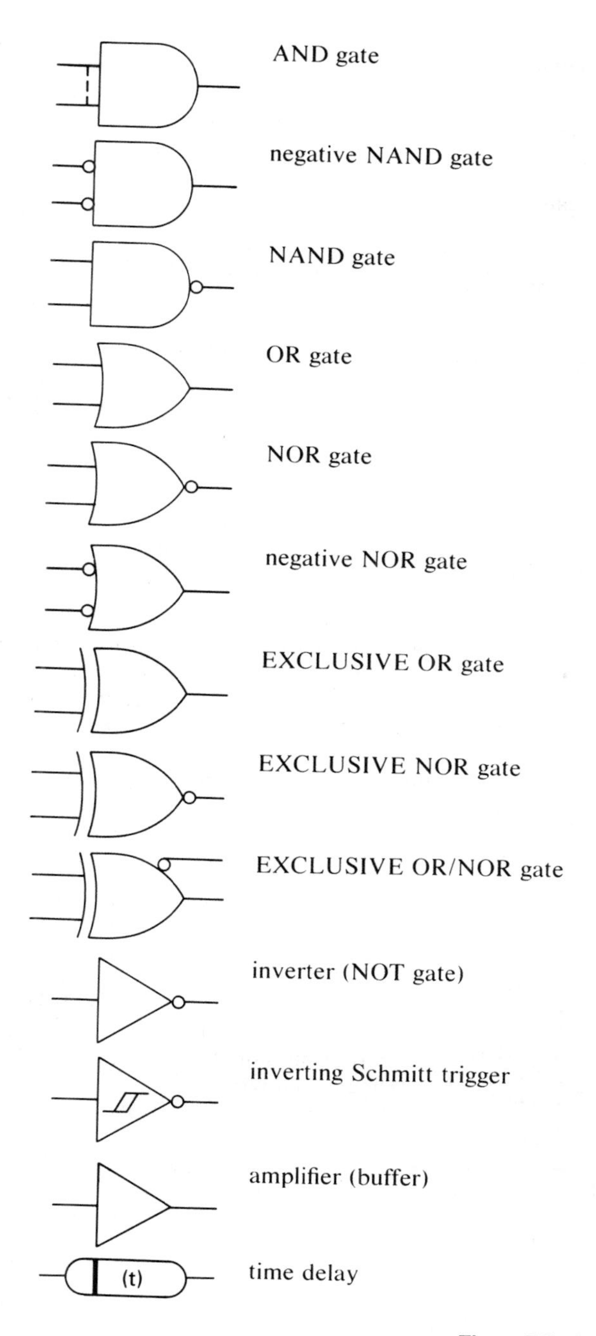

Figure 7.7 Logic symbols.

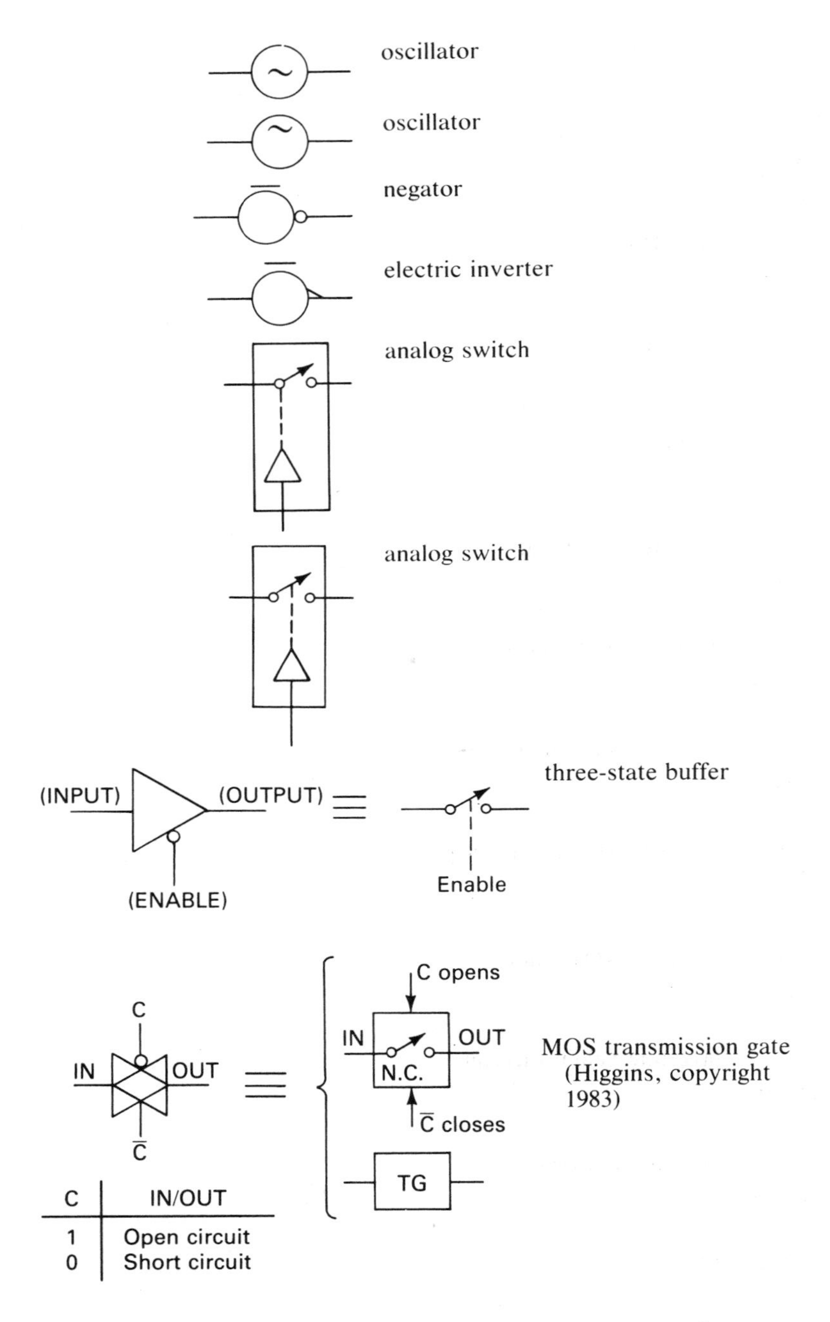

Figure 7.7 *(Continued)*

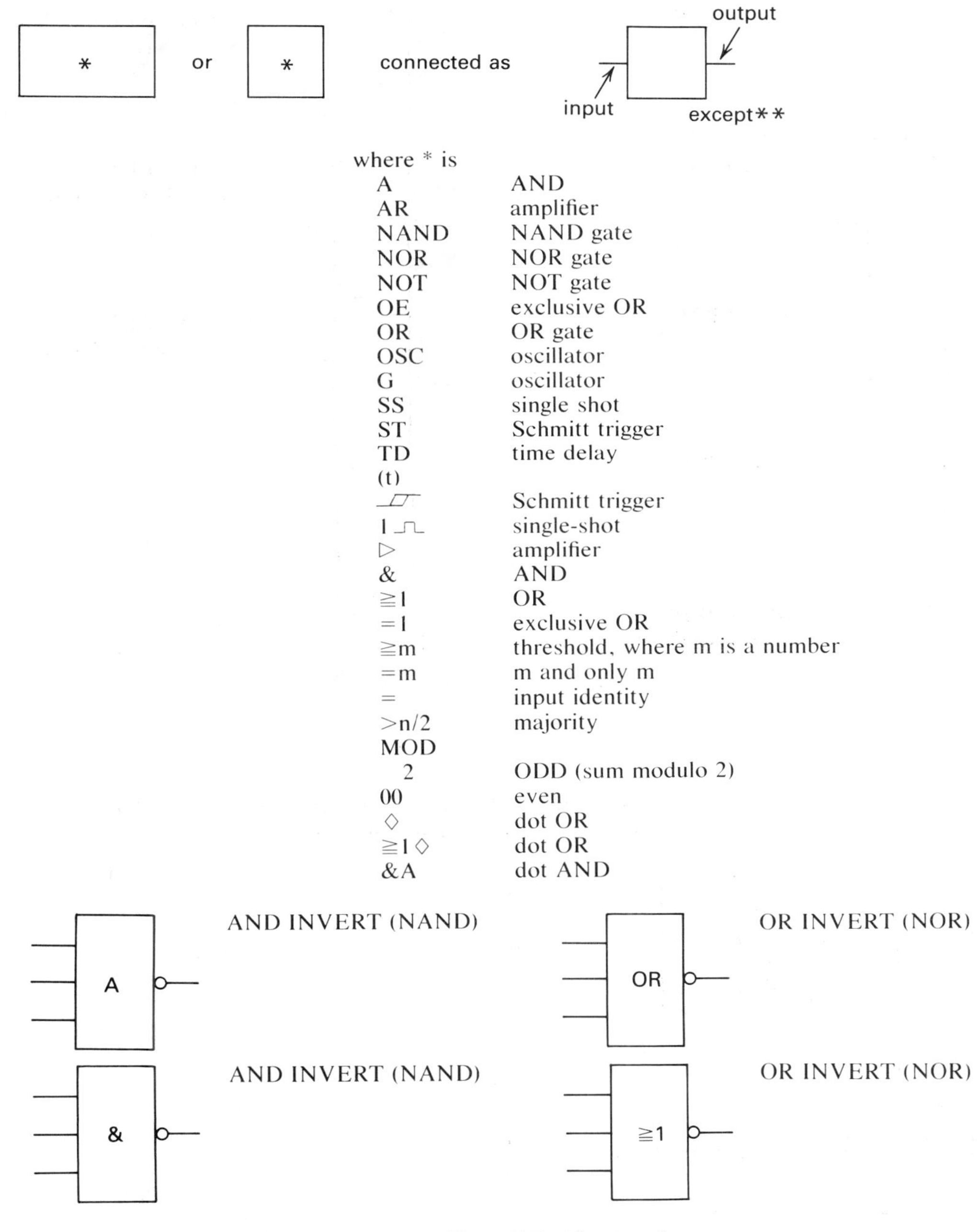

Figure 7.7 *(Continued)*

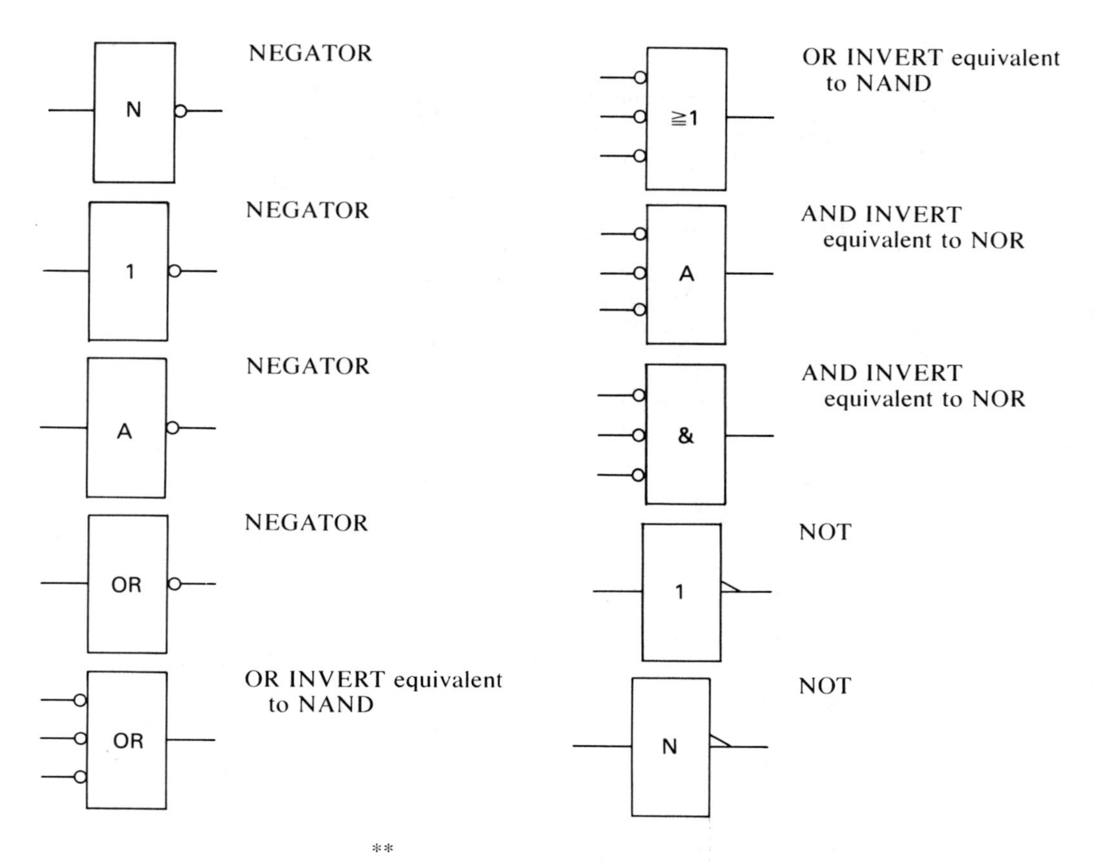

**

AND, OR, exclusive OR, dot OR, dot AND, NAND, and NOR have two or more inputs, which are shown as

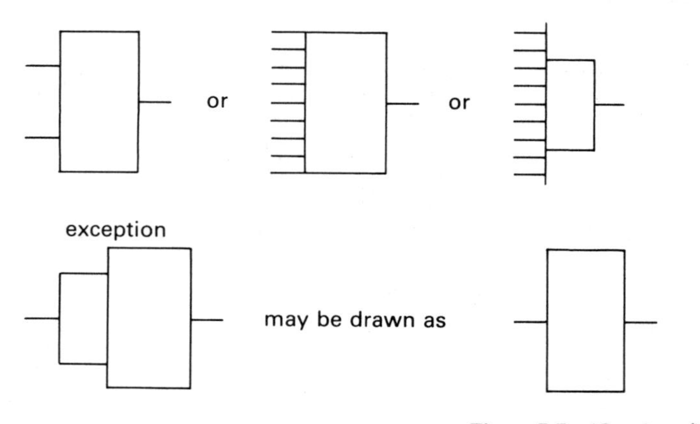

Figure 7.7 *(Continued)*

Figure 7.8 shows equivalent logic symbols. In the top illustration note that an AND gate is equivalent to a negated NOR gate. A gate is "negated" when it has negation bubbles on its inputs. The NAND gate is the same as a negated OR; a negated AND is equivalent to a NOR; and an OR is equivalent to a negated NAND.

Equivalent symbols are important because a logic diagram may use the symbol that shows the function that is being performed by a logic element. If, for instance, a negated NOR is shown on a diagram, it should be interpreted as an

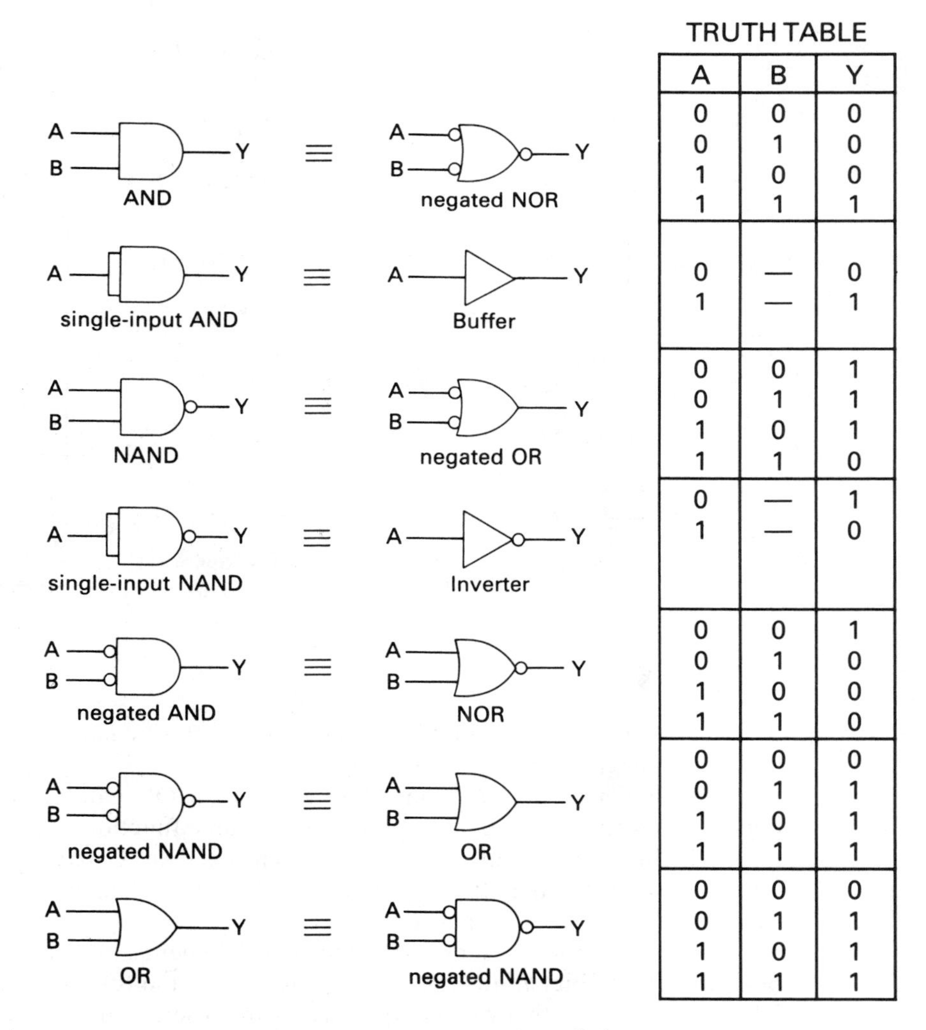

A	B	Y
0	0	0
0	1	0
1	0	0
1	1	1
0	—	0
1	—	1
0	0	1
0	1	1
1	0	1
1	1	0
0	—	1
1	—	0
0	0	1
0	1	0
1	0	0
1	1	0
0	0	0
0	1	1
1	0	1
1	1	1
0	0	0
0	1	1
1	0	1
1	1	1

Figure 7.8 Equivalent logic symbols.

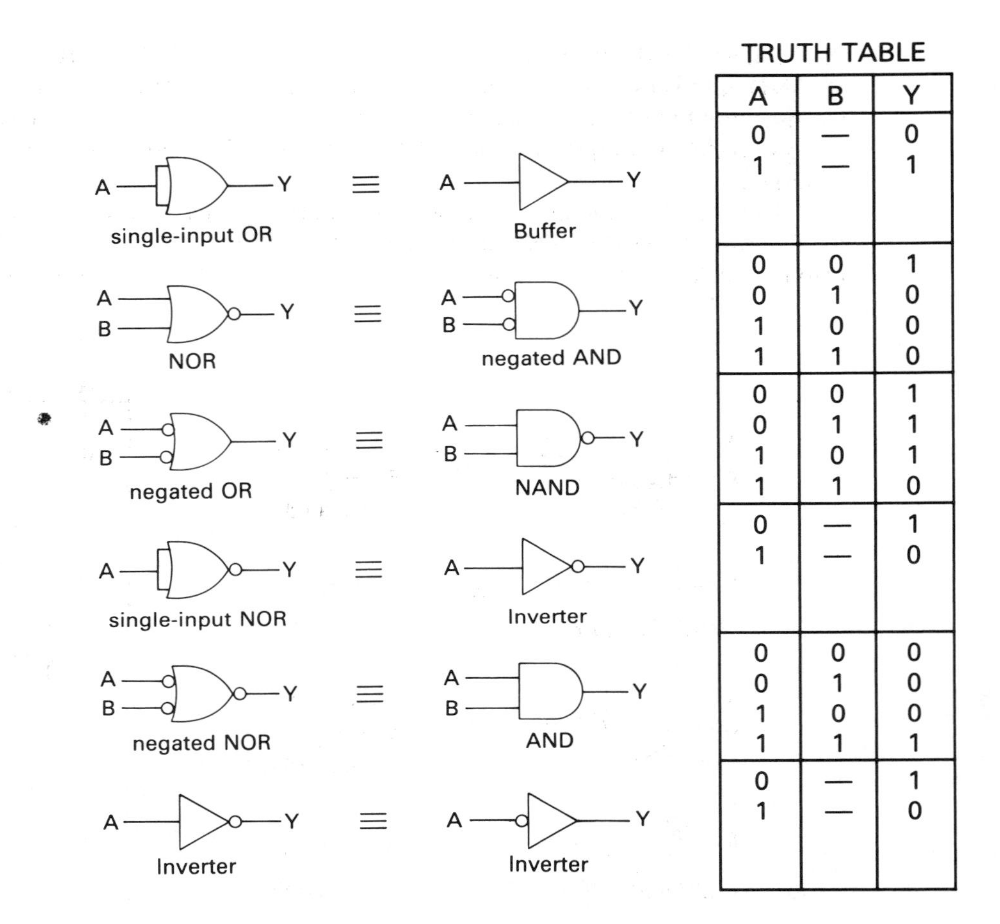

TRUTH TABLE

A	B	Y
0	—	0
1	—	1
0	0	1
0	1	0
1	0	0
1	1	0
0	0	1
0	1	1
1	0	1
1	1	0
0	—	1
1	—	0
0	0	0
0	1	0
1	0	0
1	1	1
0	—	1
1	—	0

Figure 7.8 (*Continued*)

AND gate. In fact, if you looked up the manufacturer's data sheet on this device, it would indicate that it is an AND gate.

As shown in Fig. 7.8, an AND gate or an OR gate with its inputs tied together to form only one input performs the same function as a buffer. The gate is used in this case perhaps because it was surplus. An IC may, for example, have four AND gates but only three are being used. Rather than using a separate IC for a buffer, the surplus gate is used.

Logic symbols in series indicate logic elements that are used to form a Schmitt trigger [Fig. 7.9(a)], oscillator [Fig. 7.9(b) and (c)], and time delays [Fig. 7.9(d)–(g)]. Notice that when two inverters are in series, they are equivalent to no inversion. That is, the negation effect is removed; a 0 input would result in a 0 output. Placing two buffers (or their equivalents) in series causes a desired propagation or time delay. Single buffers are also used to provide a propagation delay.

Either NOR or NAND gates can be cross-connected to form a most essential digital circuit, the flip-flop (see Fig. 7.10). In this case, the gates form an R-S flip-flop or "latch." The Q output will always be the opposite of $\overline{Q}$; if Q, for example, is 1, then $\overline{Q}$ will be 0.

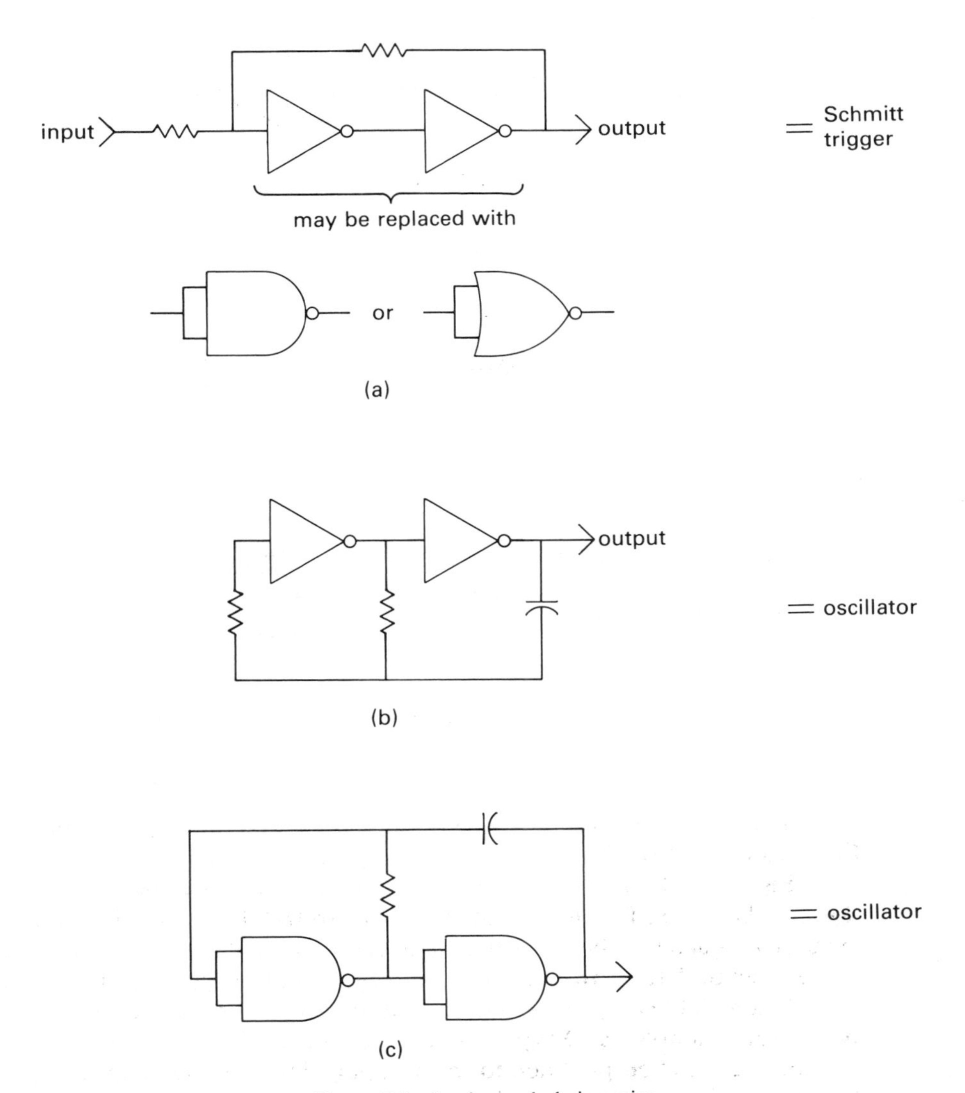

Figure 7.9 Logic symbols in series.

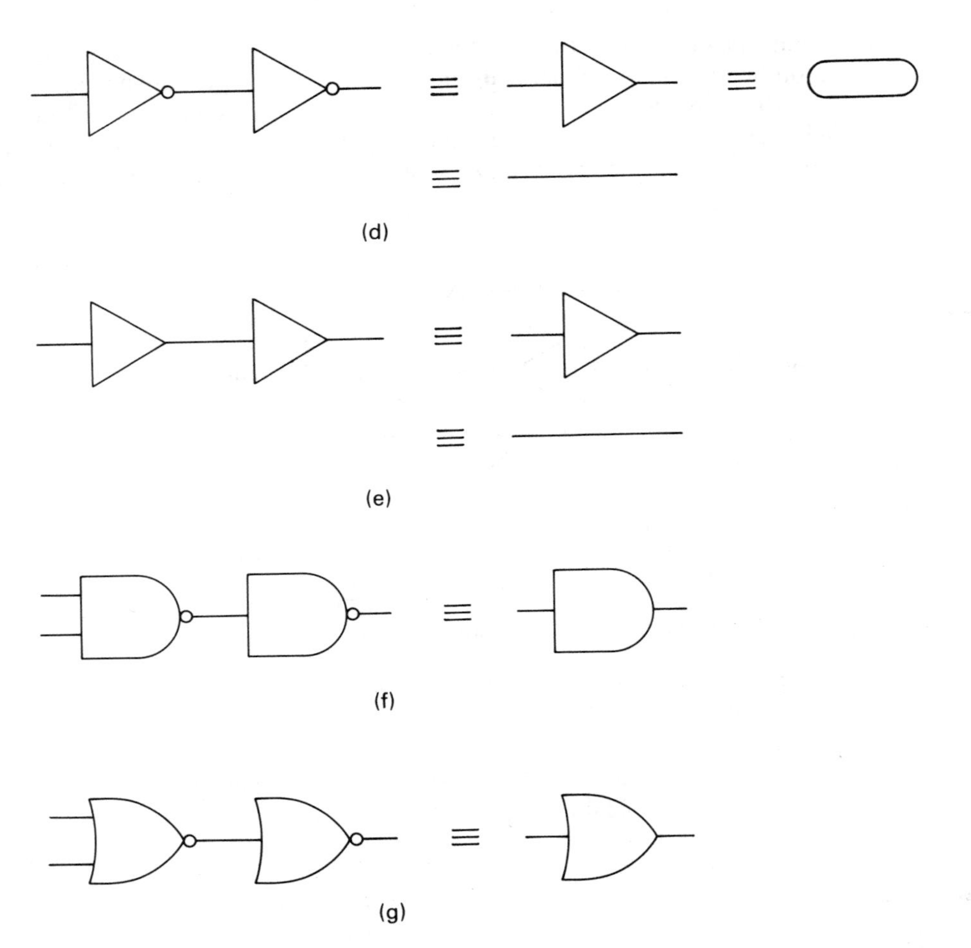

Figure 7.9 (*Continued*)

While flip-flops can be constructed from individual gates, they are more likely to be manufactured as a single integrated circuit.

Figure 7.11 shows the logic symbols for the more common types of flip-flops. Like gates, flip-flops generally have truth tables. In FF truth tables, the symbol t_n indicates the time before a clock pulse; t_{n+1}, the time after a clock pulse. Figure 7.12 shows the different types of clock input signals for flip-flops.

Figure 7.13 shows how one type of flip-flop can be wired to form a different type. For example, a D-type flip-flop can be made into a T-type flip-flop by connecting the $\overline{Q}$ output back to the D input. Thus, in interpreting logic symbols for flip-flops, particular attention must be paid to any variations in the connections.

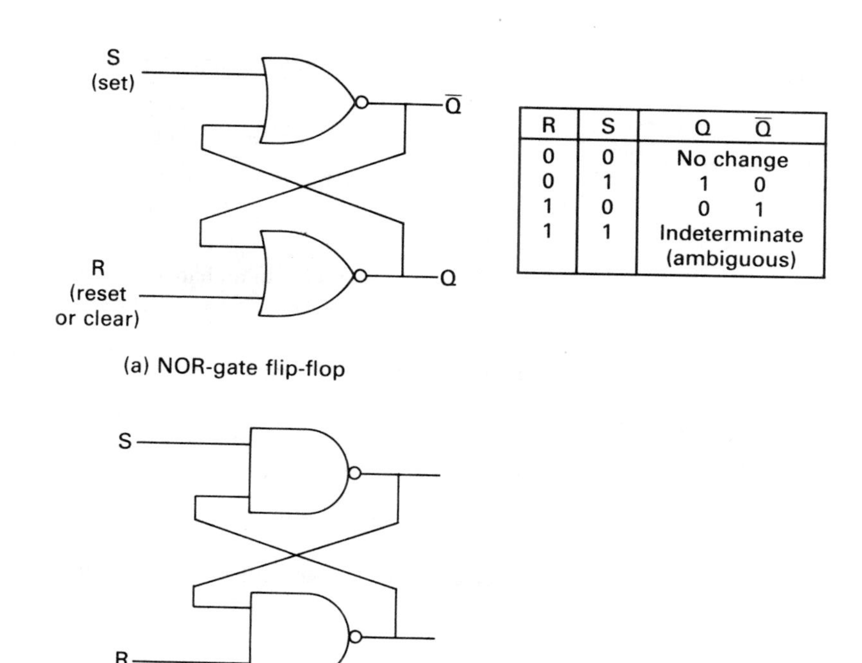

R	S	Q Q̄
0	0	No change
0	1	1 0
1	0	0 1
1	1	Indeterminate (ambiguous)

Figure 7.10 Cross-connected logic symbols: (a) NOR-gate flip-flop; (b) NAND-gate flip-flop.

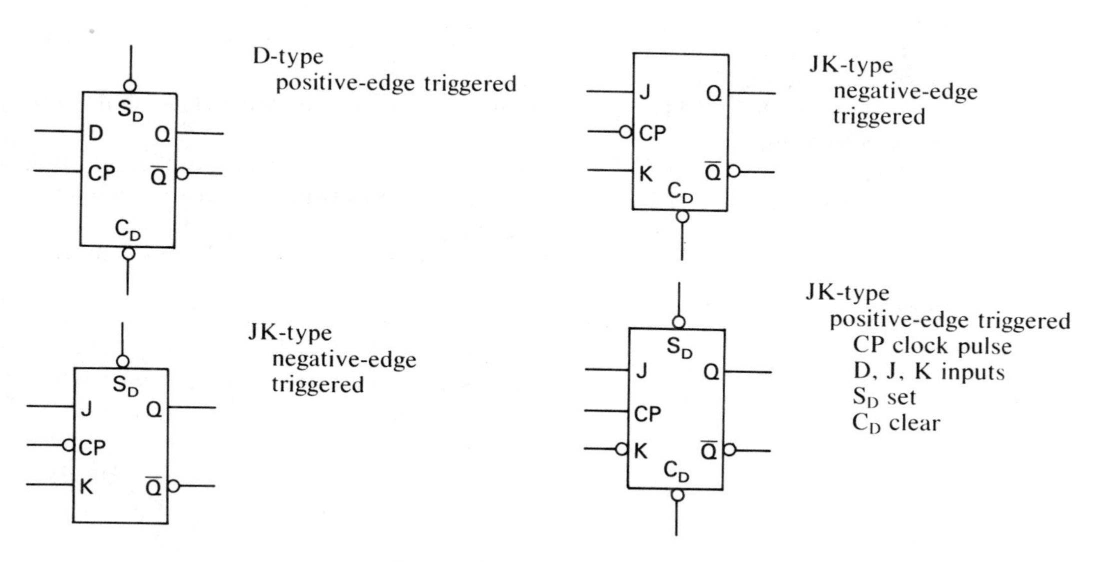

Figure 7.11 Logic symbols for flip-flops.

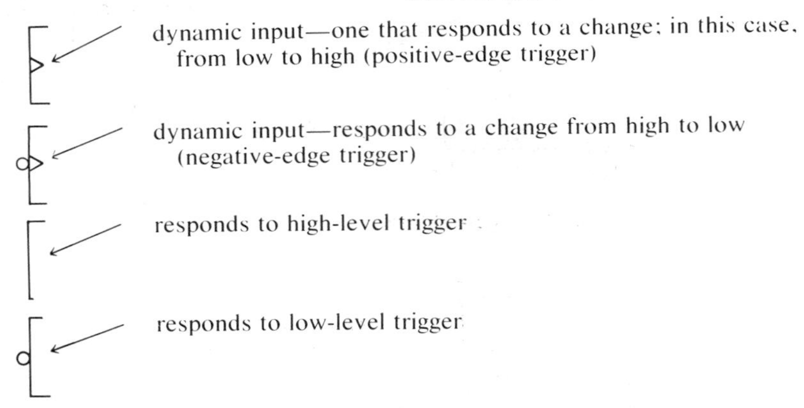

Figure 7.12 Triggers for flip-flops.

The operation of JK and RS flip-flops is shown in Fig. 7.14. Details are given because of the importance of these flip-flops in digital circuits.

Flip-flops in series form counters, as shown in Fig. 7.15. This use of flip-flops is becoming less common, however, as complete counters are being manufactured as one integrated circuit.

Typical symbols for a counter, multiplexer, and a decoder/demultiplexer are shown in Fig. 7.16. These are just a sample of the more complicated integrated circuits.

Resistors used in microcircuits are provided in integrated-circuit resistor networks, as shown in Fig. 7.17.

If an input to a TTL-type IC is not connected, it will act as if it were connected to a high level.

If a logic symbol is encountered that has not been shown in this chapter, look for explanatory notes on the diagram where the symbol appears, check the manufacturer's data sheet, or refer to Chapter 8.

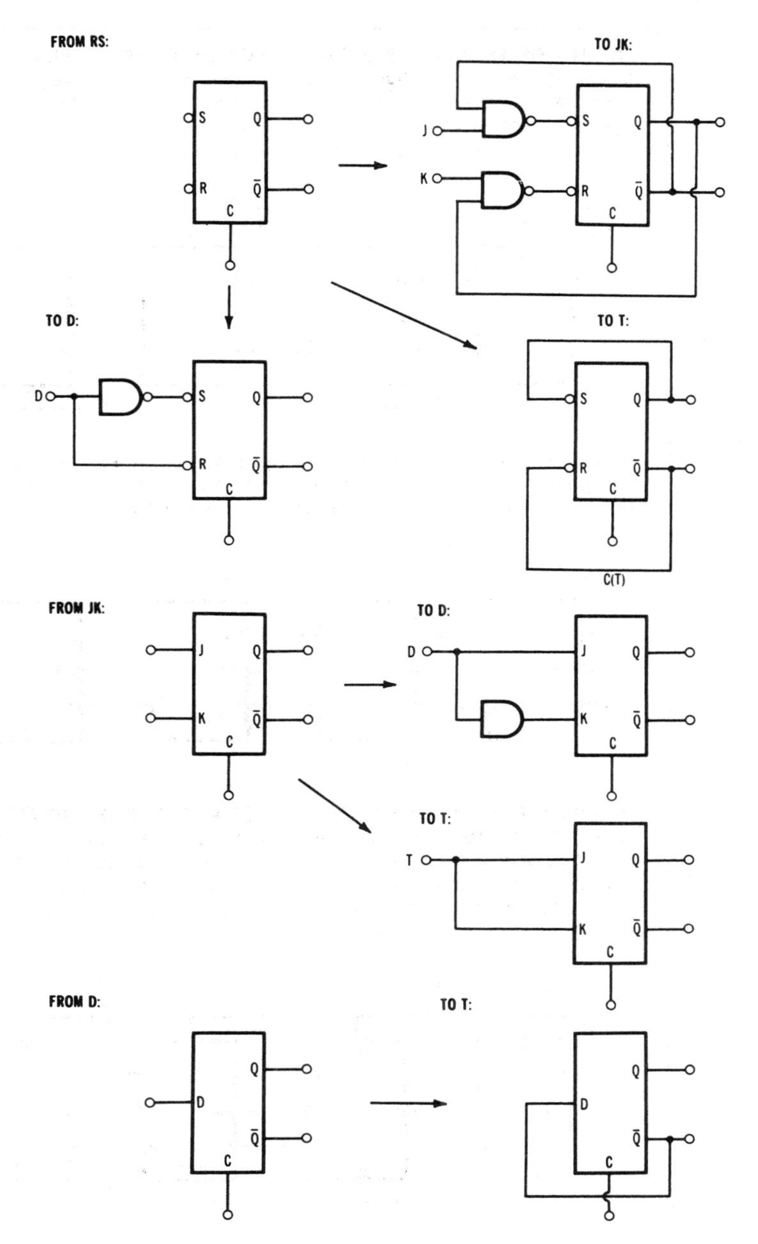

Figure 7.13 Variations of flip-flops. (Copyright 1982 by Howard W. Sams & Co., Inc.)

JK AND RS FLIP-FLOPS WITH PRESET OR CLEAR INPUT

Asserting the preset input or asserting the clear input overrides all other inputs. Note that the flip-flops shown below have active-low preset or clear inputs. In other words, the input is asserted by applying a logic-low level; if a logic-high level is applied, the input is unasserted.

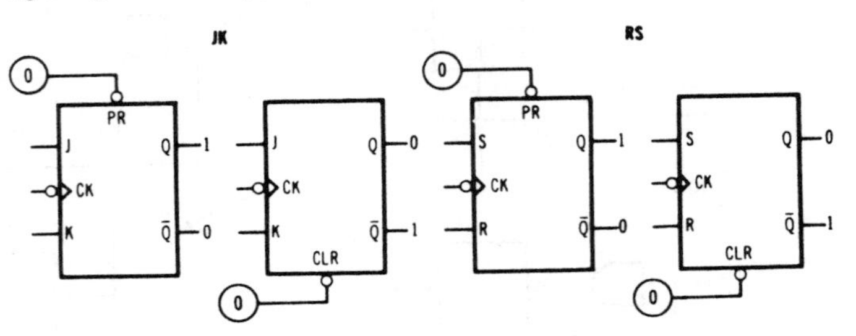

When the preset input or the clear input is unasserted, the outputs do not change until the clock pulse arrives; then, the outputs change to correspond with the input states, as in other JK or RS flip-flops.

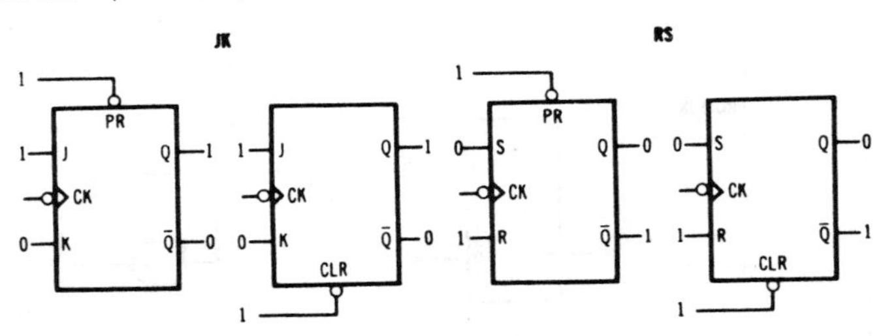

JK AND RS FLIP-FLOPS WITH PRESET AND CLEAR INPUTS

Note that the foregoing examples of flip-flops are provided with either a preset input or a clear input, but not both. The flip-flops shown below are provided with both preset and clear inputs. When the preset input is asserted (but the clear input is not asserted), the Q output goes logic-high, and all other inputs are irrelevant.

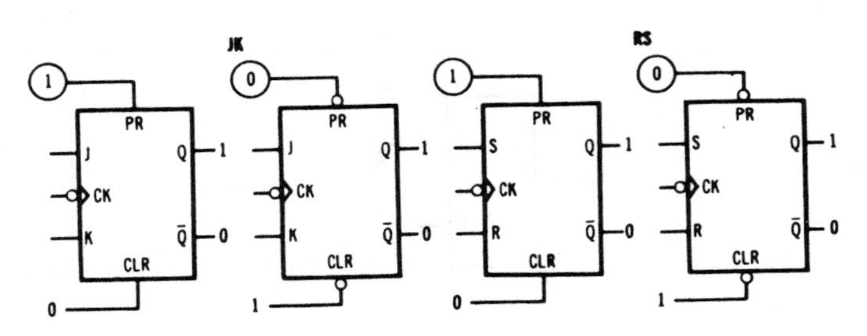

Figure 7.14 Preset and clear for J-K and R-S flip-flops. (Copyright 1982 by Howard W. Sams & Co., Inc.)

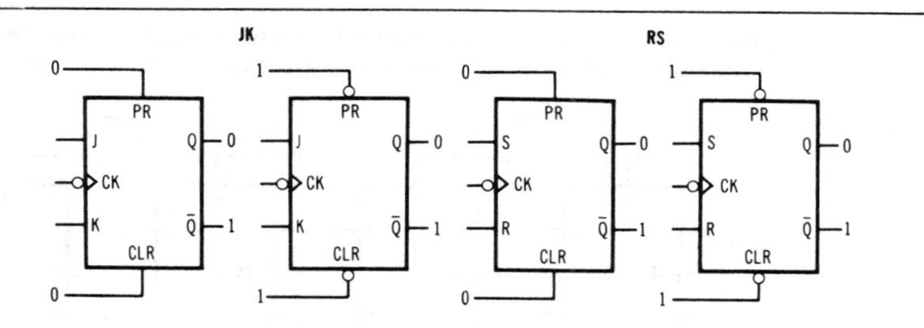

JK AND RS EDGE-TRIGGERED FLIP-FLOPS

Assuming that both the preset input and the clear input have been unasserted, if the J and K or R and S (data) inputs are in opposite logic states, the outputs change to correspond with the inputs when the trailing (negative or down-going) edge of a clock pulse arrives.

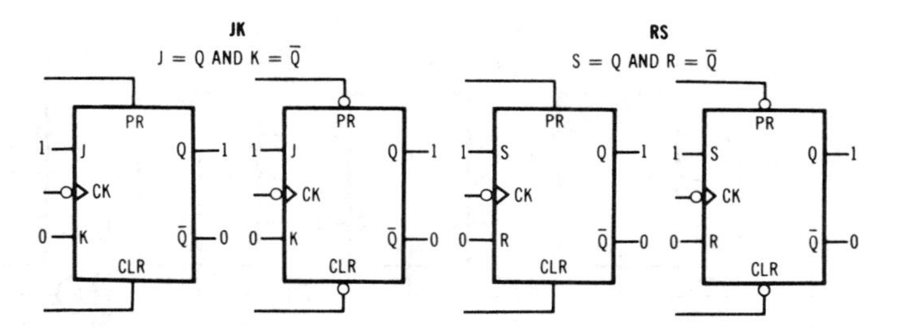

The diagrams below are the same as in the foregoing example, except that the data inputs are reversed.

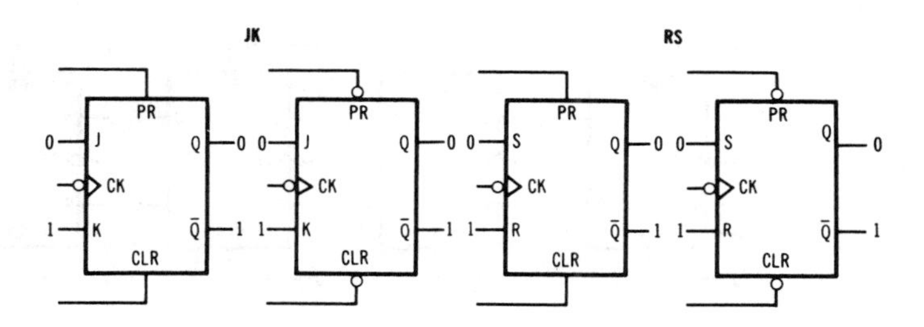

If the J and K or R and S (data) inputs are both in the logic-low state, the outputs do not change from their existing states when the down-going edge of a clock pulse arrives.

Figure 7.14 *(Continued)*

When the clear input is asserted (but the preset input is not asserted), the Q output goes logic-low, and all other inputs are irrelevant.

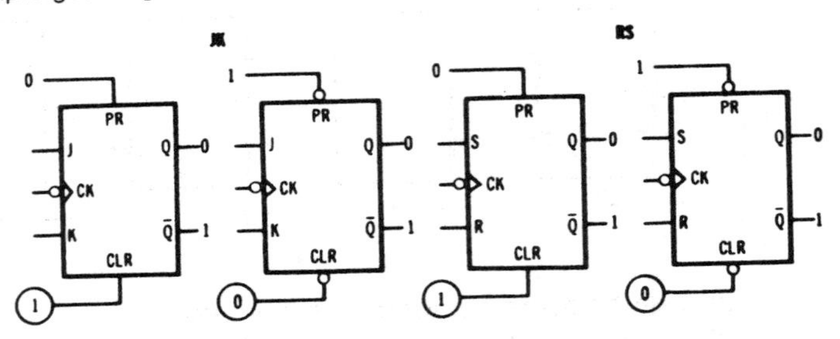

When the preset input and the clear input are both asserted, both the Q and the $\overline{Q}$ are driven logic-high but will not persist, even if both the preset input and the clear input are then unasserted at the same time (race condition occurs).

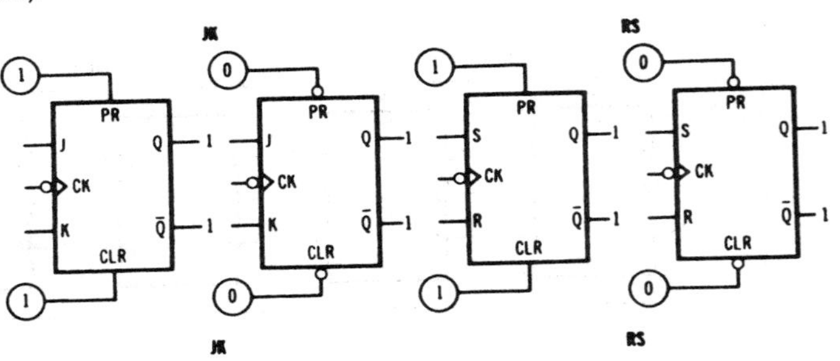

JK RS

When the preset input is unasserted after the clear input has been unasserted, the Q output will remain logic-high until a clock pulse arrives.

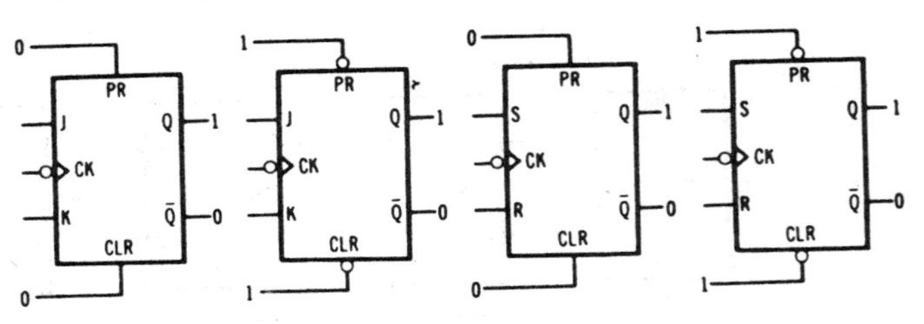

When the clear input is unasserted after the preset input has been unasserted, the Q output will remain logic-low until a clock pulse arrives.

(Continued on next page)

Figure 7.14 *(Continued)*

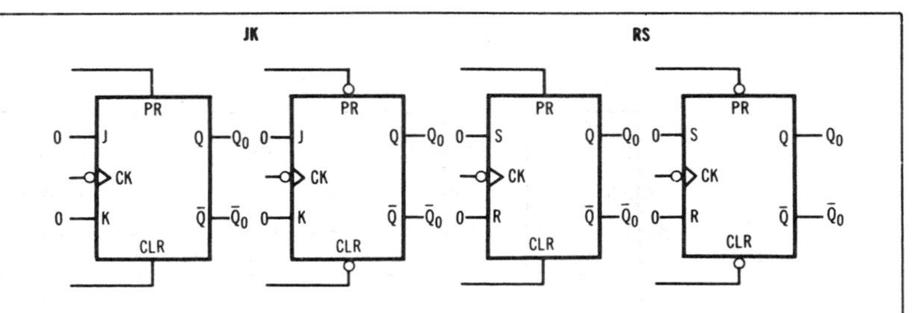

If the data inputs are both logic-high when the down-going edge of a clock pulse arrives:

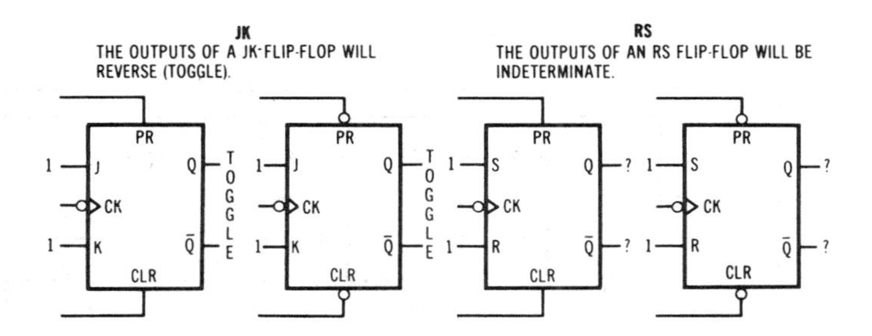

MASTER-SLAVE FLIP-FLOPS

Master-slave flip-flops recognize and store the logic states of the data inputs during and following one edge of a clock pulse, and transfer those logic states to the outputs during and following the next edge. The clock-pulse edge that causes a transfer is symbolized on the clock input line.

The logic state of a data line normally should not change during the interval between the leading edge and the trailing edge of a clock pulse. If the logic state does change during this interval, a high level probably will be stored and transferred. This is called a "ones-catching" flip-flop. Master-slave flip-flops with "data-lockout" will store only the logic level that exists during and immediately following the leading-edge clock transition.

JK AND RS PULSE-TRIGGERED (MASTER-SLAVE) FLIP-FLOPS

Assuming that the preset and clear inputs have both been unasserted, if the J and K or R and S (data) inputs were in opposite logic states during the arrival of the up-going edge of a clock pulse, the next down-going edge will transfer these stored states to corresponding outputs.

(Continued on next page)

Figure 7.14 *(Continued)*

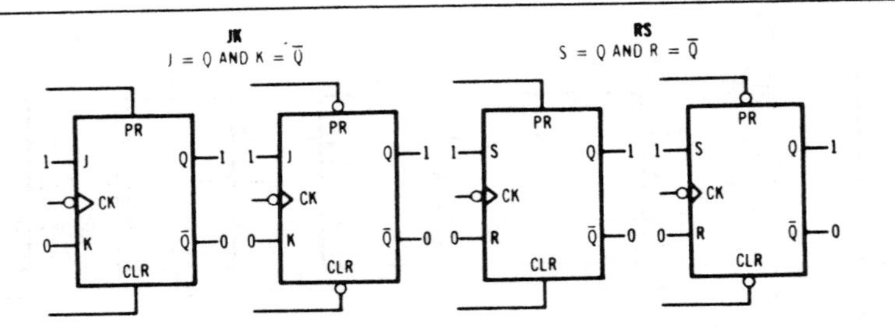

For the same conditions as above, but with the inputs reversed:

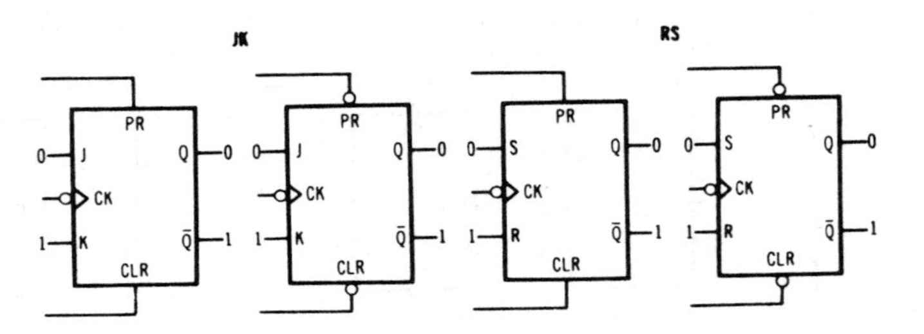

If the J and K or R and S (data) inputs were both in the logic-low state during the arrival of the up-going edge of a clock pulse, the next down-going edge will transfer no change to the outputs.

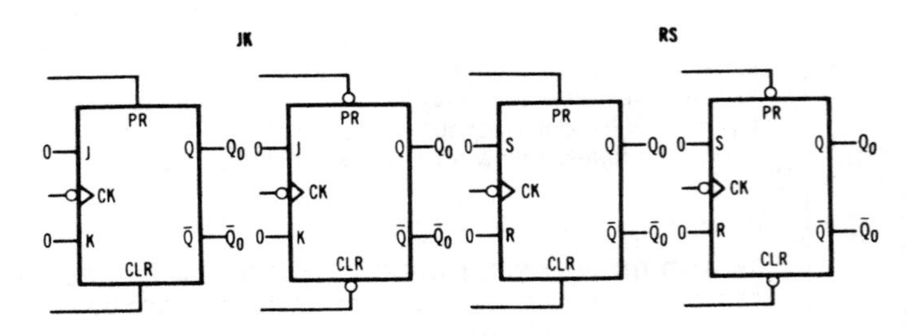

When the down-going edge of a clock pulse arrives following an up-going edge that occurred while both data inputs were logic-high:

Figure 7.14 *(Continued)*

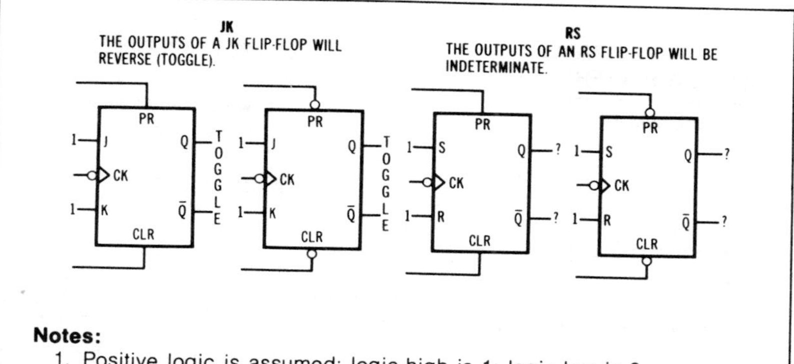

Notes:

1. Positive logic is assumed: logic-high is 1; logic-low is 0.
2. Preset (PR) is also called set (S).
3. Clock (CK) is also symbolized CP.
4. Clear (CLR) is also called reset (R).
5. A function is asserted when logic-high, unless the input line is terminated with a small circle; a small circle denotes that the function is asserted when logic-low.
6. Asserted inputs are circled, thus: ① ⓪
7. When the leading (positive, or up-going) edge of a clock pulse transfers data to the outputs, the clock input is symbolized without a small circle.

Courtesy Tektronix, Inc.

Figure 7.14 *(Continued)*

Decade Counter

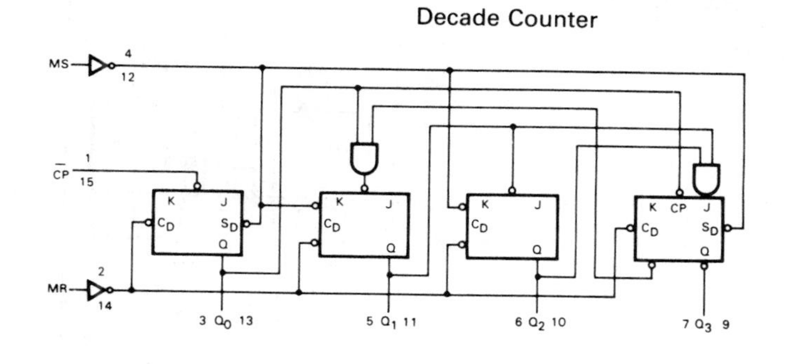

TRUTH TABLE

COUNT	OUTPUTS			
	Q_3	Q_2	Q_1	Q_0
0	L	L	L	L
1	L	L	L	H
2	L	L	H	L
3	L	L	H	H
4	L	H	L	L
5	L	H	L	H
6	L	H	H	L
7	L	H	H	H
8	H	L	L	L
9	H	L	L	H

MS	Master Set (Set to 9) Input
$\overline{MR}$	Master Reset
$\overline{CP}$	Clock Input (Active LOW Going Edge)
Q_0—Q_3	Counter Outputs

Figure 7.15 Flip-flops in series. (Copyright by Motorola, Inc. Used by permission.)

4-STAGE PRESETTABLE RIPPLE COUNTERS

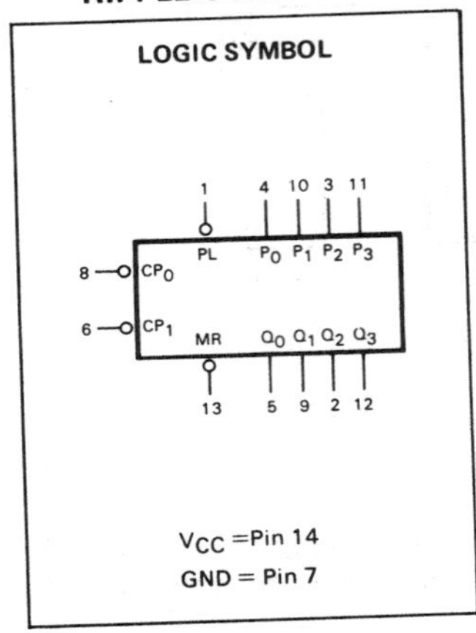

PIN NAMES

$\overline{CP}_0$	Clock (Active LOW Going Edge) Input to Divide-by-Two Section
$\overline{CP}_1$ (LS196)	Clock (Active LOW Going Edge) Input to Divide-by-Five Section
$\overline{CP}_1$ (LS197)	Clock (Active LOW Going Edge) Input to Divide-by-Eight Section
$\overline{MR}$	Master Reset (Active LOW) Input
$\overline{PL}$	Parallel Load (Active LOW) Input
P_0-P_3	Data Inputs
Q_0-Q_3	Outputs (Notes b, c)

8-INPUT MULTIPLEXER

LOW POWER SCHOTTKY

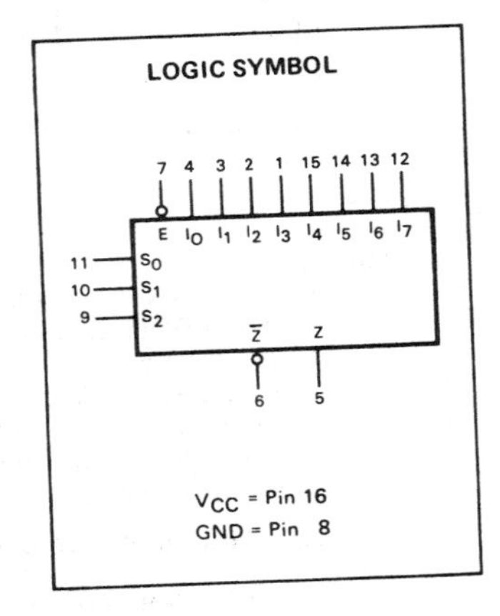

PIN NAMES

$S_0 - S_2$	Select Inputs
$\overline{E}$	Enable (Active LOW) Input
$I_0 - I_7$	Multiplexer Inputs
Z	Multiplexer Output (Note b)
$\overline{Z}$	Complementary Multiplexer Output

DUAL 1-OF-4 DECODER/ DEMULTIPLEXER

LS156-OPEN-COLLECTOR
LOW POWER SCHOTTKY

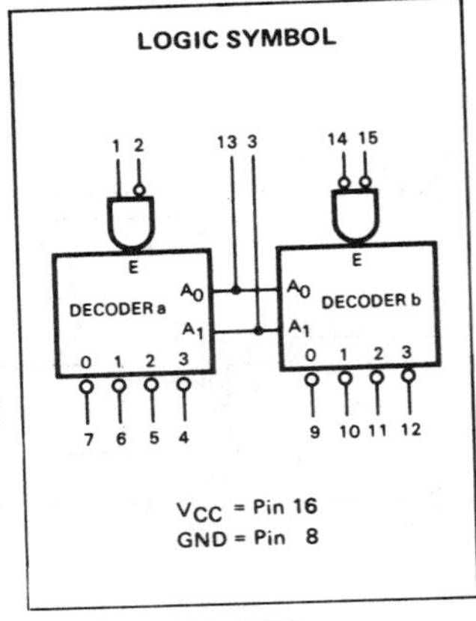

PIN NAMES

A_0, A_1	Address Inputs
$\overline{E}_a, \overline{E}_b$	Enable (Active LOW) Inputs
E_a	Enable (Active HIGH) Input
$\overline{O}_0 - \overline{O}_3$	Active LOW Outputs

Figure 7.16 Counter, multiplexer, decoder/demultiplexer. (Copyright by Motorola, Inc. Used by permission.)

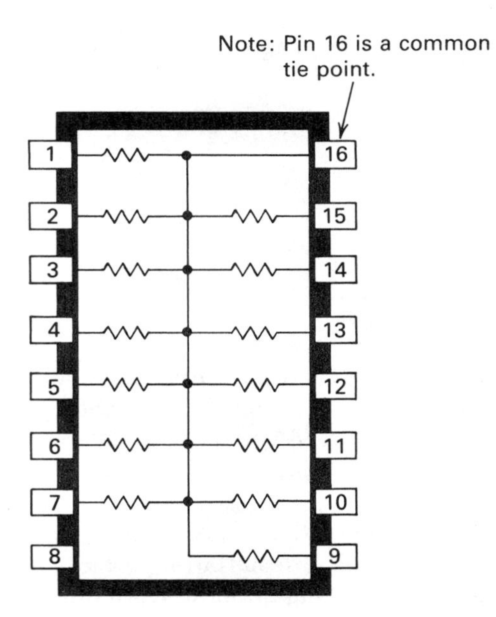

Figure 7.17 Resistor network in a microcircuit.

Chapter 8

The New Logic Symbols

8.1 A TOTALLY DIFFERENT CONCEPT

In 1984 the Institute of Electrical and Electronics Engineers published a revolutionary new way of showing functional logic symbols on a logic diagram. The new symbols have been designed so that most, if not all, of the pertinent explanations of the symbols are part of the symbols. That is, no further reference is needed to manufacturers' data sheets.

Some major semiconductor manufacturers have already adopted these new symbols, and perhaps several others will also start using these symbols. Whether or not they are universally adopted, it is important to learn how to use them, for the odds are that sooner or later you will encounter them.

The following illustrations and comments were condensed from ANSI/IEEE Std 91-1984, IEEE Standard, Graphic Symbols for Logic Functions, through the courtesy of The Institute of Electrical and Electronics Engineers, Inc. The serious student should refer to that standard for a more complete discussion.

Two symbols are used extensively in the text and illustrations: ≡, which means equivalent, and m, which indicates that a number is to be placed at that position. That is, the letter m does not appear on the symbols.

8.1.1 Definitions

A *logic state* is one of two possible abstract states that may be taken on by a logic (binary) variable. The *0-state* is the logic state represented by the binary number 0 and usually standing for an inactive or false logic condition. The *1-state* is the logic state represented by the binary number 1 and usually standing for an active or true logic condition. The *external logic state* is a logic state assumed to exist outside a symbol outline (1) on an input line prior to any external qualifying symbol at that input, or (2) on an output line beyond any external qualifying symbol at that output. The *internal logic state* is a logic state assumed to exist inside a symbol outline at an input or an output. Figure 8.1 illustrates the concept of external and internal logic states.

A *logic level* is any level within one or two nonoverlapping ranges of values of a physical quantity used to represent the logic states. *Note:* A logic variable may be equated to any physical quantity for which two distinct ranges of values can be defined. In this chapter these distinct ranges of values are referred to as logic *levels* and are denoted H and L. H is used to denote the logic level with the more positive algebraic value, and L is used to denote the logic level with the less positive algebraic value. In the case of systems in which logic states are equated with other physical properties (for example, positive or negative pulses, presence or absence of a pulse), H and L may be used to represent these properties or may be replaced by more suitable designations. A *high (H) level* is a level within the more positive (less negative) of the two ranges of the logic levels chosen to represent the logic states. A *low (L) level* is a level within the more negative (less positive) of the two ranges of logic levels chosen to represent the logic states.

The *positive-logic convention* is the representation of the 1-state and the 0-state by the high (H) and low (L) levels, respectively. The *negative-logic convention* is the representation of the 1-state and the 0-state by the low (L) and high (H) levels, respectively. A *direct polarity indication* is the designation of the internal state produced by the external level of an input, or producing the external level of an output, by the presence or absence of the polarity symbol.

A *logic function* is a definition of the relationships that hold among a set of input and output logic variables. A *combinational logic function* is a logic function in which there exists one and only one resulting combination of states of the outputs for each possible combination of input states. (The terms "combinative" and "combinatorial" have also been used to mean "combinational.") A *sequential logic function* is a logic function in which there exists at least one combination of input states for which there is more than one possible resulting combination of states at the outputs. (The outputs are functions of variables in addition to the present states of the inputs, such as time, previous internal states of the element,

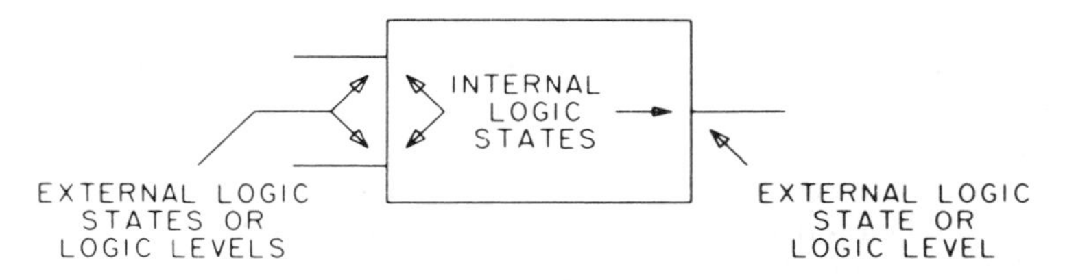

Figure 8.1 External and internal logic states. (From ANSI/IEEE Std 91-1984.)

etc.) A *bistable logic function* (*flip-flop*) is a sequential logic function that has two and only two stable internal output states.

An *element* is a representation of all or part of a logic function within a single outline, which may, in turn, be subdivided into smaller elements representing subfunctions of the overall function; alternatively, the function so represented.

A *qualifying symbol* is a symbol added to the basic outline of an element to designate the physical or logic characteristics of an input or output of the element or the overall logic characteristics of the element.

Dependency notation is a means of obtaining simplified symbols for complex elements by denoting the relationships between inputs, outputs, or inputs and outputs, without actually showing all the elements and interconnections involved.

A *distributed function* (*dot logic, wired logic*) is a logic function (either AND or OR) implemented by connecting together outputs of the appropriate type; these outputs are the inputs of the logic function thus formed; the joined connection is the output. (Definitions derived from ANSI/IEEE Std 91-1984.)

8.1.2 Applicable Documents

Two other texts may be of interest to the reader: *Using Functional Logic Symbols, 1987* (Application of IEEE Std 91-1984), by Frederic A. Mann, Texas Instruments, Dallas, Tex., and *A Practical Introduction to the New Logic Symbols*, second edition, by Ian Kampel (Butterworth, Sevenoaks, Kent, England, 1985).

8.2 SYMBOL CONSTRUCTION

8.2.1 Symbol Composition

As shown in Fig. 8.2, a logic symbol consists of an outline or combination of outlines, one or more qualifying symbols, and input and output lines.

Nonstandard information may be shown in brackets, [ ], inside the outline. For a single, unsubdivided element the outputs have identical internal logic states (except when indicated otherwise) determined by the element's function.

8.2.2 Application and Identification Information for Symbols

Reference designations, element physical identification, physical location of devices, functional use, terminal identifications, stylized waveforms, and similar information is referred to as *application information* (tagging lines) (see Fig. 8.3). It does not include qualifying symbols, indicator symbols, and control symbols.

Application and identification information for *logic* symbols is usually placed inside the symbols, as shown in Fig. 8.3(a), although it may be adjacent to the

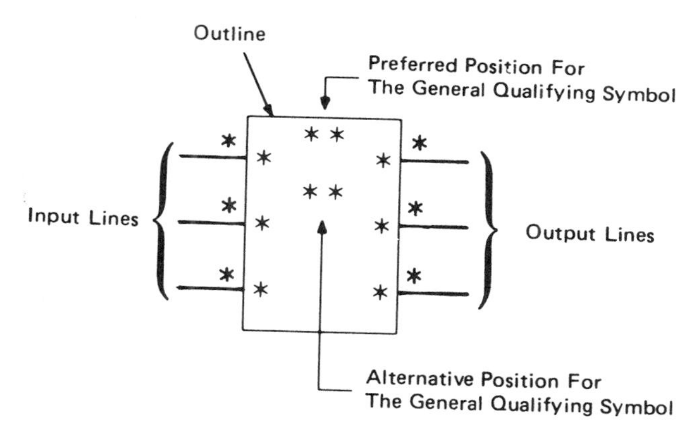

NOTE 1: The single * denotes possible positions for qualifying symbols relating to inputs and outputs.
NOTE 2: If, and only if, the function of an element is completely determined by the qualifying symbols associated with its inputs or outputs or both, no general qualifying symbol is needed.

Figure 8.2 Symbol composition. (From ANSI/IEEE Std 91-1984.)

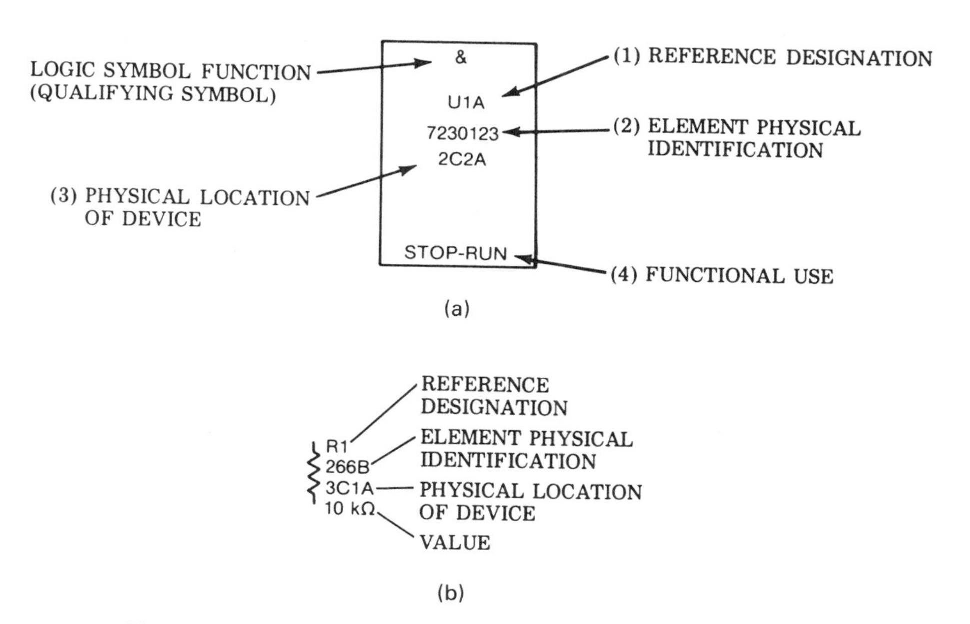

Figure 8.3 Application information. (From ANSI/IEEE Std 991-1986.)

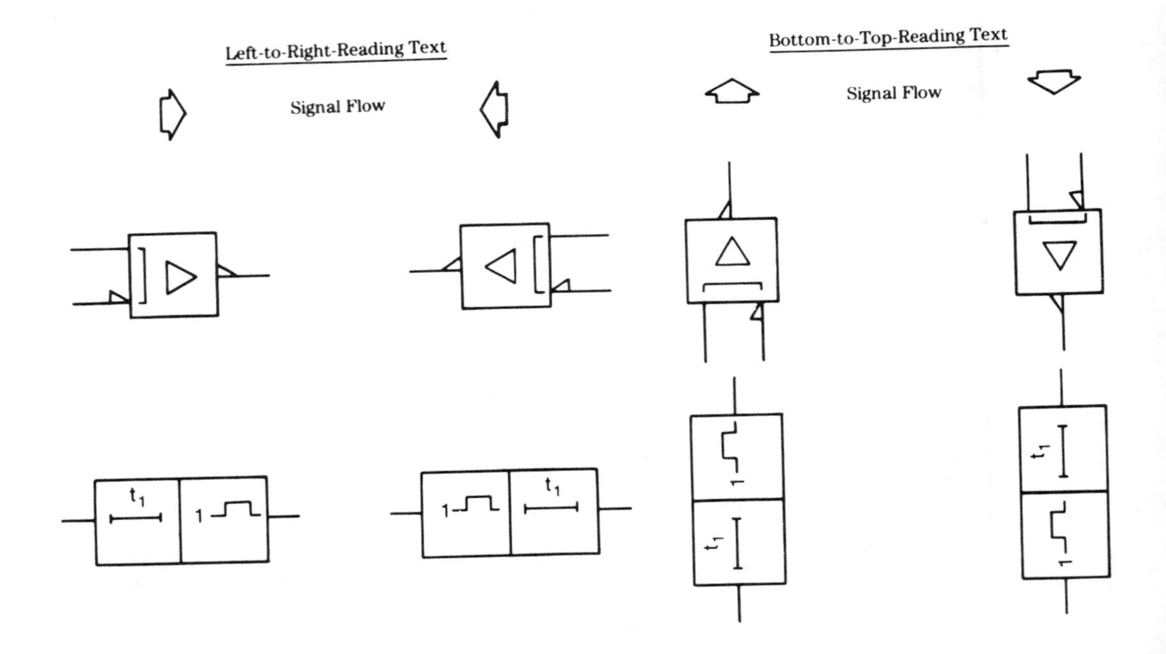

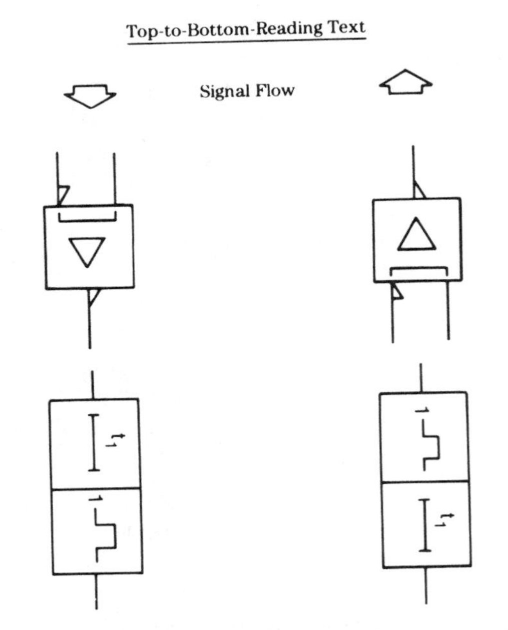

Figure 8.4 Orientation of symbols. (From ANSI/IEEE Std 991-1986.)

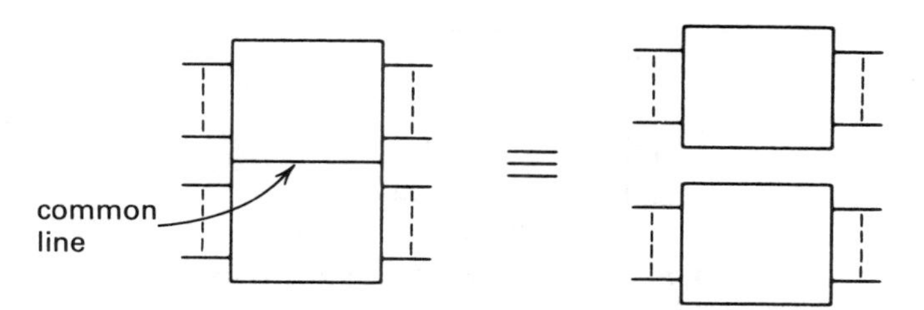

Figure 8.5 General outline for abutted element. (From ANSI/IEEE Std 91-1984.)

logic symbol. For nonlogic symbols this information may be adjacent to the symbol as shown in Fig. 8.3(b).

8.2.3 Size and Orientation of Symbols

Logic symbols are made large enough to accommodate internal annotations and input and output lines with acceptable spacing. Within a logic diagram, different sizes of symbols are sometimes used to emphasize certain aspects or to make it easier to include additional information. As a rule, line thickness does not change the meaning of a symbol. Symbols may be rotated or otherwise manipulated to simplify the circuit layout. Usually, the symbols are designed for inputs on the left and outputs on the right. However, as shown in Fig. 8.4, the orientation is different for top-to-bottom and bottom-to-top reading.

8.2.4 Embedded and Abutted Elements

To save drawing space, the outlines of associated elements may be embedded or abutted as shown in Figs. 8.5 and 8.6.

If the line common to two abutted symbols is in the direction of signal flow, as in Fig. 8.5, there is no logic connection between the elements. The exception to this rule is those arrays in which there are two or more directions of signal flow. If the line common to two outlines is perpendicular to the direction of signal flow, as shown in Fig. 8.7, there is at least one logic connection between the elements.

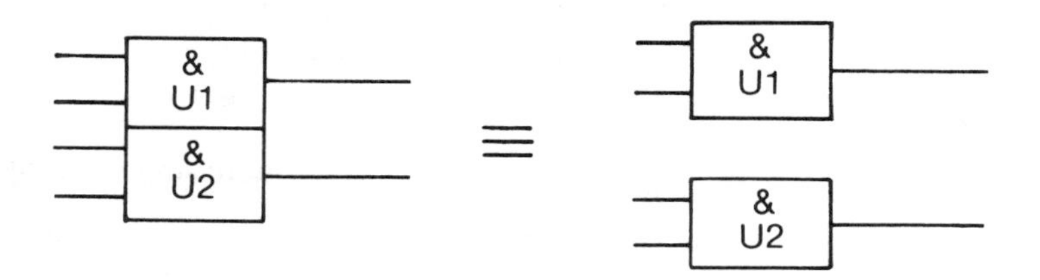

Figure 8.6 Abuttment of AND-gate symbols. (From ANSI/IEEE Std 991-1986.)

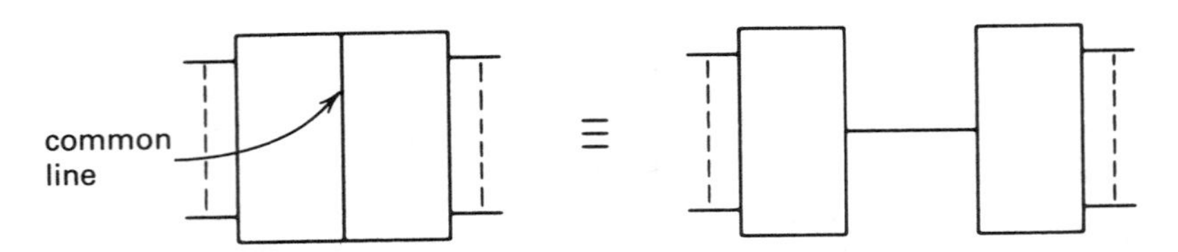

Figure 8.7 Abutted symbols with vertical common line. (From ANSI/IEEE Std 91-1984.)

Qualifying symbols on either side of the common line, as in Fig. 8.8, show each connection between elements. In an array of elements having the same qualifying symbols, the symbols inside the outline may be shown in only the first of the outlines, as in Fig. 8.9.

8.2.5 Common Control Block

As shown in Figs. 8.10 through 8.13, the common control block is the only one of the new logic symbols that has a distinctive shape. Note that it may be placed at either the top or the bottom of an array of related elements. It is used with such an array to show inputs or outputs that are common to more than one element of the array.

Unless dependency notation is indicated, an input to the block affects all elements of the array, as shown in Fig. 8.10. In the case of dependency notation, an input is common only to those elements of the array that have a matching identification number.

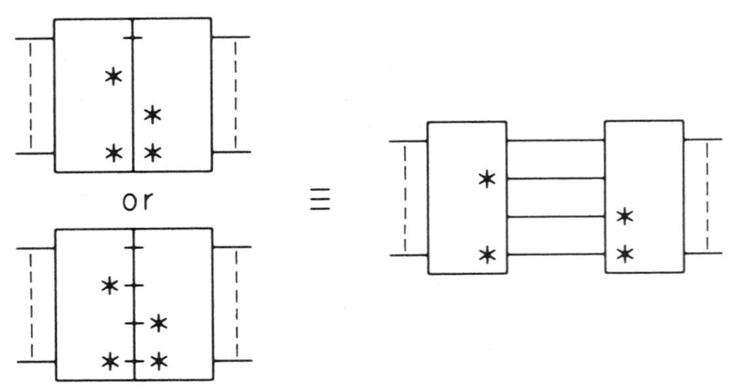

Figure 8.8 Qualifying symbols used with abutted symbols. (From ANSI/IEEE Std 91-1984.)

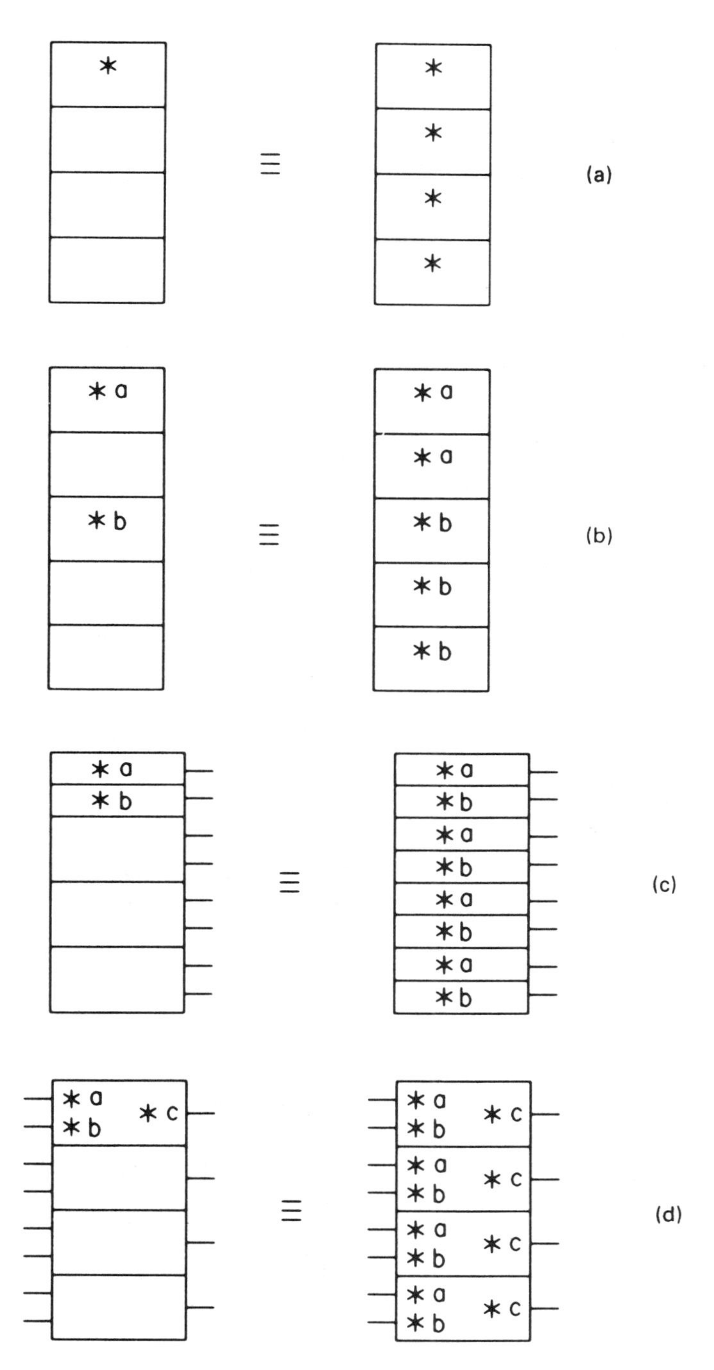

Figure 8.9 Array of elements with same qualifying symbols: (a) a group of elements with identical general qualifying symbols; (b) two successive groups of elements; (c) two interlaced groups of four elements each; (d) a group of elements with identical input and output qualifying symbols. (From ANSI/IEEE Std 91-1984.)

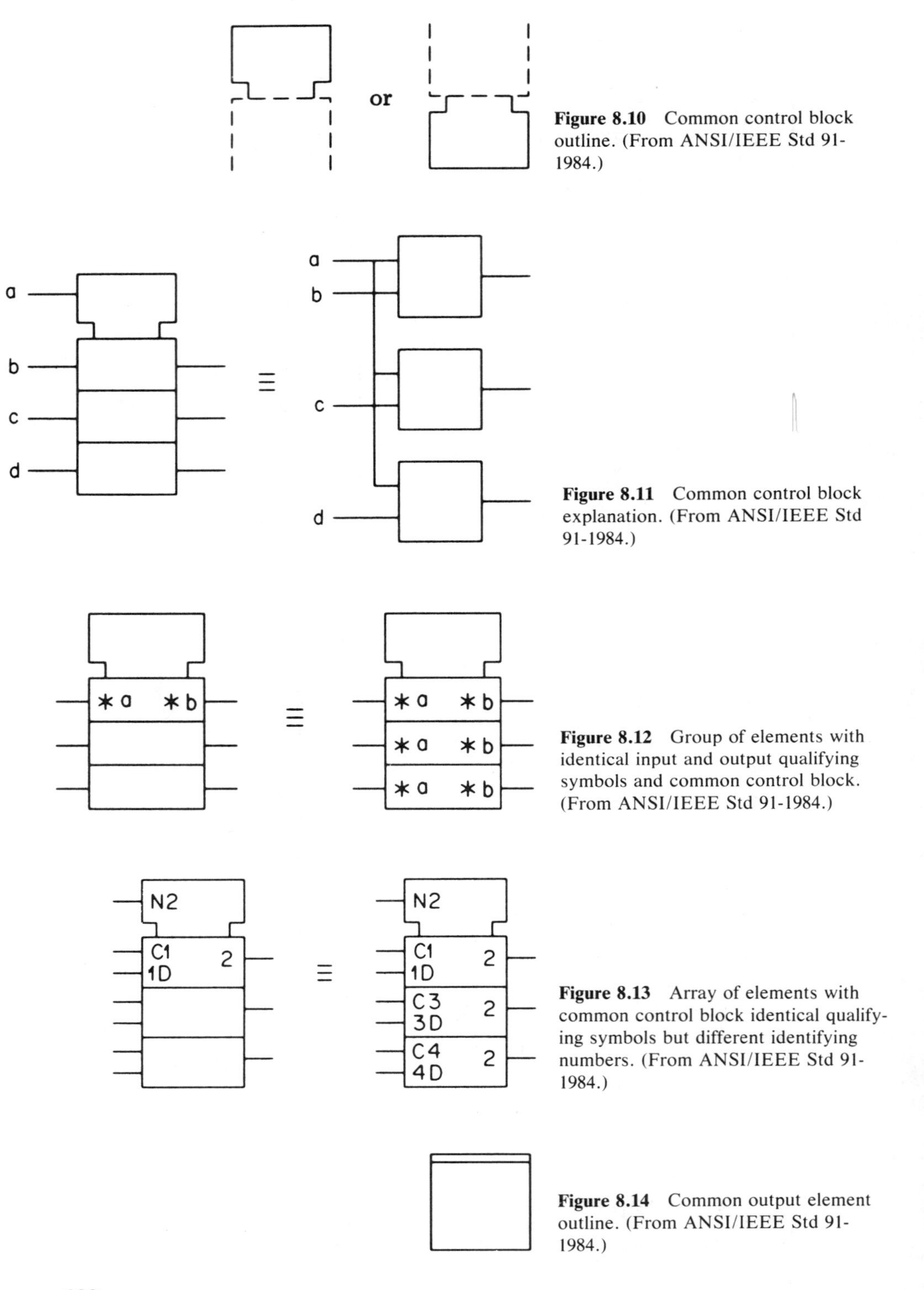

Figure 8.10 Common control block outline. (From ANSI/IEEE Std 91-1984.)

Figure 8.11 Common control block explanation. (From ANSI/IEEE Std 91-1984.)

Figure 8.12 Group of elements with identical input and output qualifying symbols and common control block. (From ANSI/IEEE Std 91-1984.)

Figure 8.13 Array of elements with common control block identical qualifying symbols but different identifying numbers. (From ANSI/IEEE Std 91-1984.)

Figure 8.14 Common output element outline. (From ANSI/IEEE Std 91-1984.)

8.2.6 Common Output Element

A common output element (Fig. 8.14) indicates that all of the similar outputs of all elements of an array form inputs to the common output element, as shown in Fig. 8.15. By similar outputs we mean that the outputs always have identical internal logic states. If one output has a different logic state, such as *d* in Fig. 8.15, that output is not connected to the common output element. The function of the element is shown by a qualifying symbol; Fig. 8.15 shows the particular case for an AND circuit. A common output element may be placed within a common control block as shown in Fig. 8.16 and at either end of the array.

In an array of common output elements, the double line may be shown only once, as illustrated in Fig. 8.17. In such an array each element has an internal connection to each of the common output elements. However, the common output elements are not connected unless so indicated.

8.2.7 Inputs and Outputs with Multiple Functions

Inputs or outputs of some logic devices may have multiple functions. In fact, an input terminal may serve as an output terminal at a different time. Multiple-

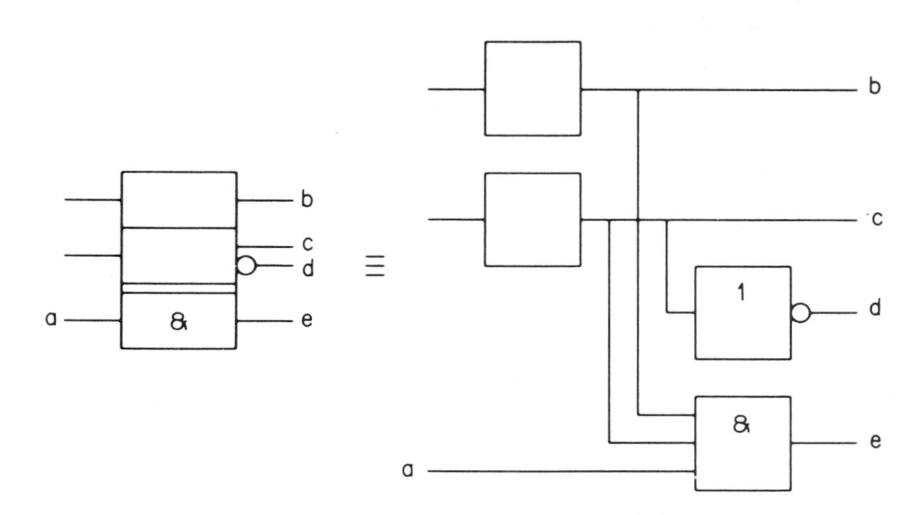

NOTE: The common output element has at least one qualifying symbol to indicate its function. The "&" is the qualifying symbol for AND.

Figure 8.15 Common output element. (From ANSI/IEEE Std 91-1984.)

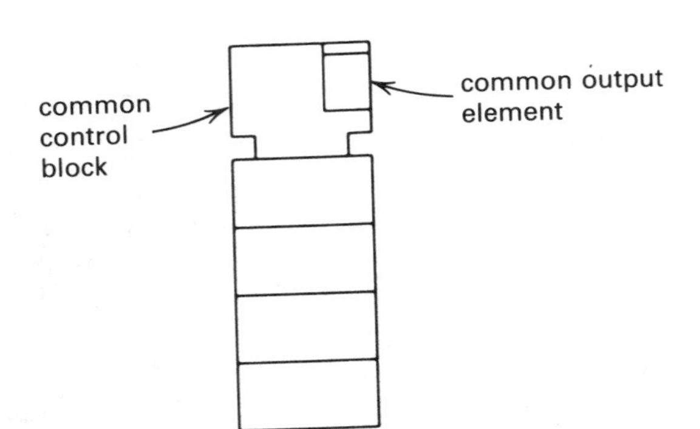

Figure 8.16 Array with common control block and common output element. (From ANSI/IEEE Std 91-1984.)

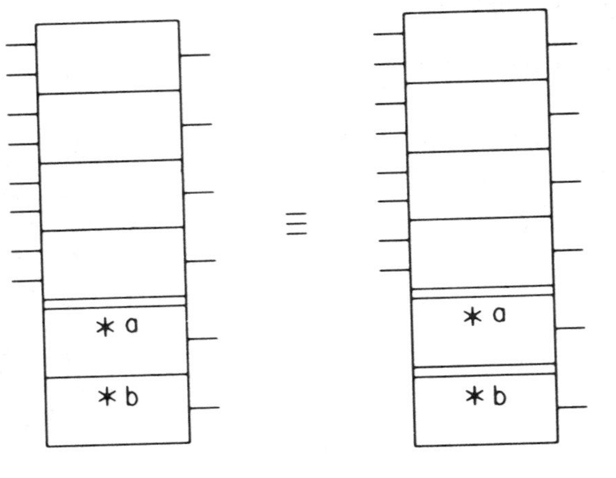

Figure 8.17 Array with two common output elements. (From ANSI/IEEE Std 91-1984.)

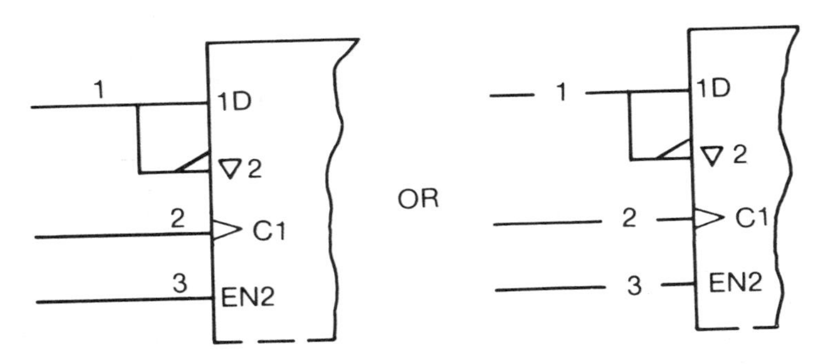

Figure 8.18 Multiple-function terminal (terminal 1) shown as separate lines. (From ANSI/IEEE Std 991-1986.)

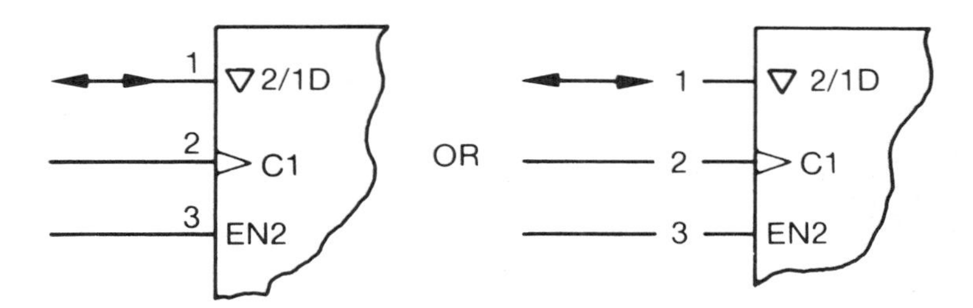

Figure 8.19 Multiple-function terminal (terminal 1) shown as a single line. (From ANSI/IEEE Std 991-1986.)

function terminals may be shown as separate inputs or outputs connected together outside the symbol outline as in Fig. 8.18.

In the method shown in Fig. 8.18, the terminal identification (pin number) is placed on or adjacent to the combined circuit path; this location shows that the connection is internal to the device.

Figure 8.19 shows how a multiple-function terminal may be drawn as a single terminal with a solidus (/) used to separate the labels associated with the separate functions.

As illustrated in Fig. 8.20, a multiple-function terminal may be shown more than once at the symbol outline with the terminal identification repeated.

Repetitive information may be identified by repeated terminal identification in parentheses.

8.2.8 Abbreviated Representation of Symbols

The symbol $-\!\!/^{n}$ indicates multiple conductors for a set of identical inputs to or outputs from an element. That is, the inputs or outputs have identical functions and labels (see Fig. 8.21). This technique is not used if terminal identifiers are used at the symbol.

As shown in Fig. 8.22, an array of identical elements may be indicated by the symbol $-\!\!/^{n}$ for multiple conductors and the notation "mX," where m is a number indicating the number of elements in the array.

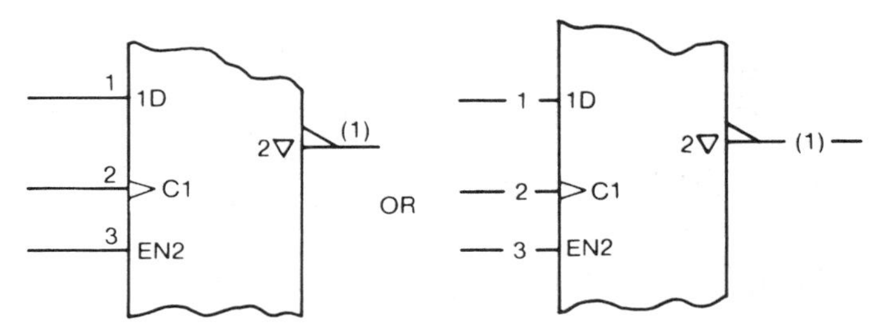

Figure 8.20 Multiple-function terminal (terminal 1) repeated on symbol outline. (From ANSI/IEEE Std 991-1986.)

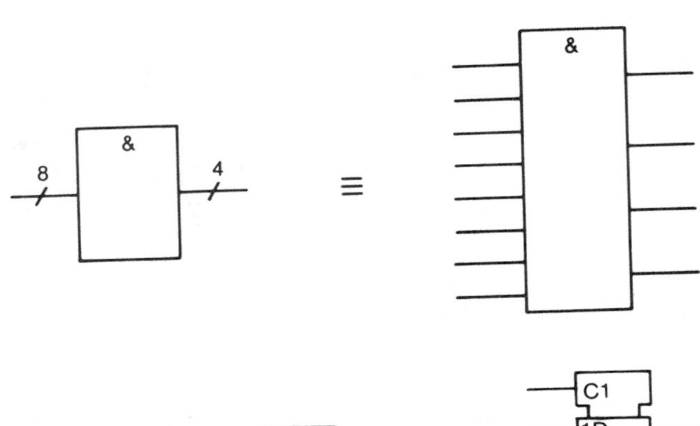

Figure 8.21 Abbreviated representation of identical inputs and outputs for an eight-input, four-output AND gate. (From ANSI/IEEE Std 991-1986.)

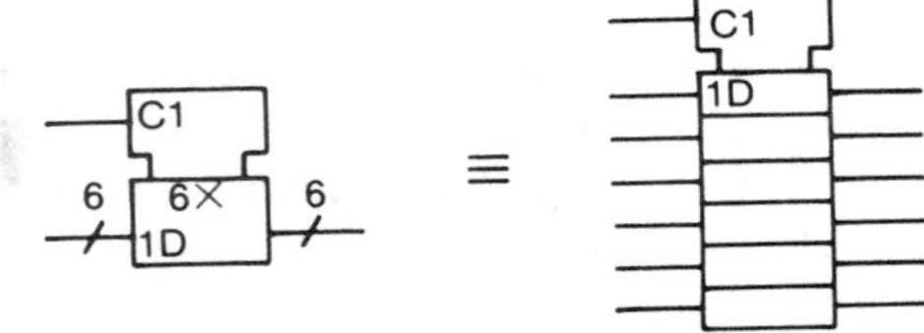

Six D-Type Latches with Common Control

Figure 8.22 Abbreviated representations of arrays of identical elements. (From ANSI/IEEE Std 991-1986.)

8.3 QUALIFYING SYMBOLS ASSOCIATED WITH INPUTS, OUTPUTS, AND OTHER CONNECTIONS

8.3.1 Negation, Polarity, and Dynamic Input Symbols

The qualifying symbols in Figs. 8.23 through 8.30 show the relationship between an internal logic state and an external logic state or level. In a diagram using positive or negative logic, the internal logic 1-state corresponds with the external logic 1-state. In a diagram using direct polarity indication, the internal logic 1-state corresponds with the logic H-level.

8.3.2 Internal Connections

A connection within an element is called an *internal* connection. The relationship between internal logic states at internal connections is indicated by qualifying symbols shown in this section. The symbols in Fig. 8.24 show the logic relationships between elements whose outlines are combined.

8.3.3 Symbols Inside the Outline

Except for bithreshold and extension inputs, if two or more inputs have the same qualifying symbol for their functions, they are in an OR relationship.

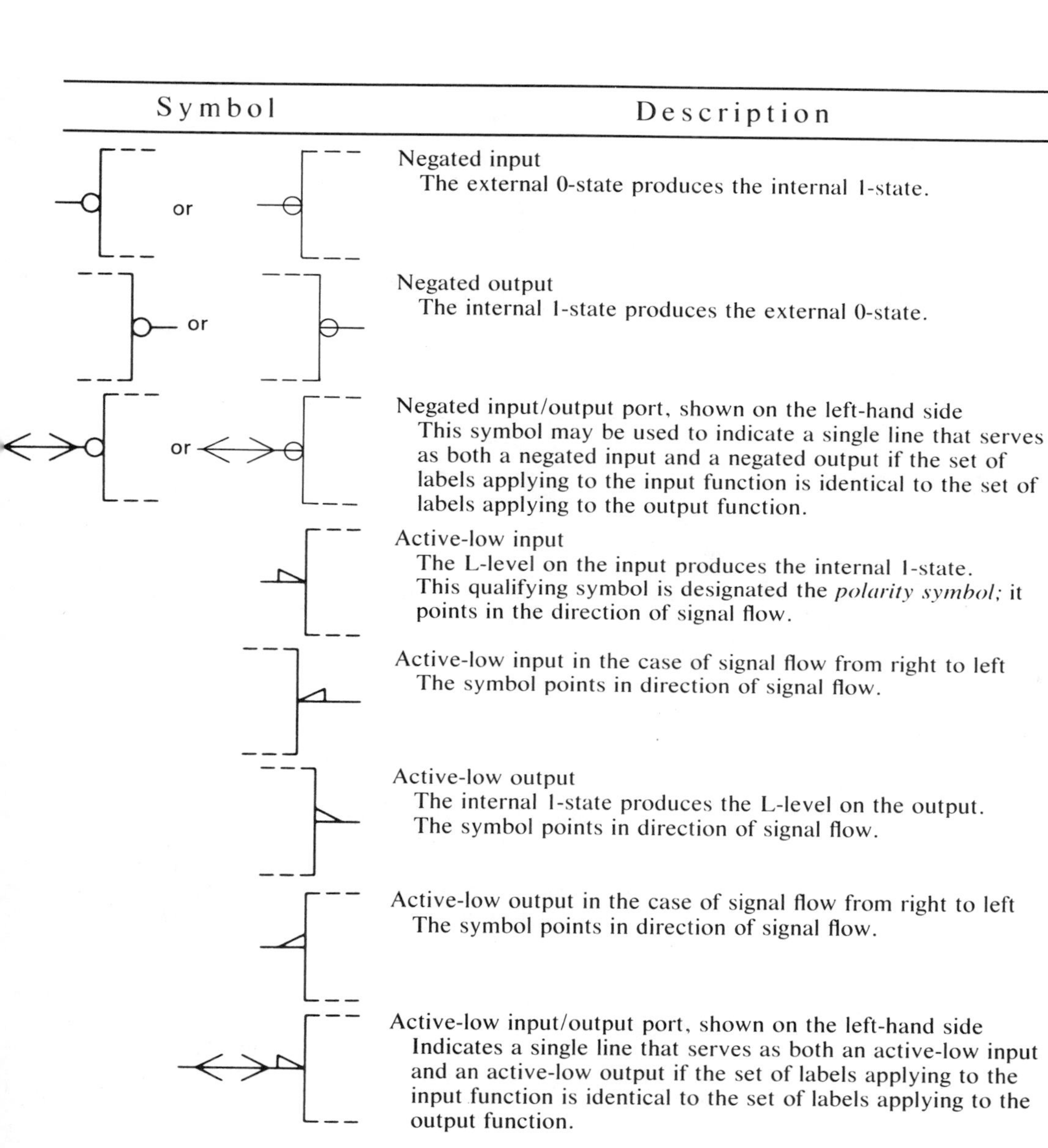

Symbol	Description
	Negated input The external 0-state produces the internal 1-state.
	Negated output The internal 1-state produces the external 0-state.
	Negated input/output port, shown on the left-hand side This symbol may be used to indicate a single line that serves as both a negated input and a negated output if the set of labels applying to the input function is identical to the set of labels applying to the output function.
	Active-low input The L-level on the input produces the internal 1-state. This qualifying symbol is designated the *polarity symbol;* it points in the direction of signal flow.
	Active-low input in the case of signal flow from right to left The symbol points in direction of signal flow.
	Active-low output The internal 1-state produces the L-level on the output. The symbol points in direction of signal flow.
	Active-low output in the case of signal flow from right to left The symbol points in direction of signal flow.
	Active-low input/output port, shown on the left-hand side Indicates a single line that serves as both an active-low input and an active-low output if the set of labels applying to the input function is identical to the set of labels applying to the output function.
	The polarity symbol should point to the right or down, relative to the orientation of the text within the symbol.
	Dynamic input The transition from the external 0-state to the external 1-state produces a transitory internal 1-state. At all other times the internal logic state is 0. On diagrams using direct polarity indication, the transition from the L-level to the H-level on the input produces a transitory internal 1-state. At all other times the internal logic state is 0.

Figure 8.23 Negation, polarity, and dynamic input symbols. (From ANSI/IEEE Std 91-1984.)

Symbol	Description
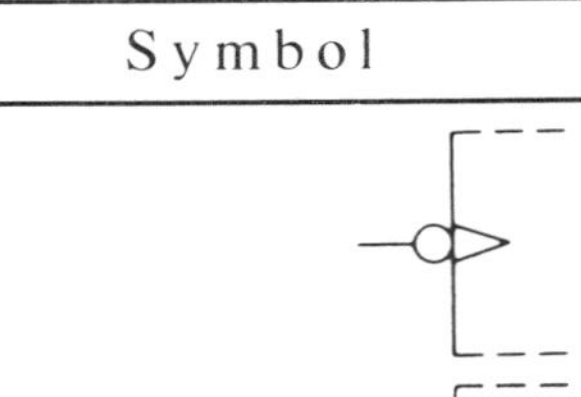	Dynamic input with negation The transition from the external 1-state to the external 0-state on the input produces a transitory internal 1-state. At all other times the internal logic state is 0.
	Dynamic input with polarity symbol The transition from the H-level to the L-level on the input produces a transitory internal 1-state. At all other times the internal logic state is 0.

Figure 8.23 *(Continued)*

Symbol	Description
	Internal connection The internal 1-state (0-state) of the output of the element on the left produces the 1-state (0-state) at the input of the element on the right. The internal connection symbol is omitted sometimes if no confusion is likely.
	Internal connection with negation The internal 1-state (0-state) of the output of the element on the left produces the 0-state (1-state) at the input of the element on the right.
	Internal connection with dynamic character Transition from the internal 0-state to the internal 1-state of the output of the element on the left produces a transitory 1-state at the input of the element on the right. At all other times the logic state at the input of the element on the right is 0.
	Internal connection with negation and dynamic character Transition from the internal 1-state to the internal 0-state of the output of the element on the left produces a transitory 1-state at the input of the element on the right. At all other times the logic state at the input of the element on the right is 0.
	Internal input (virtual input) This input always stands at its internal 1-state unless it is affected by a dependency relationship that has an overriding effect. NOTE: Internal inputs and outputs have internal logic states only.
	Internal output (virtual output) The effect of this output is indicated by dependency notation.

Figure 8.24 Internal connections. (From ANSI/IEEE Std 91-1984.)

Symbol	Description
	Postponed output The change of the internal state of this output is postponed until the input signal that initiates the change returns to its initial external logic state or logic level. The internal logic state of any input(s) affecting or affected by the *initiating* input must not change while the *initiating* input stands at its internal 1-state or the resulting output state will not be specified by the symbol. If the input signal that initiates the change appears at an internal connection, the change of state is postponed until the output at the internal connection returns to its initial logic state. NOTE: If the postponed output symbol is shown without a prefix, the output is postponed with respect to each Cm, +, →, ←, −, or T input; in all other cases the identifying numbers (or if necessary the full labels) of all inputs with respect to which the output is postponed are shown as a prefix to this symbol.

Figure 8.25 Postponed output. (From ANSI/IEEE Std 91-1984.)

When using the positive or the negative logic convention, the transition at the output takes place when the input changes:

(1) From its external 1-state to its external 0-state, or

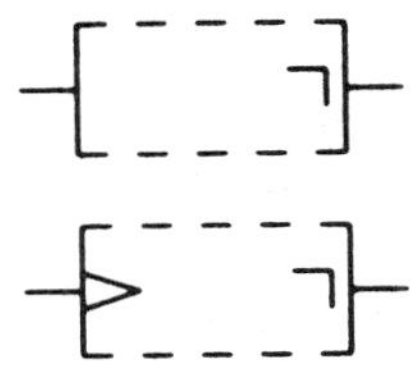

(2) From its external 0-state to its external 1-state

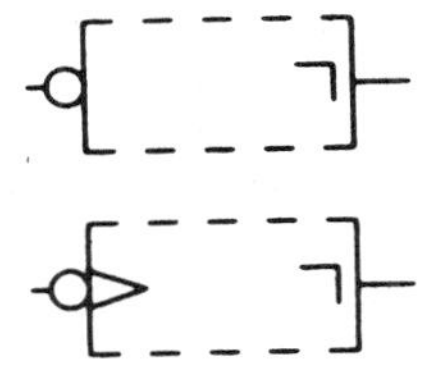

Figure 8.26 Output transition with positive or negative logic. (From ANSI/IEEE Std 91-1984.)

On a diagram using direct polarity indication, the transition at the output takes place when the input changes:

(1) From its H-level to its L-level, or

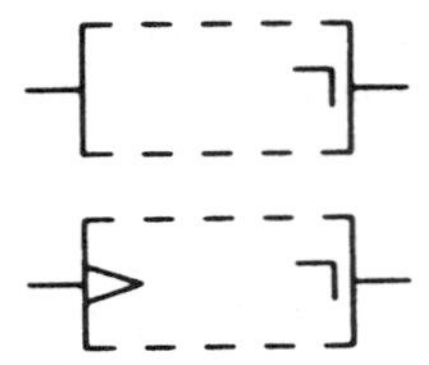

(2) From its L-level to its H-level

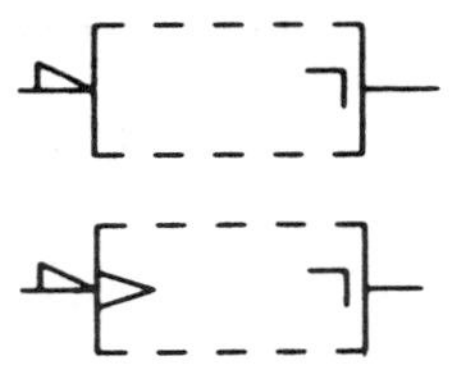

Figure 8.27 Output transition with direct polarity indication. (From ANSI/IEEE Std 91-1984.)

The internal logic state, as determined by the external logic state or level, may be modified by the state of other inputs. If the symbol ▷ is present at an input, that input has a dynamic character (see Figs. 8.25, 8.26, and 8.27). In Figs 8.26 and 8.27 transitions will occur as shown unless any input has an overriding effect.

Figure 8.28 shows the input signals that appear within the outline; Fig. 8.29 shows the output signals. Commonly used logical signal designations within the outlines are given in Table 8.1.

Bit-grouping symbols are shown in Fig. 8.30.

Symbol	Description
hysteresis symbol	Input with hysteresis Bithreshold input The input changes to its internal 1-state when the external signal level reaches a threshold value, V1. It maintains this state until the external signal level has returned through V1 and reaches another threshold value, V0. If this symbol (without the negation symbol or polarity symbol) appears on a diagram that uses either direct polarity indication or the positive-logic convention, V1 is more positive than V0. If it

Figure 8.28 Input signals within the outline. (From ANSI/IEEE Std 91-1984.)

Symbol	Description

appears on a diagram that uses the negative-logic convention, V1 is more negative than V0. If the negation symbol or polarity symbol is present, then the relationship between V1 and V0 is reversed.

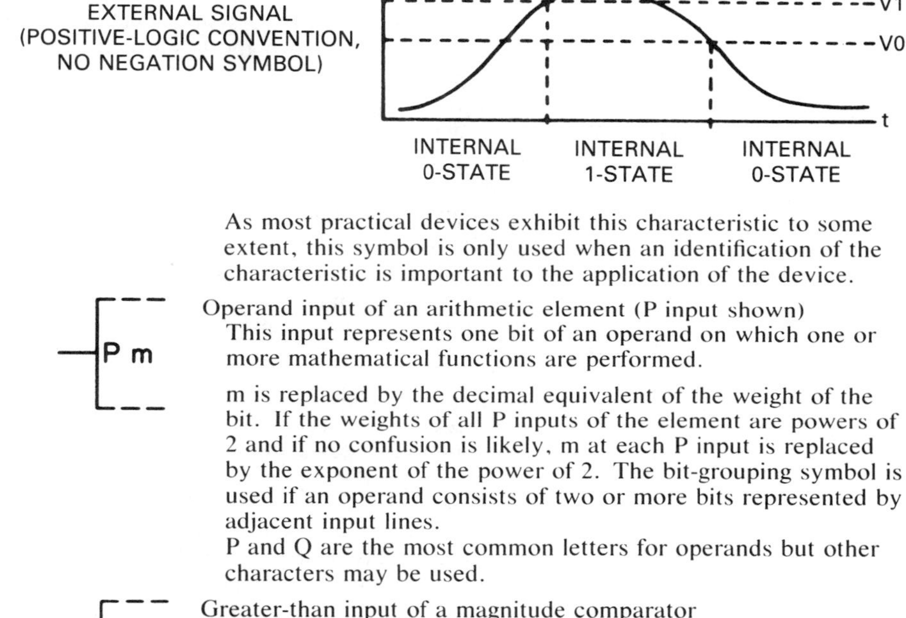

As most practical devices exhibit this characteristic to some extent, this symbol is only used when an identification of the characteristic is important to the application of the device.

P m

Operand input of an arithmetic element (P input shown)
This input represents one bit of an operand on which one or more mathematical functions are performed.

m is replaced by the decimal equivalent of the weight of the bit. If the weights of all P inputs of the element are powers of 2 and if no confusion is likely, m at each P input is replaced by the exponent of the power of 2. The bit-grouping symbol is used if an operand consists of two or more bits represented by adjacent input lines.
P and Q are the most common letters for operands but other characters may be used.

>

Greater-than input of a magnitude comparator

<

Less-than input of a magnitude comparator

=

Equal input of a magnitude comparator

T

T input of a bistable element
Each time this input takes on its internal 1-state, a single change of the internal state of the output to its complement takes place. When the input stands at its internal 0-state, it has no effect on the element.

Figure 8.28 (*Continued*)

Symbol	Description
→ m	Shifting input of a register Each time an input with a right-pointing arrow takes on its internal 1-state, the information contained in the element will be shifted once m positions from left to right or from top to bottom. When the input stands at its internal 0-state, it has no effect on the element. m in this symbol and the 3 symbols below is replaced by the relevant value. If m = 1, the 1 is usually omitted.
← m	Shifting input of a register Each time an input with a left-pointing arrow takes on its internal 1-state, the information contained in the element will be shifted once m positions from right to left or from bottom to top. When the input stands at its internal 0-state, it has no effect on the element.
+m	Count-up input Each time this input takes on its internal 1-state, the content of the element is increased once by m. When the input stands at its internal 0-state, it has no effect on the element.
-m	Count-down input Each time this input takes on its internal 1-state, the content of the element is decreased once by m. When the input stands at its internal 0-state, it has no effect on the element.
EN	Enable input When this input stands at its internal 1-state, all outputs stand at their normally defined internal logic states and have their normally defined effect on elements or distributed functions that may be connected to the outputs, provided no other inputs or outputs have an overriding and contradicting effect. When this input stands at its internal 0-state, all open-circuit outputs stand at their external high-impedance states, all passive-pulldown outputs stand at their high-impedance L-levels, all passive-pullup outputs stand at their high-impedance H-levels, all 3-state outputs stand at their normally defined internal logic states and at their external high-impedance states, and all other outputs stand at their internal 0-states. In a composite symbol this input affects all outputs that are not shown as internal connections or internal outputs. The "composite symbol" excludes outputs that are not otherwise shown to be influenced by the unsubdivided portion of the symbol containing the EN Input.

Figure 8.28 *(Continued)*

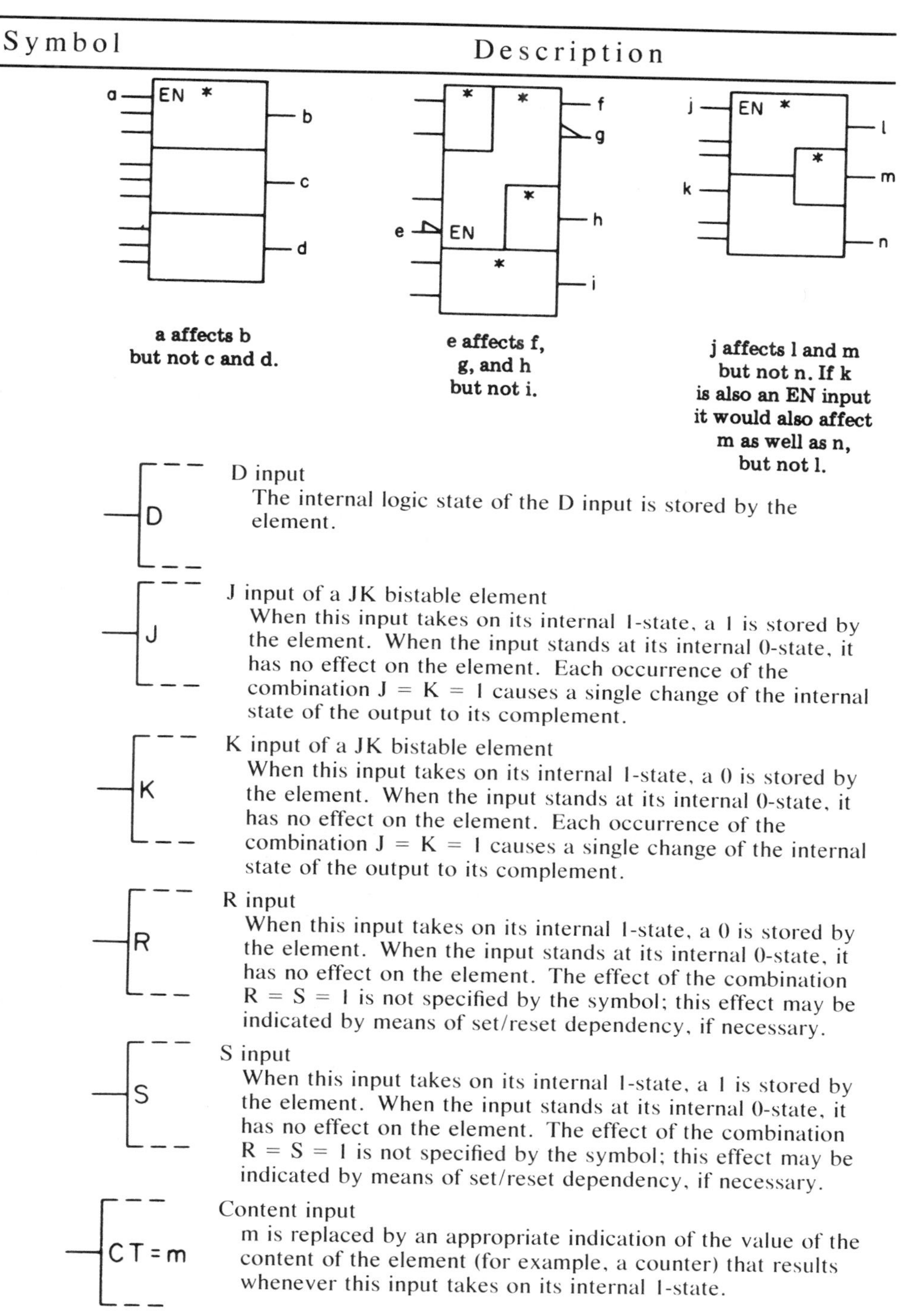

D input
The internal logic state of the D input is stored by the element.

J input of a JK bistable element
When this input takes on its internal 1-state, a 1 is stored by the element. When the input stands at its internal 0-state, it has no effect on the element. Each occurrence of the combination J = K = 1 causes a single change of the internal state of the output to its complement.

K input of a JK bistable element
When this input takes on its internal 1-state, a 0 is stored by the element. When the input stands at its internal 0-state, it has no effect on the element. Each occurrence of the combination J = K = 1 causes a single change of the internal state of the output to its complement.

R input
When this input takes on its internal 1-state, a 0 is stored by the element. When the input stands at its internal 0-state, it has no effect on the element. The effect of the combination R = S = 1 is not specified by the symbol; this effect may be indicated by means of set/reset dependency, if necessary.

S input
When this input takes on its internal 1-state, a 1 is stored by the element. When the input stands at its internal 0-state, it has no effect on the element. The effect of the combination R = S = 1 is not specified by the symbol; this effect may be indicated by means of set/reset dependency, if necessary.

Content input
m is replaced by an appropriate indication of the value of the content of the element (for example, a counter) that results whenever this input takes on its internal 1-state.

Figure 8.28 *(Continued)*

Symbol	Description

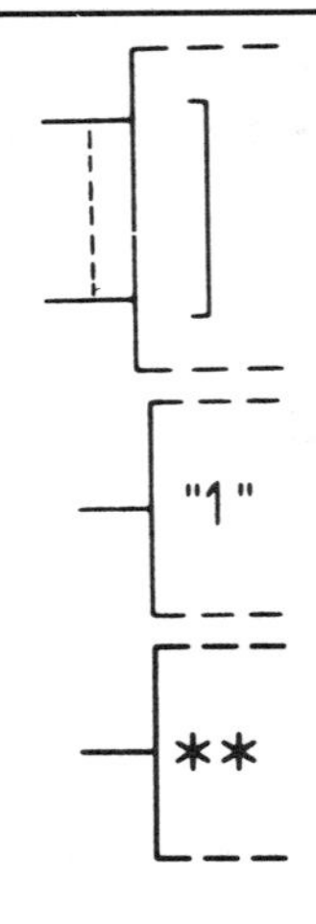

Line-grouping symbol for inputs
Indicates that two or more terminals are needed to implement a single logic output.

Fixed-mode input
Identifies an input that always must be in the internal 1-state for the element to perform the function indicated by the complete symbol.

Symbols for commonly used logic signals (shown at an input)
See Table 8.1 for a list of symbols for designating commonly used logic signals.

** is replaced by an appropriate alphabetical signal designator. If a numerical suffix is added to this alphabetical designator, it indicates the weight or position of the signal within a group of related signals.

Extension input
An input of an element to which an extender output may be connected

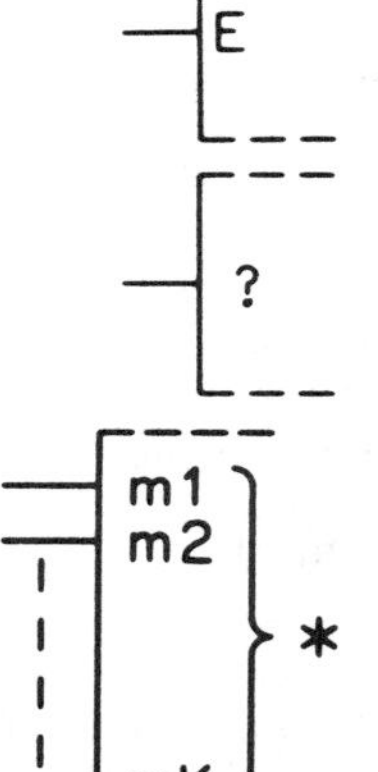

Query input (interrogate) of an associative memory
When this input takes on its internal 1-state, an interrogation of the content of the element takes place. When the input stands at its internal 0-state, it has no effect on the element.

Bit-grouping symbol for inputs (qualifying symbol for multibit input), general symbol
Inputs grouped by this symbol produce a value that is the sum of the individual weights of the inputs standing at their internal 1-states. The individual inputs are shown in ascending or descending order by weight.

This value can be regarded as:
(1) A value on which a mathematical function is performed, or
(2) Defining an identifying number in the sense of dependency notation, or
(3) A value to become the content of the element.

ml . . . mK are replaced by the decimal equivalents of the actual weights. If all the weights are powers of 2, ml . . . mK are replaced by the exponents of the powers of 2 if no confusion is likely. Labels between ml and mK are omitted if no confusion is likely.

"*" is replaced by an indication of the operand on which the mathematical function is performed (for example, P or Q), by an indication in the sense of dependency notation, by CT, or by a label in the case of a "gray box" (Section 8.6). In the case of CT the value produced by the inputs is the value that is loaded into the element.

Figure 8.28 *(Continued)*

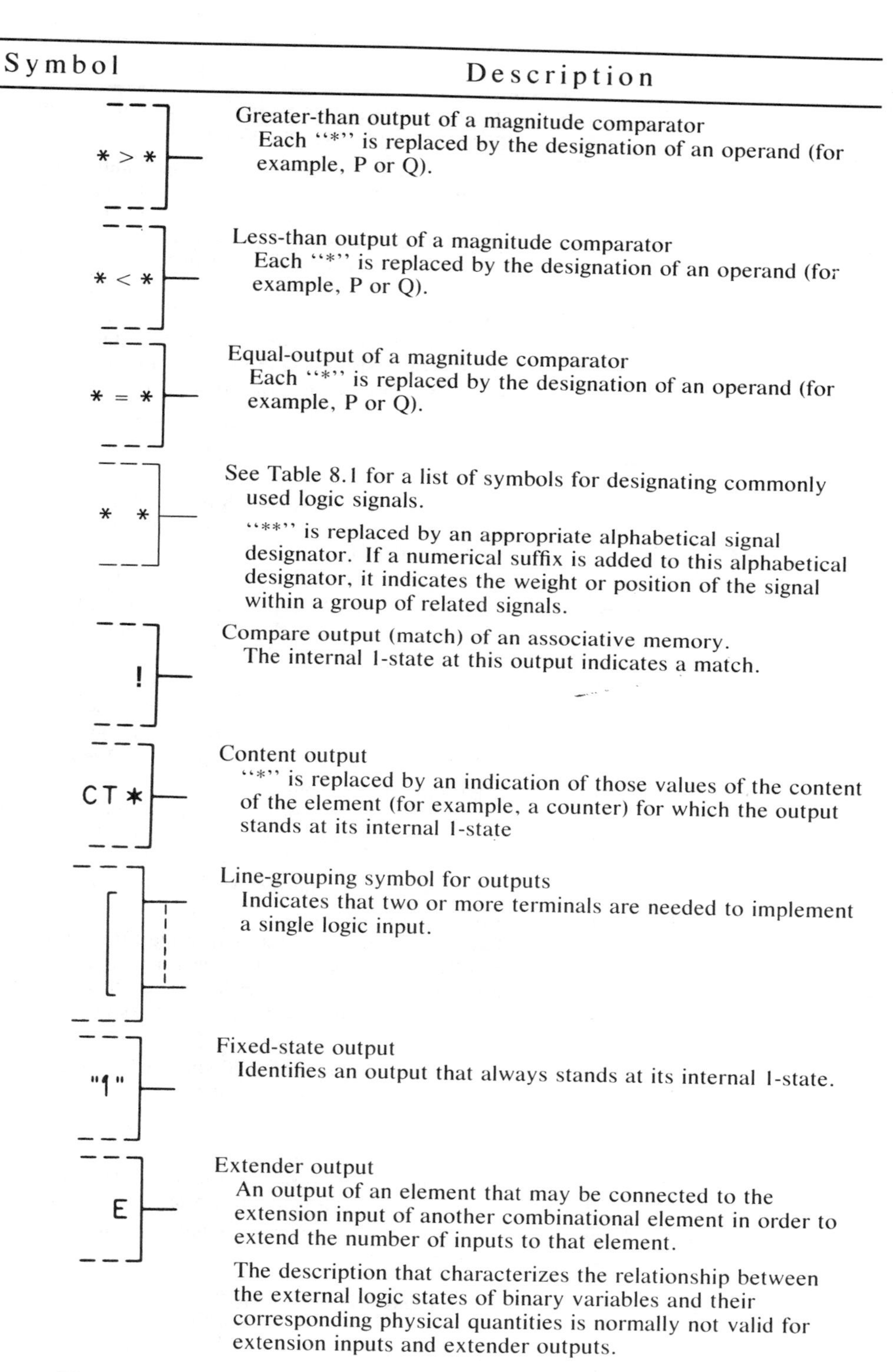

Symbol	Description
* > *	Greater-than output of a magnitude comparator Each "*" is replaced by the designation of an operand (for example, P or Q).
* < *	Less-than output of a magnitude comparator Each "*" is replaced by the designation of an operand (for example, P or Q).
* = *	Equal-output of a magnitude comparator Each "*" is replaced by the designation of an operand (for example, P or Q).
* *	See Table 8.1 for a list of symbols for designating commonly used logic signals. "**" is replaced by an appropriate alphabetical signal designator. If a numerical suffix is added to this alphabetical designator, it indicates the weight or position of the signal within a group of related signals.
!	Compare output (match) of an associative memory. The internal 1-state at this output indicates a match.
CT*	Content output "*" is replaced by an indication of those values of the content of the element (for example, a counter) for which the output stands at its internal 1-state
[	Line-grouping symbol for outputs Indicates that two or more terminals are needed to implement a single logic input.
"1"	Fixed-state output Identifies an output that always stands at its internal 1-state.
E	Extender output An output of an element that may be connected to the extension input of another combinational element in order to extend the number of inputs to that element. The description that characterizes the relationship between the external logic states of binary variables and their corresponding physical quantities is normally not valid for extension inputs and extender outputs.

Figure 8.29 Output signals within the outline. (From ANSI/IEEE Std 91-1984.)

Symbol	Description
Open-circuit output symbol (◇)	Open-circuit output (for example, open-emitter, open-collector, open-source, or open-drain output) One of the two possible internal logic states of this type of output corresponds to an external high-impedance condition. In order to produce the proper logic level in this condition, an externally connected component, often a resistor, is required. This type of output can form part of a distributed function. Although this symbol (and the 5 symbols below) is shown inside the outline, it refers to external states and levels only. The meaning of this symbol is not altered by the presence of a negation symbol (◦) or polarity symbol (▷)
Open-circuit output (H-type) symbol	Open-circuit output (H-type) (for example, PNP open-collector, NPN open-emitter, P-channel open-drain, or N-channel open-source) When not in its external high-impedance state, this type of output produces a relatively low-impedance H-level. When used as part of a distributed function, a positive-logic OR function or a negative-logic AND function is performed.
Open-circuit output (L-type) symbol	Open-circuit output (L-type) (for example, NPN open-collector, PNP open-emitter, N-channel open-drain, or P-channel open-source) When not in its external high-impedance state, this type of output produces a relatively low-impedance L-level. When used as part of a distributed function, a positive-logic AND function or a negative-logic OR function is performed.
Passive-pulldown output symbol	Passive-pulldown output Can be used as part of a distributed function to perform a positive-logic OR function or a negative-logic AND function but it produces both the H-level and the L-level without the need for an additional external component.
Passive-pullup output symbol	Passive-pullup output Can be used as part of a distributed function to perform a positive-logic AND function or a negative-logic OR function but it produces both the H-level and the L-level without the need for an additional external component.
▽	3-state output Can take on a third external state, which is a high-impedance state, having no logic significance.
▷	Output with special amplification The symbol ▷ denotes the function of amplification and points in the direction of signal flow. However, the absence of this symbol does not necessarily indicate the absence of special amplification.

Figure 8.29 (*Continued*)

Symbol	Description

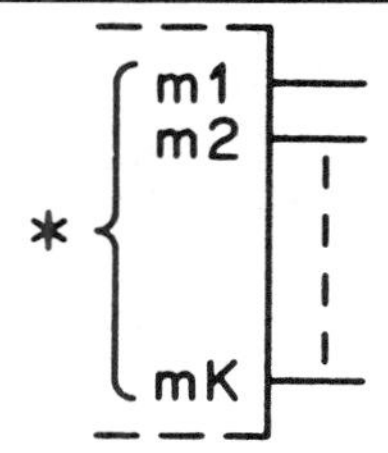

Bit-grouping symbol for outputs (qualifying symbol for multibit output), general symbol

Outputs grouped by this symbol represent a value that is the sum of the individual weights of the outputs standing at their internal 1-states. The individual outputs are shown in ascending or descending order by weight.

This value can be regarded as
(1) The result of the performance of a mathematical function, or
(2) The value of the content of the element.

ml . . . mK are replaced by the decimal equivalents of the actual weights. If all the weights are powers of 2, ml . . . mK are replaced by the exponents of the power of 2 if no confusion is likely. Labels between ml and mK are omitted if no confusion is likely.

"*" is replaced by an indication of the result of the performance of the mathematical function, CT, or an appropriate label in the case of a "gray box" (Section 8.6). In the case of CT the value represented by the outputs is the actual value of the content of the element.

Figure 8.29 (*Continued*)

Figure 8.30 Bit-grouping symbols for inputs and outputs. (From ANSI/IEEE Std 91-1984.)

TABLE 8.1 Commonly Used Logical Signal Designations

Designation	Input/Output	Description
BI (Borrow input)	Input only	If at its internal 1-state, indicates that a subtraction operation performed by a lower-order arithmetic element produces an arithmetic borrow.
BO (Borrow output)	Output only	If at its internal 1-state, indicates that a subtraction operation performed by an arithmetic element produces an arithmetic borrow (see BI signal above).
BG (Borrow generate)	Output	If at its internal 1-state, indicates that an arithmetic element performing subtraction is in the "Borrow-Generate" state, that is, that the subtrahend applied to the element is larger than the minuend, causing a borrow from that element independent of the state of the BI input to that element.
	Input	If at its internal 1-state, indicates to a borrow-acceleration element that the arithmetic element that produces the BG signal is in the "Borrow-Generate" state (see BG output above). The borrow-acceleration element uses its BG, BP, and BI input signals to determine, with reduced propagation delays, the states of the arithmetic borrow signals for a group of arithmetic elements performing binary subtraction.
BP (Borrow propagate)	Output	If at its internal 1-state, indicates that an arithmetic element performing subtraction is in the "Borrow-Propagate" state, that is, that the subtrahend and the minuend applied to the element are equal in value, so that the BO output will stand at its internal 1-state if and only if the BI input is at its internal 1-state.
	Input	If at its internal 1-state, indicates to a borrow-acceleration element that the arithmetic element that produces the BP signal is in the "Borrow-Propagate" state.
CI (Carry input)	Input only	If at its internal 1-state, indicates that an addition operation performed by a lower-order arithmetic element produces an arithmetic carry.
CO (Carry output)	Output only	If at its internal 1-state, indicates that an addition operation performed by an arithmetic element produces an arithmetic carry (see CI signal above).
CG (Carry generate)	Output	If at its internal 1-state, indicates that an arithmetic element performing addition is in the "Carry-Generate" state, that is, that the sum of its addends is sufficiently large to cause a carry from the element independent of the state of the CI input to that element.
	Input	If at its internal 1-state, indicates to a carry-acceleration element whether or not the arithmetic element that produces the CG signal is in the "Carry-Generate" state (see CG output above). The carry-acceleration element uses its CG, CP, and CI input signals to determine with reduced propagation delays, the states of the arithmetic carry signals for a group of arithmetic elements performing addition.
CP (Carry propagate)	Output	If at its internal 1-state, indicates that an arithmetic element performing addition is in the "Carry-Propagate" state, that is, that the sum of its addends is one less than the value at which the element produces an output carry. As a result, the CO output will stand at its internal 1-state if and only if its CI input is at its internal 1-state.
	Input	If at its internal 1-state, indicates to a carry-acceleration element that the arithmetic element that produces the CP signal is in the "Carry-Propagate" state.

Source: ANSI/IEEE Std 91-1984.

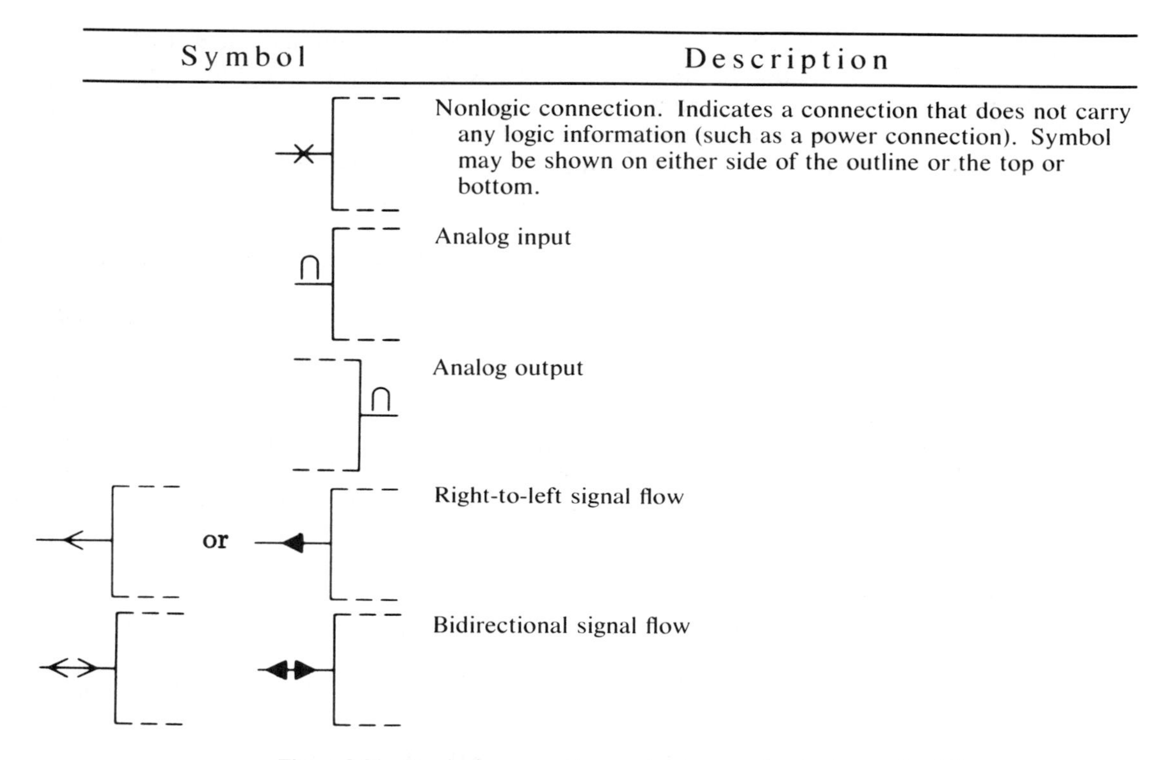

Figure 8.31 Nonlogic connection, analog input and output, and signal flow symbols. (From ANSI/IEEE Std 91-1984.)

8.3.4 Nonlogic Connections and Signal-Flow Indicators

Figure 8.31 shows nonlogic connections, analog input and output, and signal-flow symbols. The direction of signal flow is generally from left to right and from top to bottom. If arrowheads are present, they will point in the direction of signal flow.

8.4 DEPENDENCY NOTATION

8.4.1 General

Dependency notation shows the relationships between inputs, outputs, or inputs and outputs, without actually showing all the elements and interconnections involved. It is a supplement to qualifying symbols. In dependency notation, the terms "affecting" and "affected" are often used. If input 1, for example, affects input 2, then input 1 is designated as the affecting input and input 2 is called the affected input.

Because of feedback in some complex elements, outputs sometimes affect inputs and other outputs. In the following text the term "affecting inputs" is understood to refer to affecting *outputs* also.

Table 8.2 summarizes the various dependencies. In the table the word "action" is used to indicate that affected inputs have their normal effect on the element and the outputs take on the internal logic states determined by the element's function.

Dependency notation defines the relationships between internal logic states except for open-circuit outputs, passive-pullup and passive-pulldown outputs, and three-state outputs. In those cases *enable dependency* defines the relationships between the internal logic states of affecting inputs and the external states of affected outputs.

Dependency notation is indicated by a letter symbol on the input affecting other inputs or outputs. The symbol shows the relationship involved followed by an identifying number. Dependency notation is also indicated by a number on each input or output affected by the affecting input with that same number. A bar placed over the identifying number at the affected input or output indicates that it is the complement of the internal logic state of the affecting input or output that does the affecting.

The identifying number of the affected input may be used as an identifying prefix on the affected input or output to show its function. When an input or output is affected by more than one affecting input, the label of the affected one has the identifying numbers of each of the affecting inputs. The order of these identifying numbers is the same as that of the affecting relationships, as read from left to right.

When two affecting inputs have the same letter and the same identifying number, they are in an OR relationship.

If the labels indicating the functions of the affected inputs or outputs are numbers, the identifying numbers associated with both affecting inputs and affected inputs or outputs are replaced by another character, such as a Greek letter.

Affecting inputs affect only the corresponding affected inputs and outputs of the symbol.

8.4.2 Comparison of C, EN, and M Effects on Inputs

Cm, ENm, and Mm inputs have different applications even though they have the same effect on affected inputs:

1. Cm identifies an action-producing input, for example, the edge-triggered clock of a bistable circuit.
2. ENM identifies an input that produces a single preparatory effect.
3. Mm identifies one or more inputs that singly or together produce alternative preparatory effects.

TABLE 8.2 Summary of Dependency Notation

Letter[a]	Type of dependency	Effect on internal logic state of, or action of, the affected input or output: Affecting input at its 1-state	Affecting input at its 0-state
A	ADDRESS	Permits action (address selected)	Prevents action (address not selected)
C	CONTROL	Permits action	Prevents action
EN	ENABLE	Permits action	(1) Prevents action of affected inputs (2) Imposes external high-impedance state on open-circuit and three-state outputs (internal state of three-state output is unaffected) (3) Imposes high-impedance L-level on passive-pulldown outputs and high-impedance H-level on passive-pullup outputs (4) Imposes 0-state on other outputs
G	AND	Does not alter state (permits action)	Imposes 0-state
M	MODE	Permits action (mode selected)	Prevents action (mode not selected)
N	NEGATE	Complements state	Does not alter state (no effect)
R	RESET	Affected output reacts as it would to S = 0, R = 1	No effect
S	SET	Affected output reacts as it would to S = 1, R = 0	No effect
V	OR	Imposes 1-state	Does not alter state (permits action)
X	TRANSMISSION	Transmission path established	No transmission path established
Z	INTER-CONNECTION	Imposes 1-state	Imposes 0-state

Source: ANSI/IEEE Std 91-1984.

[a] These letters appear at the affecting input (or output) and are followed by a number represented in the general cases in Section 8.4 by the letter "m." Each input or output affected by that input (or output) is labeled with that same number.

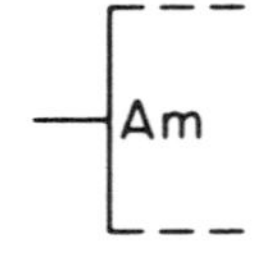

Am input

When this input stands at its internal 1-state, the inputs affected by this input (that is, the inputs of the section of the array selected by this input) have their normally defined effect on the elements of the selected section. Also, the internal logic states of the outputs affected by this input (that is, the outputs of the selected section) have their normal effect on the OR functions (or the indicated functions) determining the internal logic states of the outputs of the array.

When the input stands at its internal 0-state, the inputs affected by this input (that is, the inputs of the section selected by this input) have no effect on the elements of this section. Also, the outputs affected by this input (that is, the outputs of the section selected by this input) have no effect on the outputs of the array.

m is an identifying number

Figure 8.32 Am input. (From ANSI/IEEE Std 91-1984.)

8.4.3 A (Address) Dependency

The letter A indicates address dependency. It is used to represent those elements, such as memories, that use address control inputs to select specified sections of a multidimensional array (see Fig. 8.32). The label for an affecting address input is the letter A plus an identifying number that corresponds to the address of the particular section of the array selected by this input (see Fig. 8.33).

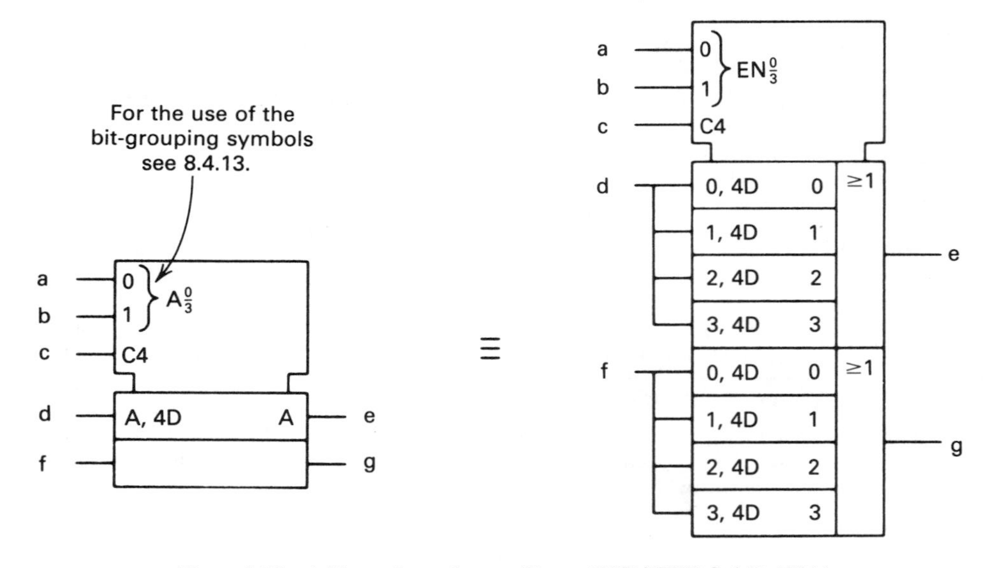

Figure 8.33 Address dependency. (From ANSI/IEEE Std 91-1984.)

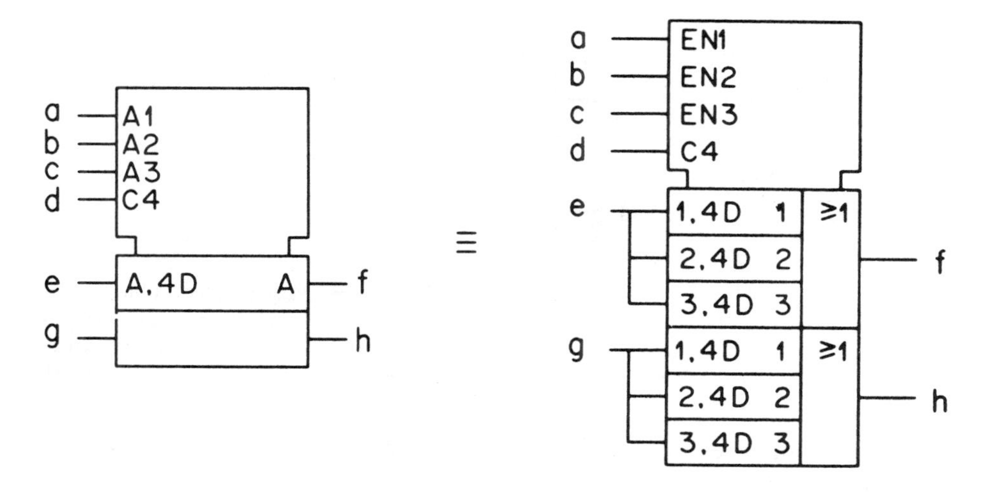

Figure 8.34 Address dependency showing affected input as an input to all selected sections and affected outputs as an OR function. (From ANSI/IEEE Std 91-1984.)

As shown in Fig. 8.34, if the address structure allows more than one section of an element to be selected at a time, an affected input of that element is an input to all the sections selected; an affected output of that element is the OR function (unless otherwise noted) of the outputs of all the sections selected.

Figure 8.35 shows an array of 16 sections, each composed of four pulse-triggered D flip-flops.

When an output label indicates that the associated output is an open-circuit, passive-pulldown, passive-pullup, three-state, or specially amplified output, this

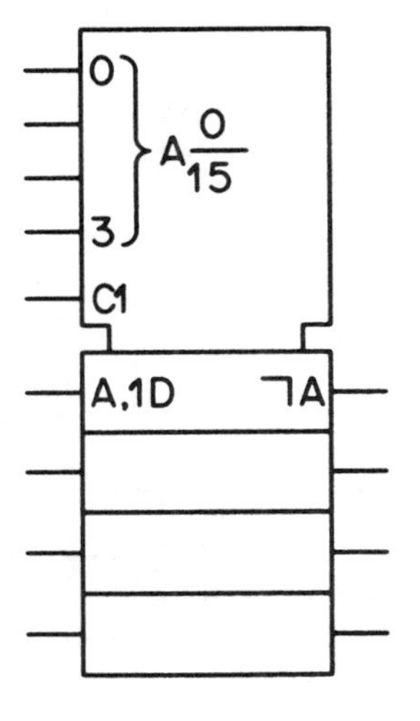

Figure 8.35 Address dependency for array of 16 sections each composed of four pulse-triggered D flip-flops. (From ANSI/IEEE Std 91-1984.)

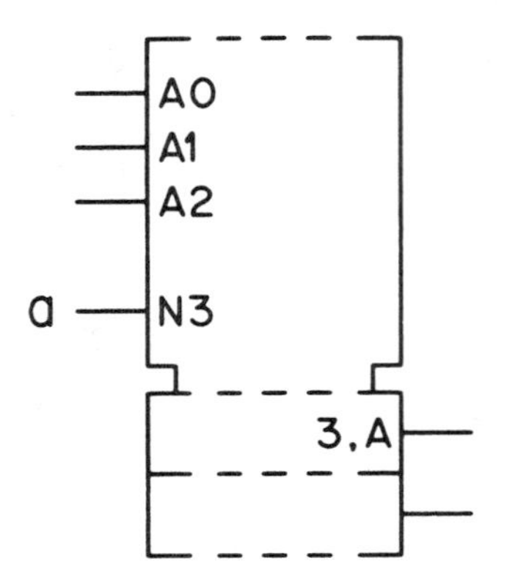

Figure 8.36 Address dependency showing internal logic state as an OR function of the complemented states. (From ANSI/IEEE Std 91-1984.)

designation refers to the output of the array, not to the individual sections of the array.

When an output affected by an Am input also has other labels, the position of the labels relative to the letter determines whether they affect the output of the selected section or the array. If the labels precede the letter A, they affect the output of the *selected section;* if the labels are placed after the letter A, they affect the output of the *array,* that is, after the application of the indicated function to the corresponding outputs of the selected sections of the array.

If $a = 1$ in Fig. 8.36, the internal logic state of each output is the result of the OR function of the complemented states of the outputs of the sections selected.

If $a = 1$ in Fig. 8.37, the internal logic state of each output is the complement of the OR function of the states of the outputs of the selected sections.

If the letter A is changed to 1A, 2A, . . . then there are several sets of affecting Am inputs for the purpose of independent access to sections of the array. This access may be simultaneous. These sets of Am inputs may have the same identifying numbers because they have access to the same sections of the array.

If two affecting address inputs have the same identifying number, there is no fixed relationship with each other. Neither is there any relationship between

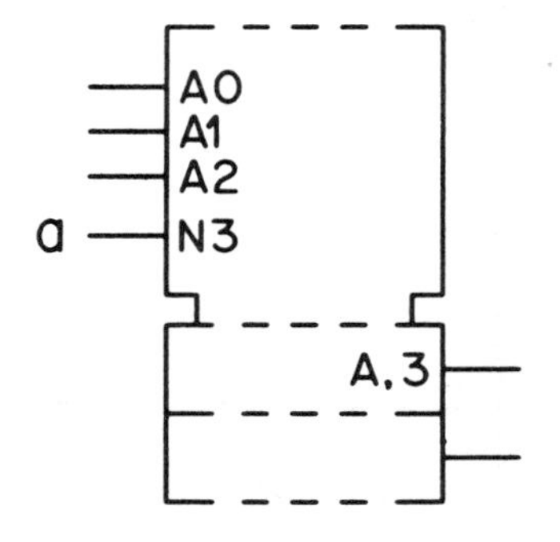

Figure 8.37 Address dependency showing internal logic state of each output as the complement of the OR function. (From ANSI/IEEE Std 91-1984.)

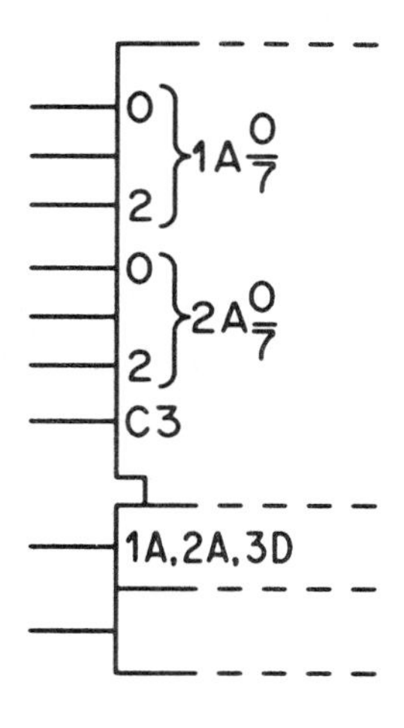

Figure 8.38 Address dependency showing selection of a particular section by both sets of Am inputs. (From ANSI/IEEE Std 91-1984.)

these inputs and any affecting dependency input (for example, G, V, N, . . .) with the same identifying number.

A particular section is selected if it is selected by *both* sets of Am inputs, as shown in Fig. 8.38.

Figure 8.39 shows that a particular section is selected if it is selected by either one or both sets of Am inputs.

8.4.4 C (Control) Dependency

The letter C indicates control dependency. Control dependency is used only for sequential elements. It indicates an action-producing input: for example, the clock input of a sequential element (see Figs. 8.40 and 8.41).

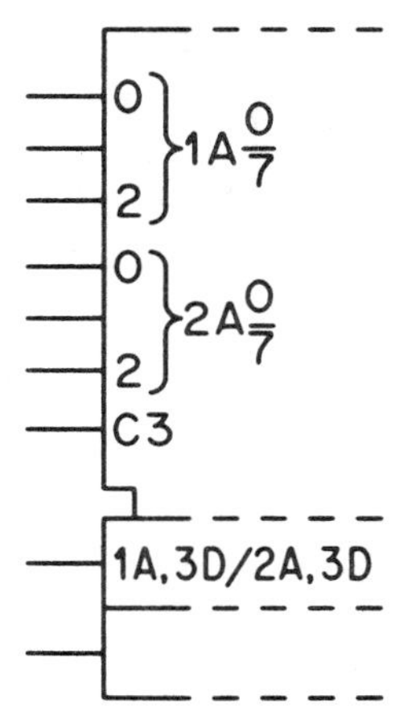

Figure 8.39 Address dependency showing selection of a particular section by either one or both sets of Am inputs. (From ANSI/IEEE Std 91-1984.)

Symbol	Description

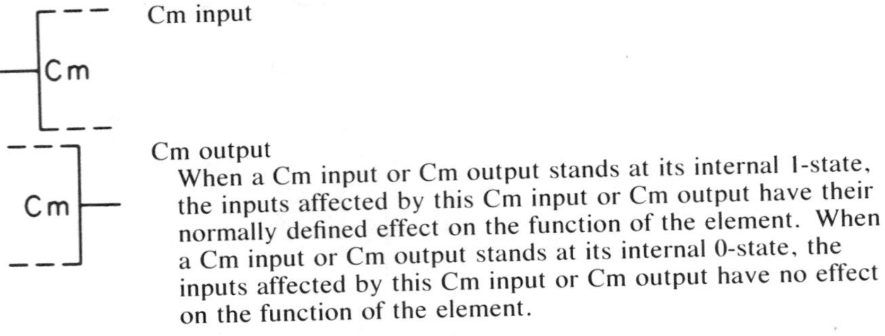

Cm input

Cm output

When a Cm input or Cm output stands at its internal 1-state, the inputs affected by this Cm input or Cm output have their normally defined effect on the function of the element. When a Cm input or Cm output stands at its internal 0-state, the inputs affected by this Cm input or Cm output have no effect on the function of the element.

m is an identifying number

Figure 8.40 Cm input and output. (From ANSI/IEEE Std 91-1984.)

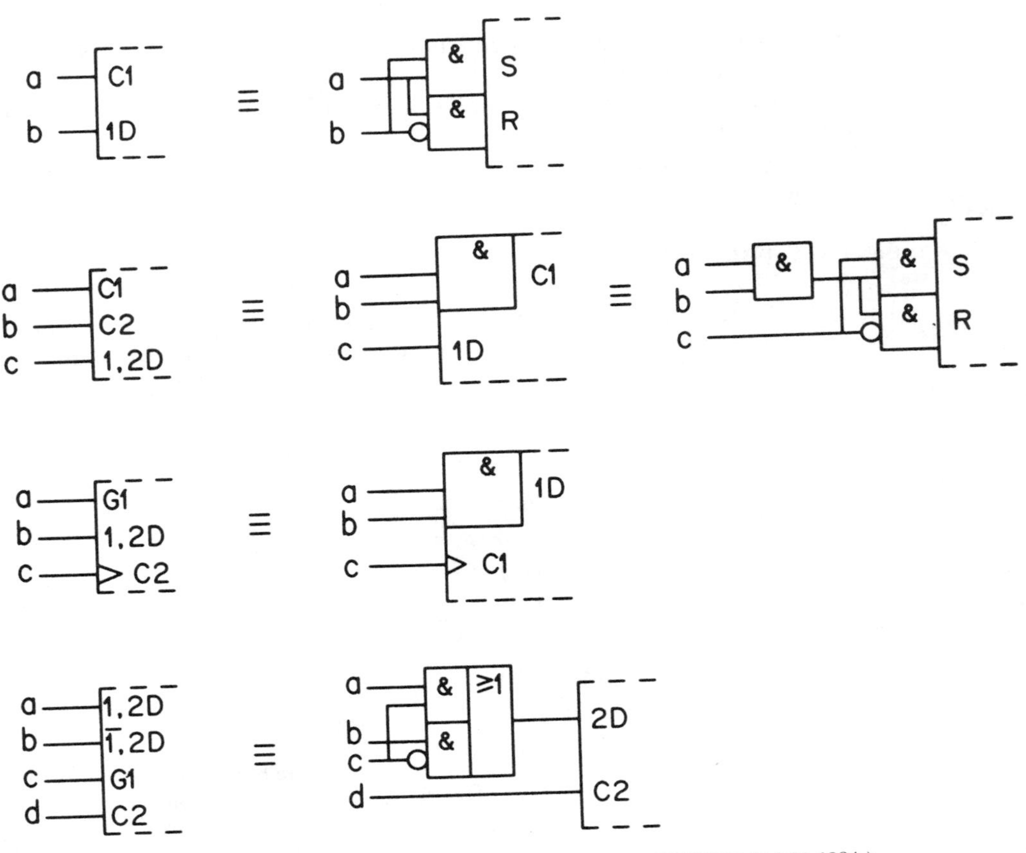

Figure 8.41 Control dependency. (From ANSI/IEEE Std 91-1984.)

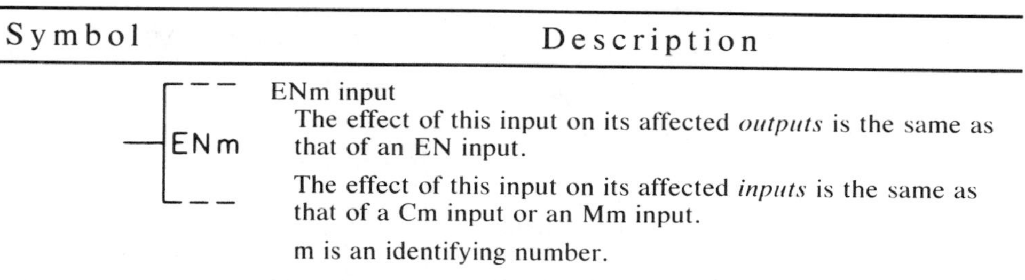

Symbol	Description
ENm	ENm input The effect of this input on its affected *outputs* is the same as that of an EN input. The effect of this input on its affected *inputs* is the same as that of a Cm input or an Mm input. m is an identifying number.

Figure 8.42 ENm input. (From ANSI/IEEE Std 91-1984.)

8.4.5 EN (Enable) Dependency

The letters EN indicate *enable* dependency. Enable dependency identifies an enable input and indicates which inputs and outputs are controlled by it (see Figs. 8.42 and 8.43).

8.4.6 G (AND) Dependency

The letter G indicates AND dependency. There is an AND (&) relationship between each Gm input or output and each input or output affected by the Gm input or output (see Figs. 8.44 and 8.45).

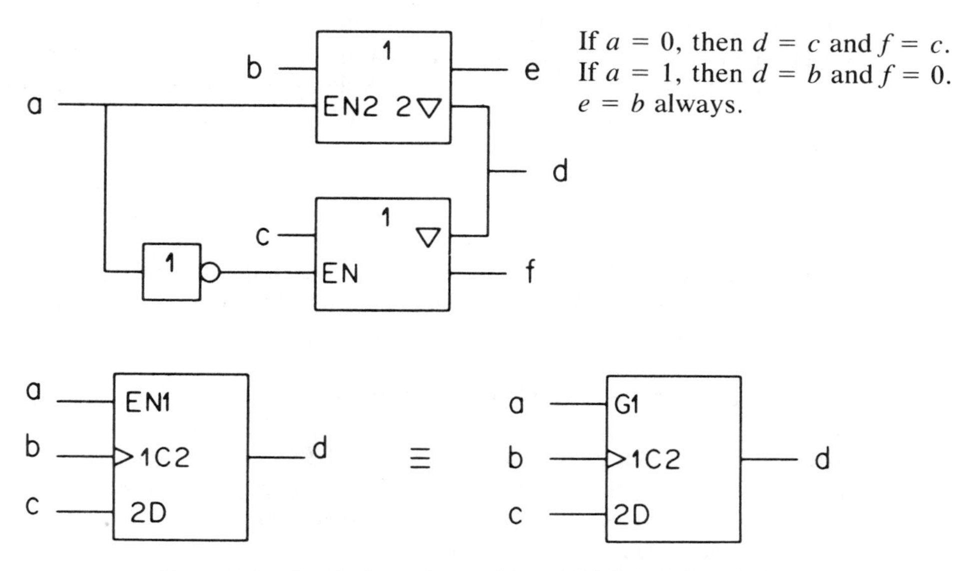

Figure 8.43 Enable dependency. (From ANSI/IEEE Std 91-1984.)

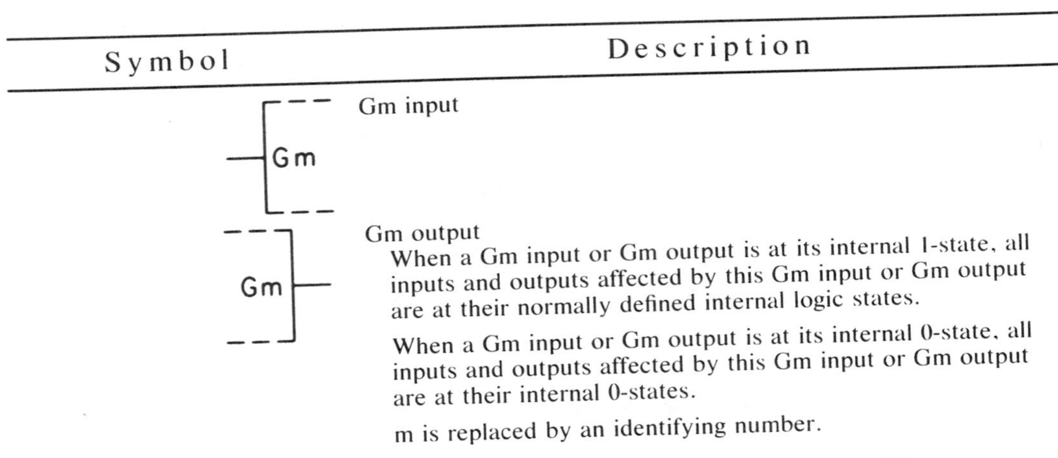

Symbol	Description
Gm	Gm input
Gm	Gm output When a Gm input or Gm output is at its internal 1-state, all inputs and outputs affected by this Gm input or Gm output are at their normally defined internal logic states. When a Gm input or Gm output is at its internal 0-state, all inputs and outputs affected by this Gm input or Gm output are at their internal 0-states. m is replaced by an identifying number.

Figure 8.44 Gm input and output. (From ANSI/IEEE Std 91-1984.)

8.4.7 M (Mode) Dependency

The letter M indicates mode dependency. *Mode* dependency identifies an input that selects the mode of operation of an element and indicates the inputs and outputs depending on that mode. It also shows that the effects of particular inputs and outputs of an element depend on the mode in which the element is operating (see Figs. 8.46 and 8.47).

The inputs and outputs affected by an affecting Mm input are labeled with the letter M in complex elements with a large number of different modes. In such diagrams there is either a table which explains the effects of these inputs in the different modes or a reference to such a table. In some cases the letter M is omitted.

8.4.8 N (Negate) Dependency

The letter N indicates negate dependency. As indicated in Figs. 8.48 and 8.49, there is an exclusive-OR relationship with an Nm input or Nm output and each input or output affected by the Nm input or Nm output.

8.4.9 S (Set) and R (Reset) Dependencies

The letter S indicates set dependency and the letter R indicates reset dependency. When R = S = 1, the set and reset dependencies specify the effect of this combination on an RS bistable element. Affecting Sm and Rm inputs affect outputs only (see Figs. 8.50 and 8.51).

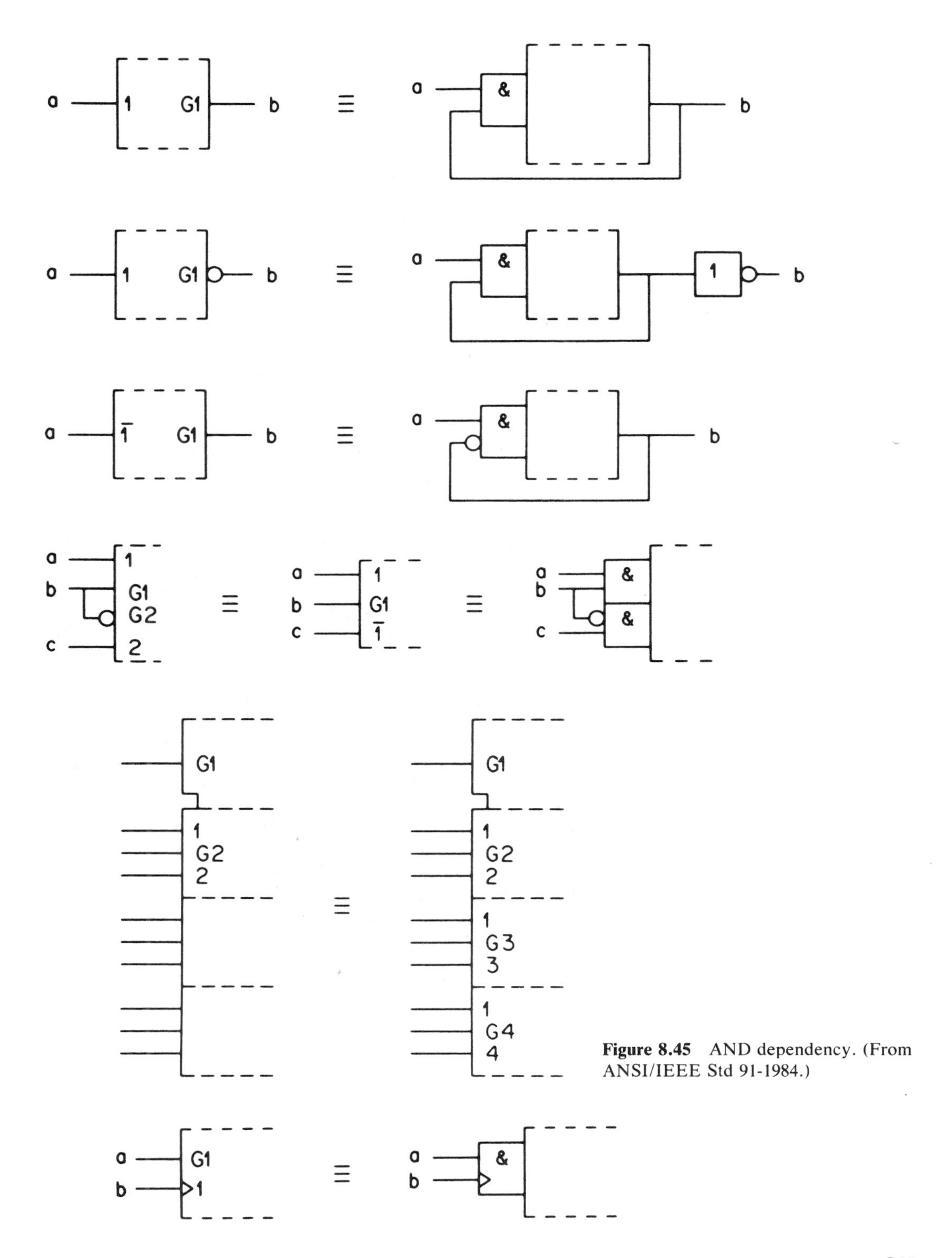

Figure 8.45 AND dependency. (From ANSI/IEEE Std 91-1984.)

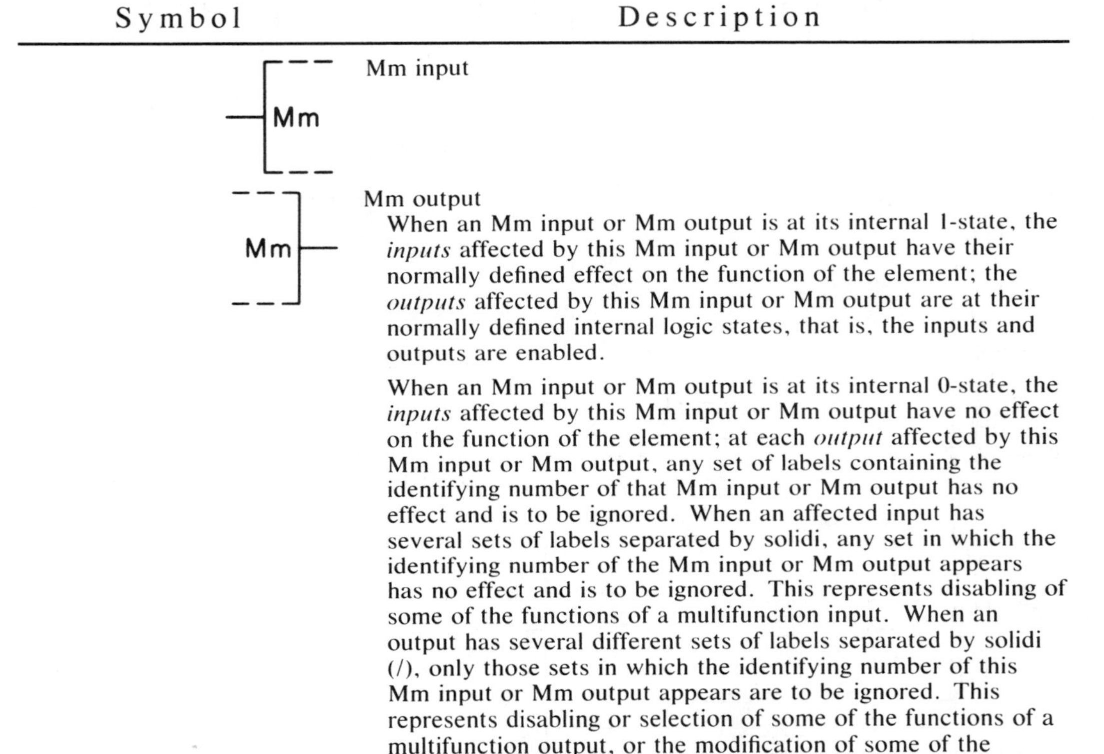

Symbol	Description
Mm	Mm input
Mm	Mm output When an Mm input or Mm output is at its internal 1-state, the *inputs* affected by this Mm input or Mm output have their normally defined effect on the function of the element; the *outputs* affected by this Mm input or Mm output are at their normally defined internal logic states, that is, the inputs and outputs are enabled. When an Mm input or Mm output is at its internal 0-state, the *inputs* affected by this Mm input or Mm output have no effect on the function of the element; at each *output* affected by this Mm input or Mm output, any set of labels containing the identifying number of that Mm input or Mm output has no effect and is to be ignored. When an affected input has several sets of labels separated by solidi, any set in which the identifying number of the Mm input or Mm output appears has no effect and is to be ignored. This represents disabling of some of the functions of a multifunction input. When an output has several different sets of labels separated by solidi (/), only those sets in which the identifying number of this Mm input or Mm output appears are to be ignored. This represents disabling or selection of some of the functions of a multifunction output, or the modification of some of the characteristics or dependent relationships of the output. m is an identifying number.

Figure 8.46 Mm input and output. (From ANSI/IEEE Std 91-1984.)

M Dependency Affecting Inputs:

For the use of the solidus and the symbol see 8.4.

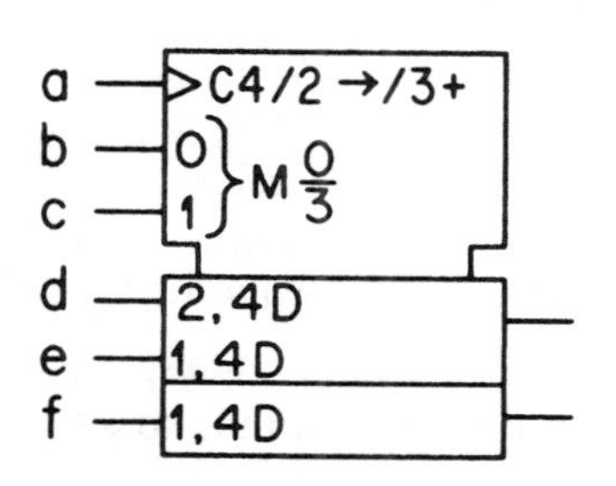

Note that all operations are synchronous.

In mode 0 ($b = 0$, $c = 0$), the outputs remain at their existing states because none of the inputs has an effect.

In mode 1 ($b = 1$, $c = 0$), parallel loading takes place through inputs *e* and *f*.

In mode 2 ($b = 0$, $c = 1$), shifting down and serial loading through input *d* take place.

In mode 3 ($b = c = 1$), counting up by increment of 1 per clock pulse takes place.

Determining the Function of an Output:

If input *a* stands at its internal 1-state establishing mode 1, output *b* will stand at its internal 1-state when the content of the register equals 15. If input *a* stands at its internal 0-state, output *b* will stand at its internal 1-state when the content of the register equals 0.

Figure 8.47 Mode dependency. (From ANSI/IEEE Std 91-1984.)

Modifying Dependent Relationships of Outputs:

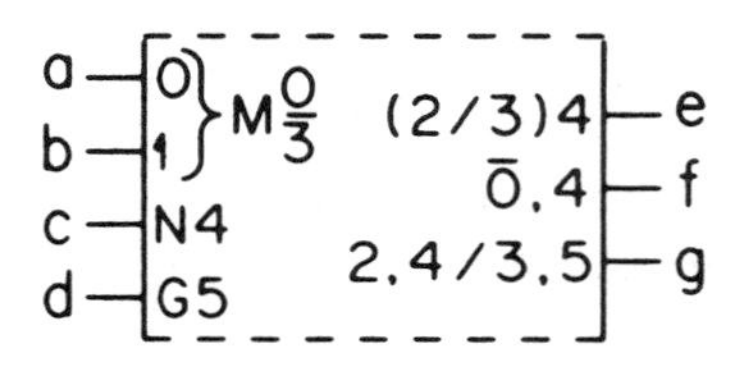

At output e the label set causing negation (if c = 1) is effective only in modes 2 and 3. In modes 0 and 1 this output stands at its normally defined state as if it had no labels.

At output f the label set has effect when the mode is not 0 so output f is negated (if c = 1) in modes 1, 2 and 3. In mode 0 the label set has no effect so the output stands at its normally defined state. In this example $\bar{0}$, 4 is equivalent to (1/2/3)4.

At output g there are two label sets. The first set, causing negation (if c = 1), is effective only in mode 2. The second set, subjecting g to AND dependency on d, has effect only in mode 3.

Note that in mode 0 none of the dependency relationships has any effect on the outputs, so e, f, and g will all stand at the same state.

Figure 8.47 (*Continued*)

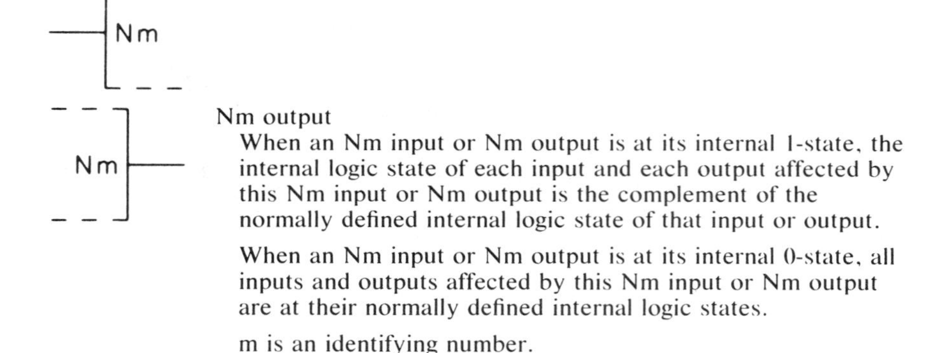

Nm input

Nm output

When an Nm input or Nm output is at its internal 1-state, the internal logic state of each input and each output affected by this Nm input or Nm output is the complement of the normally defined internal logic state of that input or output.

When an Nm input or Nm output is at its internal 0-state, all inputs and outputs affected by this Nm input or Nm output are at their normally defined internal logic states.

m is an identifying number.

Figure 8.48 Nm input and output. (From ANSI/IEEE Std 91-1984.)

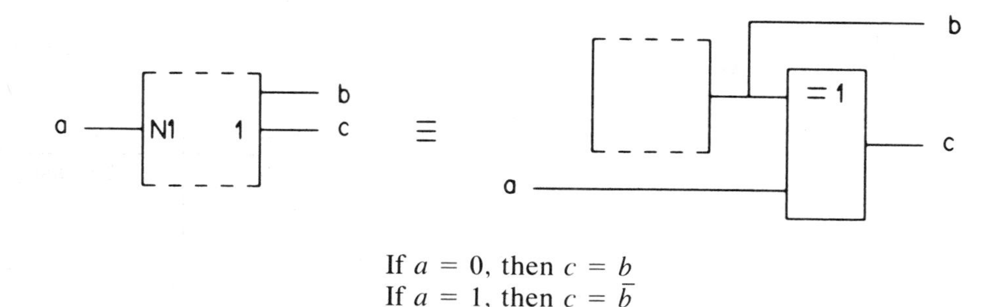

Figure 8.49 Negate dependency. (From ANSI/IEEE Std 91-1984.)

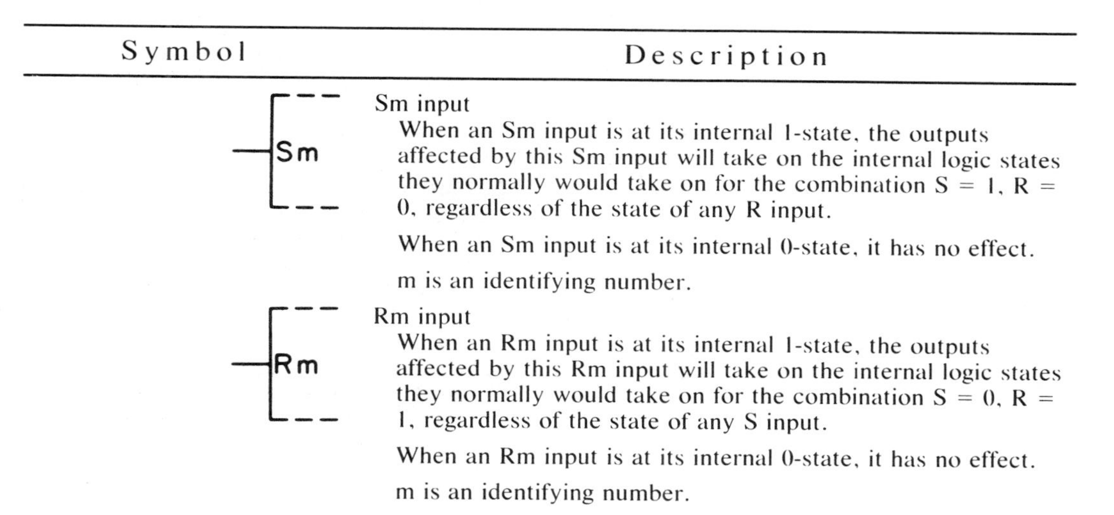

Symbol	Description
Sm	Sm input When an Sm input is at its internal 1-state, the outputs affected by this Sm input will take on the internal logic states they normally would take on for the combination S = 1, R = 0, regardless of the state of any R input. When an Sm input is at its internal 0-state, it has no effect. m is an identifying number.
Rm	Rm input When an Rm input is at its internal 1-state, the outputs affected by this Rm input will take on the internal logic states they normally would take on for the combination S = 0, R = 1, regardless of the state of any S input. When an Rm input is at its internal 0-state, it has no effect. m is an identifying number.

Figure 8.50 Sm and Rm inputs. (From ANSI/IEEE Std 91-1984.)

In the illustrations below the truth tables refer to *external* logic states.

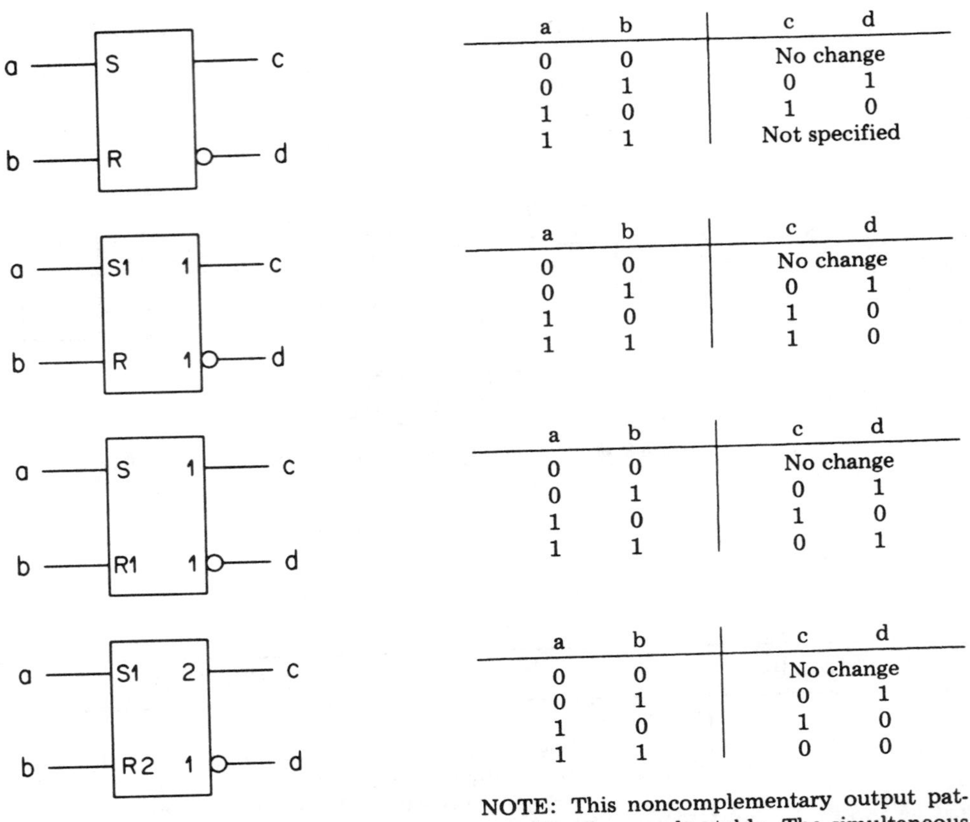

a	b	c	d
0	0	No change	
0	1	0	1
1	0	1	0
1	1	Not specified	

a	b	c	d
0	0	No change	
0	1	0	1
1	0	1	0
1	1	1	0

a	b	c	d
0	0	No change	
0	1	0	1
1	0	1	0
1	1	0	1

a	b	c	d
0	0	No change	
0	1	0	1
1	0	1	0
1	1	0	0

NOTE: This noncomplementary output pattern is only pseudo stable. The simultaneous return of *a* and *b* to 0 produces an unforseeable stable and complementary output pattern.

a
b
G1/$\bar{2}$S
G2/$\bar{1}$R
c
d

a	b	c	d
0	0	No change	
0	1	0	1
1	0	1	0
1	1	No change	

For the use of the solidus, see 8.4.13.

NOTE: This example does not make use of S and R dependencies but completes the set of alternatives to the unspecified case and shows that S and R dependencies cannot affect inputs.

Figure 8.51 Set and reset dependencies. (From ANSI/IEEE Std 91-1984.)

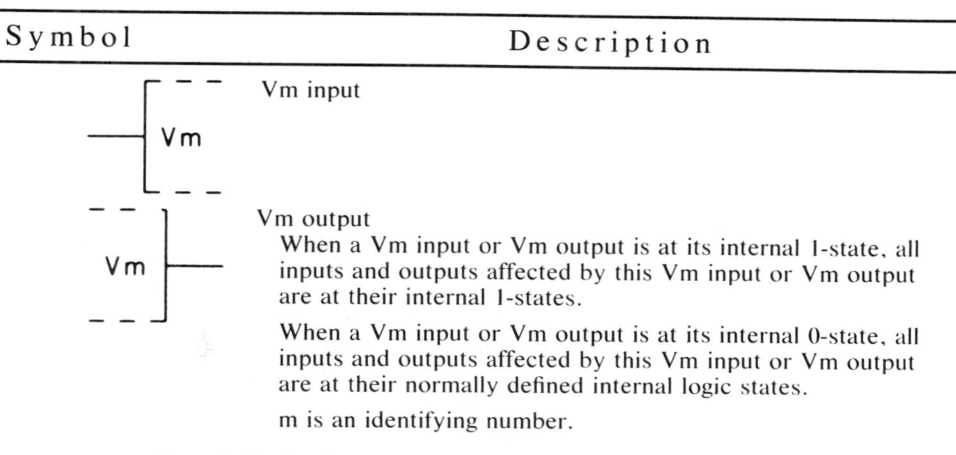

Symbol	Description
Vm	Vm input
Vm	Vm output When a Vm input or Vm output is at its internal 1-state, all inputs and outputs affected by this Vm input or Vm output are at their internal 1-states. When a Vm input or Vm output is at its internal 0-state, all inputs and outputs affected by this Vm input or Vm output are at their normally defined internal logic states. m is an identifying number.

Figure 8.52 Vm input and output. (From ANSI/IEEE Std 91-1984.)

8.4.10 V (OR) Dependency

The letter V indicates OR dependency. As shown in Figs. 8.52 and 8.53, there is an OR relationship between a Vm input or Vm output and each input or output affected by the Vm input or Vm output.

8.4.11 X (Transmission) Dependency

The letter X indicates transmission dependency, which is used to indicate controlled transmission paths between affected ports (see Figs. 8.54 and 8.55).

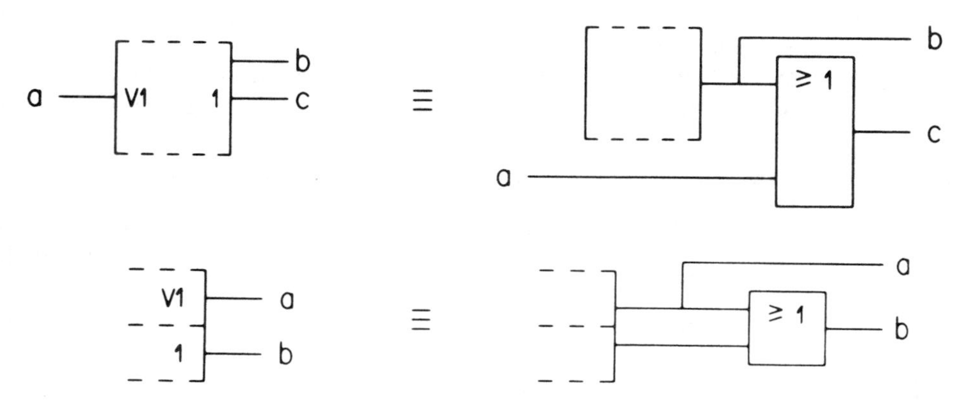

Figure 8.53 OR dependency. (From ANSI/IEEE Std 91-1984.)

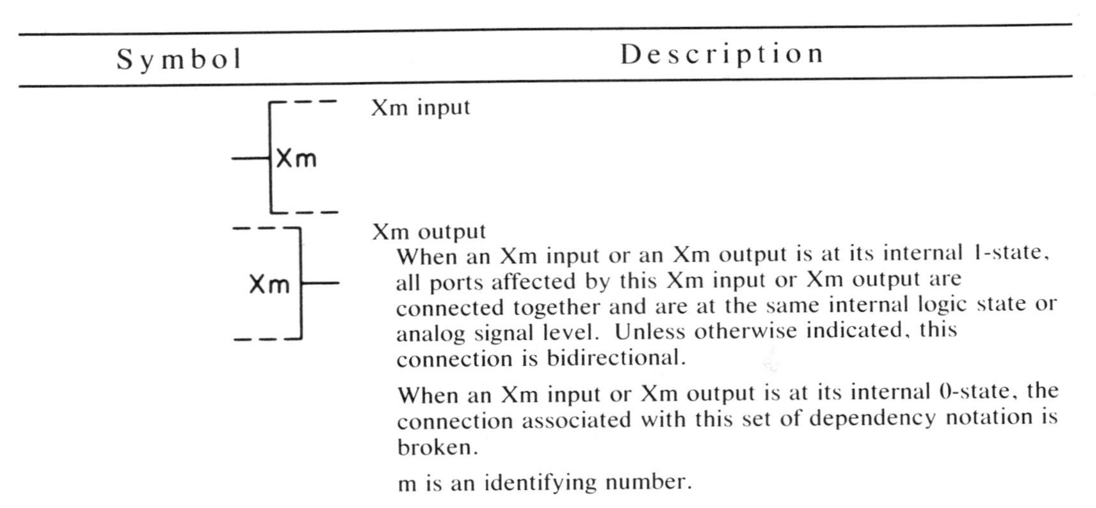

Xm input

Xm output
When an Xm input or an Xm output is at its internal 1-state, all ports affected by this Xm input or Xm output are connected together and are at the same internal logic state or analog signal level. Unless otherwise indicated, this connection is bidirectional.

When an Xm input or Xm output is at its internal 0-state, the connection associated with this set of dependency notation is broken.

m is an identifying number.

Figure 8.54 Xm input and output. (From ANSI/IEEE Std 91-1984.)

8.4.12 Z (Interconnection) Dependency

The letter Z indicates interconnection dependency. It is used to show that there are internal logic connections between inputs, outputs, internal inputs, and internal outputs. If an input or output is affected by a Zm input or output, its internal logic is identical to the internal logic state of its affecting Zm input or output, unless indicated otherwise (see Figs. 8.56 and 8.57).

8.4.13 Special Techniques

Coder. When the symbol for a coder [X/Y] is used as an embedded symbol, as in Fig. 8.58, a set of affecting inputs is produced by decoding the signals on certain inputs to an element.

When Y in qualifying symbol X/Y is replaced, as shown in Fig. 8.59, with a letter indicating the type of dependency, all the affecting inputs produced by the

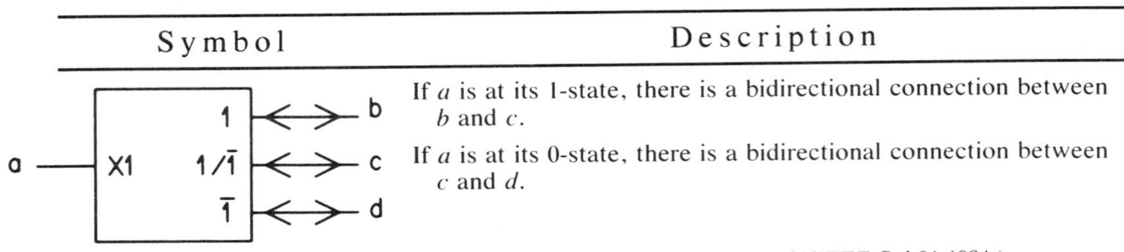

If *a* is at its 1-state, there is a bidirectional connection between *b* and *c*.

If *a* is at its 0-state, there is a bidirectional connection between *c* and *d*.

Figure 8.55 Transmission dependency. (From ANSI/IEEE Std 91-1984.)

Symbol	Description

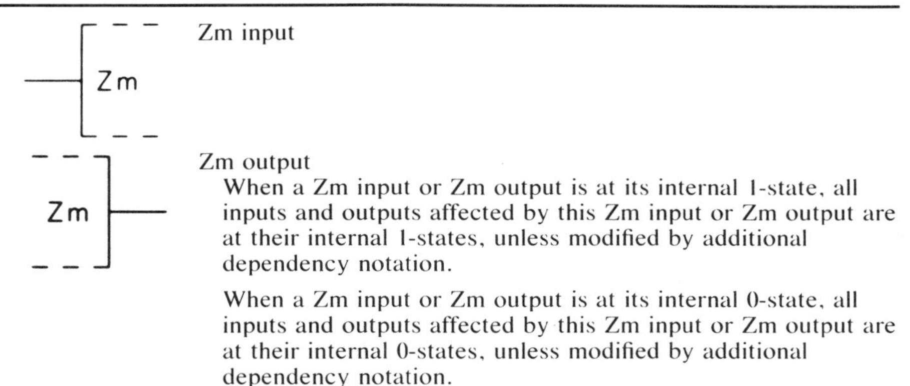

Zm input

Zm output

When a Zm input or Zm output is at its internal 1-state, all inputs and outputs affected by this Zm input or Zm output are at their internal 1-states, unless modified by additional dependency notation.

When a Zm input or Zm output is at its internal 0-state, all inputs and outputs affected by this Zm input or Zm output are at their internal 0-states, unless modified by additional dependency notation.

m is an identifying number.

Figure 8.56 Zm input and output. (From ANSI/IEEE Std 91-1984.)

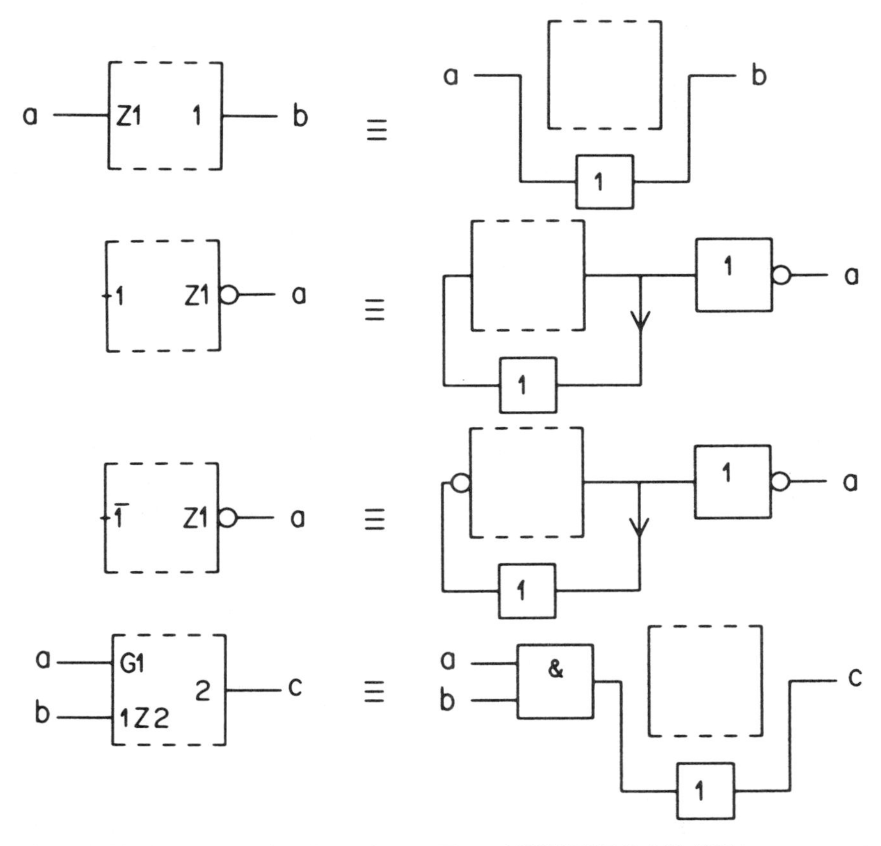

Figure 8.57 Interconnection dependency. (From ANSI/IEEE Std 91-1984.)

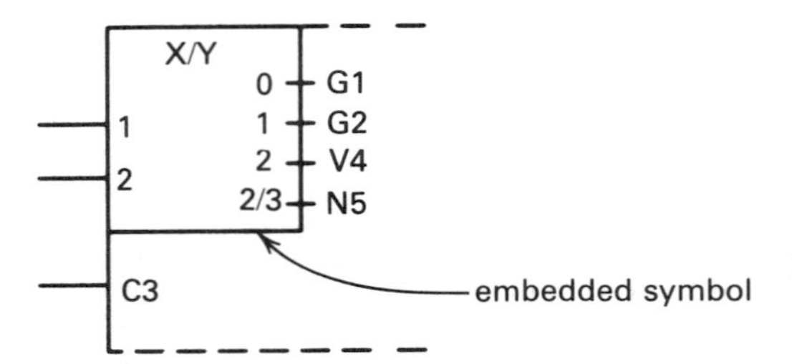

Figure 8.58 Embedded coder used to produce affecting inputs. (From ANSI/IEEE Std 91-1984.)

coder are of the same type and their identifying numbers correspond with the numbers shown at the outputs of the coder. Affecting inputs are not indicated in this case.

Bit grouping. The bit-grouping symbol, shown in Fig. 8.60, indicates that all affecting inputs produced by a coder are of the same type and have consecutive identifying numbers. In this case, the * is replaced by the letter indicating the type of dependency followed by m1/m2, where m1 is replaced by the smallest identifying number and m2 is replaced by the largest. The number of outputs of the coder is $m2 - m1 + 1$.

Input labels. If the qualifying symbol for an input having a single functional effect is preceded by labels corresponding to other inputs, that input is affected by the other inputs. The order in which the effects or modifications are applied is the left-to-right order of the preceding labels (see Fig. 8.61).

If an input has several different functional effects, these effects may be indicated on different input lines connected together outside the outline. However, in some cases the input may be shown once with the different sets of labels separated by solidi. The order of these sets of labels has no significance. When a solidus precedes the first set of labels shown, one of the functional effects of an input is that of an unlabeled input of the element (see Fig. 8.62). If all inputs of a combinational element are disabled, the symbol does not specify the internal logic states of the outputs of the element. (Disabled here means that the inputs have no effect on the function of the element.)

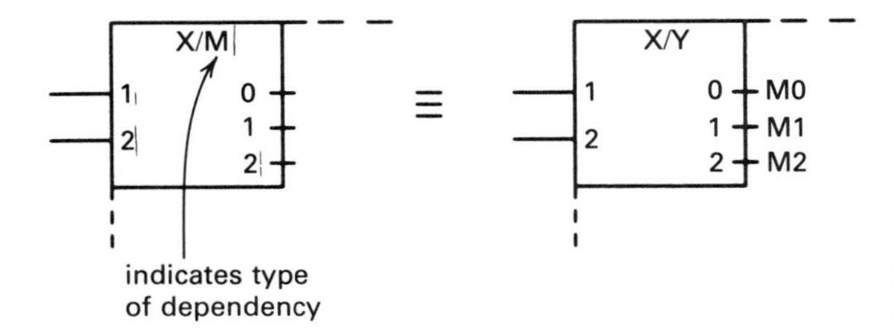

Figure 8.59 Replacement of X/Y with X/M in a coder. (From ANSI/IEEE Std 91-1984.)

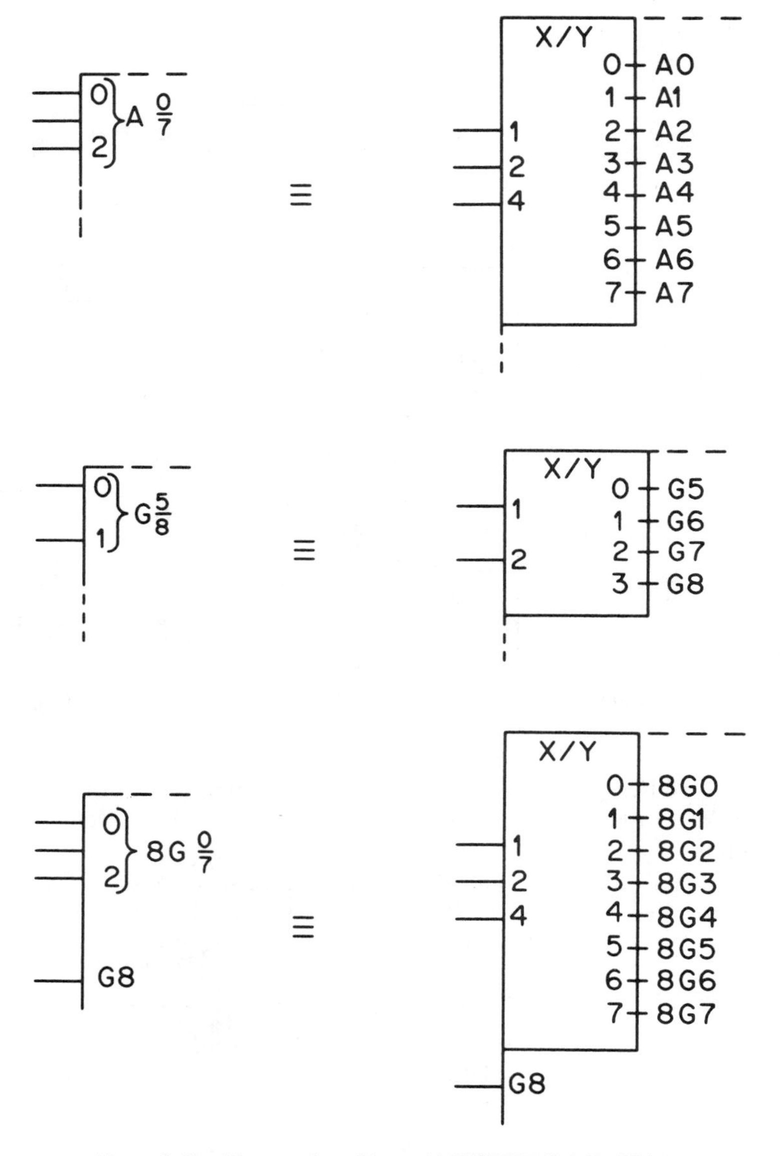

Figure 8.60 Bit grouping. (From ANSI/IEEE Std 91-1984.)

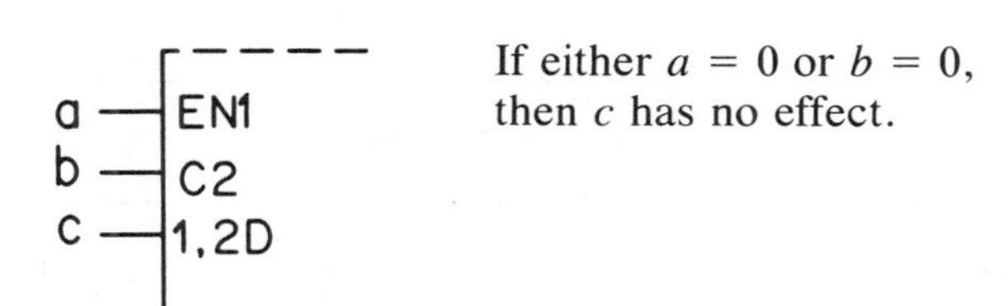

Figure 8.61 Order of input labels. (From ANSI/IEEE Std 91-1984.)

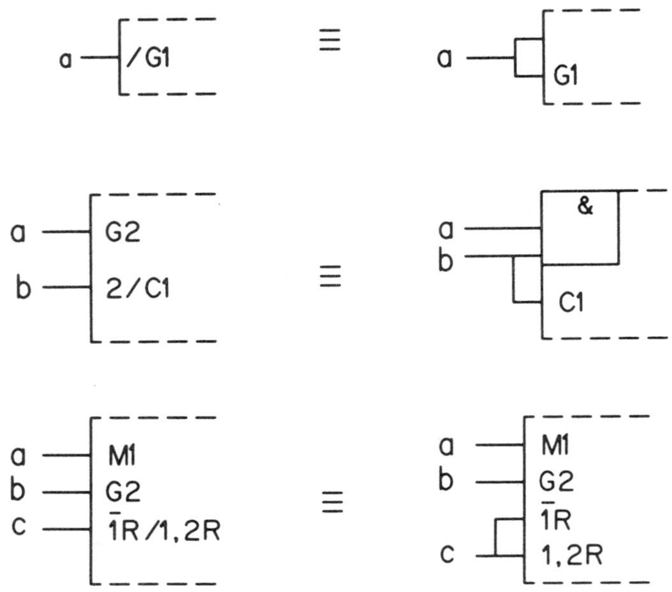

Figure 8.62 Input labels separated by a solidus (/). (From ANSI/IEEE Std 91-1984.)

When the inputs of a sequential element are disabled, the content of this element does not change and the outputs remain at their existing internal logic states.

Parentheses are used to set off a specific label or to factor labels. In this use labels can be factored as indicated in Fig. 8.63.

Note that brackets [ ] and braces { } are not the same as parentheses.

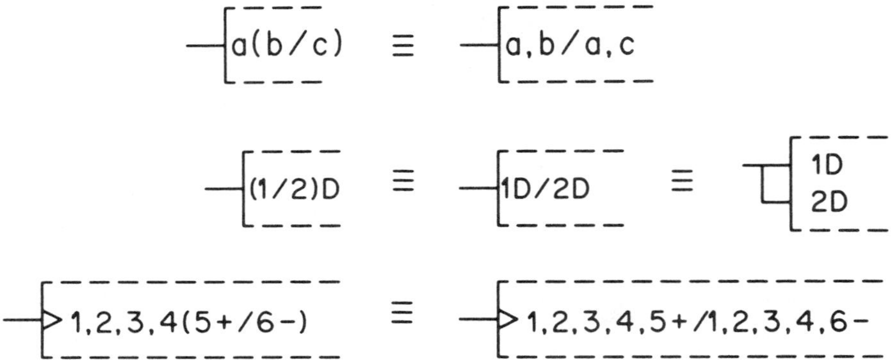

Figure 8.63 Parentheses, ( ), used to separate labels. (From ANSI/IEEE Std 91-1984.)

Output labels. Output labels are shown in the following order:

1. The postponed output symbol (⌐). In some cases the indications of the inputs to which it must be applied may precede this symbol.
2. The labels that determine or modify the internal logic state of the output. These labels are arranged to be read from left to right in the same order in which their effects must be applied.
3. A label indicating the effect of the output on the inputs and other outputs of the element.

Symbols for open-circuit, passive-pulldown, passive-pullup, tri-state, or specially amplified outputs are generally drawn adjacent to the output line.

As shown in Fig. 8.64, a comma is used to separate two adjacent identifying numbers of affecting inputs in a set of labels not already separated by a nonnumeric character.

Sets of labels separated by solidi (/) on the same output line indicate that the labels stand in an OR relationship to each other or have alternative modes of operation. This can also be indicated by sets of labels on different output lines connected together outside the outline. In either case, the output has but one internal state at a time. For OR relationships, the equivalencies in Fig. 8.65 hold.

Use of the multiline form with a negated output symbol or polarity symbol is shown in Fig. 8.66. Parentheses may also be used to separate or factor labels, as in Fig. 8.67.

As shown in Fig. 8.68, when the bit-grouping symbol for outputs is used and the sets of labels of all outputs grouped together differ only in the indications of the weights, the sets of labels are shown only once between the symbol replacing the * and the grouping symbol; in this usage the sets include the symbols for open-circuit, passive-pulldown, passive-pullup, three-state, or specially amplified outputs, but exclude the indications of the weights.

Inputs having inherent storage. The input label mD,* indicates that the input has inherent storage. (This does not apply to a D input.) In this label, "m" is the identifying numbers of the inputs that affect the storage operation and * is a symbol indicating the function of the stored input. When that symbol is a number, the comma after the "D" is omitted (see Fig. 8.69).

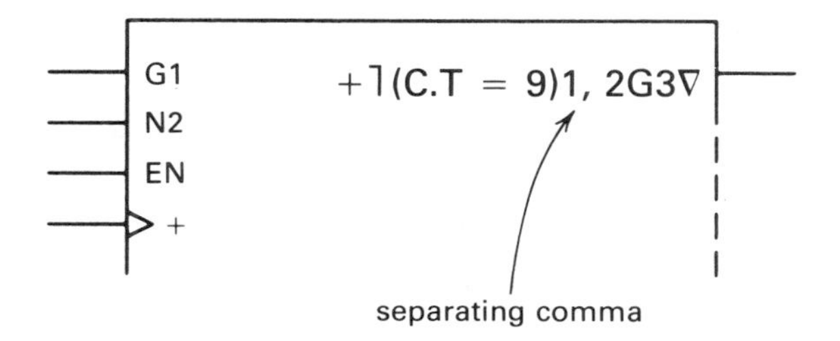

Figure 8.64 Separation of adjacent identifying numbers of affecting inputs with a comma. (From ANSI/IEEE Std 91-1984.)

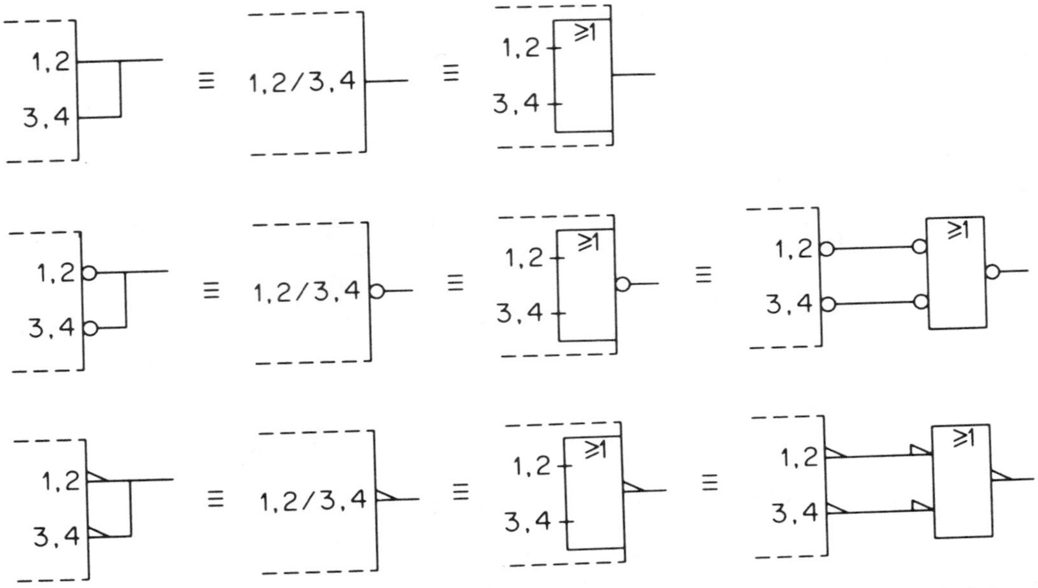

Figure 8.65 Output label equivalencies for OR relationships. (From ANSI/IEEE Std 91-1984.)

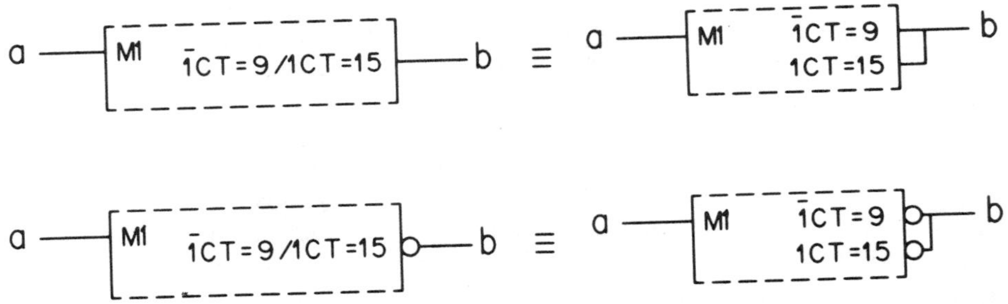

Figure 8.66 Multiline form used with negated output symbol or polarity symbol. (From ANSI/IEEE Std 91-1984.)

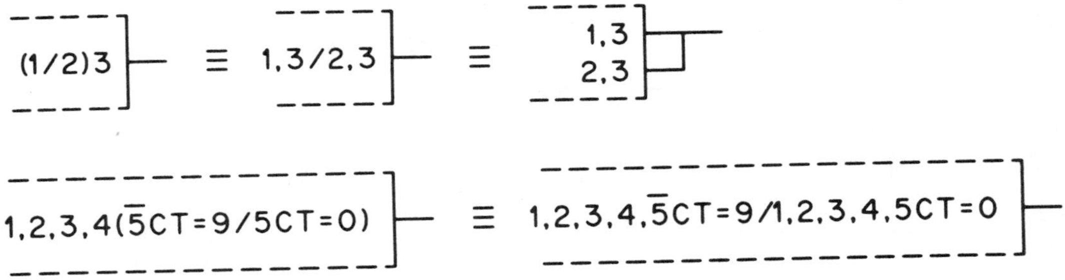

Figure 8.67 Parentheses used to separate or factor labels. (From ANSI/IEEE Std 91-1984.)

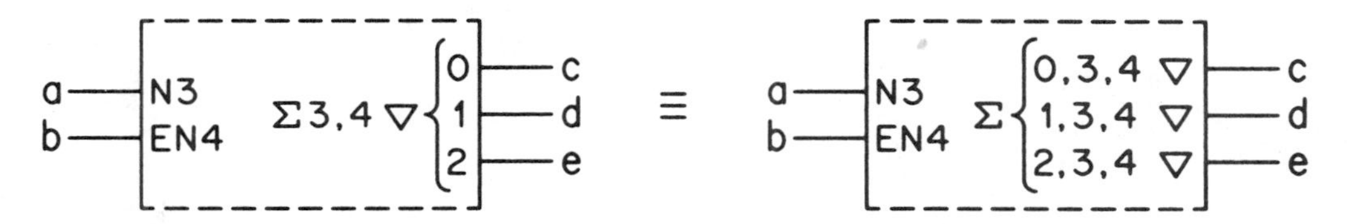

Figure 8.68 Sets of labels shown only once between the symbol replacing * and the grouping symbol. (From ANSI/IEEE Std 91-1984.)

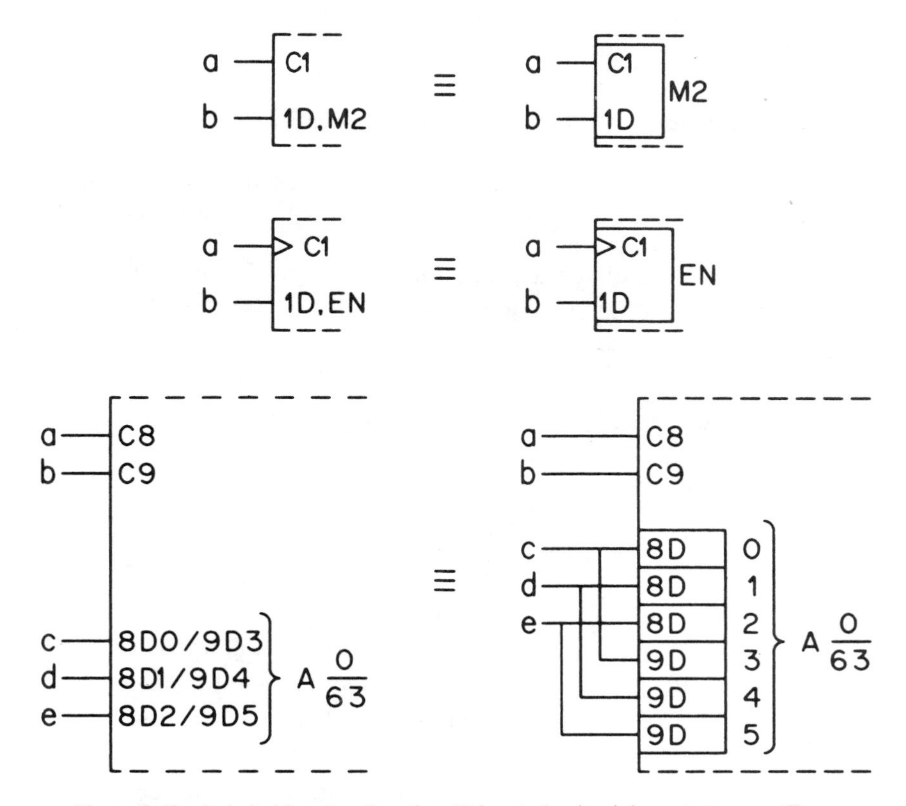

Figure 8.69 Labeled inputs other than D inputs having inherent storage. (From ANSI/IEEE Std 91-1984.)

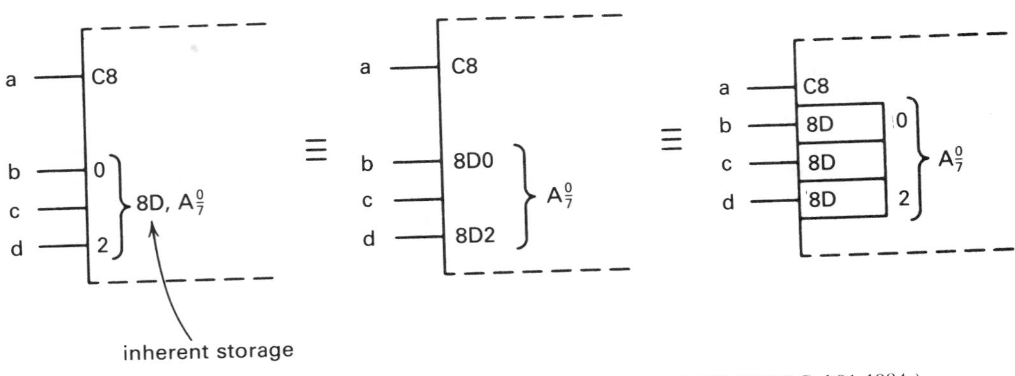

Figure 8.70 Inherent storage combined with bit grouping. (From ANSI/IEEE Std 91-1984.)

When used with the bit-grouping symbol, the inherent storage symbol (mD) is placed between the bit-grouping symbol and the letter "A" as in Fig. 8.70.

8.5 COMBINATIONAL AND SEQUENTIAL ELEMENTS

Figure 8.71 shows the qualifying symbols used inside the outline for combinational and sequential elements. These symbols are based on the internal logic states of the relevant inputs and outputs.

The symbol for a given function depends on its purpose in the system. Complementary representation is used to make diagrams easier to understand.

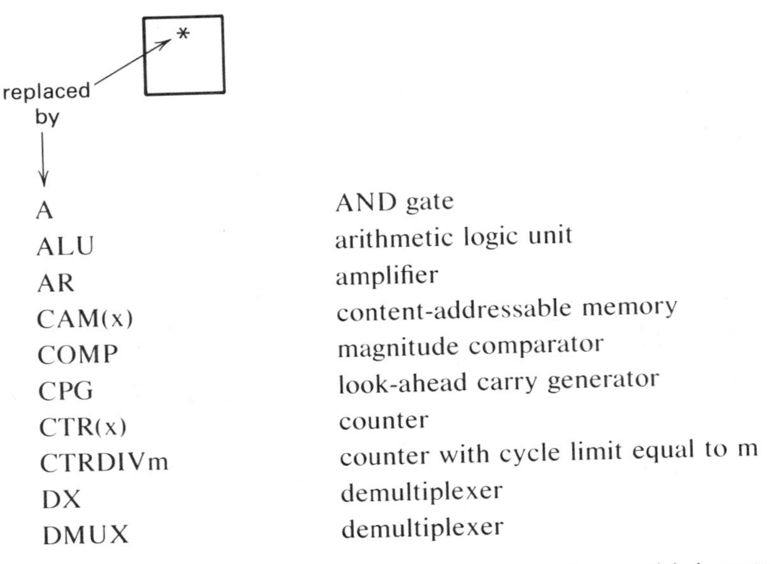

A	AND gate
ALU	arithmetic logic unit
AR	amplifier
CAM(x)	content-addressable memory
COMP	magnitude comparator
CPG	look-ahead carry generator
CTR(x)	counter
CTRDIVm	counter with cycle limit equal to m
DX	demultiplexer
DMUX	demultiplexer

Figure 8.71 Summary of combinational and sequential elements.

FF	flip flop
FIFO(x)	first-in, first-out memory
G	oscillator
MOD 2	odd function
MUX	multiplexer
NOT	negator
OE	exclusive OR
OR	OR gate
OSC	oscillator
P-Q	subtractor
PROM(x)	programmable read-only memory
RAM(x)	random-access memory
RCTR(x)	ripple counter modulo to the power (x)
RCTRDIVm	ripple counter with cycle length equal to m
ROM(x)	read-only memory
SRG(x)	shift register
SS	single-shot
ST	Schmitt trigger
TD (t)	time delay (fixed)
X/Y	signal-level converter
$X \rightarrow Y$	coder
1	buffer
&	AND gate
=	input identity (logic identity element)
=1	exclusive-OR
$\geqslant 1$	OR
Σ	adder
π	multiplier
m = number $\geqslant$ m	logic threshold element
m = number $\geqq$ m	logic threshold element (obsolete)

Figure 8.71 (*Continued*)

For example, an AND gate may be shown as an OR gate but with negated inputs and outputs.

8.5.1 Arithmetic Elements

Arithmetic elements are shown in Figs. 8.72 through 8.77.

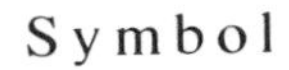

Description

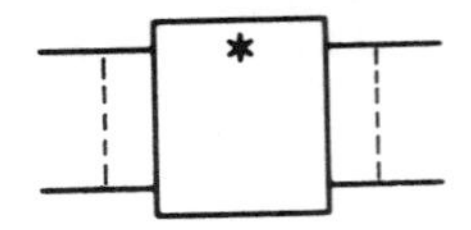

Arithmetic element, general symbol

The * is replaced by one of the following qualifying symbols:

ALU	Arithmetic-logic unit
COMP	Magnitude comparator A cascadable comparator is assumed to implement a portion of a comparison that proceeds from lower to higher order, unless otherwise indicated, for example, by "[H-L]" placed below the "COMP" qualifying symbols.
CPG	Look-ahead carry generator (carry, propagate, and generate)
P-Q	Subtractor
Σ	Adder
π	Multiplier

Figure 8.72 General symbol for arithmetic element. (From ANSI/IEEE Std 91-1984.)

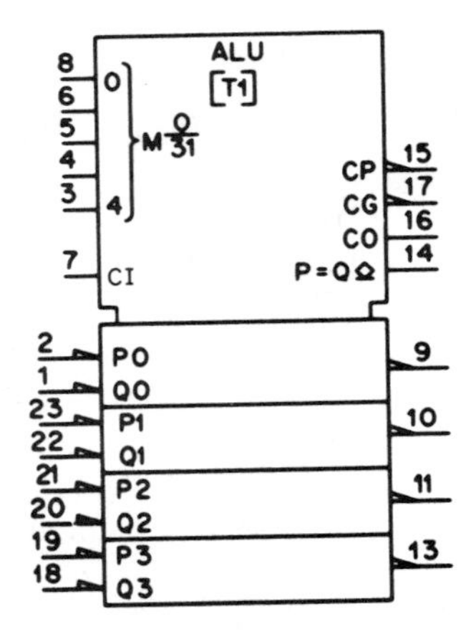

T1 refers to supplementary information.

Figure 8.73 Four-bit arithmetic logic unit. (From ANSI/IEEE Std 91-1984.)

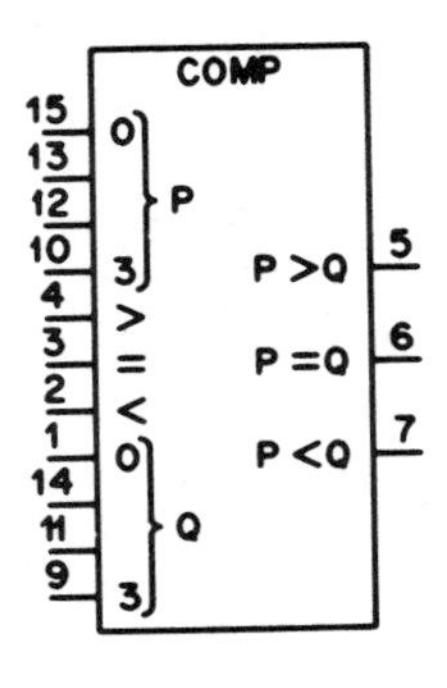

Figure 8.74 Magnitude comparator, low-order to high-order, with cascading inputs, 4-bit. (From ANSI/IEEE Std 91-1984.)

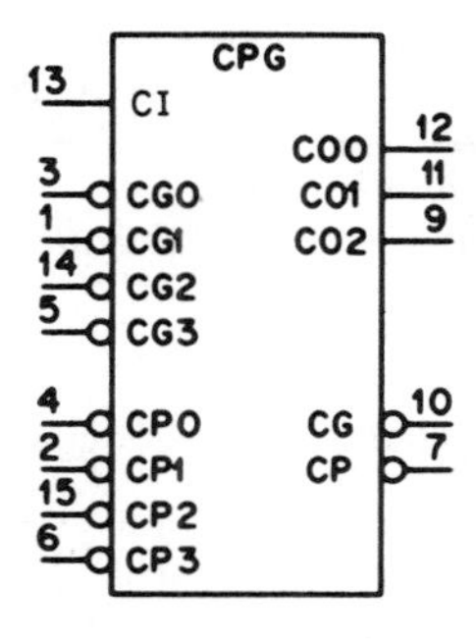

Figure 8.75 Look-ahead carry generator, 4-bit. (From ANSI/IEEE Std 91-1984.)

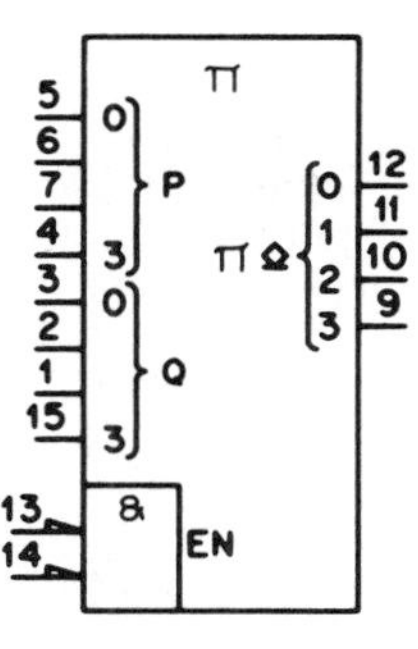

Figure 8.76 Multiplier, 4-bit parallel, generating the four most significant bits of the 8-bit product. (From ANSI/IEEE Std 91-1984.)

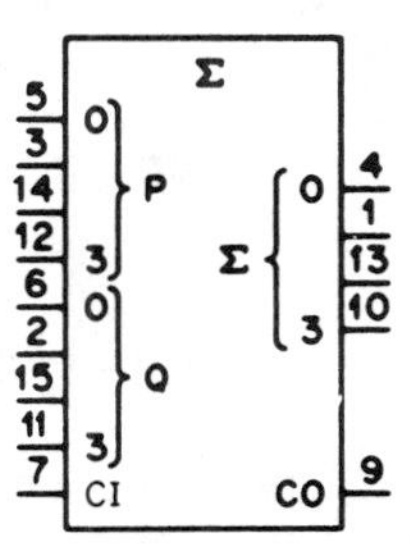

Figure 8.77 Four-bit full adder. (From ANSI/IEEE Std 91-1984.)

8.5.2 Astable Elements

The general symbols for astable elements are shown in Fig. 8.78. In these symbols, the G is the qualifying symbol for a generator. However, if the waveform is evident, the letter G may not be shown. Figure 8.79 shows an application.

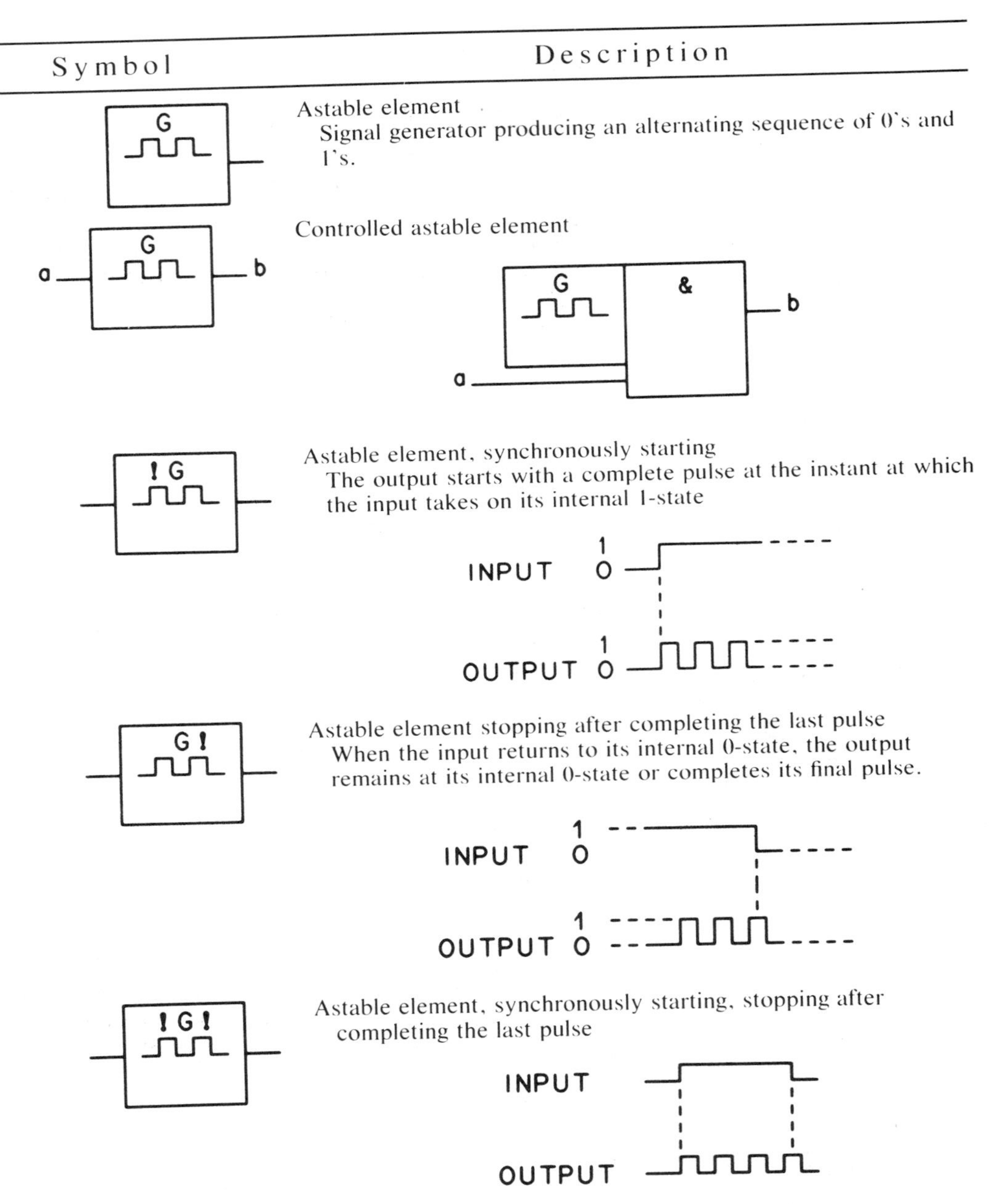

Figure 8.78 Astable elements. (From ANSI/IEEE Std 91-1984.)

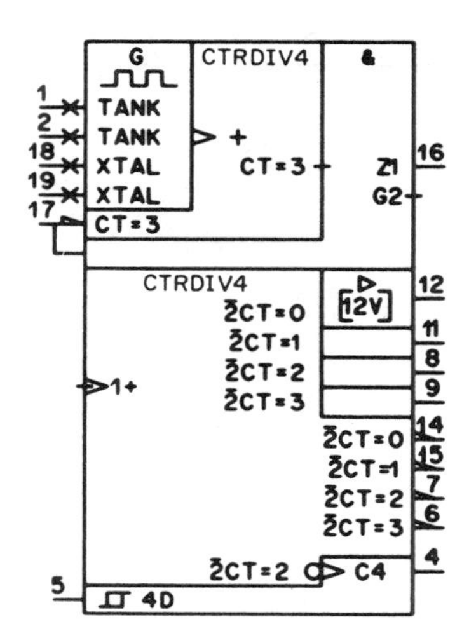

For the use of CTRDIV4, see Fig. 8.100.

Figure 8.79 Four-phase clock generator/driver. (From ANSI/IEEE Std 91-1984.)

8.5.3 Bistable Elements

Bistable elements are shown in Figs. 8.80 and 8.81. The function of a bistable element such as a flip-flop is indicated by the qualifying symbols associated with its inputs and outputs. In a bistable element the internal logic states of the outputs represent the content of the element.

For a bistable element controlled by a C input, refer to Fig. 8.80. The inputs affected by the C input in edge-triggered, pulse-triggered, and data-lock-out bistables are assumed to be stable during the time that the C input stands at its internal 1-state. The symbols in Fig. 8.80 are also used to indicate whether shift

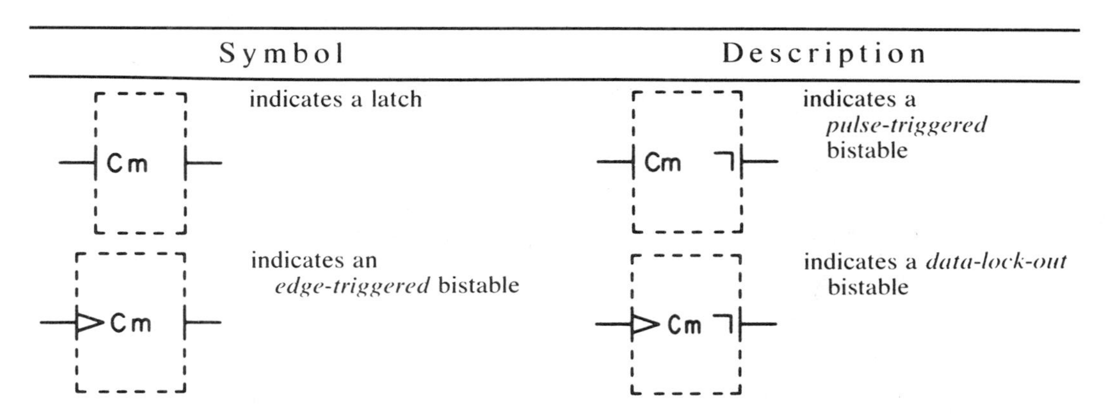

Figure 8.80 Basic C-input bistable elements. (From ANSI/IEEE Std 91-1984.)

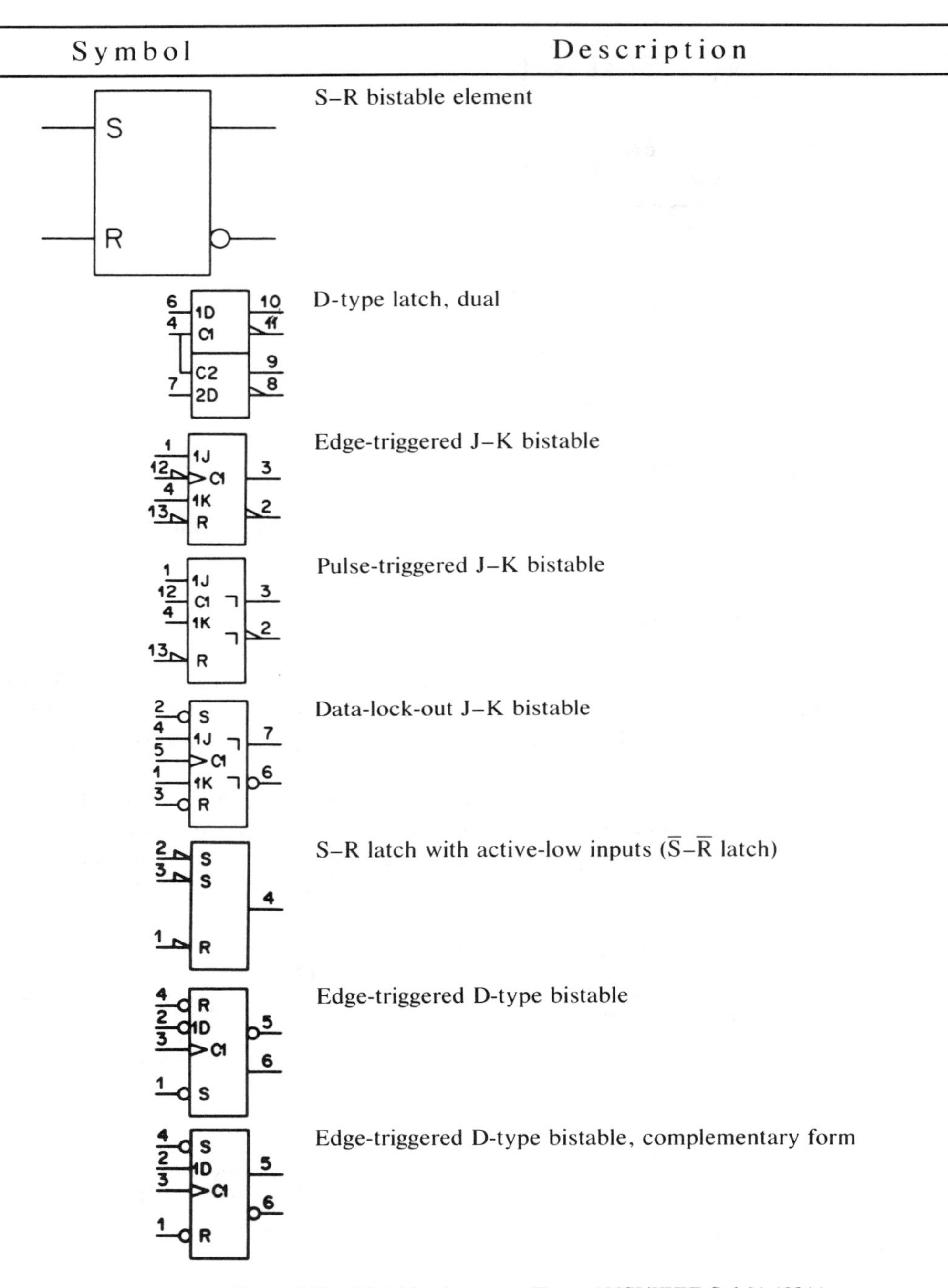

Figure 8.81 Bistable elements. (From ANSI/IEEE Std 91-1984.)

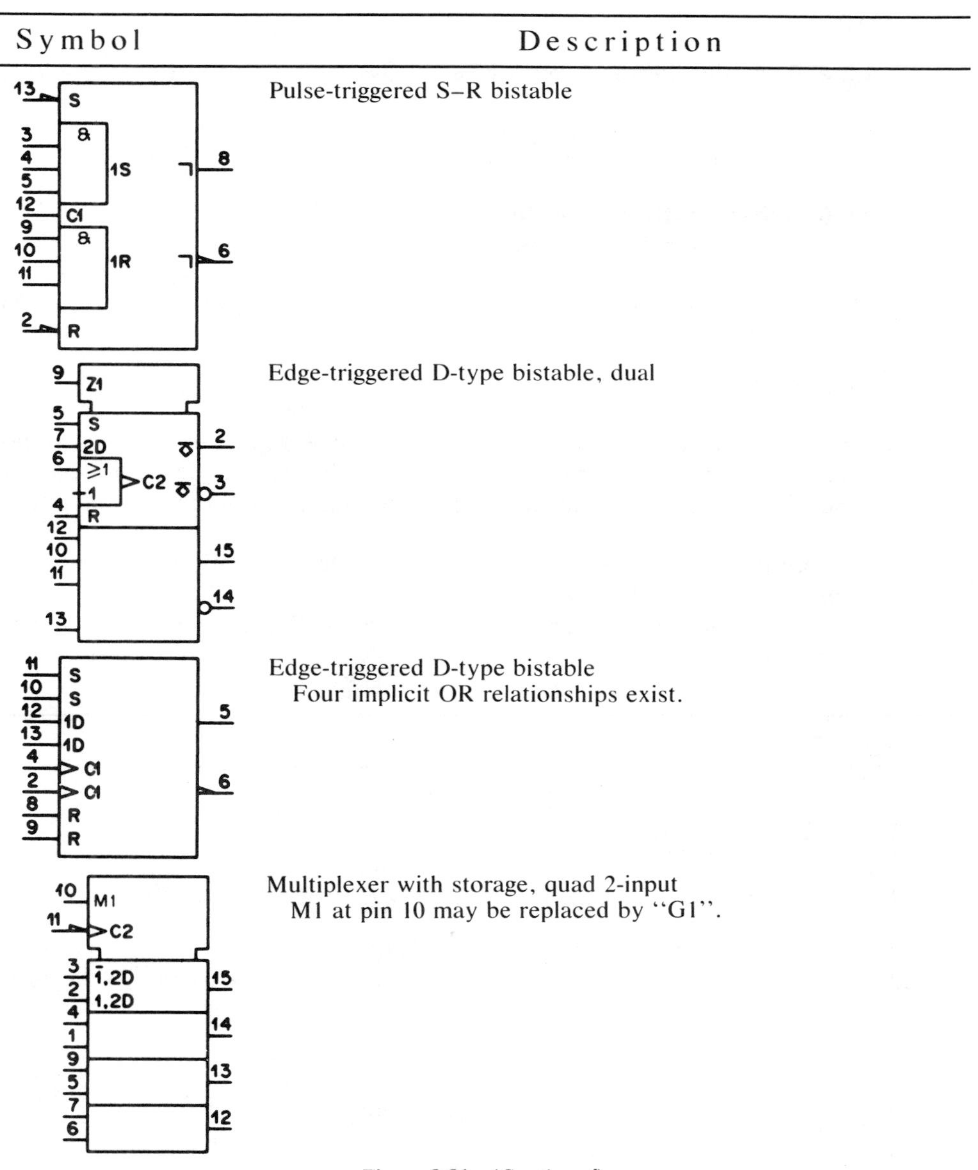

Symbol	Description
	Pulse-triggered S–R bistable
	Edge-triggered D-type bistable, dual
	Edge-triggered D-type bistable Four implicit OR relationships exist.
	Multiplexer with storage, quad 2-input M1 at pin 10 may be replaced by "G1".

Figure 8.81 *(Continued)*

registers and counters are edge-triggered, pulse-triggered, or the data-lock-out type.

8.5.4 Bistable Elements with Special Switching Properties

The symbols in Fig. 8.82 indicate what happens to the outputs of a bistable element when the supply is switched on.

8.5.5 Buffers with Special Amplification, Drivers, Receivers, and Bilateral Switches

The symbol ▷ used for amplifiers points in the direction of signal flow. Notice in Fig. 8.83 that this symbol may be combined with other qualifying symbols.

8.5.6 Coders

When code conversion is indicated (see Fig. 8.84) the internal logic states of the inputs determine an internal value based on the input code. This value is reproduced by the internal logic states of the outputs, as directed by the output code.

Inputs. Numbers on the inputs indicate the relationships between the *internal logic states* of the inputs and the *internal value*. The internal value in this case equals the sum of the weights associated with those inputs that stand at their internal 1-states.

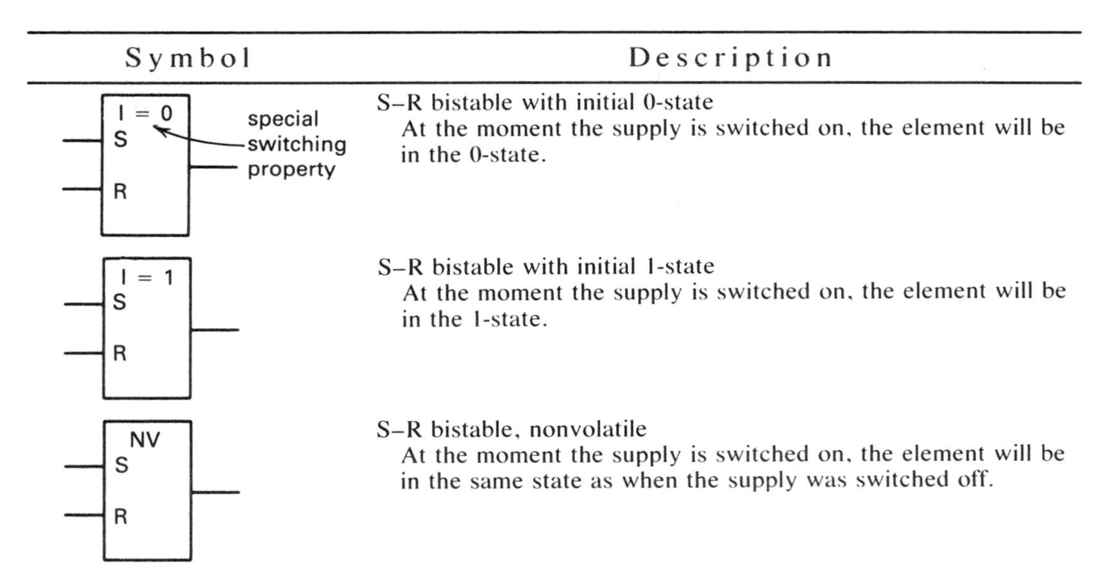

Symbol	Description
I = 0 / S / R (special switching property)	S–R bistable with initial 0-state At the moment the supply is switched on, the element will be in the 0-state.
I = 1 / S / R	S–R bistable with initial 1-state At the moment the supply is switched on, the element will be in the 1-state.
NV / S / R	S–R bistable, nonvolatile At the moment the supply is switched on, the element will be in the same state as when the supply was switched off.

Figure 8.82 Bistable elements with special switching properties. (From ANSI/IEEE Std 91-1984.)

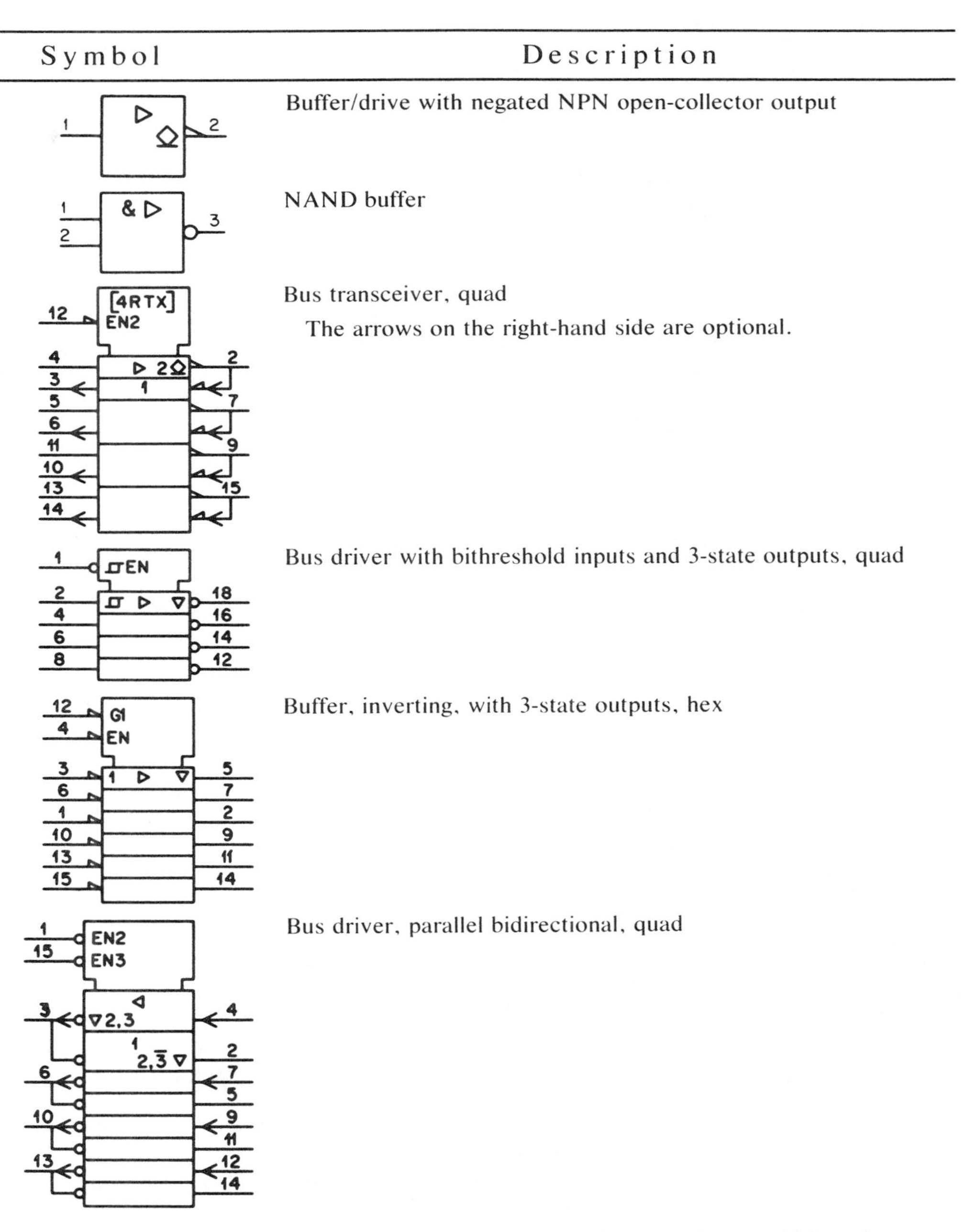

Symbol	Description
	Buffer/drive with negated NPN open-collector output
	NAND buffer
	Bus transceiver, quad The arrows on the right-hand side are optional.
	Bus driver with bithreshold inputs and 3-state outputs, quad
	Buffer, inverting, with 3-state outputs, hex
	Bus driver, parallel bidirectional, quad

Figure 8.83 Buffers with special amplification, drivers, receivers, and bilateral switches. (From ANSI/IEEE Std 91-1984.)

Symbol	Description
	Line receiver, dual
	Bus driver, bidirectional, 8-bit parallel
	Bilateral switch
	CMOS transmission gate

Figure 8.83 (*Continued*)

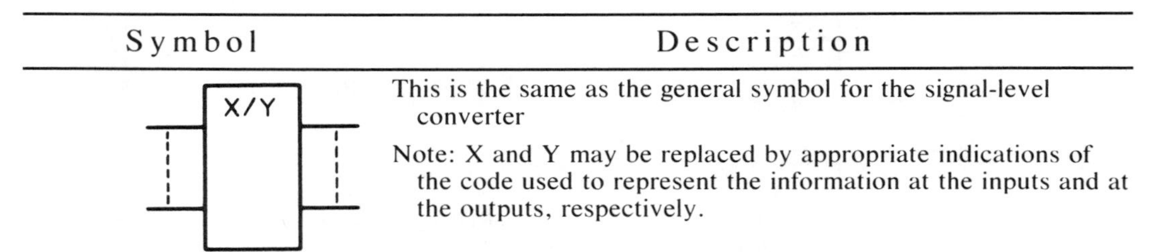

Symbol	Description
	This is the same as the general symbol for the signal-level converter Note: X and Y may be replaced by appropriate indications of the code used to represent the information at the inputs and at the outputs, respectively.

Figure 8.84 General symbol for coder or code converter. (From ANSI/IEEE Std 91-1984.)

In an alternative technique (Fig. 8.85) X is replaced by an indication of the input code; characters that refer to this code are used as labels on the inputs.

Outputs. A list of numbers on each output indicates the relationships between the *internal logic states* of the outputs and the *internal value*. The num-

Output d stands at its internal 1-state for the following combinations of internal logic states at inputs a, b, and c:

$a = 1$ $b = 0$ $c = 0$
$a = 0$ $b = 0$ $c = 1$

Output h stands at its internal 1-state for the following combination of internal states at inputs a, b, and c:

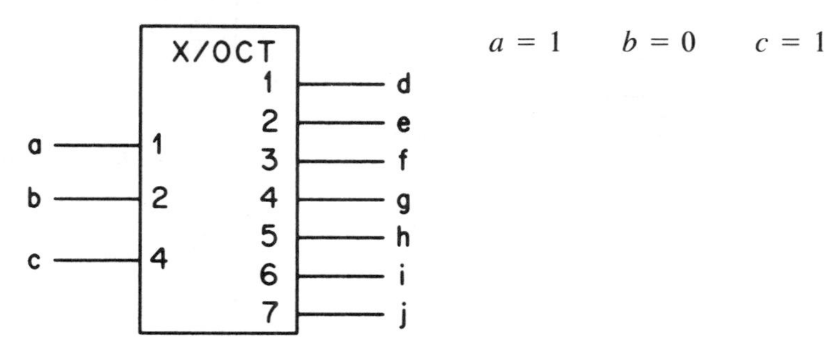

Output i stands at its internal 1-state for the following combinations of internal states at inputs a, b, c, and d:

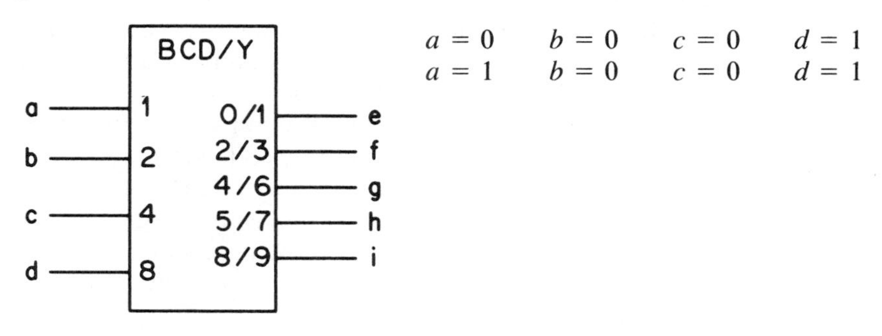

Figure 8.85 Coder or code converter. (From ANSI/IEEE Std 91-1984.)

If input j stands at its internal 1-state, outputs k and n stand at their internal 1-states.

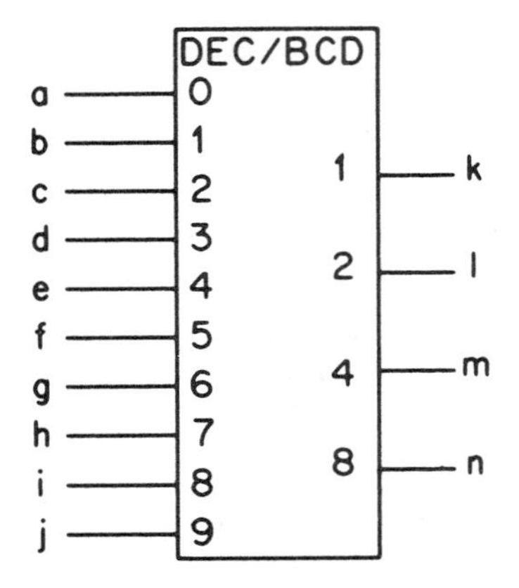

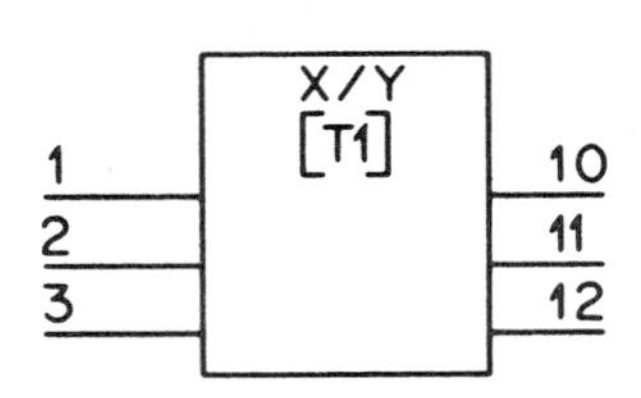

TABLE T1

INPUTS 1	2	3	OUTPUTS 10	11	12
0	0	0	1	0	0
0	0	1	0	0	0
0	1	0	0	1	0
0	1	1	0	0	0
1	0	0	0	0	0
1	0	1	0	0	0
1	1	0	0	0	1
1	1	1	0	0	0

Figure 8.85 *(Continued)*

bers, which are separated by solidi (/), represent those internal values that lead to the internal 1-state of that output. Two numbers separated by three dots, for example, 3 . . . 8, indicate that a continuous range of internal values produces the internal 1-state of an output. The two numbers are the beginning and the end of the range; in this case, 3 . . . 8 = 3/4/5/6/7/8.

In an alternative technique, Y is replaced with an indication of the output code; the characters that refer to this code are used as labels on the outputs.

Figure 8.86 shows the symbol for a BCD-to-decimal code converter.

In some cases the general symbol is used together with a reference to a table showing the relationships between the inputs and outputs. The correspondence between inputs (outputs) and the columns in the table is shown by pin numbers (see Figs. 8.87 and 8.88).

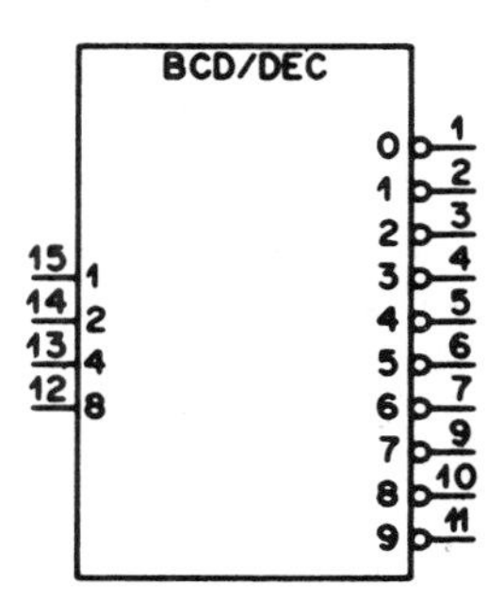

Figure 8.86 Code converter, BCD-to-decimal. (From ANSI/IEEE Std 91-1984.)

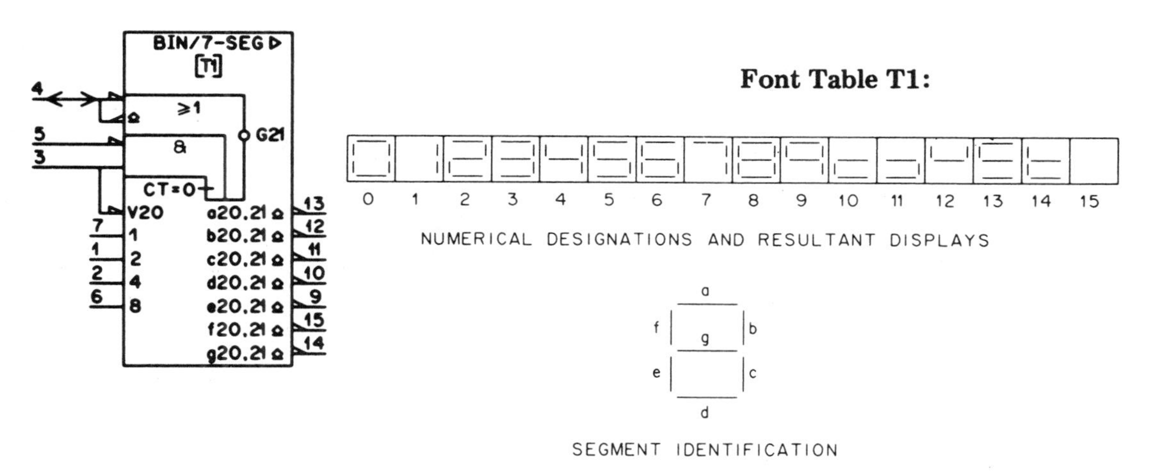

Because the direction of signal flow is implied by the polarity indicators, the arrow heads at pin 4 are optional.

Figure 8.87 Binary-to-seven segment decoder/driver. (From ANSI/IEEE Std 91-1984.)

8.5.7 Combinational Elements

The qualifying symbol for a basic combinational element indicates the number of inputs that must take on the internal 1-state in order for the outputs to take on their internal 1-states (see Fig. 8.89).

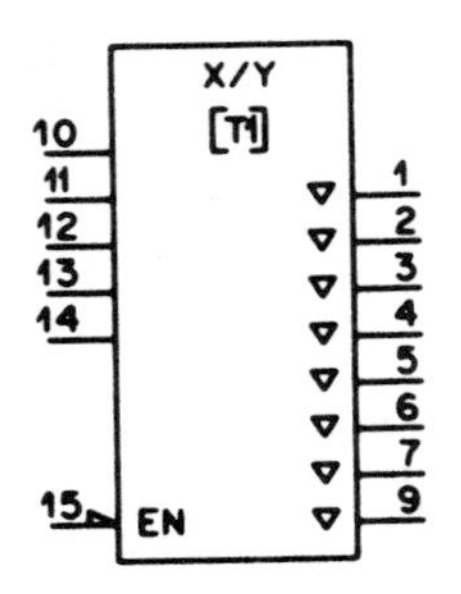

"T1" refers to a table (or to Boolean equations) showing the logic function of the device. For instance:

INPUTS					OUTPUTS (WITH DEVICE ENABLED)							
14	13	12	11	10	9	7	6	5	4	3	2	1
L	L	L	X	X	H	L	H	L	L	L	L	L
L	L	H	L	L	H	L	H	L	L	L	L	L
L	L	H	L	H	H	L	H	L	L	L	L	H
L	L	H	H	X	H	H	L	L	L	L	L	H
L	H	L	L	L	H	L	H	L	L	L	L	L
L	H	L	L	H	H	L	H	L	L	L	H	L
L	H	L	H	X	H	H	L	L	L	L	H	L
L	H	H	L	L	H	L	H	L	L	L	L	L
L	H	H	L	H	H	L	H	L	L	H	L	L
L	H	H	H	X	H	H	L	L	L	H	L	L
H	L	L	X	X	L	L	H	L	L	L	L	L
H	L	H	X	X	L	L	H	L	H	L	L	H
H	H	L	X	X	L	L	H	H	L	L	H	L
H	H	H	X	X	L	L	H	H	H	H	L	L

X indicates irrelevant (don't care)

Figure 8.88 Coder in which arbitrary combinatorial relationships between inputs and outputs are implemented in a PROM (or a ROM). (From ANSI/IEEE Std 91-1984.)

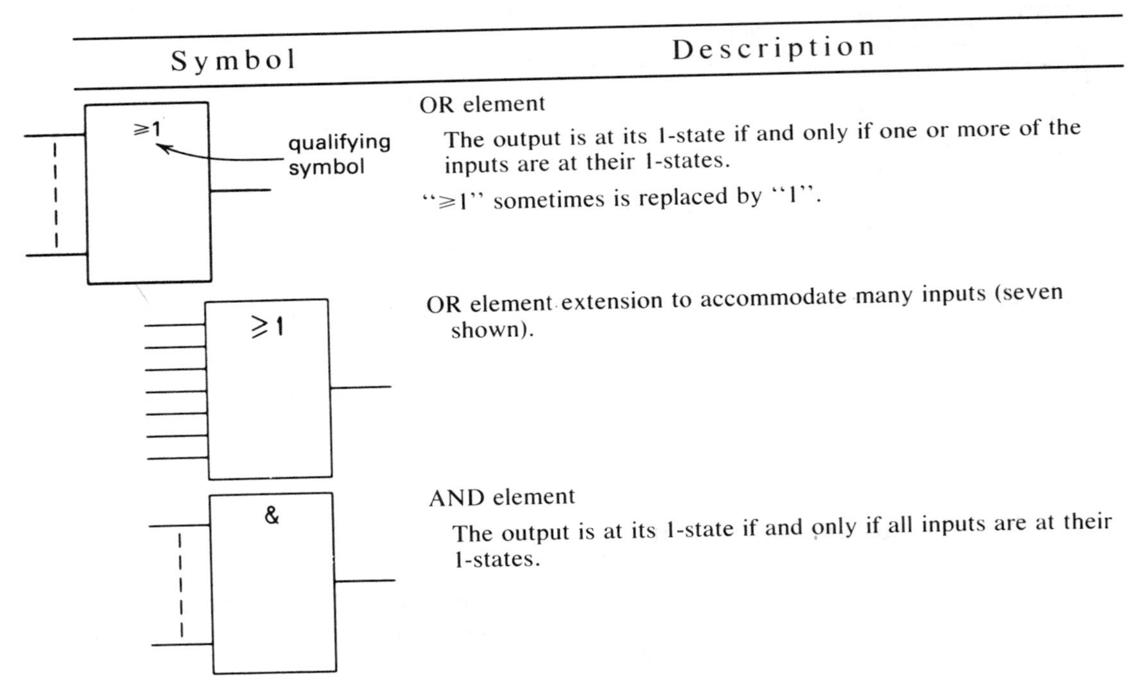

OR element

The output is at its 1-state if and only if one or more of the inputs are at their 1-states.

"≥1" sometimes is replaced by "1".

OR element extension to accommodate many inputs (seven shown).

AND element

The output is at its 1-state if and only if all inputs are at their 1-states.

Figure 8.89 Basic combinatorial elements. (From ANSI/IEEE Std 91-1984.)

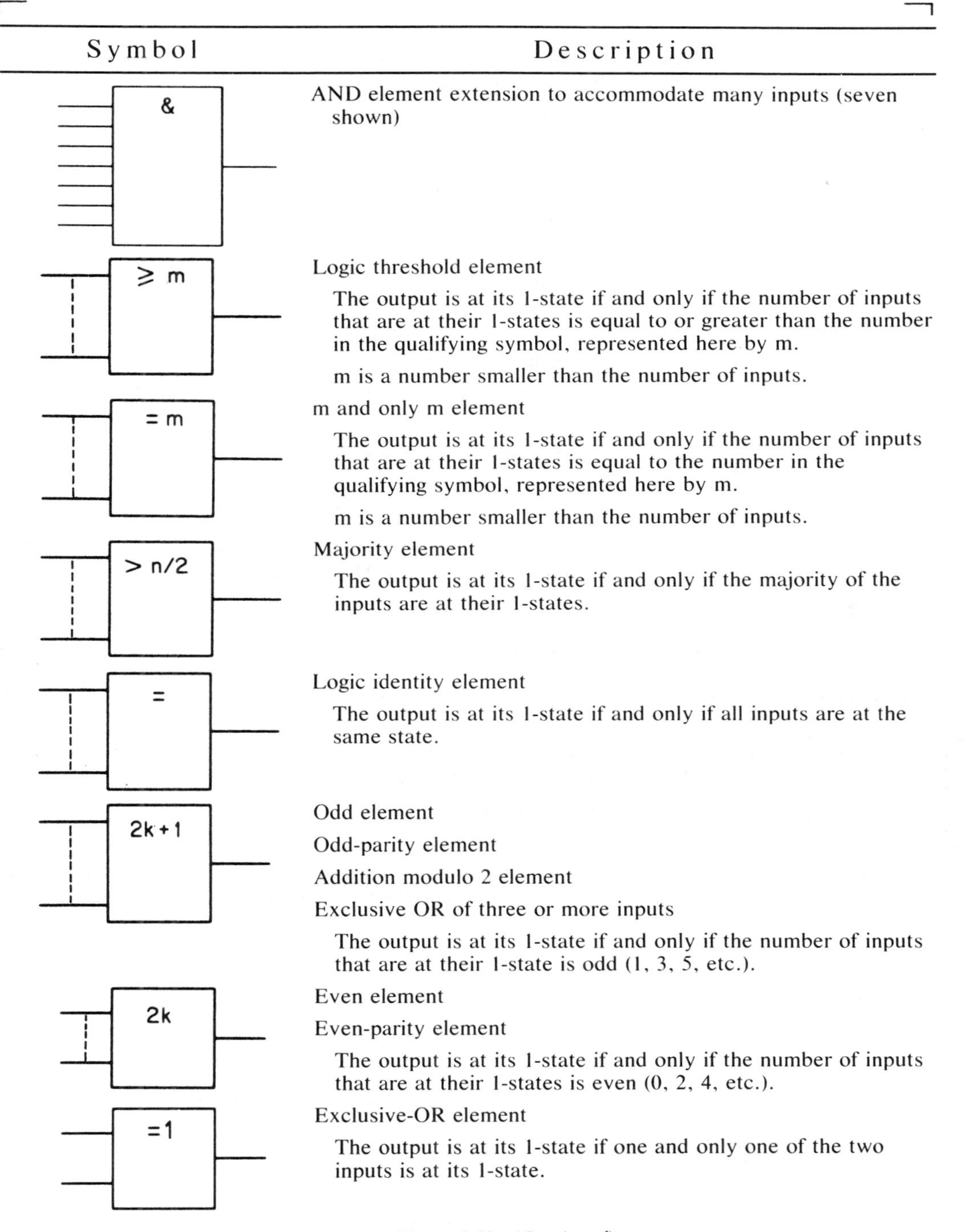

Symbol	Description
&	AND element extension to accommodate many inputs (seven shown)
⩾ m	Logic threshold element The output is at its 1-state if and only if the number of inputs that are at their 1-states is equal to or greater than the number in the qualifying symbol, represented here by m. m is a number smaller than the number of inputs.
= m	m and only m element The output is at its 1-state if and only if the number of inputs that are at their 1-states is equal to the number in the qualifying symbol, represented here by m. m is a number smaller than the number of inputs.
> n/2	Majority element The output is at its 1-state if and only if the majority of the inputs are at their 1-states.
=	Logic identity element The output is at its 1-state if and only if all inputs are at the same state.
2k + 1	Odd element Odd-parity element Addition modulo 2 element Exclusive OR of three or more inputs The output is at its 1-state if and only if the number of inputs that are at their 1-state is odd (1, 3, 5, etc.).
2k	Even element Even-parity element The output is at its 1-state if and only if the number of inputs that are at their 1-states is even (0, 2, 4, etc.).
=1	Exclusive-OR element The output is at its 1-state if one and only one of the two inputs is at its 1-state.

Figure 8.89 *(Continued)*

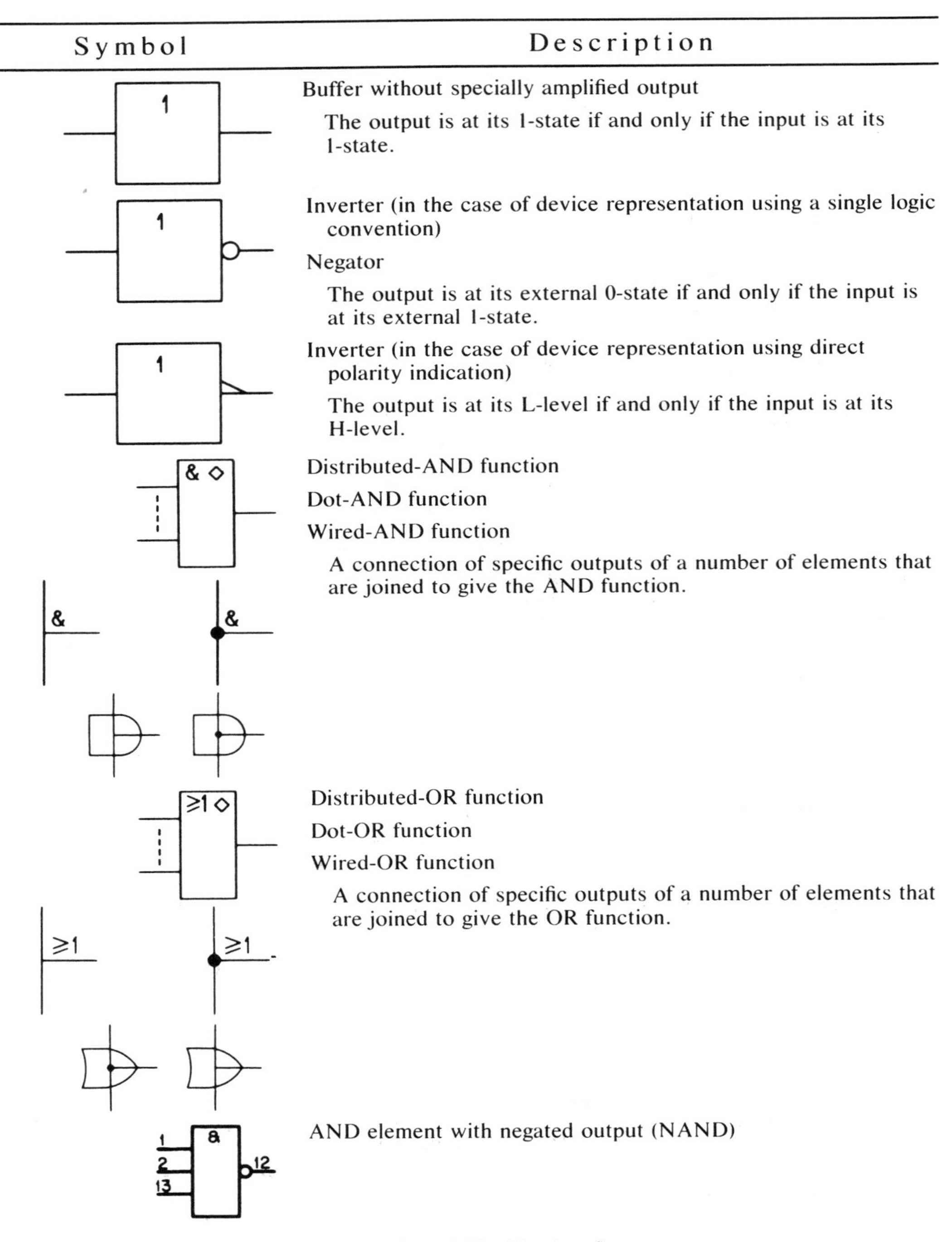

Symbol	Description
	Buffer without specially amplified output The output is at its 1-state if and only if the input is at its 1-state.
	Inverter (in the case of device representation using a single logic convention) Negator The output is at its external 0-state if and only if the input is at its external 1-state.
	Inverter (in the case of device representation using direct polarity indication) The output is at its L-level if and only if the input is at its H-level.
	Distributed-AND function Dot-AND function Wired-AND function A connection of specific outputs of a number of elements that are joined to give the AND function.
	Distributed-OR function Dot-OR function Wired-OR function A connection of specific outputs of a number of elements that are joined to give the OR function.
	AND element with negated output (NAND)

Figure 8.89 (*Continued*)

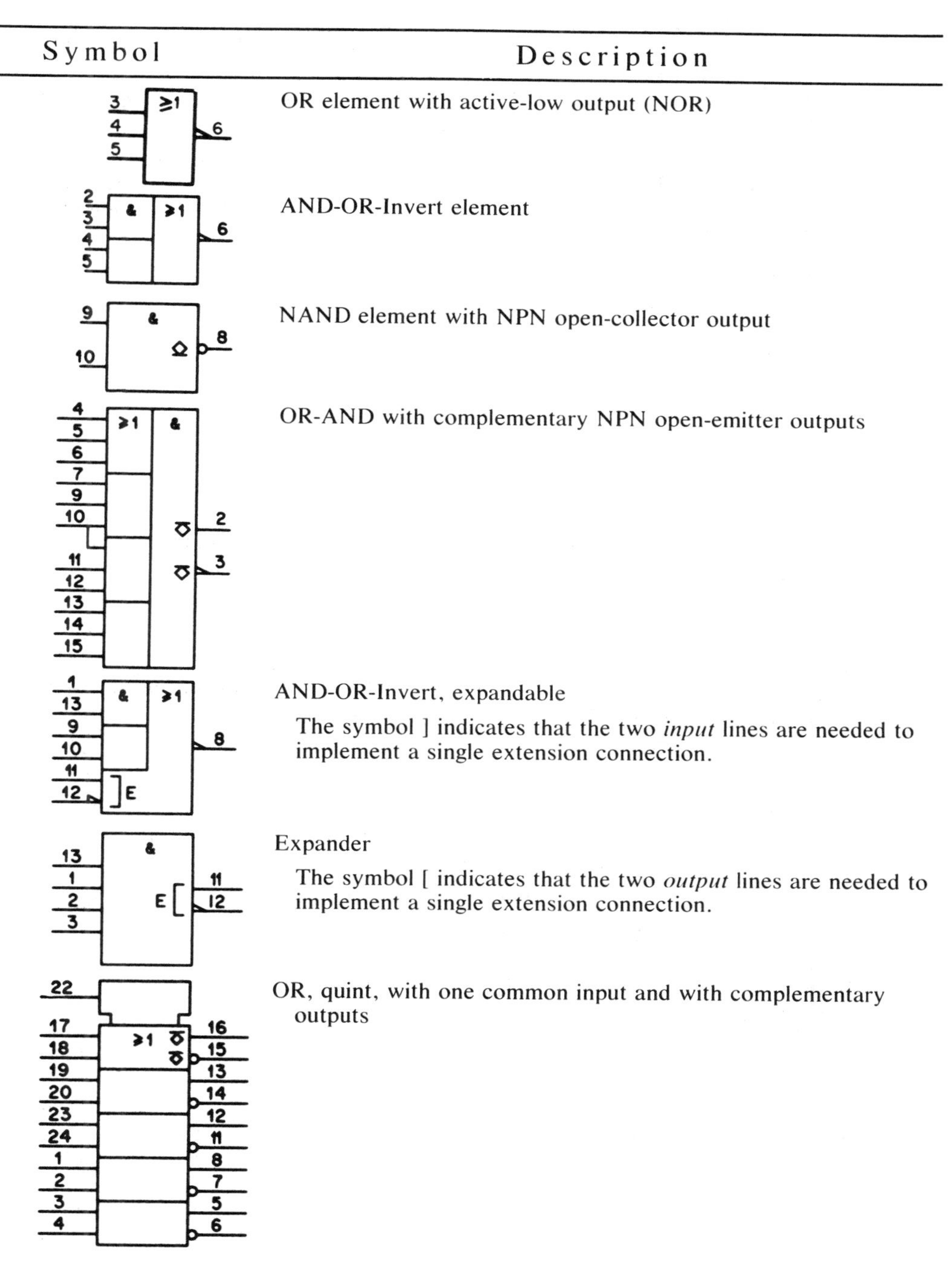

Symbol	Description
	OR element with active-low output (NOR)
	AND-OR-Invert element
	NAND element with NPN open-collector output
	OR-AND with complementary NPN open-emitter outputs
	AND-OR-Invert, expandable The symbol] indicates that the two *input* lines are needed to implement a single extension connection.
	Expander The symbol [indicates that the two *output* lines are needed to implement a single extension connection.
	OR, quint, with one common input and with complementary outputs

Figure 8.89 (*Continued*)

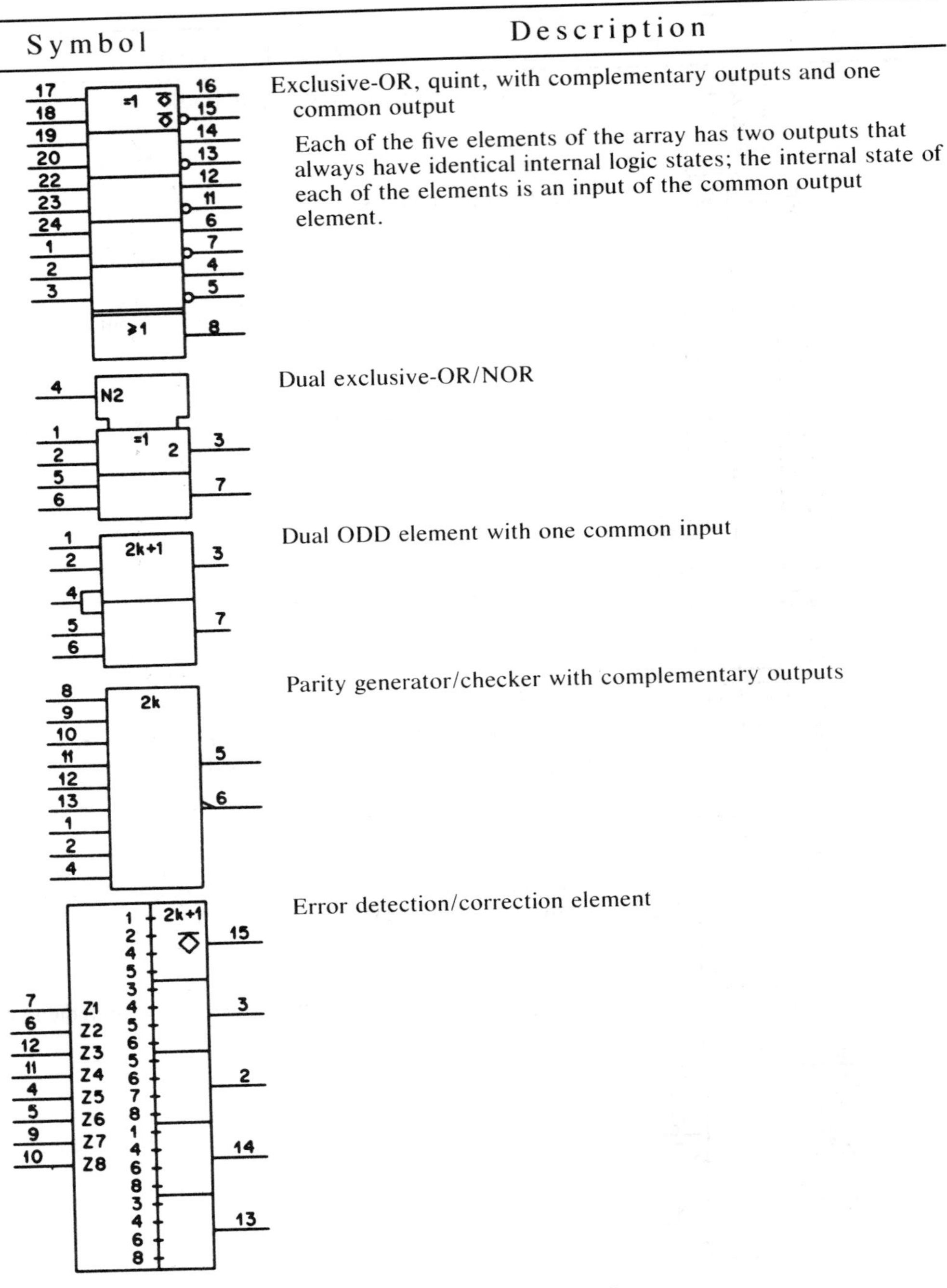

Symbol	Description
	Exclusive-OR, quint, with complementary outputs and one common output Each of the five elements of the array has two outputs that always have identical internal logic states; the internal state of each of the elements is an input of the common output element.
	Dual exclusive-OR/NOR
	Dual ODD element with one common input
	Parity generator/checker with complementary outputs
	Error detection/correction element

Figure 8.89 *(Continued)*

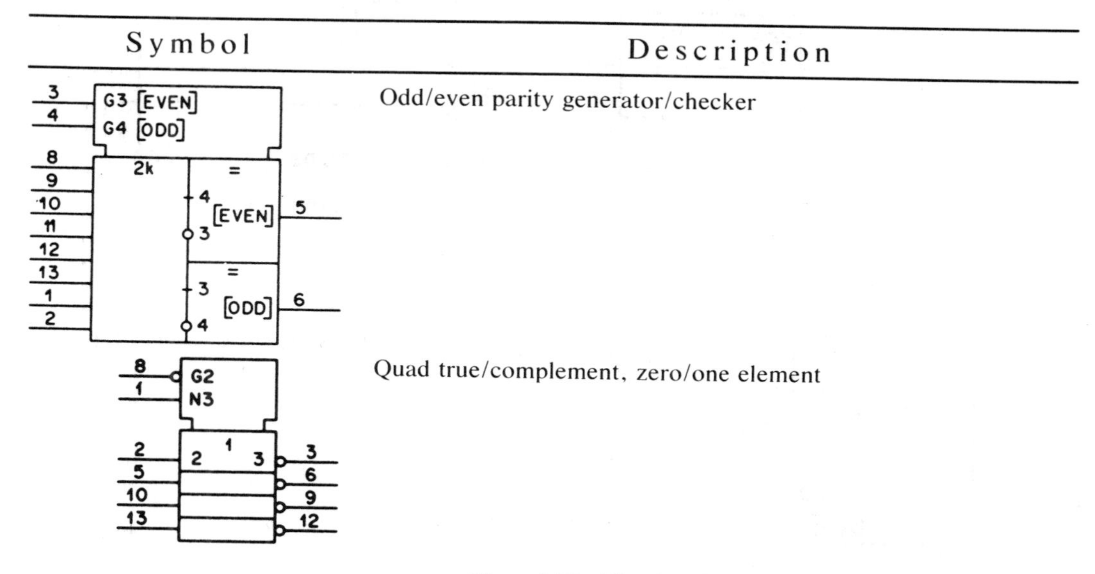

Figure 8.89 (*Continued*)

8.5.8 Delay Elements

Delay elements are shown in Figs. 8.90 and 8.91.

8.5.9 Memories

Memory symbols are shown in Figs. 8.92 through 8.96.

8.5.10 Monostable Elements

Monostable elements are shown in Fig. 8.97.

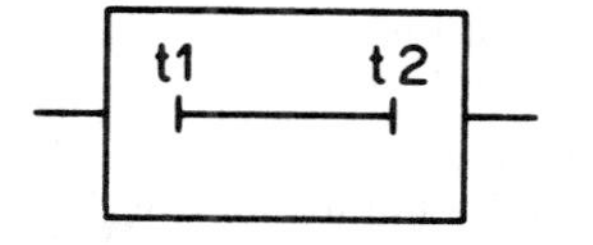

A transition from the internal 0-state to the internal 1-state at the output occurs after a delay of t1, with reference to the same transition at the input. The transition from the internal 1-state to the internal 0-state at the output occurs after a delay of t2 with reference to the same transition at the input.

Note: t1 and t2 may be replaced by the actual delays, expressed in seconds, word units, or digit units, and may be placed inside or outside the outline. If only one value is shown, the two values are equal.

Figure 8.90 General symbol for delay element with specified delay times. (From ANSI/IEEE Std 91-1984.)

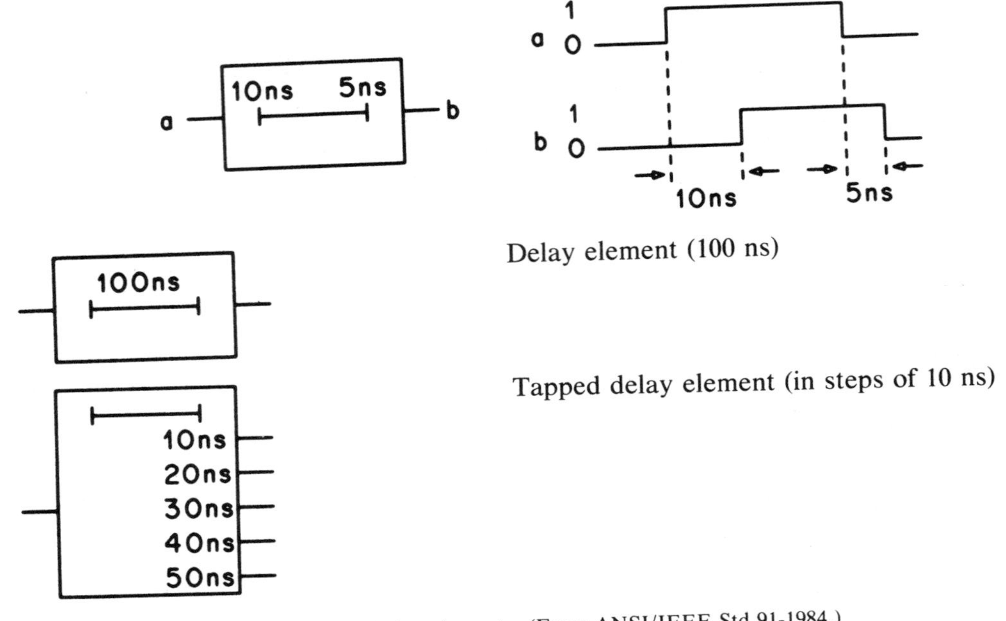

Figure 8.91 Delay elements. (From ANSI/IEEE Std 91-1984.)

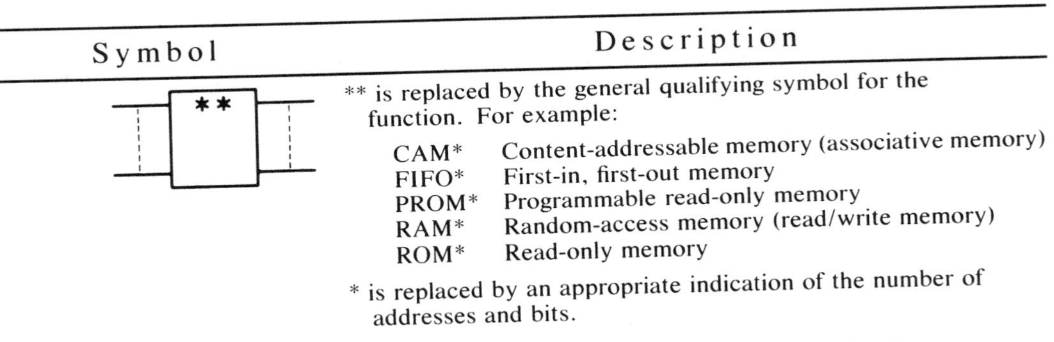

Symbol	Description
**	** is replaced by the general qualifying symbol for the function. For example: CAM* Content-addressable memory (associative memory) FIFO* First-in, first-out memory PROM* Programmable read-only memory RAM* Random-access memory (read/write memory) ROM* Read-only memory * is replaced by an appropriate indication of the number of addresses and bits.

Figure 8.92 General symbol for memory. (From ANSI/IEEE Std 91-1984.)

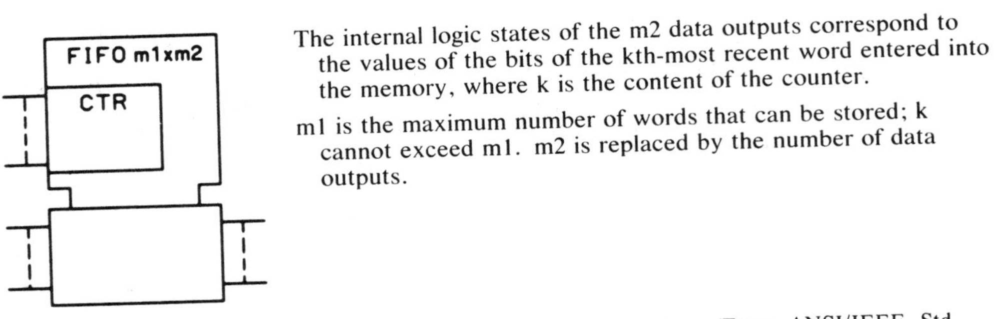

The internal logic states of the m2 data outputs correspond to the values of the bits of the kth-most recent word entered into the memory, where k is the content of the counter.

m1 is the maximum number of words that can be stored; k cannot exceed m1. m2 is replaced by the number of data outputs.

Figure 8.93 General symbol for first-in, first-out memory. (From ANSI/IEEE Std 91-1984.)

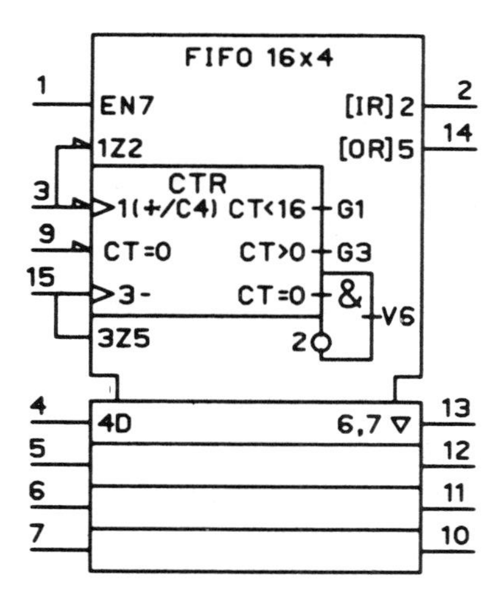

Figure 8.94 First-in, first-out memory (counter-controlled type). (From ANSI/IEEE Std 91-1984.)

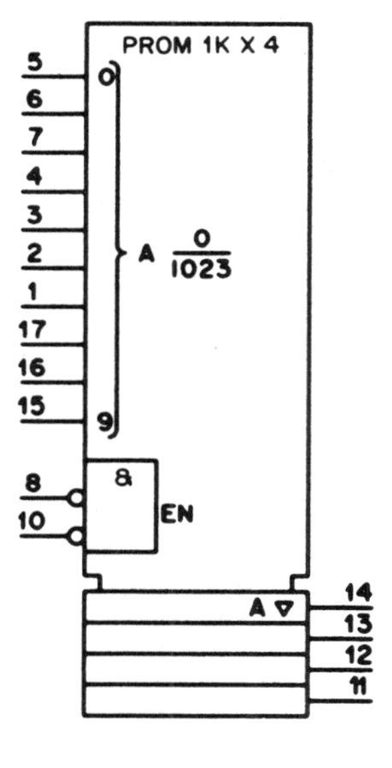

Figure 8.95 Programmable read-only memory 1024 × 4 bits. (From ANSI/IEEE Std 91-1984.)

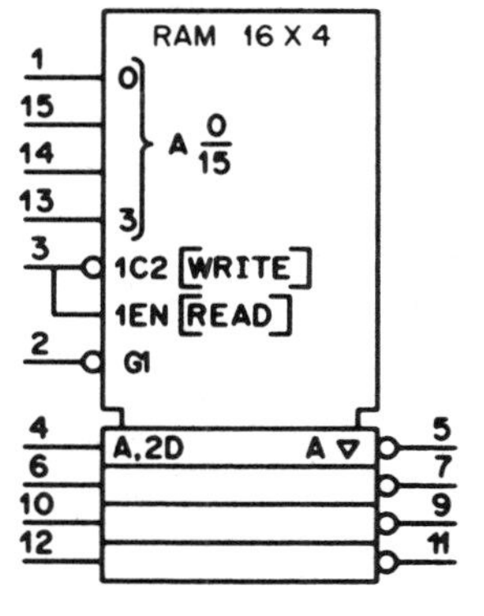

Figure 8.96 Random-access memory 16 × 4 bits. (From ANSI/IEEE Std 91-1984.)

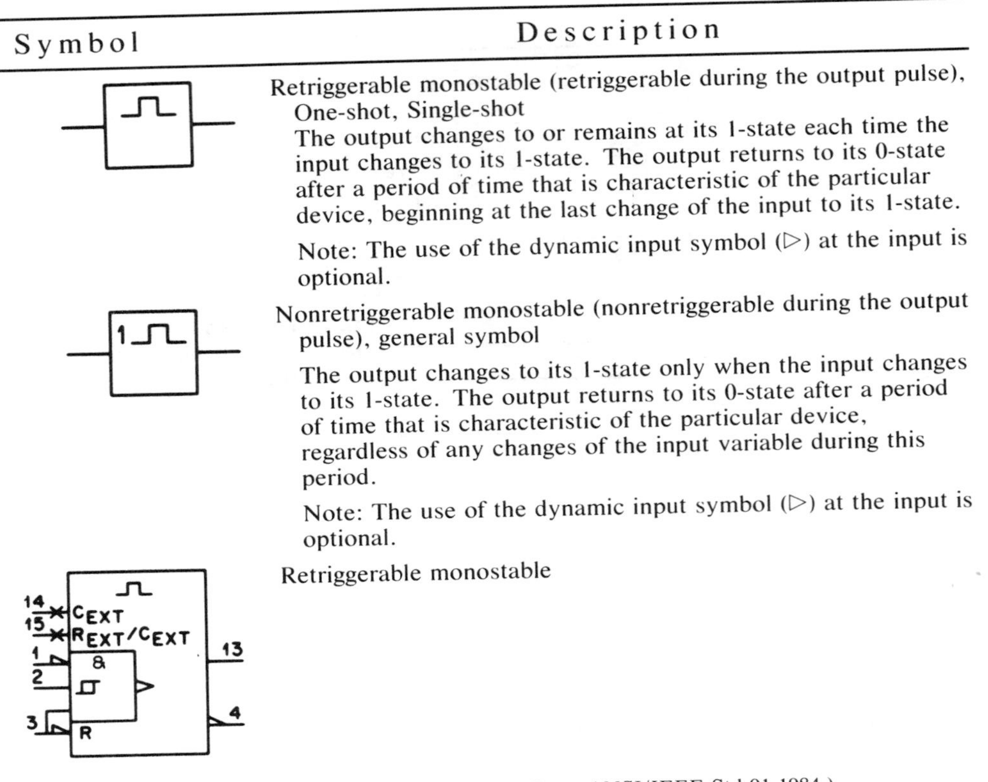

Symbol	Description
	Retriggerable monostable (retriggerable during the output pulse), One-shot, Single-shot The output changes to or remains at its 1-state each time the input changes to its 1-state. The output returns to its 0-state after a period of time that is characteristic of the particular device, beginning at the last change of the input to its 1-state. Note: The use of the dynamic input symbol (▷) at the input is optional.
1	Nonretriggerable monostable (nonretriggerable during the output pulse), general symbol The output changes to its 1-state only when the input changes to its 1-state. The output returns to its 0-state after a period of time that is characteristic of the particular device, regardless of any changes of the input variable during this period. Note: The use of the dynamic input symbol (▷) at the input is optional.
14 CEXT, 15 REXT/CEXT, 1, 2, &, 3, R, 13, 4	Retriggerable monostable

Figure 8.97 Monostable elements. (From ANSI/IEEE Std 91-1984.)

8.5.11 Multiplexers and Demultiplexers

Multiplexers and demultiplexers are shown in Fig. 8.98.

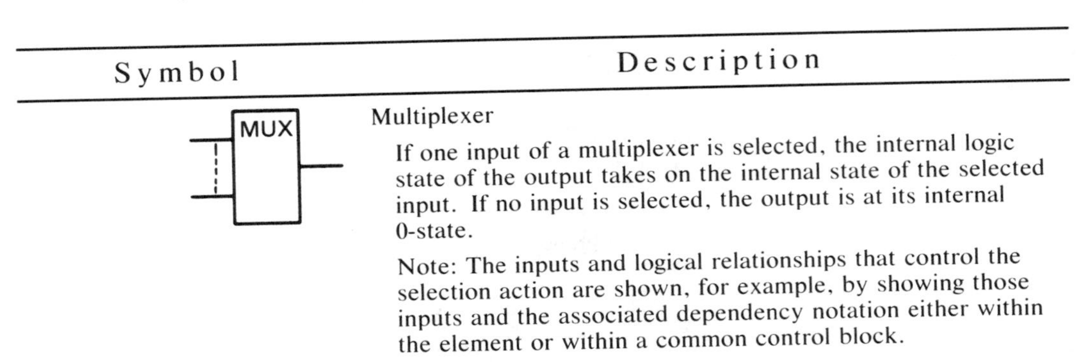

Symbol	Description
MUX	Multiplexer If one input of a multiplexer is selected, the internal logic state of the output takes on the internal state of the selected input. If no input is selected, the output is at its internal 0-state. Note: The inputs and logical relationships that control the selection action are shown, for example, by showing those inputs and the associated dependency notation either within the element or within a common control block.

Figure 8.98 Multiplexers and demultiplexers. (From ANSI/IEEE Std 91-1984.)

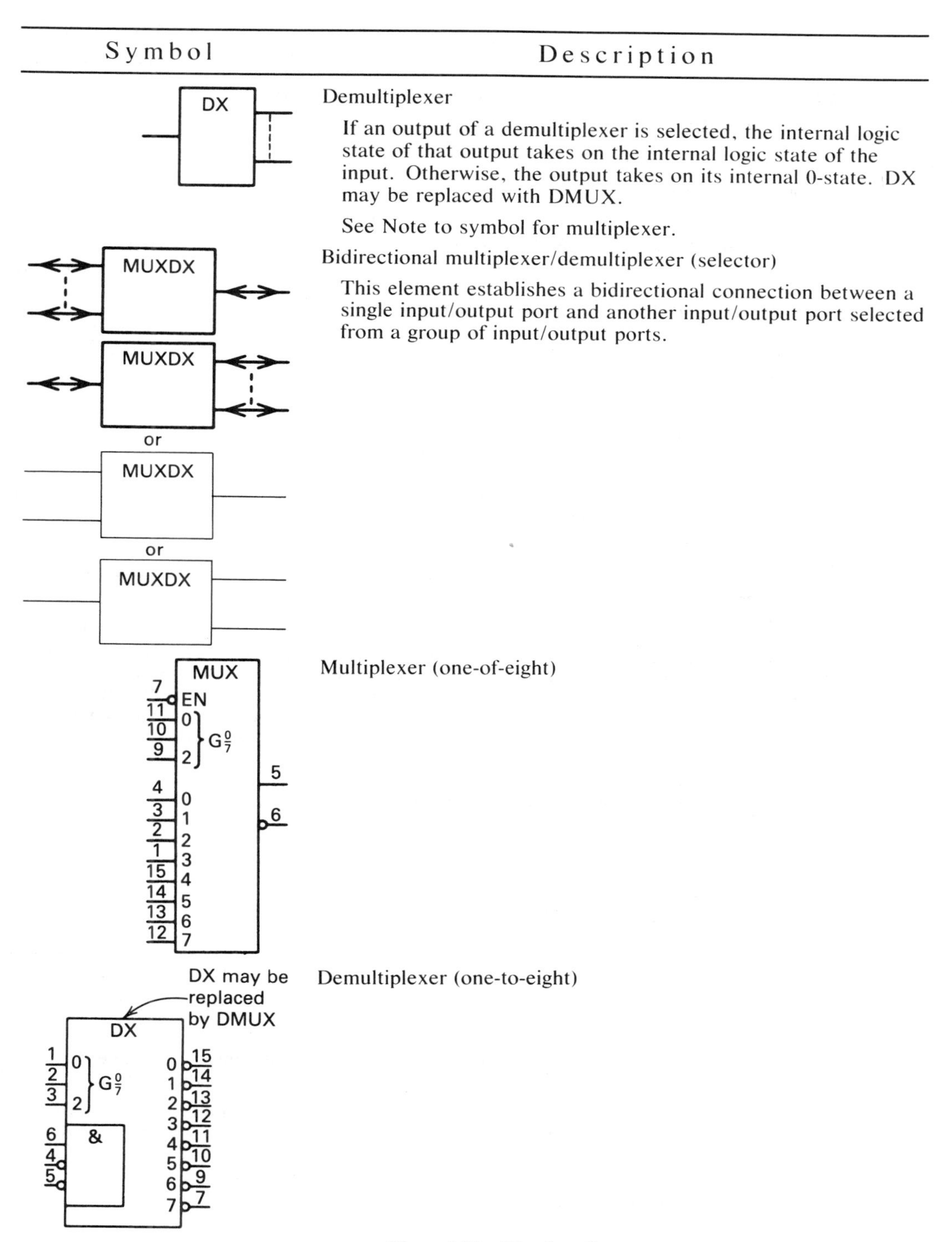

Symbol	Description
DX	Demultiplexer If an output of a demultiplexer is selected, the internal logic state of that output takes on the internal logic state of the input. Otherwise, the output takes on its internal 0-state. DX may be replaced with DMUX. See Note to symbol for multiplexer.
MUXDX / MUXDX or MUXDX or MUXDX	Bidirectional multiplexer/demultiplexer (selector) This element establishes a bidirectional connection between a single input/output port and another input/output port selected from a group of input/output ports.
MUX, EN, $G\frac{0}{7}$	Multiplexer (one-of-eight)
DX, &, $G\frac{0}{7}$; DX may be replaced by DMUX	Demultiplexer (one-to-eight)

Figure 8.98 (*Continued*)

8.5.12 Schmitt Triggers, Elements Exhibiting Hysteresis, and Bithreshold Detectors

These elements are shown in Fig. 8.99.

8.5.13 Shift Registers and Counters

Shift registers and counters are shown in Figs. 8.100, 8.101, and 8.102.

8.5.14 Signal-Level Converters

Signal-level converters are shown in Fig. 8.103.

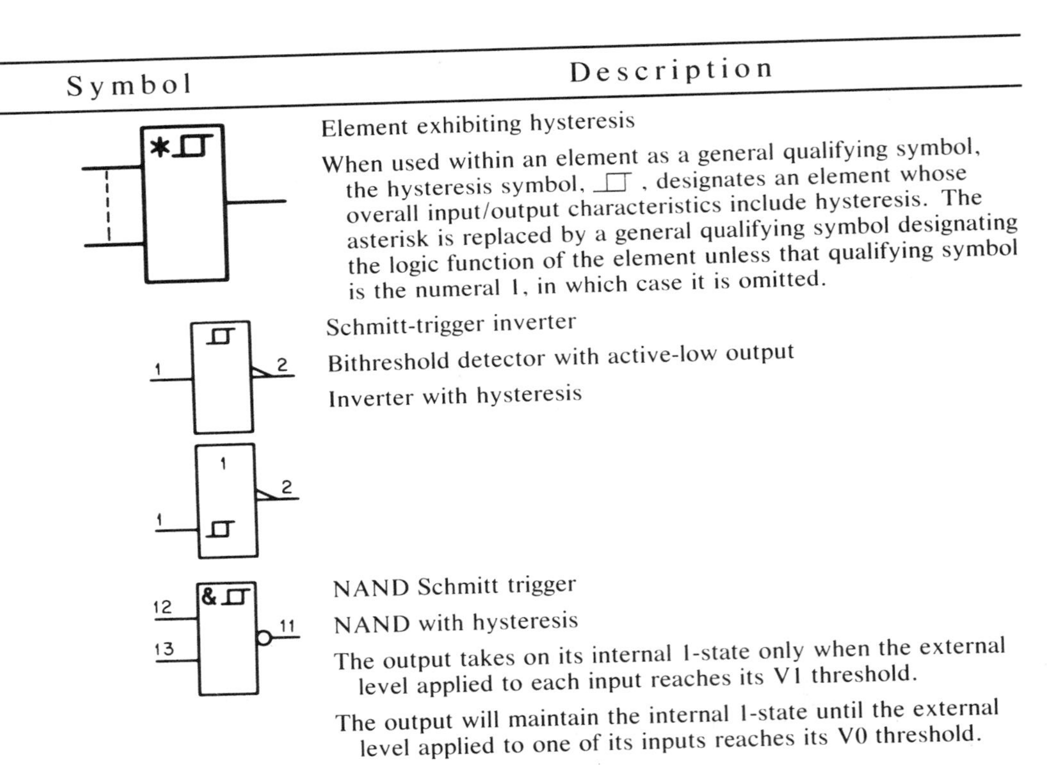

Figure 8.99 Schmitt triggers, elements exhibiting hysteresis, and bithreshold detectors. (From ANSI/IEEE Std 91-1984.)

Symbol	Description
**	Shift register or counter

"**" is replaced by the general qualifying symbol for the function, that is:

CTR*	Counter with cycle length equal to 2 to the power represented here by * (counter modulo 2 to the power *)
CTRDIVm	Counter with cycle length equal to m (counter modulo m)
RCTR*	Ripple counter with cycle length equal to 2 to the power represented here by * (ripple counter modulo to the power *)
RCTRDIVm	Ripple counter with cycle length equal to m (ripple counter modulo m)
SRG*	Shift register; "*" is replaced by the number of stages

m in CTRDIVm and RCTRDIVm is replaced by the actual value. In an array of elements having different cycle lengths, the cycle length applying to each is indicated by DIVm in each element. In such a case the letters CTR or RCTR are shown only in the common control block.

Figure 8.100 Shift registers and counters. (From ANSI/IEEE Std 91-1984.)

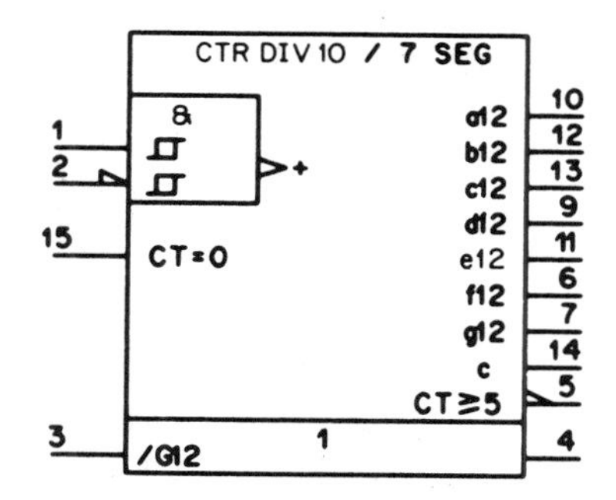

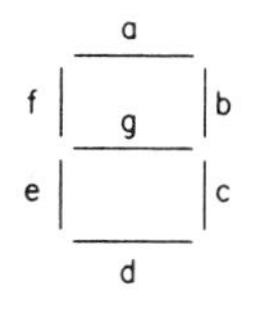

SEGMENT IDENTIFICATION

Figure 8.101 Decade counter/divider with decoded 7-segment display outputs. (From ANSI/IEEE Std 91-1984.)

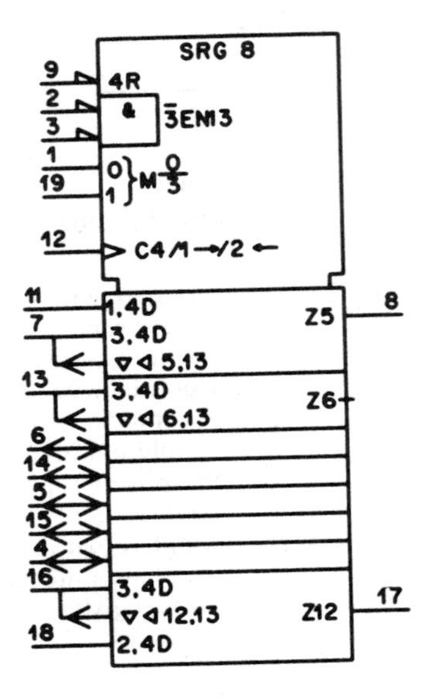

Figure 8.102 Eight-bit universal shift register. (From ANSI/IEEE Std 91-1984.)

Symbol	Description
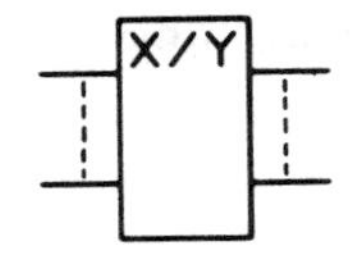	Signal-level converter Level references may be shown inside the symbol and may replace X and Y. This symbol is the same as the general symbol for a coder or code converter. If X/Y is replaced by X//Y, special isolation exists between the inputs and the outputs.
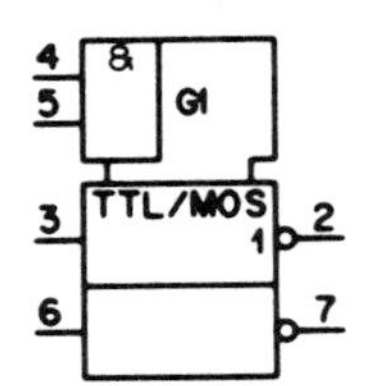	Level converter, TTL-to-MOS, dual

Figure 8.103 Signal-level converters. (From ANSI/IEEE Std 91-1984.)

8.6 HIGHLY COMPLEX FUNCTIONS

8.6.1 General

Small-scale and medium-scale integrated circuits can be represented by a combination of outlines, input/output labeling, and dependency notation. However, for

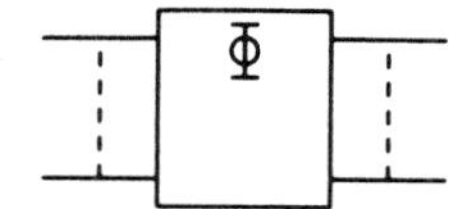

Figure 8.104 General symbol for a highly complex function (gray box). (From ANSI/IEEE Std 91-1984.)

large-scale and very-large-scale integrated circuits, a different technique, shown in Fig. 8.104, is used.

8.6.2 Input and Output Designation

Inputs and outputs inside the symbol outline are designated with the signal names appearing on the manufacturer's data sheet.

8.6.3 Negated Signal Names Designation

By using negation or polarity symbols, depending on the convention being used, negated signal names may be changed to the unnegated form inside the symbol. In the case of an input or output that has two functions activated at opposite polarities, the connecting line may use a branch to allow two separate labels to be shown, thus avoiding a negation bar, as shown in Fig. 8.105.

8.6.4 Functional Grouping

Connecting lines are grouped by function and may be separated into control and data lines.

8.6.5 Label Grouping

Inside the symbol outline a vertical line is used to group adjacent connecting lines whose labels are partially alike, as shown in Fig. 8.106.

Figure 8.105 Negated signal names. (From ANSI/IEEE Std 91-1984.)

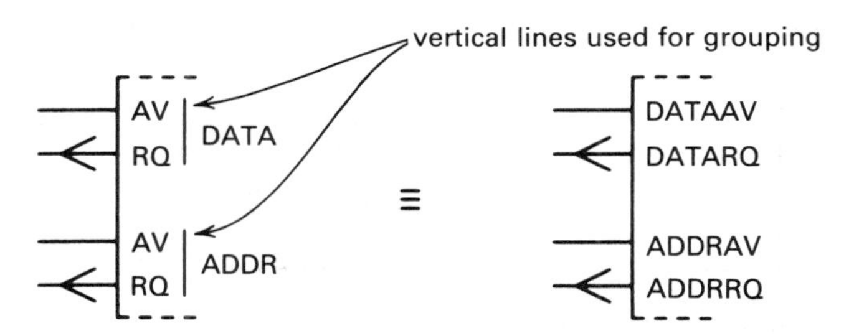

Figure 8.106 Label grouping. (From ANSI/IEEE Std 91-1984.)

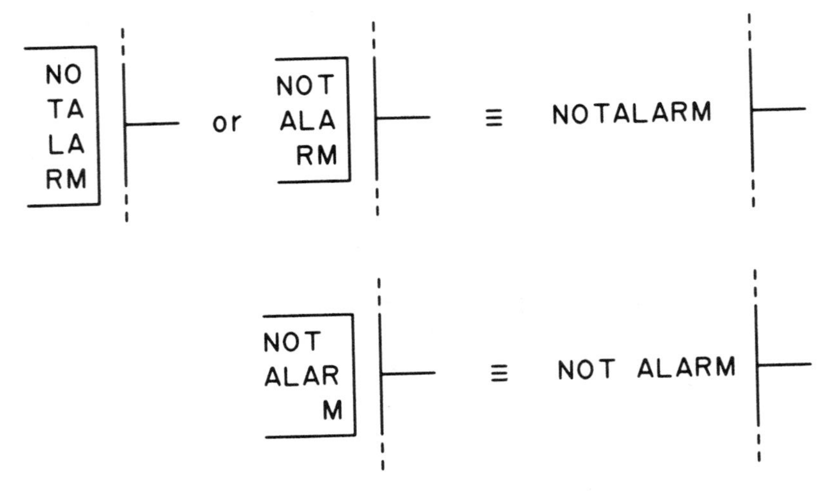

Figure 8.107 Long character strings. (From ANSI/IEEE Std 91-1984.)

8.6.6 Long Character Strings

When long character strings associated with input or output lines are too long to be printed horizontally, they may be stacked as shown in Fig. 8.107.

8.6.7 Bidirectional Signal Flow

Bidirectional signal flow is shown by a double arrow. With this arrow, a single connecting line may be used for an input or output terminal. In this usage both the input and the output have the same label and the same negation or polarity indication (see Fig. 8.108).

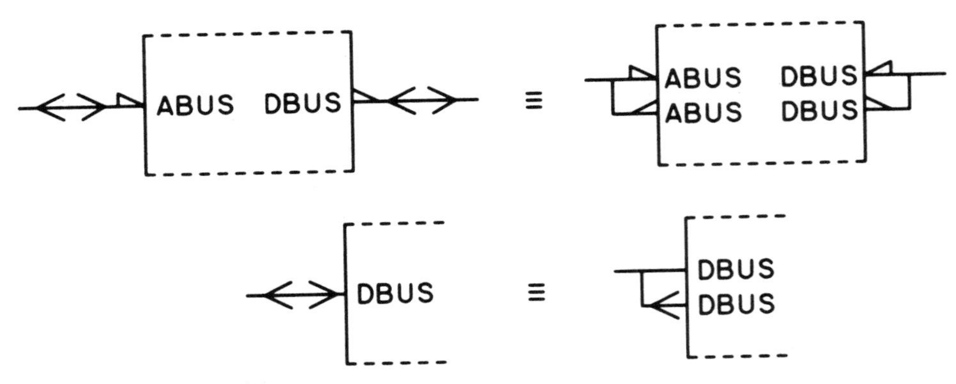

Figure 8.108 Bidirectional signal flow. (From ANSI/IEEE Std 91-1984.)

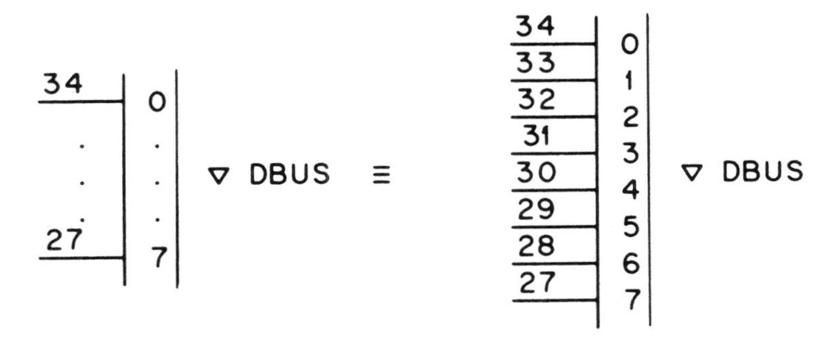

Figure 8.109 Consecutive labels and terminal numbers. (From ANSI/IEEE Std 91-1984.)

8.6.8 Consecutive Labels and Terminal Numbers

When dots separate internal labels and terminal numbers, it can be assumed that the missing labels and numbers are present in a consecutive order, as shown in Fig. 8.109.

8.6.9 Function Tables and Truth Tables

Function tables and truth tables help to explain the behavior of a circuit. If the entries to these tables refer to external logic states on a theoretical logic diagram or when using a single logic convention, note these techniques: In the table, any label derived from one appearing inside the symbol at an input or output bearing a negation symbol is modified by the addition or removal of a negation bar. However, other labels on the table appear without modification (see Fig. 8.112).

8.6.10 Internal Diagram

An *internal diagram* is a diagram drawn inside the outline of a symbol to show the behavior of a complex function. In an internal diagram notice the following:

1. A negation or polarity symbol shown at the symbol outline at those inputs (outputs) to which it applies indicates the relationship between the internal logic state of the input (output) and its external logic state or logic level.
2. When input and output labels shown inside and adjacent to the outlines of the symbols appearing on the internal diagram are repeated after the application of logic negation, they are modified by the addition (or removal) of a negation bar (see Fig. 8.110).

8.6.11 Examples

Figures 8.111, 8.112, and 8.113 are examples of complex elements.

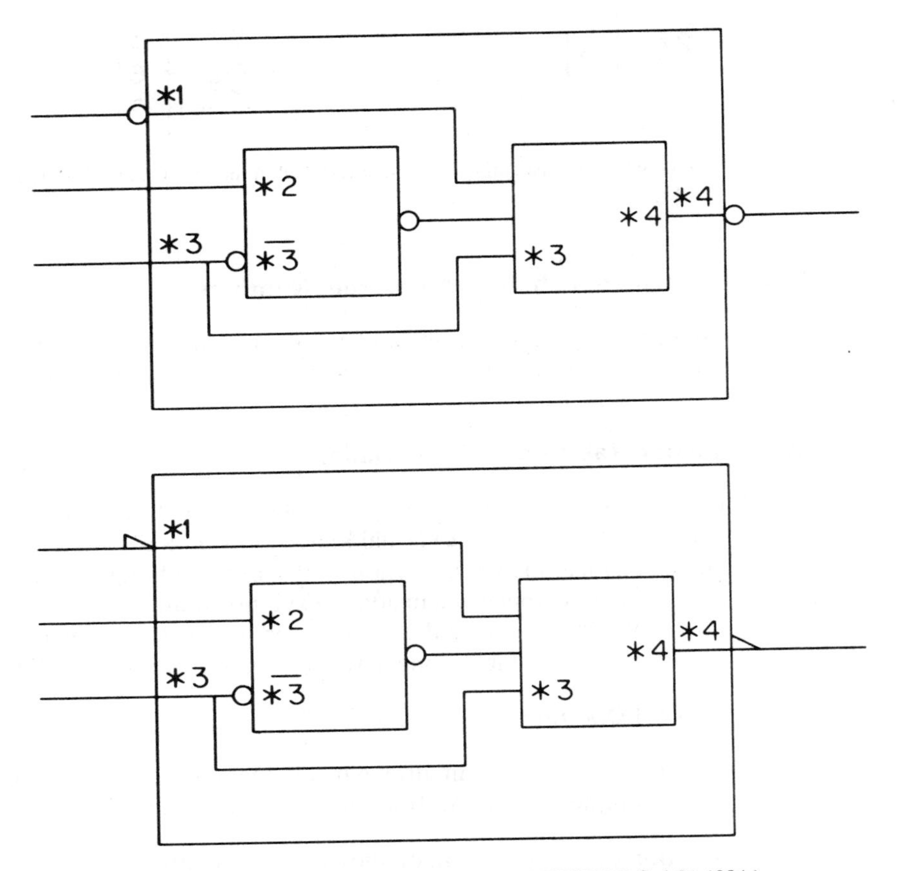

Figure 8.110 Internal diagram. (From ANSI/IEEE Std 91-1984.)

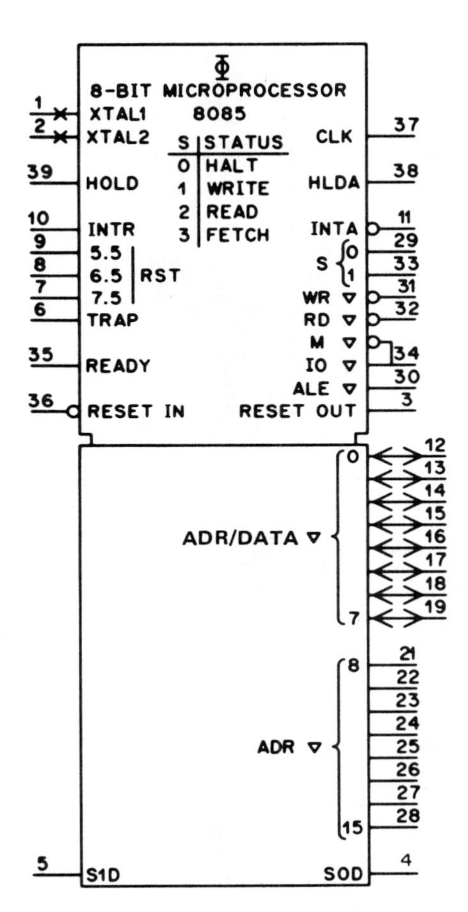

Figure 8.111 Eight-bit microprocessor. (From ANSI/IEEE Std 91-1984.)

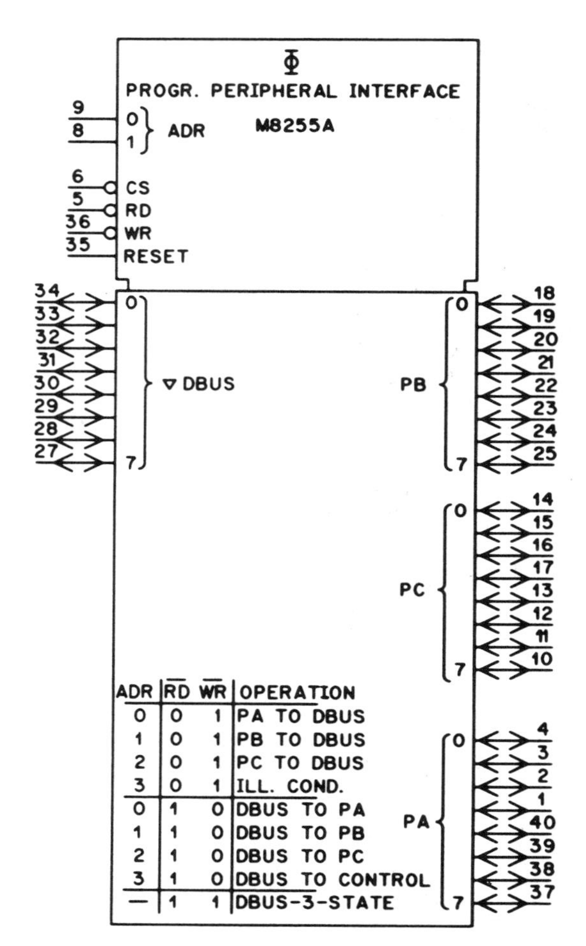

ADR	$\overline{RD}$	$\overline{WR}$	OPERATION
0	0	1	PA TO DBUS
1	0	1	PB TO DBUS
2	0	1	PC TO DBUS
3	0	1	ILL. COND.
0	1	0	DBUS TO PA
1	1	0	DBUS TO PB
2	1	0	DBUS TO PC
3	1	0	DBUS TO CONTROL
—	1	1	DBUS-3-STATE

Figure 8.112 Programmable peripheral interface. (From ANSI/IEEE Std 91-1984.)

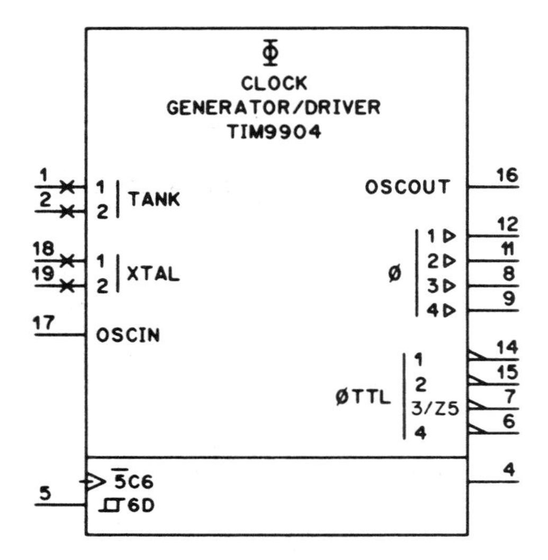

Figure 8.113 Four-phase clock generator/drive. (From ANSI/IEEE Std 91-1984.)

8.7 SUMMARY OF NEW LOGIC SYMBOLS

Figure 8.114 is a summary of the new logic symbols.

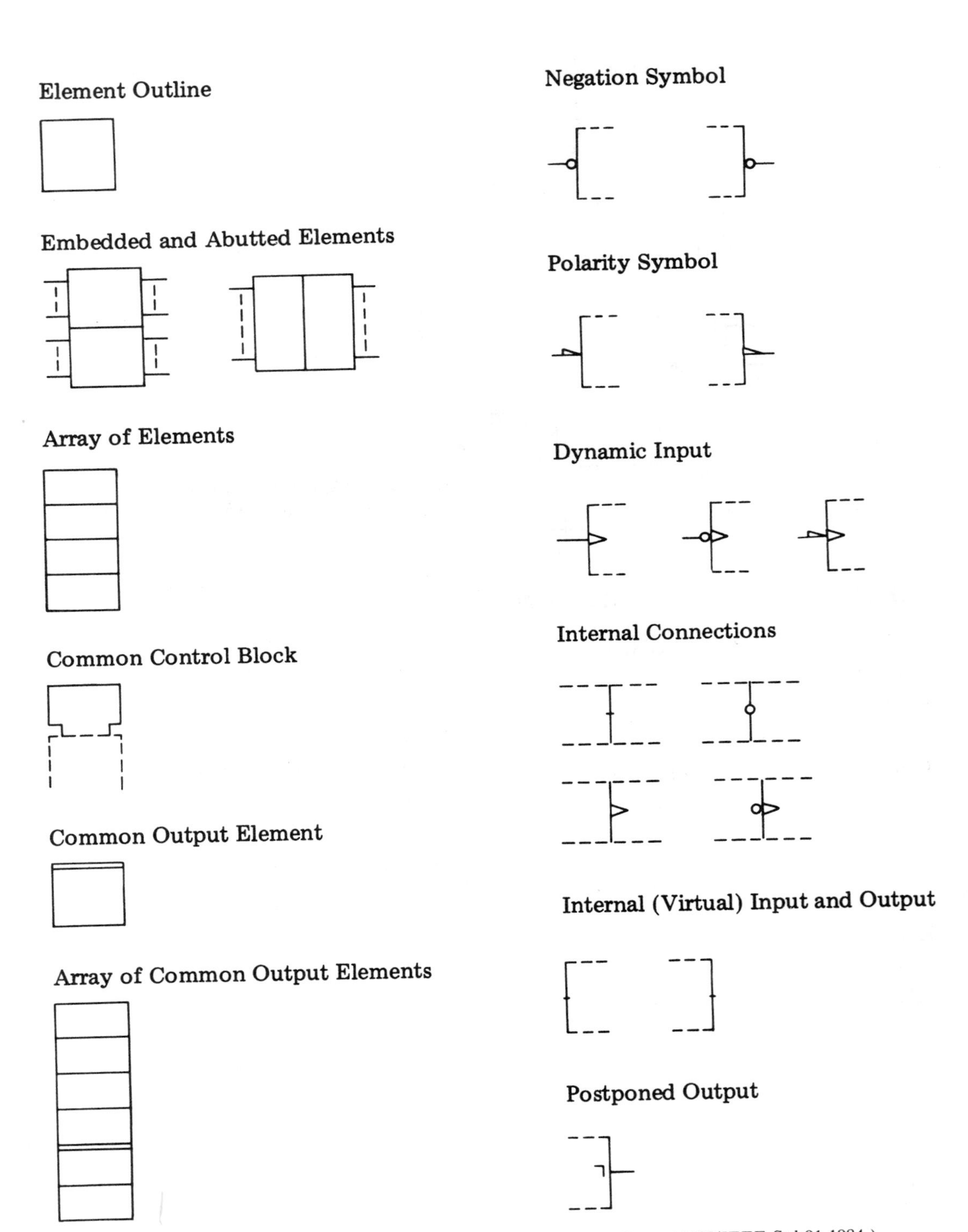

Figure 8.114 Summary of the new logic symbols. (From ANSI/IEEE Std 91-1984.)

Input with Hysteresis (Bithreshold Input)

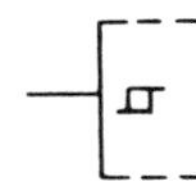

Open-Circuit Output

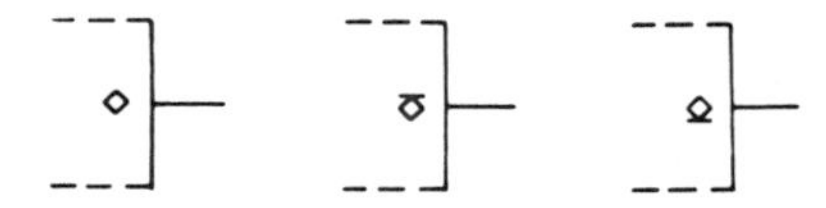

Passive-Pulldown and Passive-Pullup Outputs

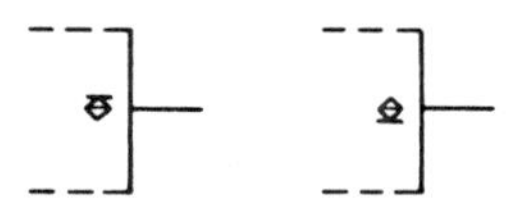

3-State Output

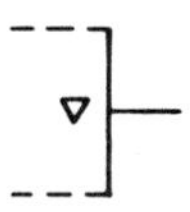

Output with Special Amplification

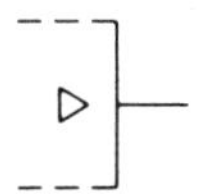

Extension Input and Extender Output

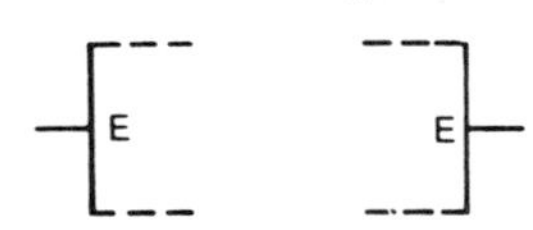

Inputs

*: EN, D, J, K, R, S, T,

−m, −m, +m, −m, ?

Compare Output

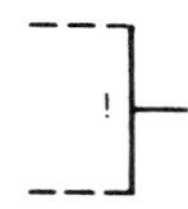

Bit-Grouping Symbol

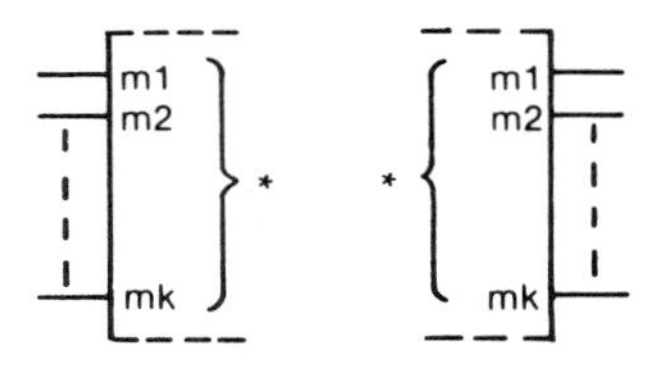

Operand Inputs and Outputs of an Arithmetic Element

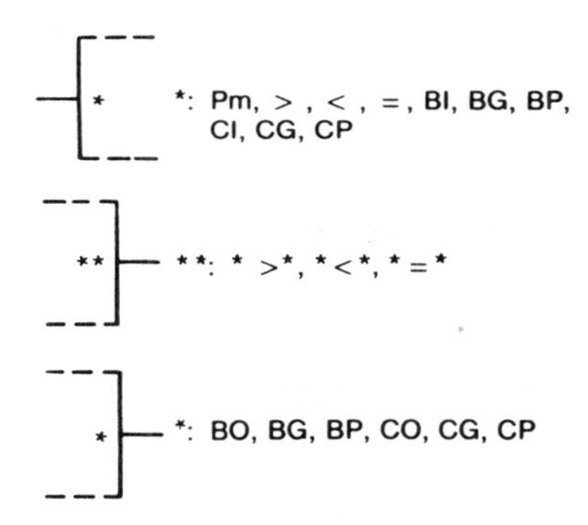

Content Input and Output

CT = m CT* *: ≤ m, = m, ≥ m, etc.

Line-Grouping Symbol

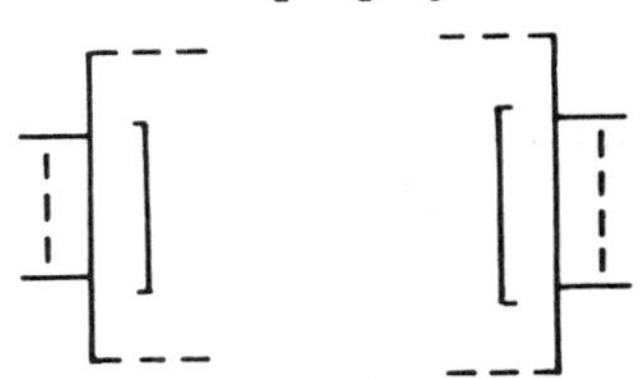

Figure 8.114 (*Continued*)

Fixed-Mode Input and Fixed-State Output

Non-Logic Connection

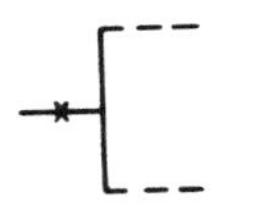

Analog Input and Output

Right-to-left Signal Flow

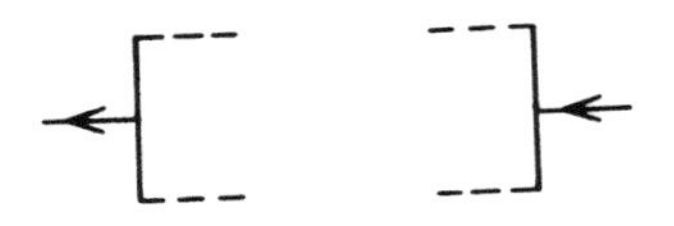

Bidirectional Signal Flow

Dependency Symbols

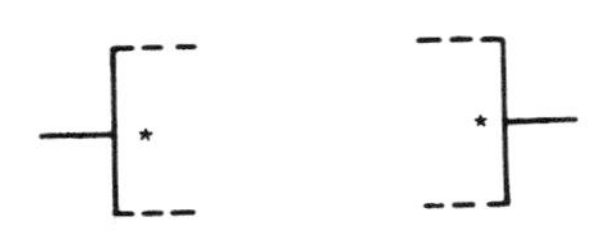

Gm	— AND	(4.3.2)
Vm	— OR	(4.3.3)
Nm	— Negate	(4.3.4)
Zm	— Interconnection	(4.3.5)
Xm	— Transmission	(4.3.6)
Cm	— Control	(4.3.7)
Sm	— Set	(4.3.8)
Rm	— Reset	(4.3.8)
ENm	— Enable	(4.3.9)
Mm	— Mode	(4.3.10)
Am	— Address	(4.3.11)

where m = a number

Basic Combinational Elements

OR Element

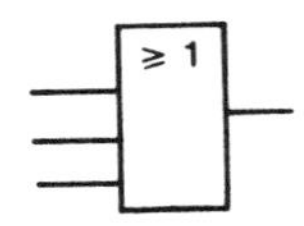

AND Element

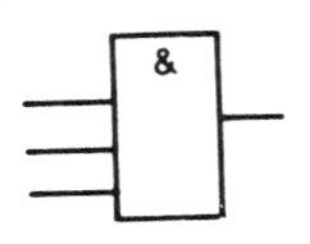

Logic Threshold Element

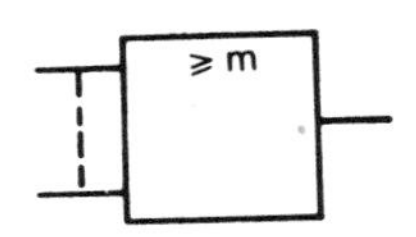

m and only m Element

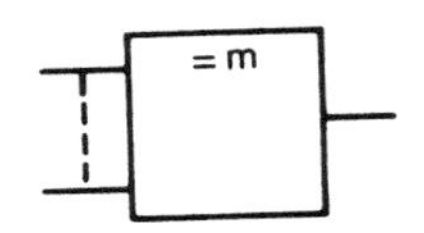

Majority Element

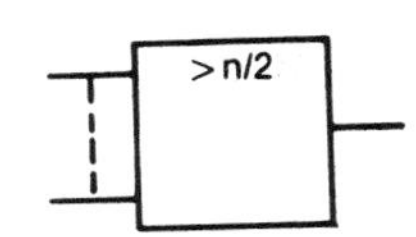

Logic Identity Element

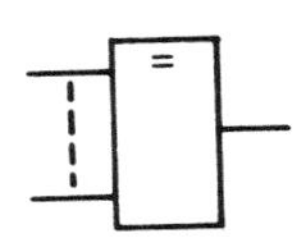

Odd Element (Addition Modulo 2 Element)

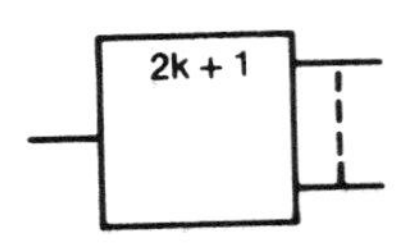

Figure 8.114 *(Continued)*

Even Element

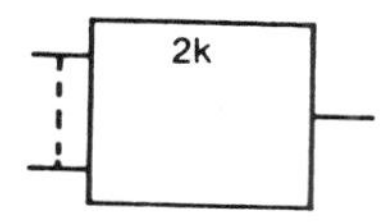

Exclusive-OR Element

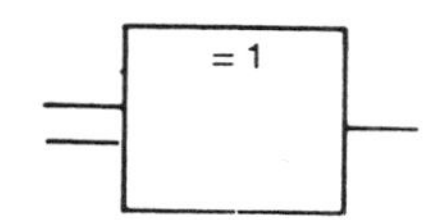

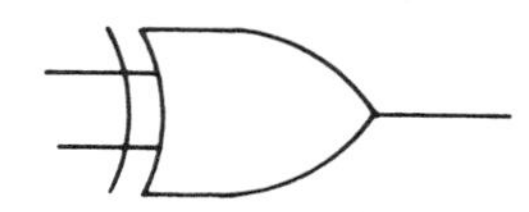

Buffer Without Specially Amplified Output

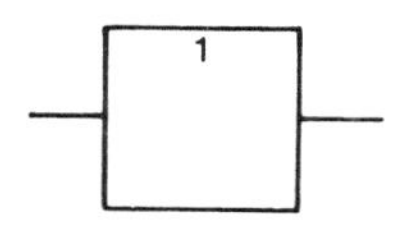

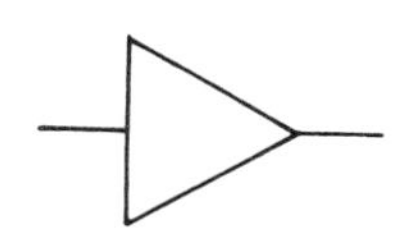

Inverter, Negators

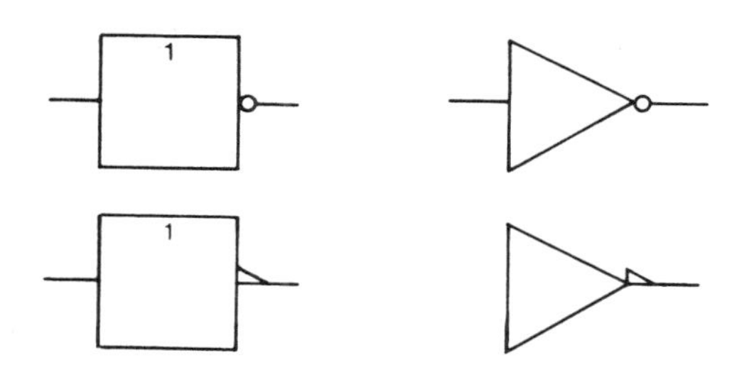

Distributed-AND Function (Dot-AND Function, Wired-AND Function)

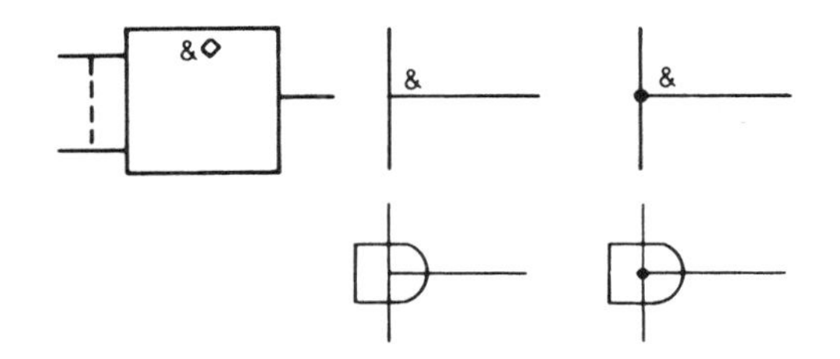

Distributed-OR Function (Dot-OR Function, Wired-OR Function)

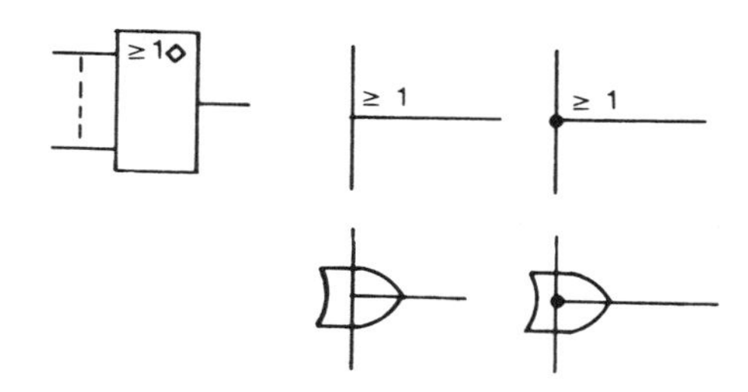

Buffer with Special Amplification

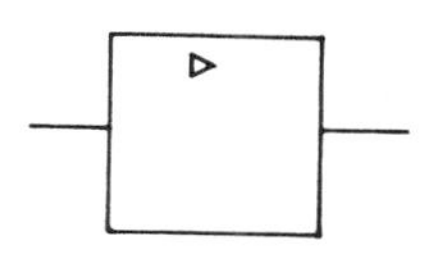

Elements Exhibiting Hysteresis (Schmitt Triggers, Bithreshold Detectors)

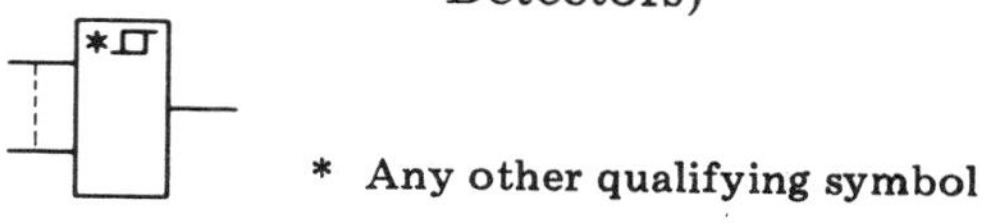

Coders

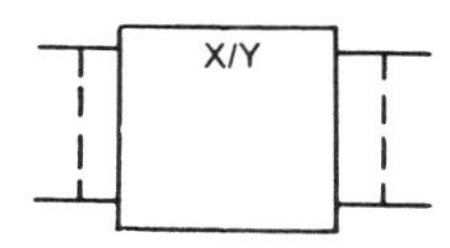

Signal-Level Converters

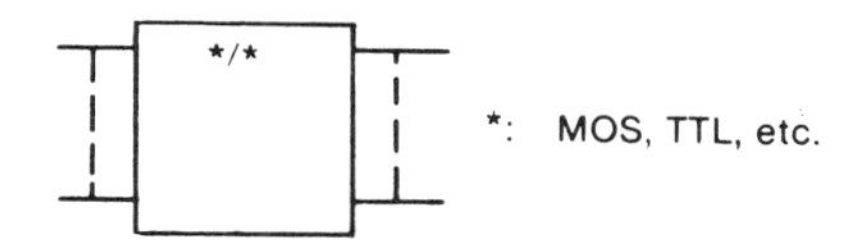

Multiplexers and Demultiplexers

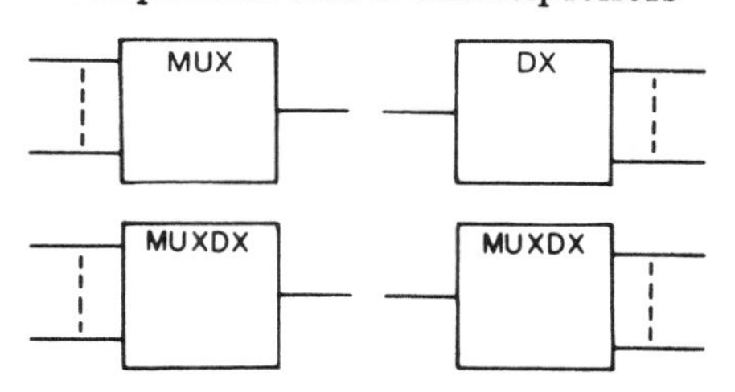

Figure 8.114 *(Continued)*

Arithmetic Elements

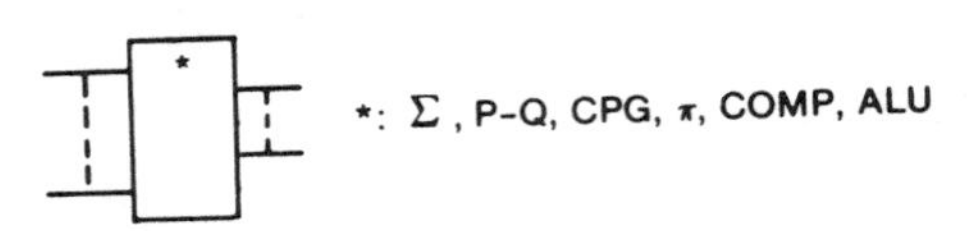

Delay Elements

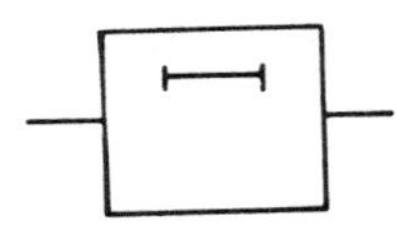

Basic Bistable Elements

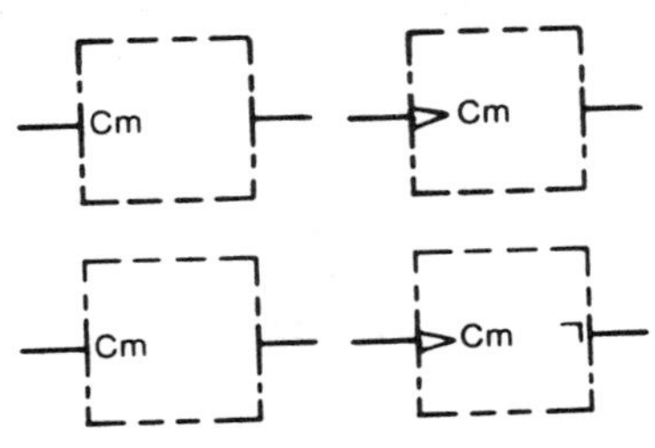

These circuits usually also have D, S, R, J or K inputs

Bistable Element with Special Switching Properties

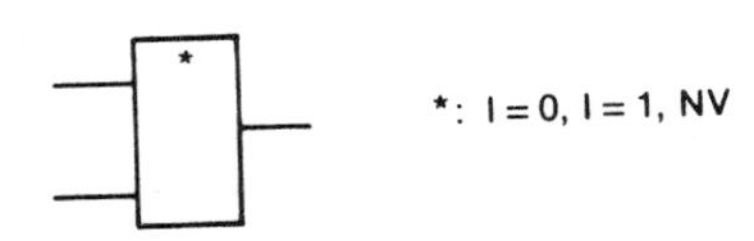

Monostable Elements

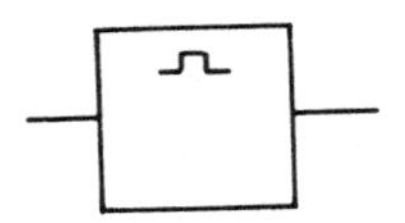

Astable Elements

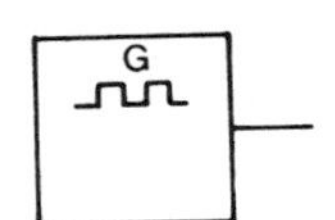

Shift Registers and Counters

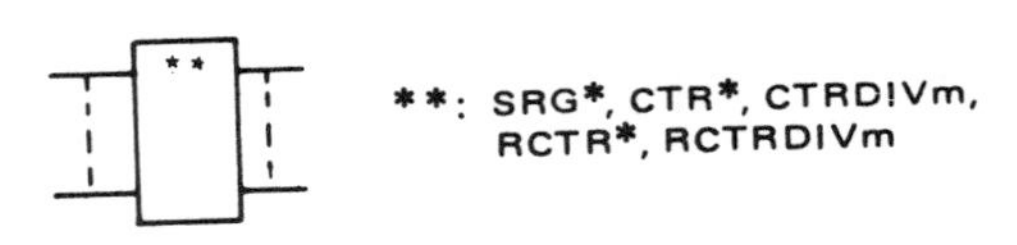

Memories

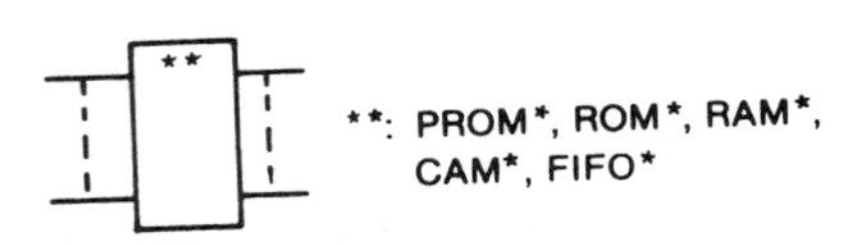

Complex-Function Element

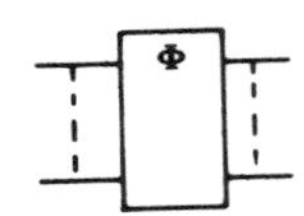

Name-Grouping Symbol for Complex Circuits

Open Box for Long Character Strings

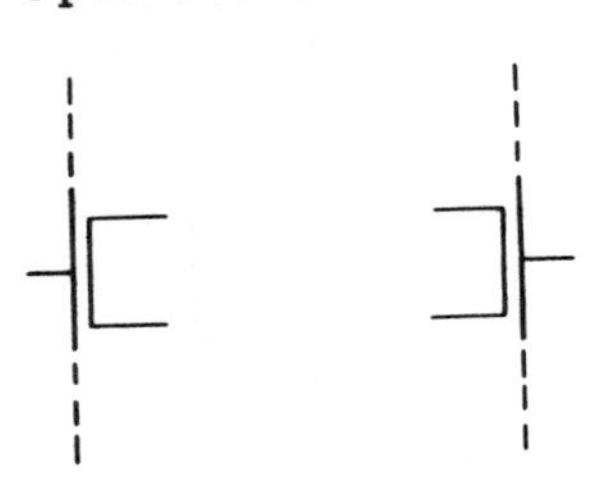

Figure 8.114 *(Continued)*

Chapter 9

Interpreting Logic Diagrams

Now that we have considered conventional schematics and logic symbols, we are ready to learn the techniques for interpreting logic diagrams. In this chapter we consider flowcharts, timing diagrams, manufacturers' data sheets, microprocessors, truth and function tables, state diagrams, and logic convention.

Note that a *basic logic diagram* shows the conceptual principles of a circuit through the use of logic symbols and other necessary functional symbols, together with their signal and major control path connections, but physical location, pin connection, and assembly level information are usually omitted (ANSI/IEEE Std 991-1986).

A *detailed logic diagram* shows the information necessary for manufacture, installation, maintenance, and training for a logic circuit or system. In addition to the items shown on the basic logic diagram, it includes power connections, reference designations, terminal identification, signal-level conventions applicable to the diagram, and information necessary to trace paths and circuits among sheets of the diagram (ANSI/IEEE Std 991-1986).

These definitions apply to diagrams that have been prepared to meet the ANSI/IEEE standard. Unfortunately, some of the diagrams you may encounter will not include all these features as they have been prepared under different or no standards.

In logic diagrams, arrows may not be shown but can be assumed, as shown in Fig. 9.1, in determining the flow of information.

Parallel signals in buses are shown either by a broad arrow or by a single line with a number, as shown in Fig. 9.2.

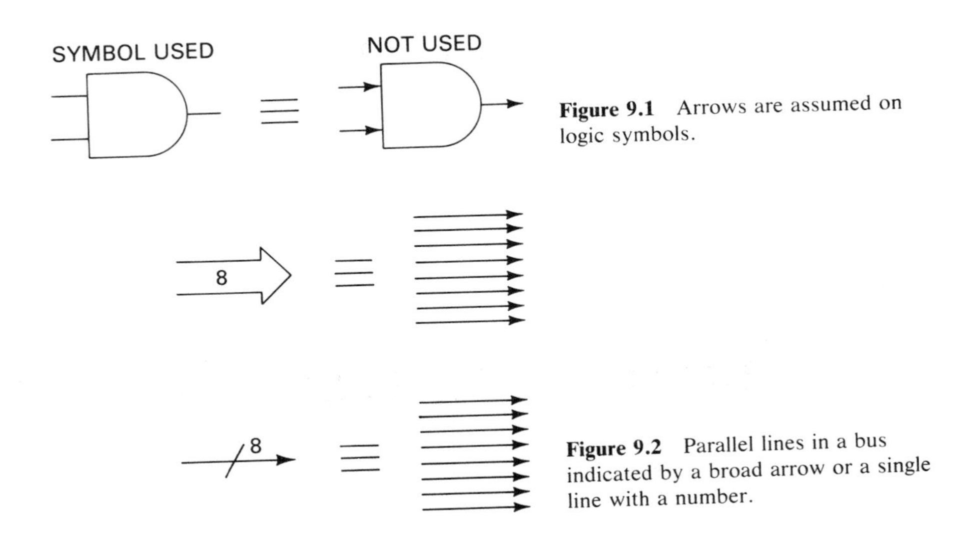

Figure 9.1 Arrows are assumed on logic symbols.

Figure 9.2 Parallel lines in a bus indicated by a broad arrow or a single line with a number.

9.1 FLOWCHARTS

A flowchart is a diagram, such as Fig. 9.3, that gives a quick picture or map of what is happening in a computer or microprocessor program. That is, it shows how a program's functions or steps are arranged in a sequential order. Although it may not be provided for each microprocessor, when it is given it is a big help in understanding the operation.

In a sense, a flowchart is a special form of the block diagram. It is generally drawn in a top-to-bottom format, but may occasionally be drawn from left to right, right to left, or some combination of these patterns. If arrowheads are shown, they will indicate the direction of flow or the sequence of steps. If they are not shown, you may assume that the flow is from top to bottom.

In any case the flow is to be followed along flow lines. Distinctive shaped symbols on these flow lines indicate the type of action or step to be taken in the program. The words within the symbols give the details of the operation.

Standard distinctive symbols used in flowcharts are shown in Fig. 9.4. Nonstandard symbols which are encountered are usually close enough in shape to the standard symbols that they can be readily interpreted.

Because of limited space within the symbols, words are frequently abbreviated and liberal use is made of mathematical symbols such as > for "greater than," < for "less than," and ≠ for "not equal to."

Referring to Fig. 9.4, the *process* function may represent an entire program or a single instruction. It is used to define any operation causing a change in value, form, or location of information.

The *comment or annotation* block provides additional description or clarification. Notice that the dashed line extends to symbols.

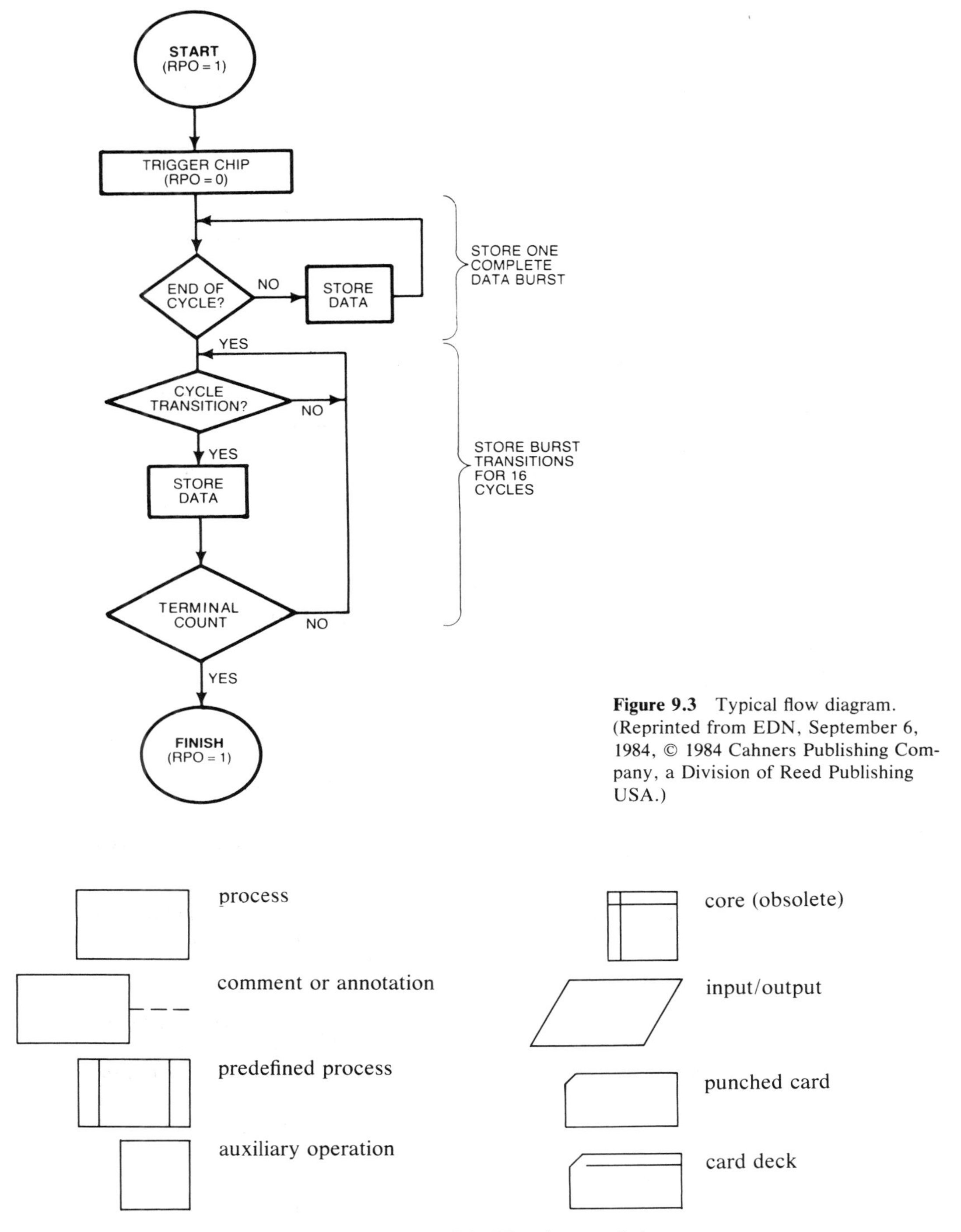

Figure 9.3 Typical flow diagram. (Reprinted from EDN, September 6, 1984, © 1984 Cahners Publishing Company, a Division of Reed Publishing USA.)

Figure 9.4 Flowchart symbols.

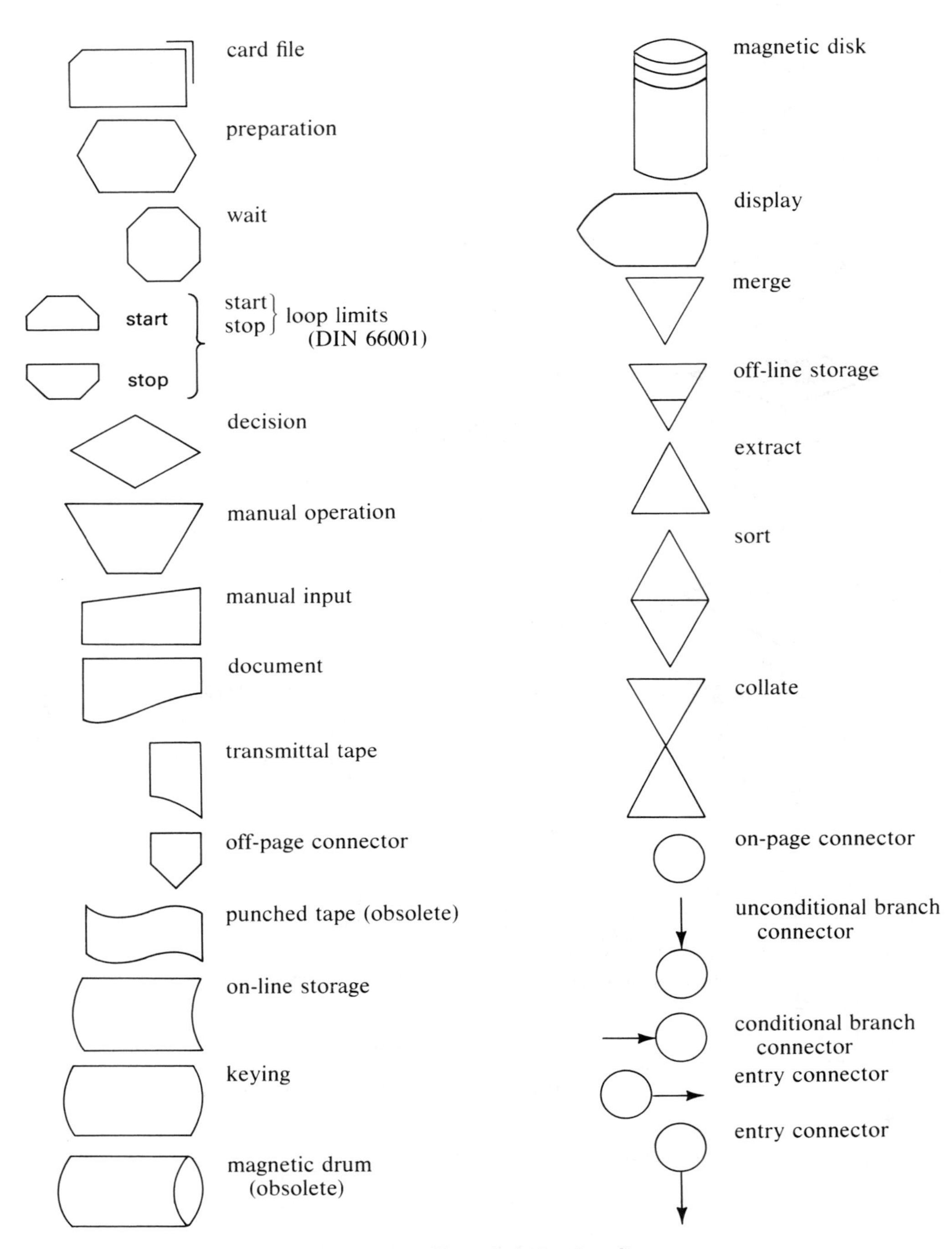

Figure 9.4 (*Continued*)

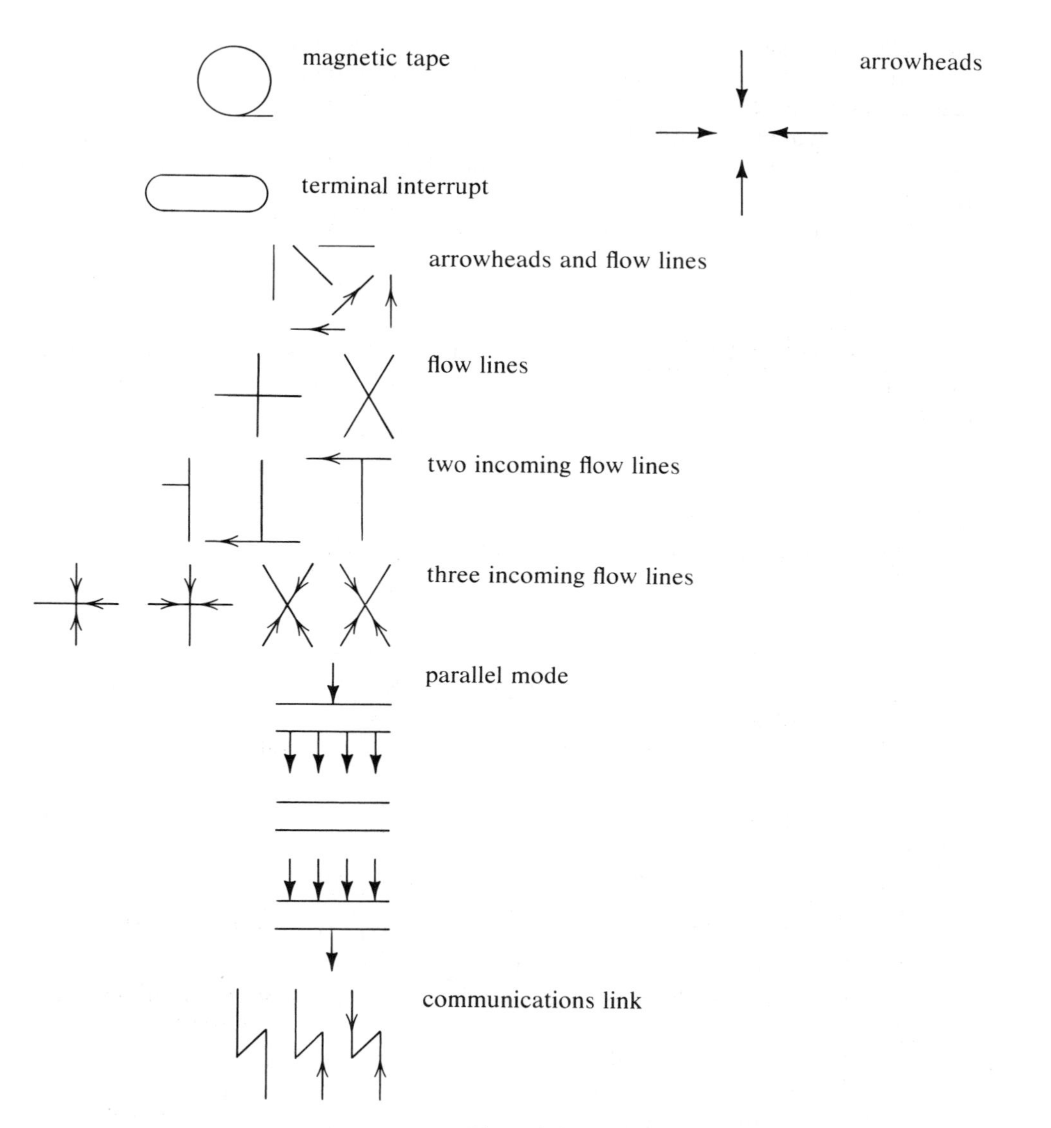

Figure 9.4 (*Continued*)

A *predefined process* is one or more named operations or program steps specified in a subroutine or another set of flowcharts.

An *auxiliary operation* is off-line performance on equipment not under direct control of the central processing unit.

The *core* symbol is used for specific media for input/output functions.

Input/output indicates information available for processing (input) or recording of processed information (output).

Punched card indicates all varieties of cards used for the input/output function. A *card deck* simply indicates a collection of punched cards; a *card file* represents a collection of related punched card records.

The *preparation* symbol represents an instruction modification to change the program: for example, to set a switch, modify an index register, or initialize a routine.

Wait indicates that another operation is occurring which delays the main routine.

Loop limits is an international symbol that you may not encounter in U.S. drawings.

A *decision* operation is a most important step that asks the microprocessor a question that can only be answered "yes" or "no." The answer determines if an alternative path is to be followed.

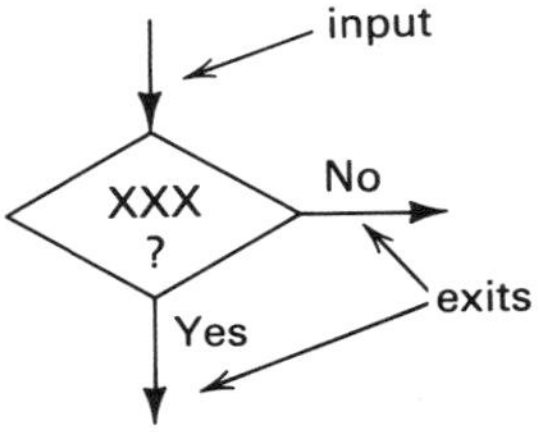

A *manual operation* is an off-line process performed at human speed with no involvement of the microprocessor.

Manual input allows information to be inputted by on-line keyboards, switch settings, or pushbuttons.

Transmittal tape is proof or adding machine tape or other batch-controlled information.

An *off-page connector* indicates that the flowchart continues on another sheet. A unique identifying letter or number on this connector will match the letter or number on the corresponding input connector on the other sheet.

On-line storage indicates input or output with magnetic disk or tape.

Keying is an operation such as punching, verifying, or typing using a key-driven device.

Magnetic disk is a specific media for input/output functions.

Display is the display of information by on-line indicators, video devices, console printers, plotters, etc.

Merge is the act of combining two or more sets of items into one set. *Off-line storage* is simply storing off-line with any recorded medium. *Extract* is the removal of one or more specific sets of items from a set. *Sort* is arranging a set of items into a sequence. *Collate* is merging with extracting, forming two or more sets of items from two or more other sets.

An *on-page connector* is used to identify common points in the flow paths when connecting lines cannot be drawn or would be confusing if drawn. It shows an exit to or entry from another part of the flowchart.

If the flowpath enters the connector from the top, the connector is called an *unconditional branch connector*. When the microprocessor reaches this instruction, it will without exception branch to another part of the flowchart. Notice that this connector is the last symbol in a sequence of steps.

A *conditional branch connector* causes the program to branch if a specific condition is met. It is always associated with a decision symbol.

Each type of branch connector has a matching *entry connector,* whether ○→ or ○↓. For the match to occur, the label on the entry connector must match exactly the label on the branch connector.

A *terminal interrupt* is a terminal point in a flowchart. It may show start, stop, halt, delay, or interrupt or it may show an exit from a closed subroutine.

Parallel mode shows the beginning or end of two or more simultaneous operations.

A *communication link* indicates the transmission of information by a telecommunication link. It may be vertical, horizontal, or diagonal. Opposing arrowheads indicate bidirectional flow.

9.2 TIMING DIAGRAMS

A *timing diagram* is a useful, if not essential, aid in the interpretation of some logic diagrams. By showing the waveforms present in a logic element or system, it allows the relative time relations of operations in parts of the element or in components of a system to be determined. It is particularly useful in showing how a particular operation affects or initiates other operations.

For signals on single lines, the waveforms can be shown as in Fig. 9.5. At times, interruptions are shown in waveforms to indicate that the waveforms have been shortened so as to allow long lines to be drawn to shorter scales, as indicated in Fig. 9.6.

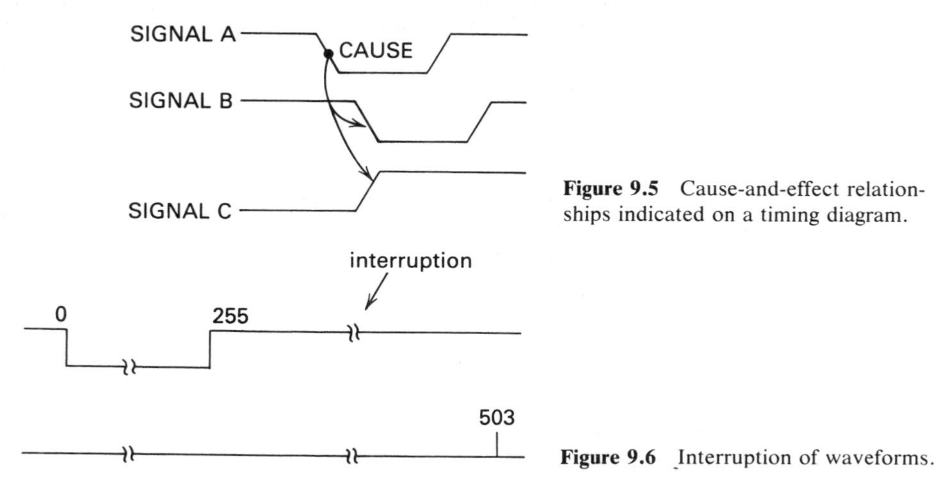

Figure 9.5 Cause-and-effect relationships indicated on a timing diagram.

Figure 9.6 Interruption of waveforms.

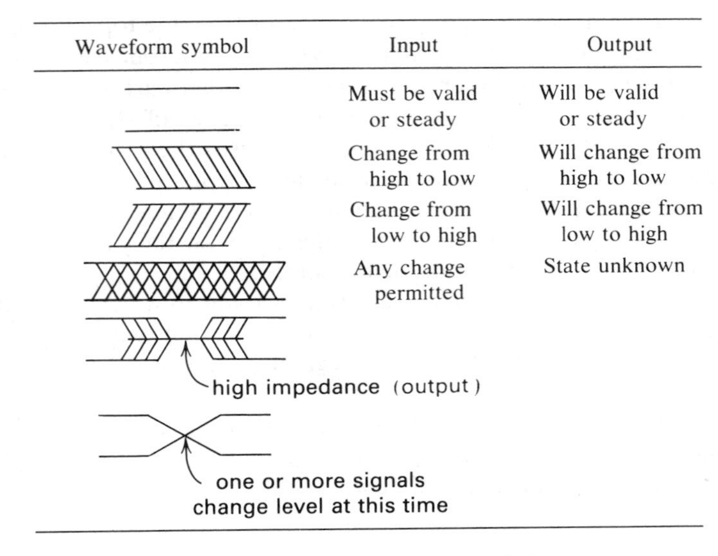

Figure 9.7 Bus waveform symbols.

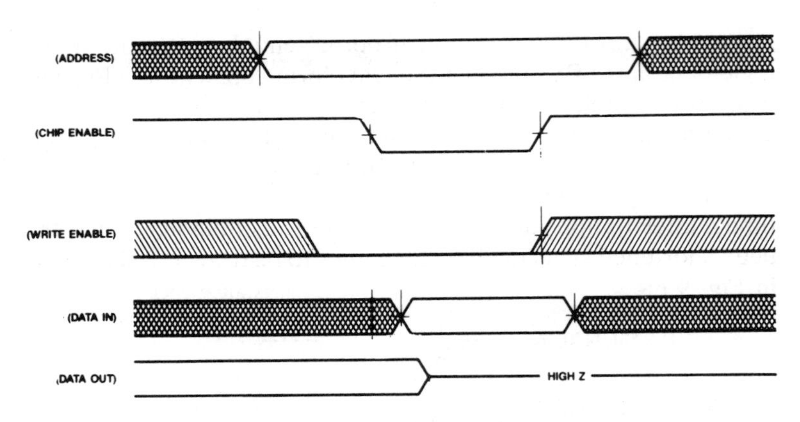

Figure 9.8 Timing diagram for a random-access memory.

Waveforms could be shown for individual lines of a bus, but it is better to group the lines and indicate what is happening by the use of parallel lines like those in Fig. 9.7. These symbols are used in Fig. 9.8, a timing diagram for a random-access memory.

9.3 MANUFACTURERS' DATA SHEETS

To understand the operation of an IC it is important to obtain as much information as possible about the IC. In the worst case, an IC on a logic diagram may have only a part number and no further information. By referring to *Towers' International Digital IC Selector,* by T. D. Towers (TAB Books Inc., Blue Ridge Summit, PA) one may determine the type of IC. For example, if the diagram states only that the IC is a 74147, by looking in this reference, you can determine that it is a 10-to-4-line priority encoder.

If you are familiar with the operation of 10-to-4-line encoders, it will not be necessary to go any further. If you need more information, you can refer to data sheets prepared by each IC manufacturer. Table 9.1 lists some addresses where this information can be obtained. Some of the data sheets, which are usually compiled in data books, will be supplied free of charge. For others there will be a charge. To obtain data on all the current ICs will cost you considerably. Figure 9.9 shows one of two pages of one manufacturer's data book catalog. Obviously, there is a lot of data available.

Another approach to data on ICs is the publication *IC Master,* (Garden City, New York: Hearst Business Communications) a two-volume set that includes hundreds of manufacturers' data sheets and application notes on more than 38,000 integrated circuits. However, it is an expensive book and does not describe the function of some of the ICs.

The *Encyclopedia of Integrated Circuits,* by Walter H. Buchshaum (Englewood Cliffs, N.J.: Prentice Hall, Inc., 1981), describes "families" of analog and digital ICs. For example, for an eight-input multiplexer it gives a description of its operation, a function block diagram, key parameters, applications, and other information.

Figure 9.10 is a sample of a GE/RCA data sheet for a logic element—a dynamic shift register. All the information presented in the four sheets is impor-

TABLE 9.1 IC MANUFACTURERS

Advanced Micro Devices, Inc. P.O. Box 3453 Sunnyvale, CA 94088	National Semiconductor 2900 Semiconductor Drive Santa Clara, CA 95051
Analog Devices Two Technology Way Norwood, MA 02062-9106	RCA Box 3200 Somerville, NJ 08876
Fairchild Semiconductor Corp. 10400 Ridgeview Court Cupertino, CA 95014	Siemens Corporation 186 Wood Avenue South Iselin, NJ 08830
Intel Books P.O. Box 58130 Santa Clara, CA 95052-08130	Standard Microsystems Corporation 35 Marcus Boulevard Hauppauge, NY 11788
Gould AMI Semiconductors 3800 Homestead Road Santa Clara, CA 95051	Texas Instruments Incorporated Data Book Sales P.O. Box 655012, M/S 54 Dallas, TX 75265
Motorola, Inc. 5005 East McDowell Road Phoenix, AZ 85008	

Texas Instruments Semiconductor Book Descriptions

Interface Circuits Data Book — 1981, 700 pages LCC5921
Includes specifications and applications information on TTL logic interface circuits, line drivers/receivers and peripheral drivers.

TMS4500A Dynamic Ram Controller User's Manual — 1985, 39 pages SCG690A
Describes functional operation of the TMS4500A and how it can be used in a microprocessor system. Includes introduction, basic operation, typical implementation, design criteria and application examples.

High Speed CMOS Logic Data Book — 1984, 800 pages SCLD001A
Detailed specifications and application information on the TI family of HCMOS (High-speed CMOS) logic devices. Includes product selection guide, glossary and alphanumeric index.

Telecommunications Circuits Data Book — 1986, 380 pages SCTD001
Detailed specifications and application information on the TI family of telecommunications products. This includes switching and transmission, codecs, filters, combos, FSK modems, subscriber line control, and subscriber products. Contains product selection guide, glossary and alphanumeric index.

TTL Data Book Vol. III — 1984, 729 pages SDAD001A
Replaced by SDAD001B

TTL Data Book Vol. III Supplement — 1984, 255 pages SDAD003
Included in SDAD001B

ALS/AS Logic Data Book (TTL Vol. III) — 1986, 1100 pages SDAD001B
Detailed specifications and applications information on the TI family of Advanced Low-power Schottky (ALS) & Advanced Schottky (AS) logic devices, and includes a functional index of all TI bipolar digital devices. Supercedes SDAD001A and SDAD003 (TTL Vol. III and Vol. III Supplement).

SN74AS888/890 Bit-Slice User's Guide (8-Bit) — 1985, 222 pages SDBU001A
Introduces the AS888 and AS890 8-bit bit-slice processors, provides detailed specifications on their operation, and outlines support tools available for system development.

Texas Instruments Incorporated
Data Book Sales
P.O. Box 655012, M/S 54
Dallas, TX 75265

SN74AS897 16-Bit Barrel Shifter User's Guide — 1985, 60 pages SDBU003
Detailed specifications and application information on the SN74AS897 16-bit barrel shifter. Includes architecture description, 16-bit/32-bit shift operation examples, instruction set, multiple cycle operation and electrical characteristics.

74AS-EVM-8 Bit-Slice Evaluation Module User's Guide — 1985, 78 pages SDBU004
Describes the functional operation of the 74AS-EVM-8 bit-slice evaluation module. This bit-slice development and evaluation system consists of a single board, dual-processor computer, extensive monitor software, nonvolatile memory and software for TI or IBM PCs.

TTL Data Book Vol. II — 1985, 1400 pages SDLD001
Detailed specifications and application information on the TI family of Low-power Schottky (LS), Schottky (S), and standard TTL logic devices.

LSI Logic Data Book — 1986, 872 pages SDVD001
Detailed specifications and application information on LSI special functions that add to AS, ALS, and LS families. Includes 8-bit bit-slice devices, FIFOs, EDAC, memory mapping units, and 8-, 9-, & 10-bit registers.

TTL Data Book Vol. I — 1984, 336 pages SDYD001
Product guide for all TI TTL devices, functional indexes, alphanumeric index, and general information.

Overview of IEEE Standard 91 — 1984, 32 pages SDYZ001
A brief condensed overview of the symbolic language for digital logic circuits by the IEEE and the IEC Technical Committee TC-3 Incorporated in IEEE Std. 91-1984.

TTL Data Book IV — 1985, 416 pages SDZD001A
Replaced by SDZD001B

TTL Data Book Vol. IV — 1986, 486 pages SDZD001B
Detailed specifications and application information on the TI family of bipolar field-programmable logic (FPLA & PAL), programmable read-only memories (PROM), random-access memories (RAM), microprocessors, and support circuits. Supercedes SDZD001A.

Power Products Data Book — 1985, 624 pages SLPD001
Detailed specification on standard JEDEC devices, advanced planar and darlington TI devices, BD, BDX, BU, BUX, and BUY, TO-3 and plastic packaged devices, and TIC triacs and SCRs.

Linear and Interface Circuits Applications Vol. I — 1985, 160 pages SLYA001
Containing basic theory, key characteristics, and applications, the prime objective of the book is to assist the user of the broad range of linear and interface integrated circuits to understand the operation principles. Such circuits are normally used to sense or activate circuits, subsystems, or systems that interface to complex digital systems. Covers amplifiers, comparators, timers and voltage regulators.

Linear and Interface Circuits Applications Vol. II — 1985, 136 pages SLYA002
Continues with the same content as volume I. Emphasis is on circuits used for data transmission and for the display of information at the human-machine interface. LED, plasma (both dc and ac), gas discharge, vacuum fluorescent, and thin-film electroluminescent display drivers are included. For data transmission, all the various RS-232C, 423A, 422A, 485, and IEEE-488 line drivers, receivers and transceivers are included.

Linear and Interface Circuits Application (Vol. I & II), SLYX001
Combines SLYA001 and SLYA002 in a binder (see above).

Linear Circuits Data Book — 1984, 792 pages SLYD001
Detailed specifications on operational amplifiers, voltage comparators, voltage regulators, data-acquisition devices, A/D converter, timers, switches, amplifiers, and special functions. Includes LinCMOS™ functions. Contains product guide, interchangeability guide, glossary, and alphanumeric index.

MOS Memory Data Book — 1986, 804 pages SMYD006
Detailed specifications and application information on dynamic RAMs, static RAMs, EPROMs, ROMs, cache address comparators, and memory controllers. Contains product guide, interchangeability guide, glossary, and alphanumeric index. Also, chapters on testing and reliability.

Figure 9.9 Sample of IC data information available. (Courtesy of Texas Instruments Incorporated.)

CD4062A Types

CMOS 200-Stage Dynamic Shift Register

MAXIMUM RATINGS, *Absolute-Maximum Values:*

STORAGE TEMPERATURE RANGE (T_{stg}) .... -65 to +150°C
OPERATING-TEMPERATURE RANGE (T_A):
PACKAGE TYPES K, T, H .... -55 to +125°C
DC SUPPLY-VOLTAGE RANGE, (V_{DD})
(Voltages referenced to V_{SS} Terminal) .... -0.5 to +15 V
POWER DISSIPATION PER PACKAGE (P_D):
For T_A = -55 to +100°C (PACKAGE TYPES K, T) .... 500 mW
For T_A = +100 to +125°C (PACKAGE TYPES K, T) .... Derate Linearly at 12 mW/°C to 200 mW
DEVICE DISSIPATION PER OUTPUT TRANSISTOR
For T_A = FULL PACKAGE-TEMPERATURE RANGE (All Package Types) .... 100 mW
INPUT VOLTAGE RANGE, ALL INPUTS .... -0.5 to V_{DD} +0.5 V
LEAD TEMPERATURE (DURING SOLDERING):
At distance 1/16 ± 1/32 inch (1.59 ± 0.79 mm) from case for 10 s max. .... +265°C

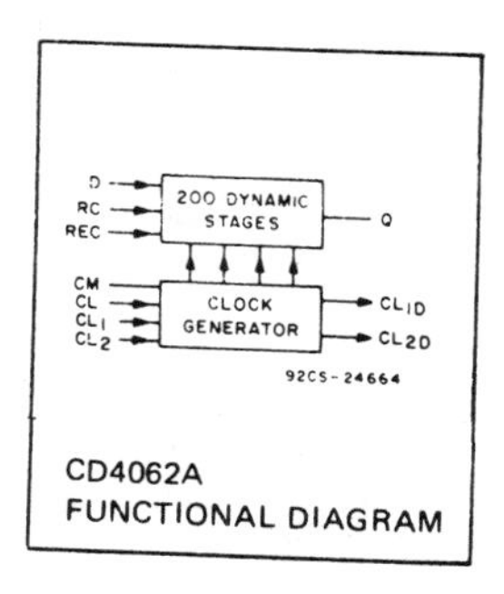

CD4062A
FUNCTIONAL DIAGRAM

The RCA-CD4062A is a 200-stage dynamic shift register with provision for either single- or two-phase clock input signals. Single-phase-clocked operation is intended for low-power, low clock-line capacitance requirements. Single-phase clocking is specified for medium-speed operation (< 1 MHz) at supply voltages up to 10 volts. Clock input capacitance is extremely low (< 5 pF), and clock rise and fall times are non-critical. The clock-mode signal (CM) must be low for single-phase operation.

Two-phase clock-input signals may be used for high-speed operation (up to 5 MHz) or to further reduce clock rise and fall time requirements at low speeds. Two-phase operation is specified for supply voltages up to 15 volts. Clock input capacitance is only 50 pF/phase. The clock-mode signal (CM) must be high for two-phase operation. The single-phase-clock input has an internal pull-down device which is activated when CM is high and may be left unconnected in two-phase operation.

The logic level present at the data input is transferred into the first stage and shifted one stage at each positive-going clock transition for single-phase operation, and at the positive-going transition of CL_1 for two-phase operation.

The CD4062A-Series types are supplied in 12-lead hermetic TO-5 packages (T suffix), 16-lead ceramic flat packages (K suffix), and in chip form (H suffix).

Features:

- **Minimum shift rates over full temperature range—**

  Single-phase clock: 3 V ≤ V_{DD} ≤ 10 V; f_{min} = 10 kHz; -55° C ≤ T_A ≤ +125° C (f_{min} = 1 kHz up to T_A ≤ 75° C)

  Two-phase clock: 3 V ≤ V_{DD} ≤ 15 V; f_{min} = 10 kHz; -55° C ≤ T_A ≤ +125° C (f_{min} = 1 kHz up to T_A ≤ 75° C)

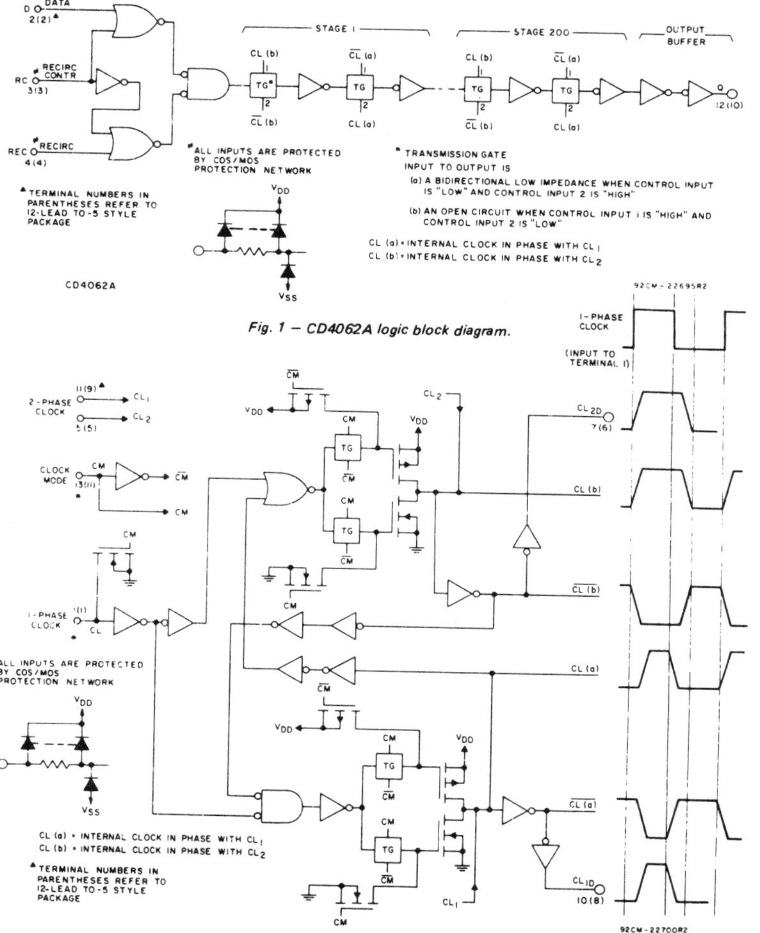

Fig. 1 – CD4062A logic block diagram.

Fig. 2 – Clock circuit logic diagram.

Figure 9.10 Sample of GE/RCA data sheet. (Copyright 1983 by GE/RCA Corporation.)

RECOMMENDED OPERATING CONDITIONS at T_A = 25° C, Except as Noted.
For maximum reliability, nominal operating conditions should be selected so that operation is always within the following ranges:

CHARACTERISTIC	V_{DD} (V)	LIMITS MIN.	LIMITS MAX.	UNITS
Supply-Voltage Range (For T_A = Full Package-Temperature Range): Single-Phase Clock		3	10	V
Two-Phase Clock		3	12	V
Clock Input Frequency, f_{CL}*	5	0.15	500	kHz
	10	0.15	1000	
Clock Pulse Width, t_W*	5	250	66.7 X 10⁶	ns
	10	500	66.7 X 10⁶	
Clock Rise or Fall Times, t_rCL or t_fCL*	5	–	10	μs
	10	–	1	
Data Hold Time, t_H*	5	150	–	ns
	10	50	–	

* For single-phase clock, 50% duty cycle

Two-Phase Clock Operation (CL_1, CL_2); Clock Mode (CM) = High; 3 V ≤ V_{DD} ≤ 15 V. See Figure 4.

CHARACTERISTIC	TEST CONDITIONS V_{DD} V	LIMITS MIN.	LIMITS TYP.	LIMITS MAX.	UNITS
Maximum Clock Input Frequency, f_{CL}	5	1.25	2.5	–	MHz
	10	2.5	5	–	
Minimum Clock Input Frequency, f_{CL}	5	150	10	–	Hz
	10	150	10	–	
Clock Overlap Time (CL2, CL1; 90%, 10%; td1, td2)		40	–	–	ns
Average Input Capacitance, C_I CL_1, CL_2		–	50	–	pF
Propagation Delays; t_{PHL}, t_{PLH} CL_1 to Q	5	–	250	500	ns
	10	–	100	200	
CL_1 to CL_{1D}	5	–	250	500	
CL_2 to CL_{2D}	10	–	100	200	
Minimum Data Setup Time t_S (CL_2, D)	5	–	150	300	ns
	10	–	50	100	
Minimum Data Hold Time t_H (CL_2)	5	–	–	0	ns
	10	–	–	0	
Clock Rise and Fall Times t_rCL_1, CL_2 t_fCL_1, CL_2		No Restrictions If Clock Overlap Requirement Is Met			

Features (Cont'd):

- Low power dissipation
  0.3 mW/bit at 1 MHz and 10 V
  0.04 mW/bit at 0.5 MHz and 5V
  (alternating 1-0 data pattern)
- Data output TTL-DTL compatible
- Recirculating capability
- Delayed two-phase clock outputs available for cascading registers
- Asynchronous ripple-type presettable to all 1's or 0's
- Ultra-low-power-dissipation standby operation
- Quiescent current specified to 15 V
- Maximum input leakage current of 1 μA at 15 V (full package-temperature range)
- 1-V noise margin (full package-temperature range)

Applications:

- Serial shift registers
- Time-delay circuits
- CRT refresh memory
- Long serial memory

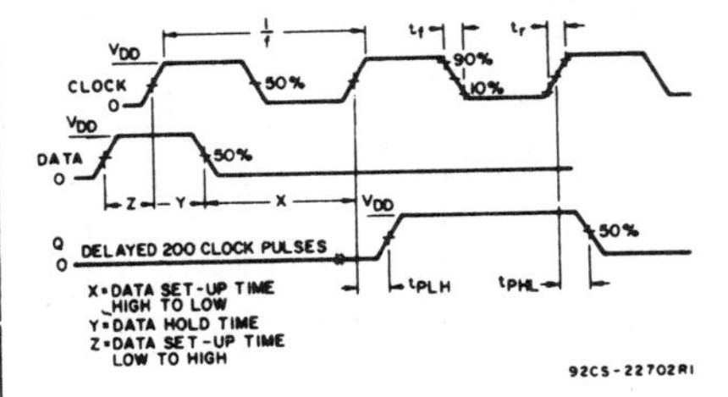

Fig. 3 – Timing diagram–single-phase clock.

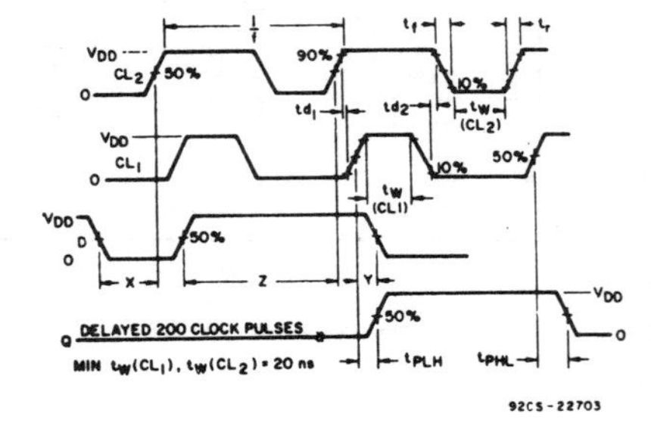

Fig. 4 – Timing diagram–two-phase clock.

Figure 9.10 (*Continued*)

CD4062A Types

STATIC ELECTRICAL CHARACTERISTICS

CHARACTERISTICS	CONDITIONS			LIMITS AT INDICATED TEMPERATURES (°C)				UNITS
	V_O (V)	V_{IN} (V)	V_{DD} (V)	-55	+25 TYP.	+25 LIMIT	+125	
Quiescent Device Current, I_L Max. CM=High, CL_1=High, CL_2=Low	–	–	5	12	0.5	12	720	μA
	–	–	10	25	1	25	1500	μA
	–	–	15	50	1	50	2000	μA
Output Voltage: Low Level, V_{OL}	–	5	5	0 Typ.; 0.05 Max				V
	–	10	10	0 Typ.; 0.05 Max				V
High Level V_{OH}	–	0	5	4.95 Min.; 5 Typ.				V
	–	0	10	9.95 Min.; 10 Typ.				V
Noise Immunity: Inputs Low, V_{NL}	4.2	–	5	1.5 Min.; 2.25 Typ.				V
	9	–	10	3 Min.; 4.5 Typ.				V
Inputs High V_{NH}	0.8	–	5	1.5 Min.; 2.25 Typ.				V
	1	–	10	3 Min.; 4.5 Typ.				V
Noise Margin: Inputs Low, V_{NML}	4.5	–	5	1 Min.				V
	9	–	10	1 Min.				V
Inputs High, V_{NMH}	0.5	–	5	1 Min.				V
	1	–	10	1 Min.				V
Output Drive Current: N-Channel (Sink), I_DN Min. Q Output	0.4	–	4.5	1.6	2.6	1.3	0.91	mA
	0.5	–	10	5	8*	4	3.2	mA
CL_{1D}, CL_{2D}	0.5	–	5	0.87	1.4	0.7	0.49	mA
	0.5	–	10	2.2	3.6	1.8	1.26	mA
P-Channel (Source): I_DP Min. Q Output	4.5	–	5	-0.31	-0.5	-0.25	-0.17	mA
	2.5	–	5	-0.93	-1.5	-0.75	-0.52	mA
	9.5	–	10	-0.87	-1.4	-0.7	-0.49	mA
CL_{1D}, CL_{2D}	4.5	–	5	-0.43	-0.7	-0.35	-0.24	mA
	9.5	–	10	-1.1	-1.8	-0.9	-0.63	mA
Input Leakage Current, I_{IL}, I_{IH}	Any Input –	–	15	$\pm10^{-5}$ Typ., ±1 Max.				μA

* Maximum power dissipation rating ⩽ 200 mW.

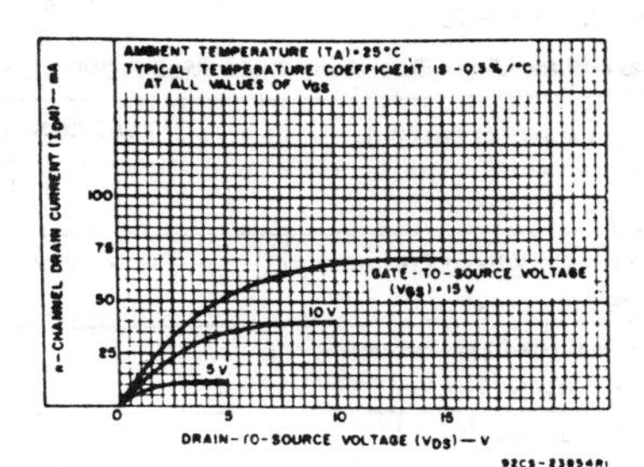

Fig. 5— Typical n-channel drain characterstics for Q output.

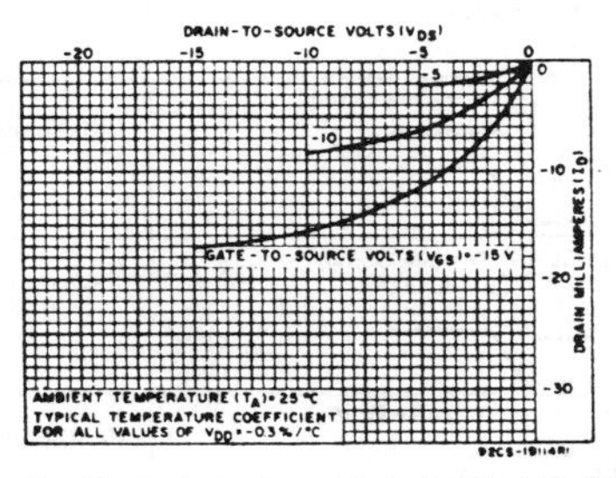

Fig. 6— Typical p-channel drain characteristics for Q output.

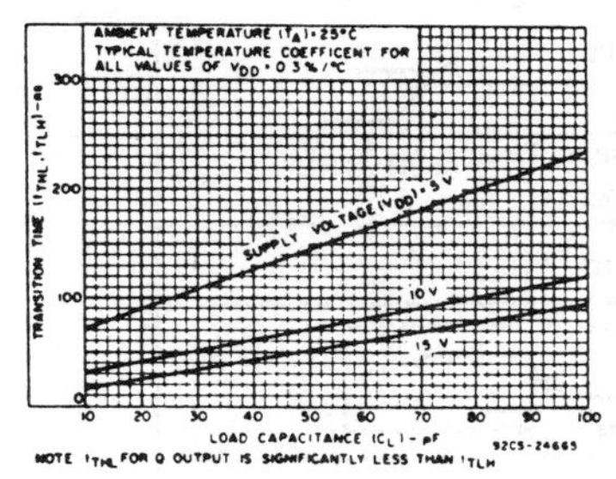

Fig. 7— Typical transition time vs. C_L for data outputs.

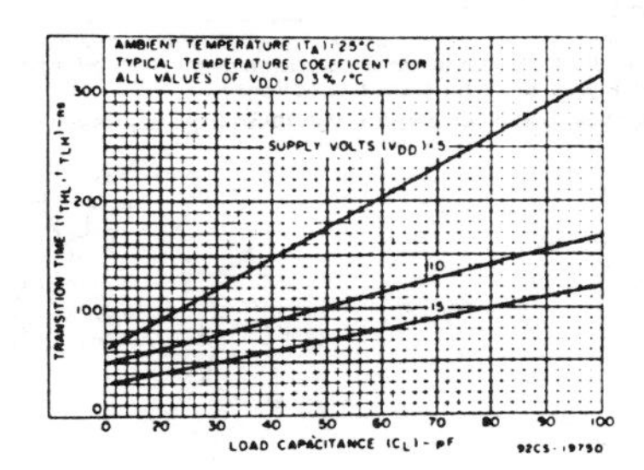

Fig. 8— Typical transition time vs. C_L for delayed clock output.

Figure 9.10 (*Continued*)

DYNAMIC CHARACTERISTICS AT T_A = 25°C, V_{SS} = 0 V, C_L = 50 pF, Input t_r, t_f = 20 ns, except t_rCL and t_fCL

Single-Phase-Clock Operation; Clock Mode (CM) = Low; 3 V < V_{DD} < 10 V (See Figure 3)

CHARACTERISTIC	TEST CONDITIONS	V_{DD} V	LIMITS MIN.	TYP.	MAX.	UNITS
Maximum Clock Input Frequency, f_{CL} (50% Duty Cycle)	t_r, t_f=20 ns	5	0.5	1	–	MHz
		10	1	2	–	
Minimum Clock Input Frequency, f_{CL} (50% Duty Cycle)		5	150	10	–	Hz
		10	150	10	–	
Clock Rise and Fall Times** t_rCL, t_fCL		5	–	–	10	μs
		10	–	–	1	
Average Input Capacitance, C_I	All Inputs Except CL_1 and CL_2		–	5	–	pF
Propagation Delays: CL to Q		5	–	1000	2000	ns
		10	–	400	800	
CL to CL_{1D} (Positive Going)	(50% Points)	5	–	750	1500	ns
		10	–	300	600	
CL to CL_{2D} (Positive Going)	(50% Points)	5	–	500	1000	
		10	–	200	400	
CL to CL_{1D} (Negative Going)	(50% Points)	5	–	450	900	
		10	–	175	350	
CL to CL_{2D} (Negative Going)	(50% Points)	5	–	750	1500	
		10	–	300	600	
Transition Time: t_{TLH}, t_{THL} Q Output		5	–	100	200	ns
		10	–	50	100	
CL_{1D}, CL_{2D}		5	–	200	400	
		10	–	100	200	
Data Set-Up Time t_S		5	–	–	0	ns
		10	–	–	0	
▲Data Hold Time t_H		5	–	–	150	ns
		10	–	–	150	

** If more than one unit is cascaded in single-phase parallel clocked application, t_rCL should be made less than or equal to the sum of the propagation delay at 15 pF, and the transition time of the output driving stage. (See Figs. 5 and 7 for cascading options.)

▲ Use of delayed clock permits high-speed logic to precede CD4062A register (see cascade register operation).

CL, CM, D, Q, RC, CL1, REC, CL1D, CL2, VSS, CL2D, VDD — TOP VIEW

92CS-22693

CD4062AT

TERMINAL DIAGRAM

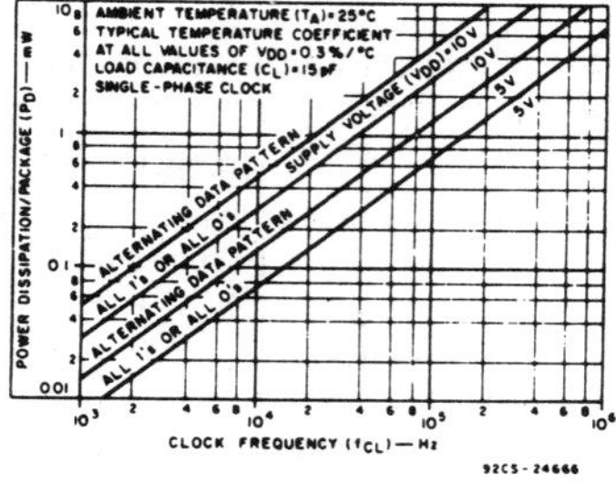

Fig. 9— Typical power dissipation vs. frequency.

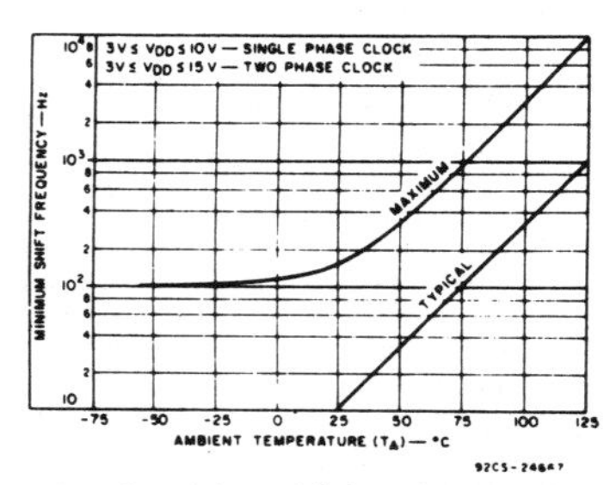

Fig. 10— Minimum shift frequency vs. ambient temperature.

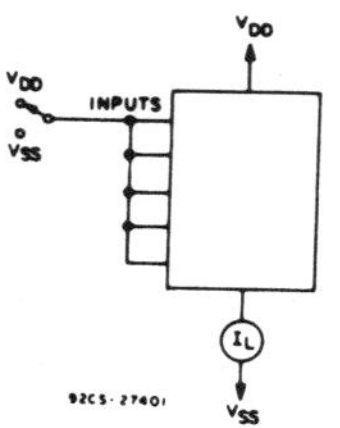

Fig. 11— Quiescent-device-current test circuit.

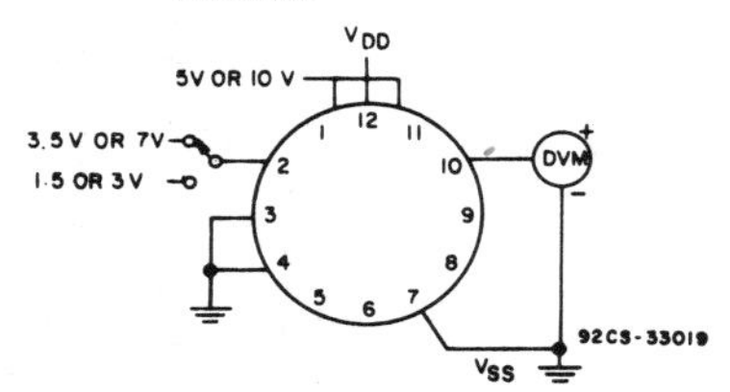

Fig. 12— Noise-immunity test circuit.

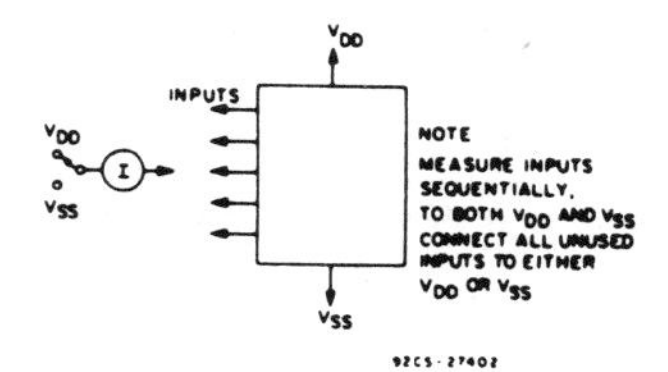

Fig. 13—Input-leakage-current test circuit.

Figure 9.10 (*Continued*)

TYPES SN54160 THRU SN54163, SN54LS160A THRU SN54LS163A, SN54S162, SN54S163, SN74160 THRU SN74163, SN74LS160A THRU SN74LS163A, SN74S162, SN74S163 SYNCHRONOUS 4-BIT COUNTERS

OCTOBER 1976—REVISED DECEMBER 1983

'160,'161,'LS160A,'LS161A . . . SYNCHRONOUS COUNTERS WITH DIRECT CLEAR
'162,'163,'LS162A,'LS163A,'S162,'S163 . . . FULLY SYNCHRONOUS COUNTERS

- **Internal Look-Ahead for Fast Counting**
- **Carry Output for n-Bit Cascading**
- **Synchronous Counting**
- **Synchronously Programmable**
- **Load Control Line**
- **Diode-Clamped Inputs**

SERIES 54', 54LS', 54S' . . . J OR W PACKAGE
SERIES 74' . . . J OR N PACKAGE
SERIES 74LS', 74S' . . . D, J OR N PACKAGE
(TOP VIEW)

Pin	Name	Pin	Name
1	$\overline{CLR}$	16	V_{CC}
2	CLK	15	RCO
3	A	14	Q_A
4	B	13	Q_B
5	C	12	Q_C
6	D	11	Q_D
7	ENP	10	ENT
8	GND	9	$\overline{LOAD}$

NC—No internal connection

TYPE	TYPICAL PROPAGATION TIME, CLOCK TO Q OUTPUT	TYPICAL MAXIMUM CLOCK FREQUENCY	TYPICAL POWER DISSIPATION
'160 thru '163	14 ns	32 MHz	305 mW
'LS162A thru 'LS163A	14 ns	32 MHz	93 mW
'S162 and 'S163	9 ns	70 MHz	475 mW

SERIES 54LS', 54S' . . . FK PACKAGE
SERIES 74LS', 74S' . . . FN PACKAGE
(TOP VIEW)

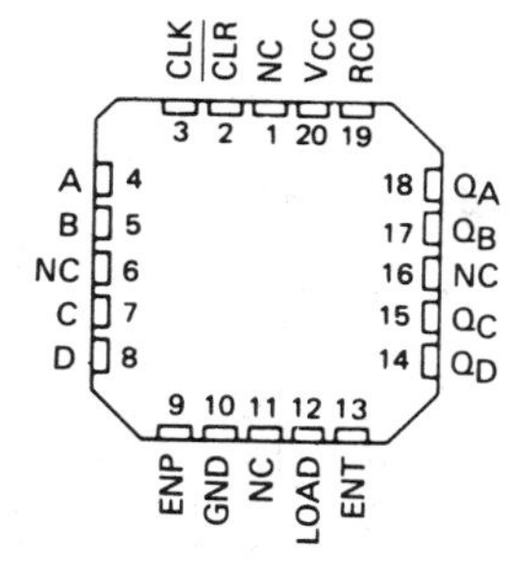

NC—No internal connection

description

These synchronous, presettable counters feature an internal carry look-ahead for application in high-speed counting designs. The '160,'162,'LS160A,'LS162A, and 'S162 are decade counters and the '161,'163,'LS161A,'LS163A, and 'S163 are 4-bit binary counters. Synchronous operation is provided by having all flip-flops clocked simultaneously so that the outputs change coincident with each other when so instructed by the count-enable inputs and internal gating. This mode of operation eliminates the output counting spikes that are normally associated with asynchronous (ripple clock) counters, however counting spikes may occur on the (RCO) ripple carry output. A buffered clock input triggers the four flip-flops on the rising edge of the clock input waveform.

These counters are fully programmable; that is, the outputs may be preset to either level. As presetting is synchronous, setting up a low level at the load input disables the counter and causes the outputs to agree with the setup data after the next clock pulse regardless of the levels of the enable inputs. Low-to-high transitions at the load input of the '160 thru '163 should be avoided when the clock is low if the enable inputs are high at or before the transition. This restriction is not applicable to the 'LS160A thru 'LS163A or 'S162 or 'S163. The clear function for the '160, '161,'LS160A, and 'LS161A is asynchronous and a low level at the clear input sets all four of the flip-flop outputs low regardless of the levels of clock, load, or enable inputs. The clear function for the '162,'163,'LS162A,'LS163A, 'S162, and 'S163 is synchronous and a low level at the clear input sets all four of the flip-flop outputs low after the next clock pulse, regardless of the levels of the enable inputs. This synchronous clear allows the count length to be modified easily as decoding the maximum count desired can be accomplished with one external NAND gate. The gate output is connected to the clear input to synchronously clear the counter to 0000 (LLLL). Low-to-high transitions at the clear input of the '162 and '163 should be avoided when the clock is low if the enable and load inputs are high at or before the transition.

Figure 9.11 Sample of TI's data sheet using new logic symbol. (Courtesy of Texas Instruments Incorporated.)

TYPES SN54160 THRU SN54163, SN54LS160A THRU SN54LS163A, SN54S162, SN54S163, SN74160 THRU SN74163, SN74LS160A THRU SN74LS163A, SN74S162, SN74S163 SYNCHRONOUS 4-BIT COUNTERS

The carry look-ahead circuitry provides for cascading counters for n-bit synchronous applications without additional gating. Instrumental in accomplishing this function are two count-enable inputs and a ripple carry output. Both count-enable inputs (P and T) must be high to count, and input T is fed forward to enable the ripple carry output. The ripple carry output thus enabled will produce a high-level output pulse with a duration approximately equal to the high-level portion of the Q_A output. This high-level overflow ripple carry pulse can be used to enable successive cascaded stages. High-to-low-level transitions at the enable P or T inputs of the '160 thru '163 should occur only when the clock input is high. Transitions at the enable P or T inputs of the 'LS160A thru 'LS163A or 'S162 and 'S163 are allowed regardless of the level of the clock input.

'LS160A thru 'LS163A,'S162 and 'S163 feature a fully independent clock circuit. Changes at control inputs (enable P or T, or load) that will modify the operating mode have no effect until clocking occurs. The function of the counter (whether enabled, disabled, loading, or counting) will be dictated solely by the conditions meeting the stable setup and hold times.

The 'LS160A thru 'LS163A are completely new designs. Compared to the original 'LS160 thru 'LS163, they feature 0-nanosecond minimium hold time and reduced input currents I_{IH} and I_{IL}.

N-BIT SYNCHRONOUS COUNTERS

This application demonstrates how the look-ahead carry circuit can be used to implement a high-speed n-bit counter. The '160, '162, 'LS160A, 'LS162A, or 'S162 will count in BCD and the '161, '163, 'LS161A, 'LS163A or 'S163 will count in binary. Virtually any count mode (modulo-N, N_1-to-N_2, N_1-to-maximum) can be used with this fast look-ahead circuit.

logic symbols

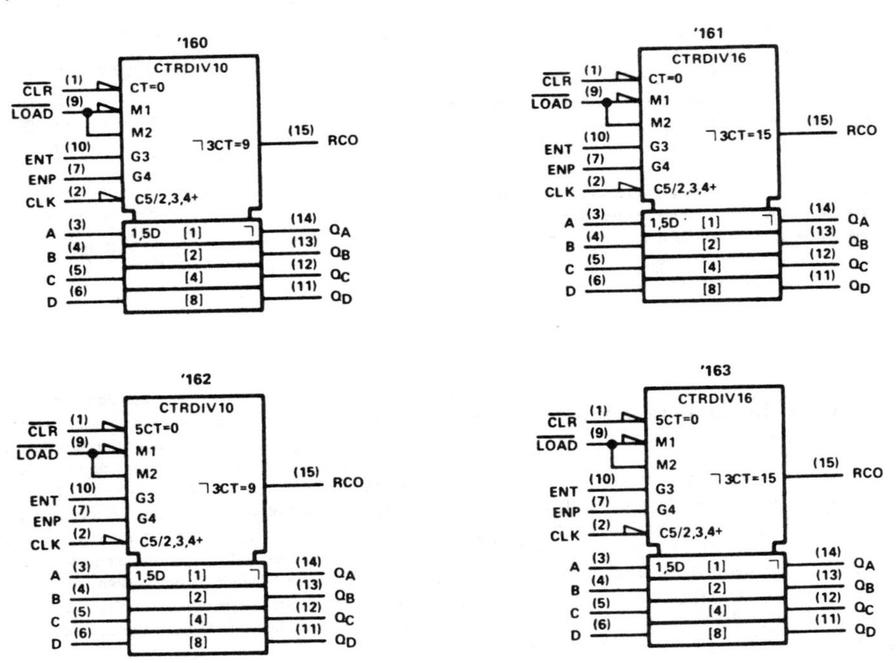

Pin numbers shown on logic notation are for D, J or N packages.

Figure 9.11 *(Continued)*

logic diagram

SN54160, SN74160 SYNCHRONOUS DECADE COUNTERS

SN54162, SN74162 synchronous decade counters are similar; however the clear is synchronous as shown for the SN54163, SN74163 binary counters at right.

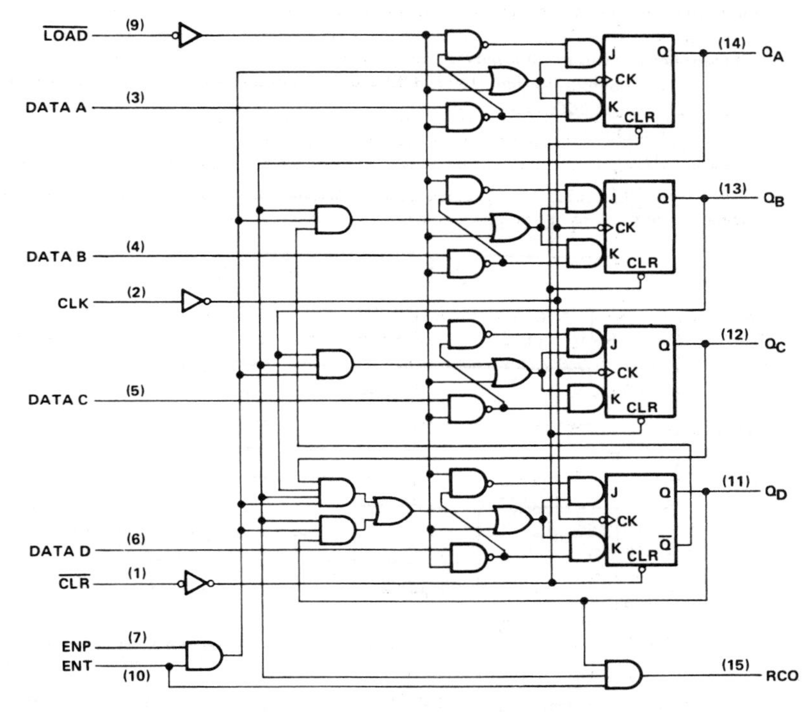

Pin numbers shown on logic notation are for D, J or N packages.

Figure 9.11 (*Continued*)

tant to the design engineer. But for a person trying to understand the operation of this shift register, some of the information is irrelevant and should be ignored. On the first sheet, the description and the diagrams are vital. The "maximum ratings" on sheet 1, the "static electrical characteristics" on sheet 3, and the "dynamic characteristics" on sheet 4 can safely be ignored. The "applications," timing diagrams, and some of the characteristics provide useful information in interpreting the operation of the shift register.

Figure 9.11 gives Texas Instruments data sheets, using the new logic symbol, for a synchronous 4-bit counter. Notice that these data sheets cover several types of this 4-bit counter. For purposes of illustration we picked the type SN54160 and deleted those pages that had no information on this type. The first

'160, '162, 'LS160A, 'LS162A, 'S162 DECADE COUNTERS

typical clear, preset, count, and inhibit sequences

Illustrated below is the following sequence:

1. Clear outputs to zero ('160 and 'LS160A are asynchronous; '162, 'LS162A, and 'S162 are synchronous)
2. Preset to BCD seven
3. Count to eight, nine, zero, one, two, and three
4. Inhibit

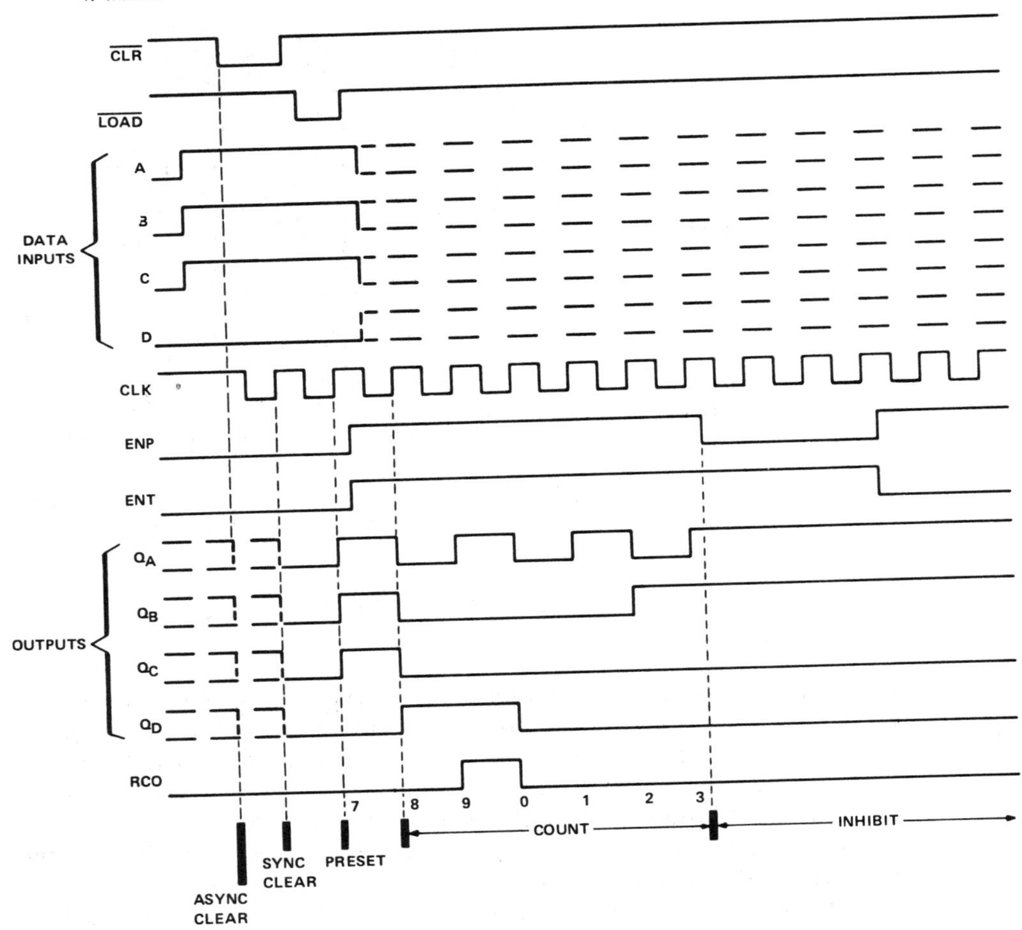

Figure 9.11 *(Continued)*

schematics of inputs and outputs

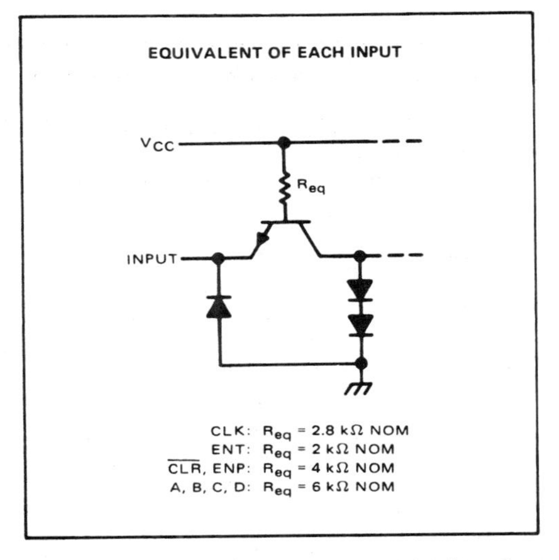

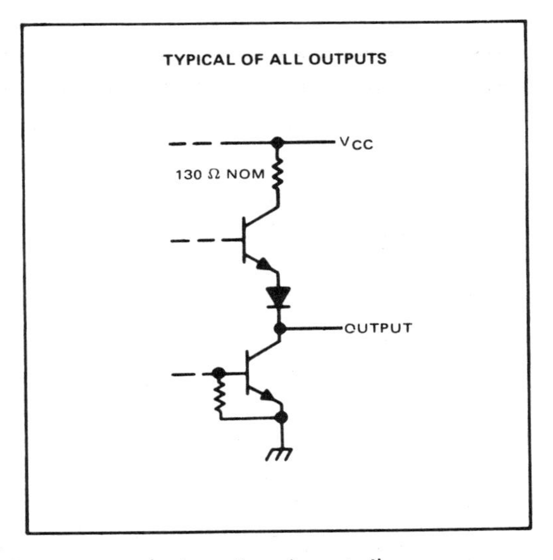

absolute maximum ratings over operating free-air temperature range (unless otherwise noted)

Supply voltage, V_{CC} (see Note 1) . . . 7 V
Input voltage . . . 5.5 V
Interemitter voltage (see Note 2) . . . 5.5 V
Operating free-air temperature range: SN54' Circuits . . . −55°C to 125°C
SN74' Circuits . . . 0°C to 70°C
Storage temperature range . . . −65°C to 150°C

NOTES: 1. Voltage values, except interemitter voltage, are with respect to network ground terminal.
2. This is the voltage between two emitters of a multiple-emitter transistor. For these circuits, this rating applies between the count enable inputs P and T.

recommended operating conditions

		SN54160, SN54161 SN54162, SN54163			SN74160, SN74161 SN74162, SN74163			UNIT
		MIN	NOM	MAX	MIN	NOM	MAX	
Supply voltage, V_{CC}		4.5	5	5.5	4.75	5	5.25	V
High-level output current, I_{OH}				−800			−800	μA
Low-level output current, I_{OL}				16			16	mA
Clock frequency, f_{clock}		0		25	0		25	MHz
Width of clock pulse, $t_{w(clock)}$		25			25			ns
Width of clear pulse, $t_{w(clear)}$		20			20			ns
Setup time, t_{su} (see Figures 1 and 2)	Data inputs A, B, C, D	20			20			ns
	ENP	20			20			
	$\overline{LOAD}$	25			25			
	$\overline{CLR}$ †	20			20			
Hold time at any input, t_h		0			0			ns
Operating free-air temperature, T_A		−55		125	0		70	°C

†This applies only for '162 and '163, which have synchronous clear inputs.

Figure 9.11 *(Continued)*

electrical characteristics over recommended operating free-air temperature range (unless otherwise noted)

PARAMETER			TEST CONDITIONS†	SN54160, SN54161 SN54162, SN54163			SN74160, SN74161 SN74162, SN74163			UNIT
				MIN	TYP‡	MAX	MIN	TYP‡	MAX	
V_{IH}	High-level input voltage			2			2			V
V_{IL}	Low-level input voltage					0.8			0.8	V
V_{IK}	Input clamp voltage		V_{CC} = MIN, $I_I = -12$ mA			−1.5			−1.5	V
V_{OH}	High-level output voltage		V_{CC} = MIN, V_{IH} = 2 V, V_{IL} = 0.8 V, $I_{OH} = -800\ \mu A$	2.4	3.4		2.4	3.4		V
V_{OL}	Low-level output voltage		V_{CC} = MIN, V_{IH} = 2 V, V_{IL} = 0.8 V, I_{OL} = 16 mA		0.2	0.4		0.2	0.4	V
I_I	Input current at maximum input voltage		V_{CC} = MAX, V_I = 5.5 V			1			1	mA
I_{IH}	High-level input current	CLK or ENT	V_{CC} = MAX, V_I = 2.4 V			80			80	μA
		Other inputs				40			40	
I_{IL}	Low-level input current	CLK or ENT	V_{CC} = MAX, V_I = 0.4 V			−3.2			−3.2	mA
		Other inputs				−1.6			−1.6	
I_{OS}	Short-circuit output current§		V_{CC} = MAX	−20		−57	−18		−57	mA
I_{CCH}	Supply current, all outputs high		V_{CC} = MAX, See Note 3		59	85		59	94	mA
I_{CCL}	Supply current, all outputs low		V_{CC} = MAX, See Note 4		63	91		63	101	mA

†For conditions shown as MIN or MAX, use the appropriate value specified under recommended operating conditions.
‡All typical values are at V_{CC} = 5 V, T_A = 25°C.
§Not more than one output should be shorted at a time.
NOTES: 3. I_{CCH} is measured with the load input high, then again with the load input low, with all other inputs high and all outputs open.
4. I_{CCL} is measured with the clock input high, then again with the clock input low, with all other inputs low and all outputs open.

switching characteristics, V_{CC} = 5 V, T_A = 25°C

PARAMETER¶	FROM (INPUT)	TO (OUTPUT)	TEST CONDITIONS	MIN	TYP	MAX	UNIT
f_{max}			C_L = 15 pF, R_L = 400 Ω, See Figures 1 and 2 and Note 5	25	32		ns
t_{PLH}	CLK	RCO			23	35	ns
t_{PHL}					23	35	
t_{PLH}	CLK ($\overline{LOAD}$ input low)	Any Q			13	20	ns
t_{PHL}					15	23	
t_{PLH}	CLK ($\overline{LOAD}$ input high)	Any Q			17	25	ns
t_{PHL}					19	29	
t_{PLH}	ENT	RCO			11	16	ns
t_{PHL}					11	16	
t_{PHL}	$\overline{CLR}$	Any Q			26	38	ns

¶f_{max} = Maximum clock frequency
t_{PLH} = propagation delay time, low-to-high-level output
t_{PHL} = propagation delay time, high-to-low-level output
NOTE 5: Propagation delay for clearing is measured from the clear input for the '160 and '161 or from the clock input transition for the '162 and '163.

Figure 9.11 *(Continued)*

four sheets provide needed information for interpretation. The last two sheets can be ignored.

9.4 MICROPROCESSORS

As the ultimate in complex circuitry, the microprocessor requires a good bit of study if its operation in a circuit is to be understood completely. Notice in Table 9.2 that entire books have been written on individual microprocessors. A short description of individual types is contained in *Encyclopedia of Integrated Circuits,* by Walter H. Buchsbaum, as shown in Fig. 9.12.

In any logic circuit with a memory, it will never be possible to understand the circuit completely unless you know what is in the memory. If the information in a ROM is considered proprietary by the manufacturer, it will not be possible to obtain this information.

If the program in the memory is available, it may be in machine code (1's and 0's). If so, it must be converted to assembly language, which can be interpreted with the aid of the instruction set for the microprocessor.

Figure 9.13 shows the assembly language program for a vending machine. To understand this program, note that

> the label is a name of the location of an instruction in the microcomputer memory and is used on any instruction that may be jumped to from another instruction. The mnemonic is an abbreviation of the operation being performed. The operand field indicates which storage location is involved in the instruction. The comment field is simply a verbal explanation for what the instruction is doing. (Copyright 1983 Texas Instruments Incorporated. All Rights Reserved.)

TABLE 9.2 TEXTBOOKS ON MICROPROCESSORS

Programming the 65816 Microprocessor: Including 6502 and 65C02, by David Eyes and Ron Lichty (Englewood Cliffs, N.J.: Prentice Hall, Inc., 1985).

The 8085A Cookbook by Jonathan Titus and David Larsen

Practical Microprocessors: Hardware, Software, and Troubleshooting, by Hewlett-Packard Corporation, Palo Alto, Calif.

Intel 80386, by Intel Books, Santa Clara, Calif.

The 68000: Principles and Programming, by Leo J. Scanlon (Indianapolis, Indiana: Howard W. Sams, 1981).

M68000 16/32-Bit Microprocessor Programmer's Reference Manual, by Motorola, Inc., Phoenix, Ariz.

Using Microprocessors and Microcomputers: The 6800 Family, by Joseph D. Greenfield and William C. Wray (New York: John Wiley & Sons, 1981).

The Motorola MC 68000 Microprocessor Family: Assembly Language, Interface Design and System Design, by Thomas L. Harmon and Barbara Lawson (Englewood Cliffs, N.J.: Prentice Hall, Inc., 1985).

The 8086 Microprocessor: Architecture, Software & Interfacing Techniques by Walter Triebel and Avtar Singh (Englewood Cliffs, N.J.: Prentice Hall, 1985).

The 9900 Microprocessor: Architecture, Software & Interface Techniques by Walter Triebel and Avtar Singh (Englewood Cliffs, N.J.: Prentice Hall, 1984).

D.12.4 8-Bit Microprocessor (MP)

DESCRIPTION

The 8-bit microprocessor illustrated in Figure D.12.4 acts as controller in a microcomputer system. An 8-bit MP provides faster execution times and higher performances than a 4-bit MP. A 4-bit MP would require more than one add instruction when adding numbers greater than 16, but an 8-bit MP can handle numbers up to 256. The fetch and execute cycles of the 8-bit MP are essentially the same as described in D.12.3 for the 16-bit MP.

In addition to the fetch and execute cycle, the following control functions are included in most 8 and 16-bit microprocessors:

Reset Signal: All control programs have a starting point when power is applied. As a general guide, the starting address is location 0. When the input reset line is activated, the MP resets the program counter to "0" and enters a fetch cycle. Information from location 0 of the external memory is read into the microprocessor and executed. This information could represent a linking address to another section of the control program or an instruction.

Load Signal: The load signal operates in a similar fashion to the reset signal. The load signal is used to force the MP to another section of the

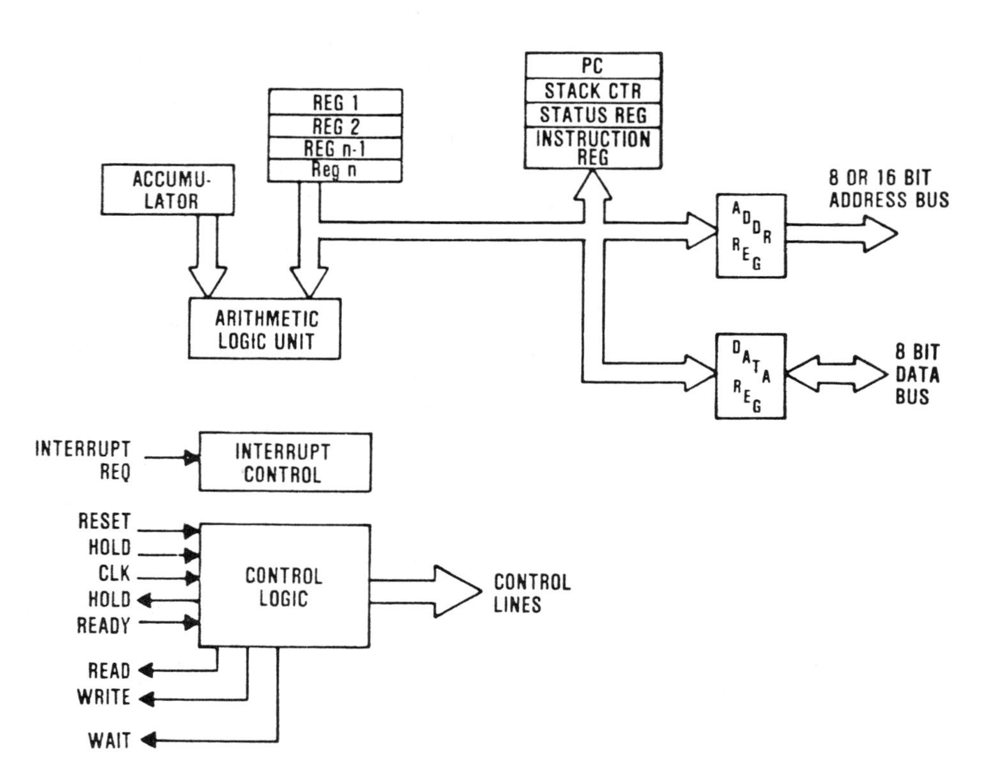

Figure 9.12 Microprocessor summary sheet. (From Walter H. Buchsbaum, *Encyclopedia of Integrated Circuits,* Prentice-Hall, Inc., Englewood Cliffs, N.J., 1981.)

control program. When the load input is activated, the program counter (PC) is set to a specific value and connected to the address bus. The value on the address bus is used to specify the memory location to be read back into the PC.

Based on the design of the specific MP, this information is used as a link or starting address for the program.

Hold/Hold A: These signals are used to stop the MP from executing the control program in order to execute direct memory access (DMA) transfer. The DMA operation transfers binary information between the memory and other parts of the system. DMA is used because the MP cannot operate at the speeds required by the incoming data, such as information from a disc or a high-speed data link. Activation of a DMA transfer occurs when the hold signal is applied. The MP stops at the end of the next memory cycle and puts the address and data bus, write and read lines into a high impedance state. This allows the DMA controller to utilize these lines without interference from the MP. The DMA data and address word is then connected to the memory bus and a write signal is generated.

Stack: This section of the MP is used to store the address of routines which have been interrupted until the interrupt is completed.

KEY PARAMETERS

a) *Direct addressing.* The number of memory locations directly accessible by the MP. This varies from 2K bytes to 128K bytes.
b) *Interrupts.* 8-bit MPs have interrupt requests but no provisions to process multiple interrupts. (See Section D.12.3.) External hardware must be included in the system to provide a multiple interrupt function.
c) *Instruction set.* The number of instructions goes from 8 to 96.
d) *Clock.* The highest frequency at which the MP operates. For 8-bit microprocessors the maximum is between 2 and 10 MHz.
e) *Internal registers.* The number of general purpose registers available within the MP. They vary from 0 to 128.

APPLICATIONS

8-bit MPs are used in the majority of microcomputer systems on the market. They find application in electronic games, appliance controls, personal computers, automotive controls, and industrial controls.

COMMENTS

Some manufacturers of the 8-bit MPs are designing their 16-bit MPs to be instructional compatible with the 8-bit control programs.

Figure 9.12 (*Continued*)

Label	Mnemonic	Operand	Comment
S0	CLR	R1	Clear TOTAL register
	CLR	R2	Clear VALUE register
	LI	12,>3C0	Set up I/O Base Address
	SBZ	16	Clear SELECT flip-flop
	SBO	16	
S3	SBZ	17	Clear COIN flip-flop
	SBO	17	
	LIIM	3	Activate S ($\overline{S}$) and C ($\overline{C}$) interrupts
S1	IDLE		Wait for interrupt
S2	TB	8	Check for nickel entry
	JEQ	NL	
	TB	9	Test for dime entry
	JEQ	DI	
	TB	10	Test for quarter entry
	JNE	S3	
	JMP	QT	
NL	AI	1, 5	Add 5¢ to TOTAL
	DCA		
	JMP	S3	Go To State 3
DI	AI	1,>10	Add 10¢ to TOTAL
	DCA		
	JMP	S3	Go To State 3
QT	AI	1,>25	Add 25¢ to TOTAL
	DCA		
	JMP	S3	Go To State 3
S4	STCR	3, 5	Get SELECT code
	MOV	VAL(3),2	Get item VALUE from table into R2
	C	2,1	Compare item VALUE to TOTAL
	JLE	VD	Go to VEND if VALUE ≤ TOTAL
	SBO	18	Generate DUMP
	SBZ	18	
	JMP	S0	Go to State 0 (S0) if VALUE > TOTAL
VD	SBZ	19	Generate VEND
	SBO	19	
	JEQ	S0	If VALUE = TOTAL go to State 0 (S0)
S5	S	2,1	else subtract VALUE from TOTAL
	LI	4,1	
	COC	4,1	If difference is odd
	JEQ	NC	go to NC to deliver nickel
DC	SBO	21	else deliver dime
	SBZ	21	by generating DC
D?	LI	3,>10	Subtract 10¢ from difference
	S	3,1	
	DCS		
	JEQ	S0	If 0, go to State 0 (S0)
S6	SBO	21	else deliver another dime
	SBZ	21	
	JMP	S0	go to State 0 (S0)
NC	SBO	20	Deliver nickel by
	SBZ	20	generating NC
	LI	3,>5	Subtract 5¢ from difference
	S	3,1	
	DCS		
	CI	1,0	
	JNE	56	
	JMP	50	Check for need for dime in change

Figure 9.13 Assembly language program for a vending machine. (Copyright 1983 Texas Instruments Incorporated. All Rights Reserved.)

9.5 TRUTH AND FUNCTION TABLES

As we have seen in Chapter 7, a *truth table* shows how a particular logic element reacts to all possible input logic levels. The truth tables shown earlier were for relatively simple functions. However, truth tables are also used to describe complicated logic functions and circuits.

As the inputs and outputs increase, the truth table also becomes more complex. A truth table for a 1-of-16 decoder, for example, has 22 columns. Whether a table is simple or complex, the entries for the table will be H (high) or L (low), 1 or 0, X (don't care or immaterial), or waveform symbols indicating a transition from low to high or high to low (see Fig. 9.14).

In a multiple-output table, a particular output is shown to be active when some combination of inputs is active or inactive. If that output label has a negation bar, such as $\overline{O}_1$, then $\overline{O}_1$ is an active low. Look for a 0 or L in the column beneath $\overline{O}_1$. Opposite the 0 or L there will be, on the same line, an indication of the particular inputs (some combination of highs and lows) that will give that output. If the output label does not have a negation bar, such as O_1, look for a 1 or H in the column beneath the O_1 label, and note the input combination that will produce that output.

Truth tables are commonly supplied in manufacturers' data sheets for logic elements. For logic circuits, however, it will be necessary for readers to construct their own truth tables. These tables will make it easier to analyze the logic circuit.

A *function table* is similar to a truth table in that it shows certain outputs when certain inputs are present. The symbols in a function table represent more conditions than simply high or low, as indicated in Fig. 9.15.

Texas Instruments explains the operation of a function table in this manner:

> If, in the input columns, a row contains only the symbols H, L, and/or X, this means that the output indicated is valid whenever the input configuration is achieved and regardless of the sequence in which it is achieved. The output persists as long as the input configuration is maintained.
>
> If, in the input columns, a row contains H, L, and/or X together with ↑ and/or ↓, this means the output is valid whenever the input configuration is achieved but the transition(s) must occur following the achievement of the steady-state levels. If the

INPUTS			OUTPUT
$\overline{E}$	CP	D_n	Q_n
H	_⌐	X	No change
L	_⌐	H	H
L	_⌐	L	L

H = HIGH Voltage Level
L = LOW Voltage Level
X = Immaterial

Figure 9.14 Truth table for a parallel D register with enable. (Copyright by Motorola, Inc. Used by permission.)

H	=	high level (steady state)
L	=	low level (steady state)
↑	=	transition from low to high level
↓	=	transition from high to low level
→	=	value/level or resulting value/level is routed to indicated destination
↶	=	value/level is re-entered
X	=	irrelevant (any input, including transitions)
Z	=	off (high-impedance) state of a 3-state output
a . . h	=	the level of steady-state inputs at inputs A through H respectively
Q_0	=	level of Q before the indicated steady-state input conditions were established
$\overline{Q}_0$	=	complement of Q_0 or level of $\overline{Q}$ before the indicated steady-state input conditions were established
Q_n	=	level of Q before the most recent active transition indicated by ↓ or ↑
⊓	=	one high-level pulse
⊔	=	one low-level pulse
TOGGLE	=	each output changes to the complement of its previous level on each active transition indicated by ↓ or ↑.

Figure 9.15 Symbols used in function tables on TI data sheets. (Courtesy of Texas Instruments Incorporated.)

output is shown as a level (H, L, Q_0, or $\overline{Q}_0$), it persists as long as the steady-state input levels and the levels that terminate indicated transitions are maintained. Unless otherwise indicated, input transitions in the opposite direction to those shown have no effect at the output. (If the output is shown as a pulse, ⊓ or ⊔, the pulse follows the input transition indicated and persists for an interval dependent on the circuit.)

Among the most complex functions are those of the shift registers. These embody most of the symbols used in any of the function tables, plus more. Figure 9.16 is the

INPUTS										OUTPUTS			
CLEAR	MODE		CLOCK	SERIAL		PARALLEL				Q_A	Q_B	Q_C	Q_D
	S1	S0		LEFT	RIGHT	A	B	C	D				
L	X	X	X	X	X	X	X	X	X	L	L	L	L
H	X	X	L	X	X	X	X	X	X	Q_{A0}	Q_{B0}	Q_{C0}	Q_{D0}
H	H	H	↑	X	X	a	b	c	d	a	b	c	d
H	L	H	↑	X	H	X	X	X	X	H	Q_{An}	Q_{Bn}	Q_{Cn}
H	L	H	↑	X	L	X	X	X	X	L	Q_{An}	Q_{Bn}	Q_{Cn}
H	H	L	↑	H	X	X	X	X	X	Q_{Bn}	Q_{Cn}	Q_{Dn}	H
H	H	L	↑	L	X	X	X	X	X	Q_{Bn}	Q_{Cn}	Q_{Dn}	L
H	L	L	X	X	X	X	X	X	X	Q_{A0}	Q_{B0}	Q_{C0}	Q_{D0}

Figure 9.16 Function table. (Courtesy of Texas Instruments Incorporated.)

function table of a 4-bit bidirectional universal shift register, for example, type SN74194.

The first line of the table represents a synchronous clearing of the register and says that if clear is low, all four outputs will be reset low regardless of the other inputs. In the following lines, clear is inactive (high) and so has no effect.

The second line shows that as long as the clock input remains low (while clear is high), no other input has any effect and the outputs maintain the levels they assumed before the steady-state combination of clear high and clock low was established. Since on other lines of the table only the rising transition of the clock is shown to be active, the second line implicitly shows that no further change in the outputs will occur while the clock remains high or on the high-to-low transition of the clock.

The third line of the table represents synchronous parallel loading of the register and says that if S1 and S0 are both high, then, without regard to the serial input, the data entered at A will be at output Q_A, data entered at B will be at Q_B, and so on, following a low-to-high clock transition.

The fourth and fifth lines represent the loading of high- and low-level data, respectively, from the shift-right serial input and the shifting of previously entered data one bit; data previously at Q_A is now at Q_B, the previous levels of Q_B and Q_C are now at Q_C and Q_D, respectively, and the data previously at Q_D is no longer in the register. This entry of serial data and shift takes place on the low-to-high transition of the clock when S1 is low and S0 is high and the levels at inputs A through D have no effect.

The sixth and seventh lines represent the loading of high- and low-level data, respectively, from the shift-left serial input and the shifting of previously entered data one bit; data previously at Q_B is now at Q_A, the previous levels of Q_C and Q_D are now at Q_B and Q_C, respectively, and the data previously at Q_A is no longer in the register. This entry of serial data and shift takes place on the low-to-high transition of the clock when S1 is high and S0 is low and the levels at inputs A through D have no effect.

The last line shows that as long as both mode inputs are low, no other input has any effect and, as in the second line, the outputs maintain the levels they assumed before the steady-state combination of clear high and both mode inputs low was established. (Courtesy of Texas Instruments Incorporated.)

9.6 STATE DIAGRAMS

The *state* diagram is a drawing that shows the logic states of a counter. Notice in Fig. 9.17 that the normal, proper, legal states for a decade counter are 0 through 9. The states above 9 (10 through 15) are called *illegal* or *improper states*. The counter can get in such states by being preset to them or by the power being turned on or by parallel loading. The diagram shows how the counter returns to the desired sequence within two counts.

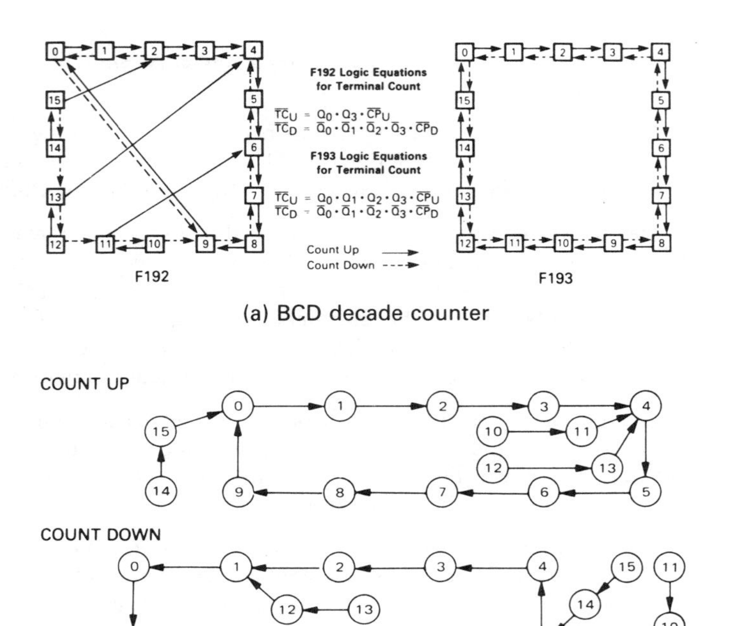

Figure 9.17 Examples of state diagrams: (a) BCD decade counter; (b) Universal decade counter. (Copyright by Motorola, Inc. Used by permission.)

9.7 LOGIC CONVENTIONS AND POLARITY INDICATION*

9.7.1 Logic States and Logic Levels

The relationship between logic states and logic levels is expressed by either the logic negation symbol (○) or the direct polarity symbol (◺). When the negation symbol is used, the whole diagram must use a single logic convention, either positive or negative. When the polarity symbol is used, its presence or absence shows the relationship between the logic level and the internal logic state at each input and output.

9.7.2 Positive- or Negative-Logic Convention

The negation symbol gives the relationship between the *external* logic state and the *internal* logic state. When it is present at an input or output, the internal and external logic states are opposites at that terminal. If there is no negation symbol

* Sections 9.7 through 9.11 were derived from ANSI/IEEE Standard 991-1986.

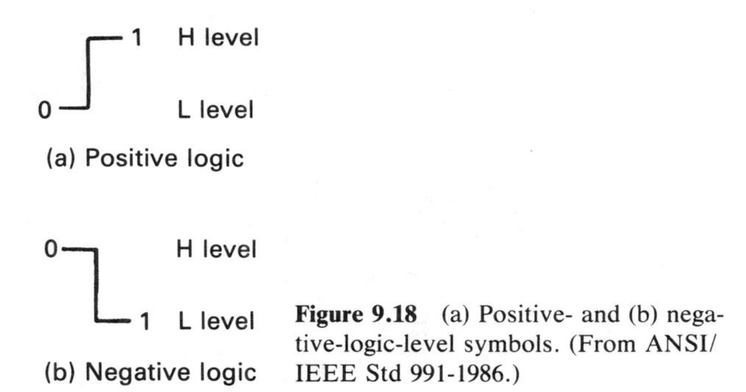

Figure 9.18 (a) Positive- and (b) negative-logic-level symbols. (From ANSI/IEEE Std 991-1986.)

at a terminal, the internal state is the same as the external state at that input or output. When the negation symbol is used on a diagram, the logic polarity symbol is not used.

Hopefully, the logic diagram or technical manual will state whether positive logic or negative logic is being used and will provide other useful information about the logic states.

When positive logic is being used, for every logic signal, the more positive voltage (H-level) corresponds to the external 1-state as shown in Fig. 9.18(a). The less positive voltage (L-level) corresponds to the external 0-state.

When negative logic is used, for every logic signal, the less positive voltage (L-level) corresponds to the external 1-state, as shown in Fig. 9.18(b). The more positive voltage (H-level) corresponds to the external 0-state.

See Fig. 9.19 for a sample logic diagram using positive logic.

9.7.3 Direct Polarity Indication

When the logic polarity symbol (⊾) is present at an input or output of a logic element, an external *low* level corresponds to the internal 1-state for that terminal. If the symbol is absent, an external *high* level corresponds to the internal 1-state for that terminal.

In the past, direct polarity indication has been called *mixed logic* but this is misleading and should not be used. When direct polarity indication is used there is no fixed relationship between logic levels and external logic states as in positive or negative logic.

See Fig. 9.20 for a sample of a logic diagram using direct polarity indication.

9.8 INTERCONNECTION OF SYMBOLS

Interconnecting lines between symbols may be horizontal, vertical, or oblique. Just as in schematics, these lines may be shown crossing or connected (Fig. 5.20)

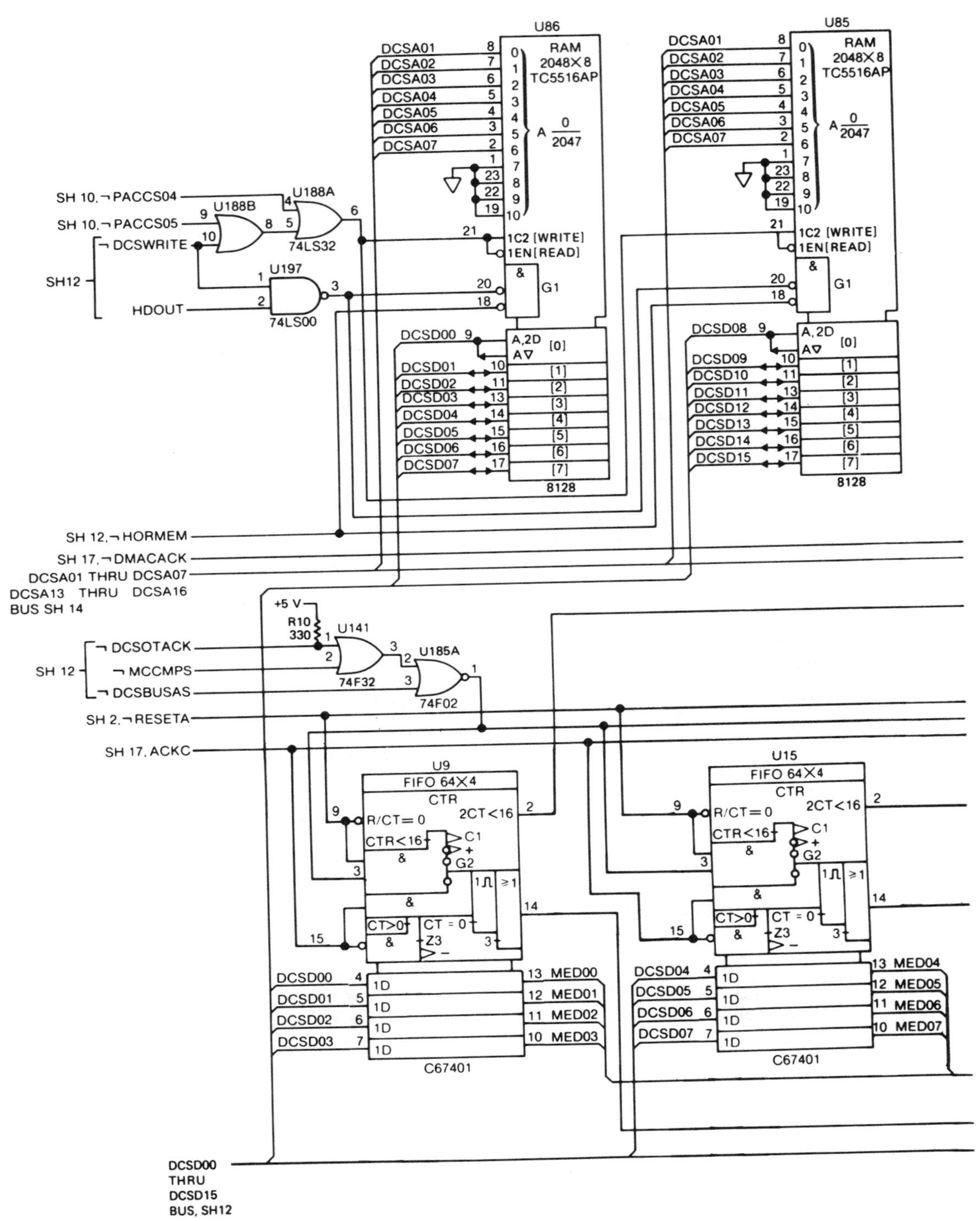

Figure 9.19 Sample of a logic diagram using positive logic. (From ANSI/IEEE Std 991-1986.)

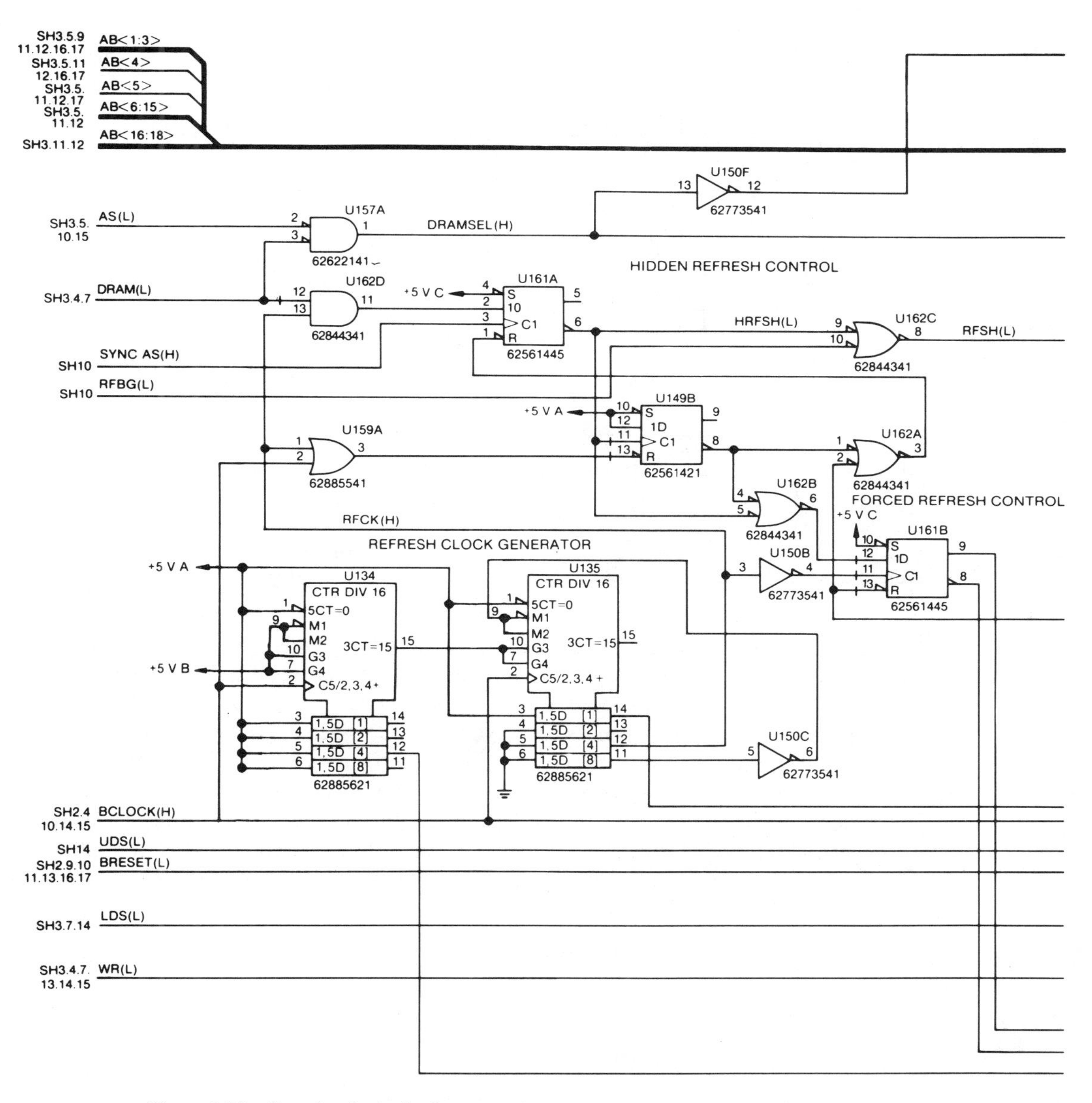

Figure 9.20 Sample of a logic diagram using direct polarity indication. (From ANSI/IEEE Std 991-1986.)

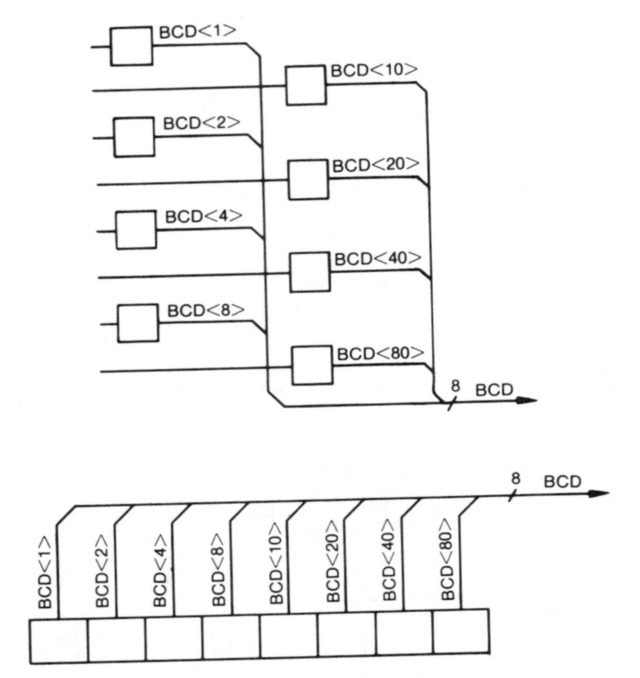

Figure 9.21 Grouping of lines. (From ANSI/IEEE Std 991-1986.)

and interrupted (Figs. 5.25 through 5.29). Ideally, symbols are arranged for maximum clarity, but they may be arranged simply because of a drafter's preference.

Groups of similar logic signal lines can be combined in highway or cable diagrams just as other signals can be combined, as discussed in Chapter 6. Figure 9.21, for example, shows the combining of BCD lines 1, 2, 4, 8, 10, 20, 40, and 80.

When the polarity or negation indication of a signal at its source is not the same as at its destination, you must invert the internal logic state of the source before using it as the internal logic state of the next input. As these mismatches can make you misinterpret a logic circuit, make sure that you do not overlook

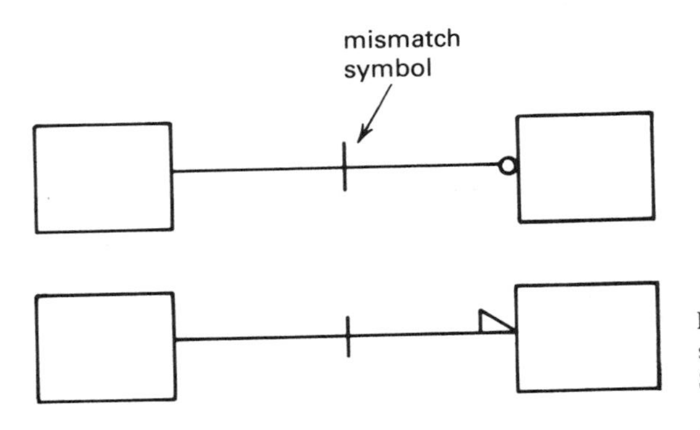

Figure 9.22 Signal mismatch between source and load. (From ANSI/IEEE Std 991-1986.)

intentional mismatches. These mismatches may be shown by a short perpendicular line across the connecting line as in Fig. 9.22.

If power connections are not shown to logic devices, the information may be given in a table or a note on the diagram.

9.9 SIGNAL NAMES

A signal name generally indicates what the signal does or what information it carries by means of mnemonic names and standard abbreviations. If you are lucky, the mnemonics and abbreviations will be explained on the diagram or in a related technical manual. If not, refer to Chapter 2, Table 9.3, and Fig. 9.23. Table 9.3 is a partial list of suggested mnemonics for use in signal names. Other meanings may be encountered for the mnemonics listed here, and different mnemonics may be found for the same meanings.

TABLE 9.3 SIGNAL NAMES—ALPHABETICALLY BY MNEMONIC

Mnemonic	Meaning	Mnemonic	Meaning
ACC	accept	CLR	clear
ACC	accumulator	CMD	command
ACK	acknowledge	CNT	count
ACT	activate	CNTL	control
ADD	adder	CO	carry output
ADR	address	COL	column
ALI	alarm inhibit	COMP	compare
ALU	arithmetic-logic unit	CORR	corrected
ASYNC	asynchronous	CP	carry propagate, compare
		CPU	central processing unit
BCD	binary-coded decimal	CRC	cycle redundancy check
BCTR	bit counter	CRY	carry
BG	borrow generate	CS	chip select
BI	borrow input	CTR	counter
BIN	binary	CYC	cycle
BIT	bit		
BLK	block	D	data
BO	borrow output	DEC	decimal
BP	borrow propagate	DEV	device
BUF	buffer, buffered	DIS	disable
BUS	bus	DISK	disk, disc
BUSY	busy	DLY	delay
BYT	byte	DMA	direct memory access
		DRAM	dynamic ram
CE	chip enable	DRV	driver
CG	carry generate	DWN	down
CHK	check		
CI	carry input	EN	enable
CK	clock	END	end
CLK	clock	EOF	end of file

TABLE 9.3 (CONTINUED)

Mnemonic	Meaning
EOL	end of line
EOT	end of tape, end of transmission
ERR	error
ERS	erase
EXOR	exclusive OR
EXT	external
FF	flip-flop
FIFO	first in, first out
FLD	field
FLT	fault
FNC	function
G	gate
GEN	generate
GND	ground
HEX	hexadecimal
HLD	holding
HORZ	horizontal
ID	identification
IN	in, input
INH	inhibit
INT	internal, interrupt
INTFC	interface
INTRPT	interrupt
I/O	input/output
IRQ	interrupt request
KYBD	keyboard
LCH	latch, latched
LD	load
LFT	left
LOC	location
LRC	longitudinal redundancy check
LSB	least-significant bit
LSBYT	least-significant byte
LT	light
MAR	memory address register
MEM	memory
MOT	motor
MPX	multiplex
MSB	most-significant bit
MSBYT	most-significant byte
MSK	mask
MSTR	master
MTR	motor
MULT	multiply, multiplier
NACK	negative acknowledge
NC	normally closed
NEG	negative
NO	normally open
OCT	octal
OFF	off
ON	on
OUT	out, output
OVFL	overflow
PAR	parity
PC	program counter
PCI	program-controlled interrupt
PE	parity error
POS	positive
PRGM	program
PROS	process, processor
PU	pull-up
PWR	power
RAM	random-access memory
RCVR	receiver
RD	read
RDY	ready
REG	register
REJ	reject
REQ	request
RES	reset
RFSH	refresh
ROM	read-only memory
ROW	row
RST	restart
RT	right
RTN	return
RTZ	return to zero
SEL	select
SET	set
SFT	shift
SIM	simulation
SLV	slave
SPLY	supply
SRQ	service request
START	start
STAT	status
STDBY	standby
STK	stack
STOP	stop
STRB	strobe
SW	switch
SYNC	synchronization
SYS	system

TABLE 9.3 (CONTINUED)

Mnemonic	Meaning	Mnemonic	Meaning
TERM	terminate, terminal	VIRT	virtual
TG	toggle	VLD	valid
TRIG	trigger		
TST	test	WR	write
		WRD	word
UP	up		
UTIL	utility	XCVR	transceiver
		XMIT, XMT	transmission, transmit
VERT	vertical	XMTR	transmitter
VID	video	XOR	exclusive-OR

Source: ANSI/IEEE Std 991-1986.

In an ideal situation, the name of a signal will reveal the function it performs, not the names of the signals that produced it. For example, if a signal RES is gated with a second signal STRB to produce a signal that clears a counter called CTR, its function is obvious if the output signal is named CLRCTR.

Signal names generally consist of capital letters (A to Z) and digits (0 to 9) but may also include negation characters (—, ~, ¬, ', /) and special characters (!, '', %, &).

Watch for identical names applied to different signals when they have similar functions. Expect a signal name to be changed whenever anything happens to the signal. A serial number or letter suffix may be added to a signal that is generated more than once or is amplified or level shifted. For example, if the signal RUN drives two buffers, the outputs of those buffers may be labeled RUN1 and RUN2.

On a diagram using single logic convention, when a logic signal is inverted, its basic name does not change; however, a negation bar is added to the name if it did not have a bar before the inversion. For example, if the signal RUN is inverted, it becomes $\overline{\text{RUN}}$. If the signal name had a negation bar before the inversion, the bar is removed for the inverted signal name. For example, $\overline{\text{RUN}}$ after inversion becomes RUN.

When direct polarity indication is used on a diagram, the signal level indicated (see Section 9.9.2) may be changed instead of using a negation bar. If a signal is inverted more than once, different versions of the signal are indicated by serial numbers or letters.

Logic signals have only two states; these states correspond to two ranges (called "levels") of physical values for the signal.

9.9.1 Signal State

For logic signals, the signal name is usually a condensation of a true-or-false statement. For example, in Table 9.4 the name ALARM is a condensation of the

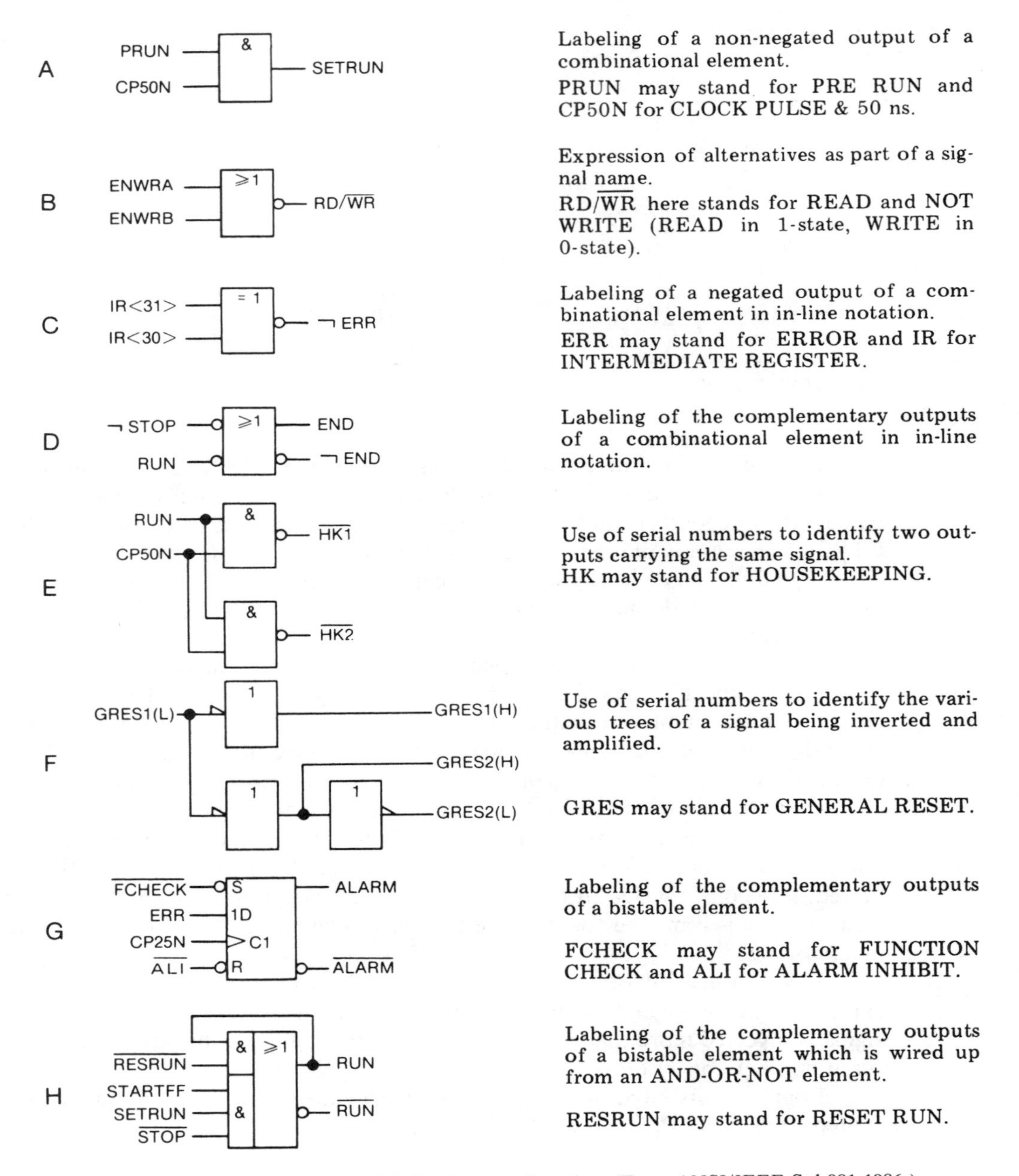

Figure 9.23 Examples of signal name allocation. (From ANSI/IEEE Std 991-1986.)

TABLE 9.4 RELATIONSHIPS AMONG STATES AND SIGNAL NAMES (SINGLE LOGIC CONVENTION)

				Relationship defined by presence or absence of negation symbol	
Row	Input (or output)	System condition	Signal state (truth value)[a]	External logic state	Internal logic state
1	ALARM	Alarm	True = 1	1	1
		No alarm	False = 0	0	0
2	ALARM	Alarm	True = 1	1	0
		No alarm	False = 0	0	1
3	$\overline{\text{ALARM}}$	Alarm	False = 0	0	0
		No alarm	True = 1	1	1
4	$\overline{\text{ALARM}}$	Alarm	False = 0	0	1
		No alarm	True = 1	1	0

Source: ANSI/IEEE Std 991-1986.

[a] The signal state being true always corresponds to the external logic state being 1. The signal state being false always corresponds to the external logic state being 0.

statement ALARM IS ACTIVE. When the statement represented by a signal name is evaluated, the truth value obtained is called the *signal state*.

When a signal is at its 1-state, the statement represented by the signal name is true. If the signal is at its 0-state, the statement will be false. For example, when the signal RUN is at its 1-state, RUN IS ACTIVE is true; it is false when the signal is in its 0-state. See Table 9.4, rows 1 and 2, for an example using ALARM.

Negated signals. Logic negation may be indicated by either a negation bar ($\overline{\quad}$), an in-line mathematical symbol (¬), a tilde (~), a prime (′), or a solidus (/).

A negation bar over a signal name indicates that that portion of the name representing the expression is to be negated. For example, $\overline{\text{RUN}}$ corresponds to the statement RUN IS NOT ACTIVE; the statement is true when the signal is in its 1-state and false when the signal is in its 0-state. It follows, then, that RUN IS ACTIVE is true when the signal $\overline{\text{RUN}}$ is in its 0-state and false when it is in its 1-state. See Table 9.4, rows 3 and 4, for a similar condition for the signal ALARM.

The in-line mathematical symbol for logic negation (¬) performs the same function as the negation bar. For example, ¬ CLR is the same as $\overline{\text{CLR}}$. Figure

$$\neg\,\text{RAS EN} = \overline{\text{RAS EN}}$$
$$(\neg\,\text{RAS})\ \text{EN} = \overline{\text{RAS}}\ \text{EN}$$
$$(\neg\,\text{RAS})\neg\,\text{EN} = \overline{\text{RAS}}\ \overline{\text{EN}}$$
$$(\neg\,\text{RAS})\text{EN} = \overline{\overline{\text{RAS}}\ \text{EN}}$$
$$\neg\,\text{RAS})\text{EN} = \overline{\text{RAS/}}\text{EN}$$
$$\neg(\text{RAS/EN}) = \overline{\text{RAS/EN}}$$
$$\neg(\neg(\text{RAS/EN})/\text{CAS}) = \overline{\overline{\text{RAS/EN}}/\text{CAS}}$$

Figure 9.24 Use of in-line negation symbol (¬). (From ANSI/IEEE Std 991-1986.)

9.24 shows how the in-line negation symbol is used in various constructions. (The tilde symbol can be used in the same position for the same result.)

Negated signals can also be indicated by a prime or a solidus (/): for example,

$$\overline{\text{BUSY}} = \text{BUSY}' = \text{BUSY/}$$

Arithmetic and logic expressions. While the plus sign (+) can indicate the OR function in signal names, most often it simply indicates algebraic addition. The minus sign (−) shows algebraic subtraction. For example, SFT − 1 may be the mnemonic for SHIFT MINUS 1.

A dot (•) or an asterisk (*) in a mnemonic indicates a logic AND relationship between the two signals. For example, "SET•CTR" may be the mnemonic for "SET" ANDed with "COUNTER." At times the dot or the asterisk may be deleted from the mnemonic. For example, "HK" may mean "H ANDed with K," but it may also mean "HOUSEKEEPING."

Parentheses in a mnemonic also indicate an AND relationship. For example, (SRQ)PWR should be interpreted as SERVICE REQUEST ANDed with POWER.

Bus signals. Within a bus the bits label may include a numeric suffix (enclosed in angle brackets) to the bus name. This suffix may represent the actual weights of the signals for buses with an inherent weighting of the signals within. For example, the eight lines of a storage register may be labeled SRBUS ⟨00⟩ to SRBUS ⟨08⟩.

The following labels may be used for connecting lines that represent entire buses: SRBUS ⟨0 : 8⟩ ≡ SRBUS ⟨0⟩, SRBUS ⟨1⟩ . . . SRBUS ⟨8⟩.

Clock signals. Numbers in clock signal names indicate the period or frequency. For example, CP50N may indicate that the basic clock period is 50 nanoseconds.

9.9.2 Signal Level

With either positive or negative logic, there is a fixed relationship between the external logic states of the signals and the corresponding logic levels. If positive logic is being used, for example, the 1-state of a signal (the true state of the signal name) corresponds to the high (H) level. For negative logic, the 1-state corresponds to the low (L) level.

When *direct polarity* indications are used, however, each logic signal name has an indication of the logic level corresponding to the 1-state (true state) of the signal. When there is either an H or L in parentheses at the end of a signal name, the signal has either a high (H) or low (L) logic level. For example, RUN(H) means that RUN IS ACTIVE is true when the logic level of the signal is high but is false when the logic level is low. $\overline{\text{RUN}}$(H) means that RUN IS NOT ACTIVE is true when the logic level is high but is false when the logic level is low.

From this we can conclude that RUN IS ACTIVE is true when the logic level of the signal is low and false when the logic level is high. See Table 9.5 for combinations of the signal ALARM.

RUN(L) means RUN IS ACTIVE is true when the logic level of the signal is low; it is false when the logic level is high.

When the *true* state of a signal corresponds with a high level, the signal is called a *true when high* signal; if the true state corresponds with a low level, the signal is called a *true when low* signal. If there are no logic level indications on the signal names in a diagram, all the signals are true when high.

When both logic negation and level inversion are applied to a signal name, there is no change in the signal: for example,

$$\text{RUN(L)} = \overline{\text{RUN}}\text{(H)}$$

As shown in Fig. 9.25 the level indication of a signal name (here FIRE OUT(L)) matches the polarity indication at the signal's source.

As shown in Fig. 9.26, when mismatched polarity is indicated, the polarity indication of a signal name matches the polarity indication on the closest device.

9.9.3 Names for Power Connections

Power supply connections are most often indicated by an abbreviation such as VCC for TTL supply voltage but may also be indicated by a unit of measure (for example, 0 V or +5.2 V).

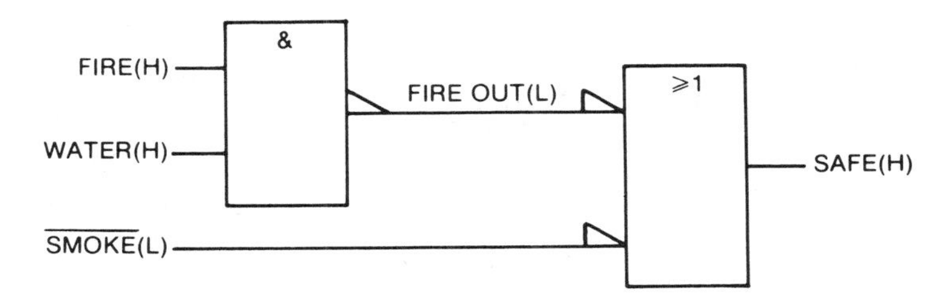

Figure 9.25 Signal-name-level indication and polarity indication. (From ANSI/IEEE Std 991-1986.)

TABLE 9.5 RELATIONSHIPS AMONG STATES, LEVELS, AND SIGNAL NAMES (DIRECT POLARITY INDICATION)

				Relationship defined by presence or absence of negation symbol	
Row	Input (or output)	System condition	Signal state (truth value)[a]	External logic state	Internal logic state
1	ALARM (H)	Alarm No alarm	True = 1 False = 0	H L	1 0
2	ALARM (L)	Alarm No alarm	True = 1 False = 0	L H	1 0
3	ALARM (L)	Alarm No alarm	True = 1 False = 0	L H	0 1
4	ALARM (H)	Alarm No alarm	True = 1 False = 0	H L	0 1
5	$\overline{\text{ALARM}}$ (H)	Alarm No alarm	False = 0 True = 1	L H	0 1
6	$\overline{\text{ALARM}}$ (L)	Alarm No alarm	False = 0 True = 1	H L	0 1
7	$\overline{\text{ALARM}}$ (L)	Alarm No alarm	False = 0 True = 1	H L	1 0
8	$\overline{\text{ALARM}}$ (H)	Alarm No alarm	False = 0 True = 1	L H	1 0

Source: ANSI/IEEE Std 991-1986.

[a] The signal state being true corresponds to the external logic level being that level specified in the signal name. The signal state being false corresponds to the logic level being the opposite of the level specified in the signal name.

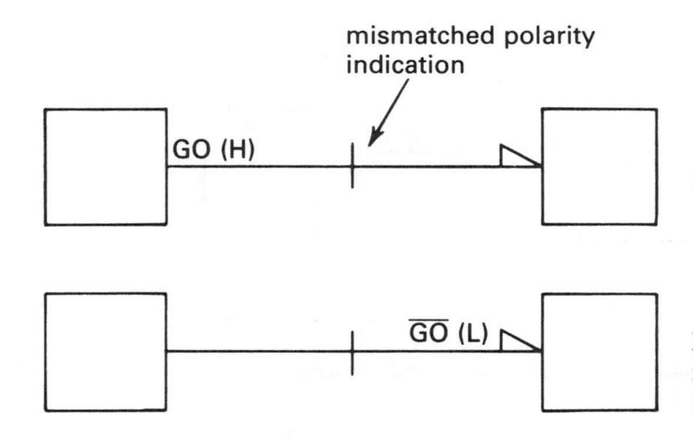

Figure 9.26 Signal names on mismatched portions. (From ANSI/IEEE Std 991-1986.)

9.10 WAVEFORMS

Waveforms are useful for testing and for explaining how a circuit works. They may be adjacent to or directly on a signal line.

While the wave shapes may be exact representations as they would appear on an oscilloscope, as shown in Fig. 9.27, they are most often approximations, as shown in Fig. 9.28. The stylized waveforms have perpendicular leading and fol-

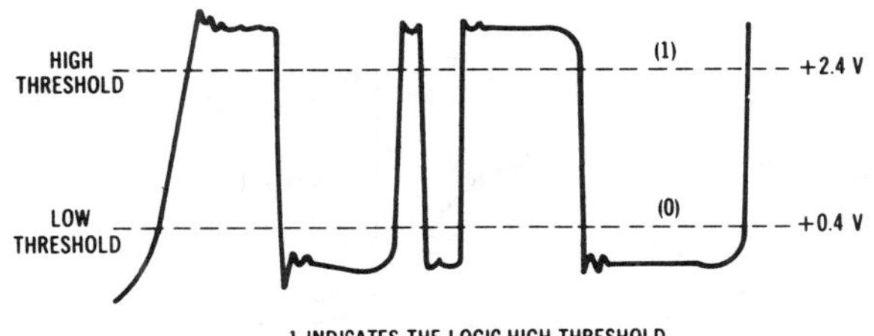

1 INDICATES THE LOGIC-HIGH THRESHOLD.
0 INDICATES THE LOGIC-LOW THRESHOLD.
ANY VOLTAGE BETWEEN THE 0 AND 1 THRESHOLDS IS TERMED A "BAD LEVEL" VOLTAGE.

Figure 9.27 Actual waveform. (Copyright 1982 by Howard W. Sams & Co., Inc.)

A

Level starting at pulse time A and ending at pulse time B.

Pulse occurring at pulse time C.

B

In the left-hand figure: going from the more positive level (+5 V) to the less-positive level (−2 V), and then returning to +5 V.

In the right-hand figure: going from the less-positive level (−2 V) to the more-positive level (+5 V), and then returning to −2 V.

C

+7 V

-7 V

+7 V

-7 V

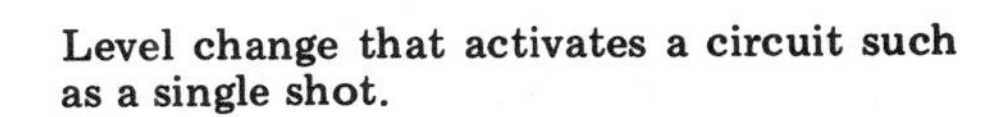

Level change that activates a circuit such as a single shot.

D

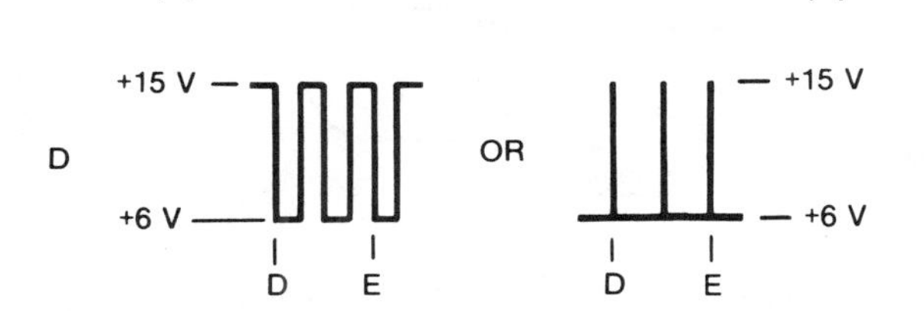

Pulse train with first pulse occurring at pulse time D and last pulse at pulse time E (In the case of an information-bearing pulse train, not all pulses are necessarily shown).

Figure 9.28 Stylized waveforms. (From ANSI/IEEE Std 991-1986.)

lowing edges, square corners, and flat tops and bottoms which could never occur in fact because they would require an infinite bandwidth. Note, too, that a single line may represent a narrow pulse if it is not necessary to indicate the pulse duration. Figure 9.29 shows how pulse characteristics can be indicated in a simplified manner.

Standard pulse characteristics

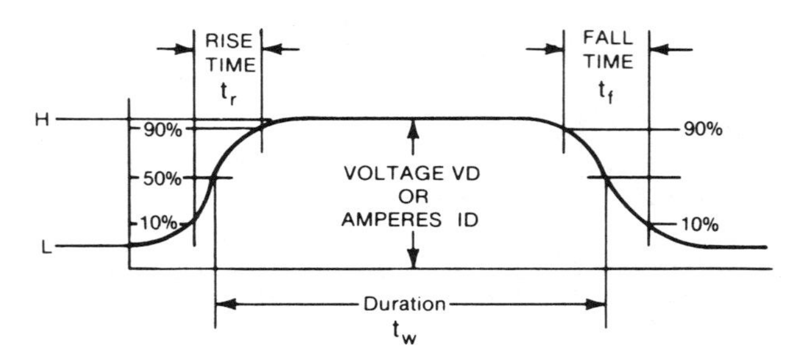

May be indicated by the following convention:

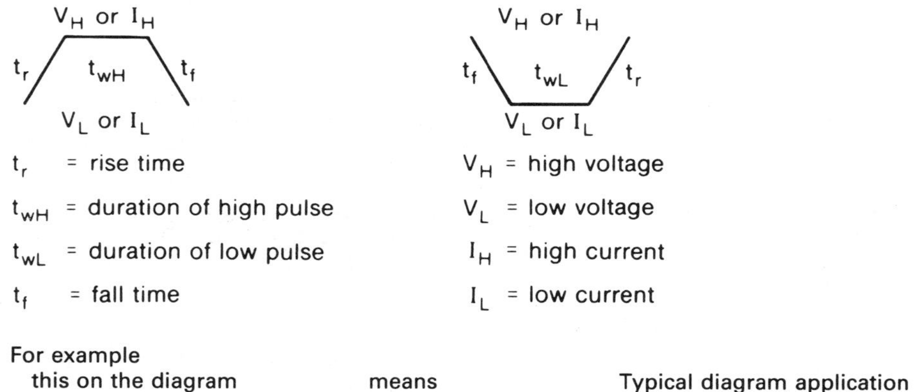

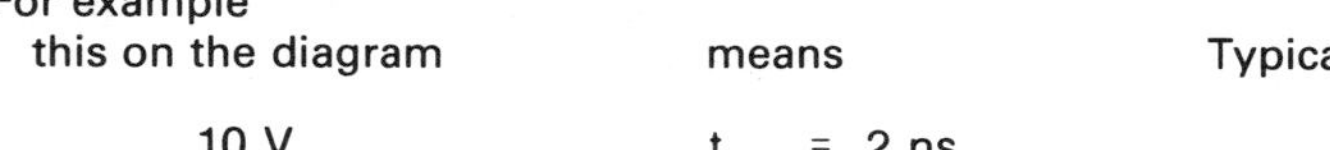

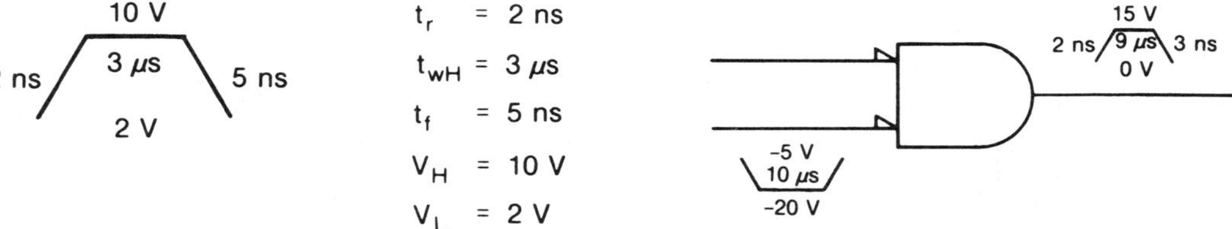

Figure 9.29 Simplified waveform notations. (From ANSI/IEEE Std 991-1986.)

9.11 DIAGRAM SIMPLIFICATION TECHNIQUES

Parts of diagrams may be simplified in order to put more information on the complete diagram or to reduce clutter by eliminating repetitive details. In some cases when simplification techniques are used, they will be explained on the diagram, but if not, they can be interpreted by the techniques in the following paragraphs.

The three most common techniques are repeated symbol simplification, repeated circuit patterns, and connections paired with ground.

9.11.1 Repeated Symbol Simplification

If a logic symbol for a specific device is shown more than once on a diagram, the full symbol may be shown only once. On subsequent use, the symbol may be shown by a rectangle that contains all relevant application and identification information and an appropriate reference to the fully delineated symbol. A box is added to the upper-left corner of the fully delineated symbol, on the outside, with a unique reference identification, as shown in Fig. 9.30. The same reference identification is shown in the same way on each corresponding simplified symbol. If a device is repeated on more than one sheet of a diagram, look for a sheet cross reference to the fully delineated symbol at the lower-left corner of the repeated pattern box, as shown in Fig. 9.31.

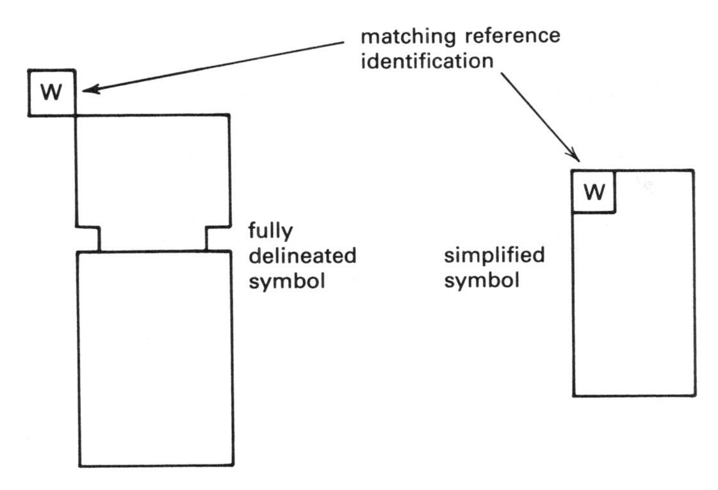

Figure 9.30 Fully delineated symbol versus simplified symbol. (From ANSI/IEEE Std 991-1986.)

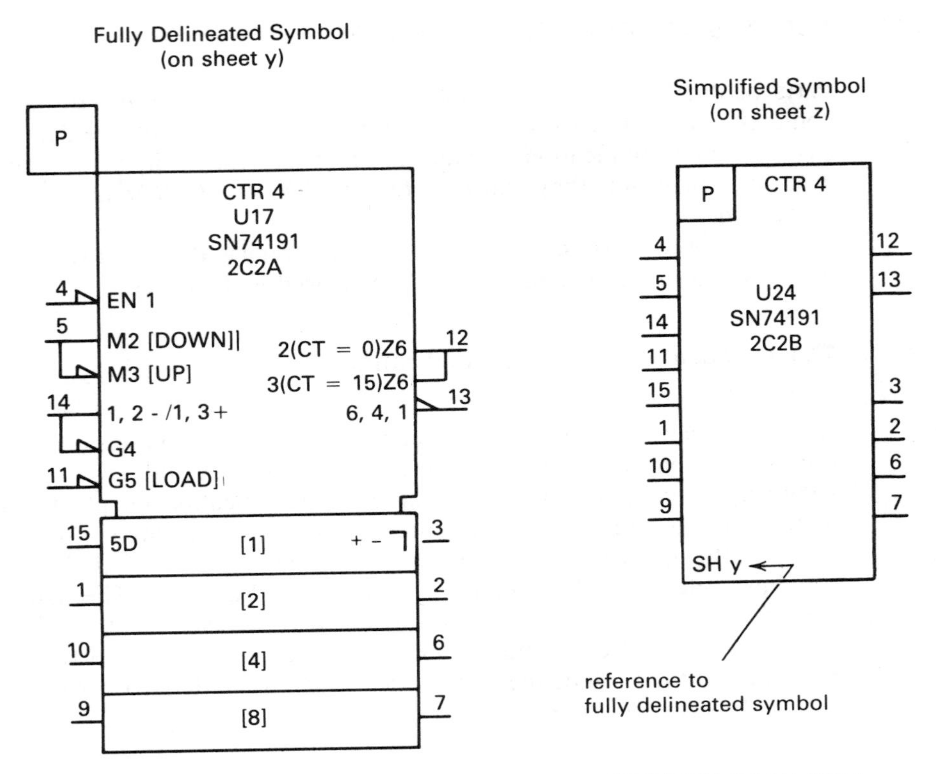

Figure 9.31 Repeated symbol simplification. (From ANSI/IEEE Std 991-1986.)

9.11.2 Repeated Circuit Patterns

If a portion of circuitry is used more than once in a diagram, additional applications of that portion may be shown in simplified form, as illustrated in Fig. 9.32. Solid single-line boxes are used to enclose the original pattern and the repeated pattern enclosures. Note that a box for repeated-pattern enclosure may be smaller than the original since circuit detail is omitted. If a pattern is repeated on more than one sheet of a diagram, look for a sheet cross reference to the fully delineated pattern at the lower-left corner of the repeated pattern. In interpreting repeated circuit patterns, note that the connections are arranged in the same order and direction as on the fully delineated circuit patterns unless otherwise noted.

Each different circuit pattern has a unique identification which is placed in the upper-left corner of all like patterns in a diagram.

Any differences between the repeated pattern and the original are indicated in the succeeding repeated-pattern enclosure. Repeated patterns may be parts of a higher-order repeated patterns.

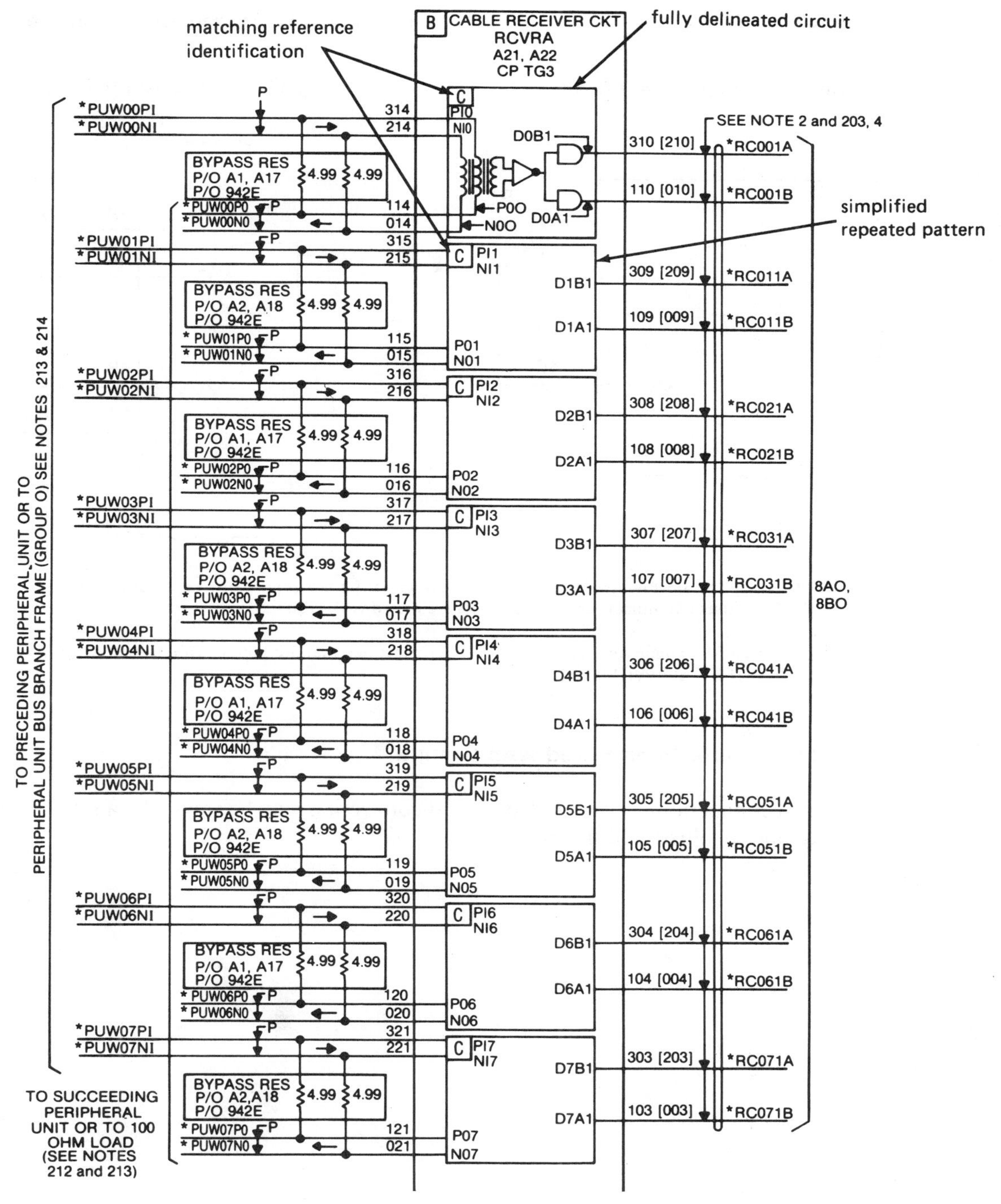

Figure 9.32 Typical diagram with repeated circuit pattern. (From ANSI/IEEE Std 991-1986.)

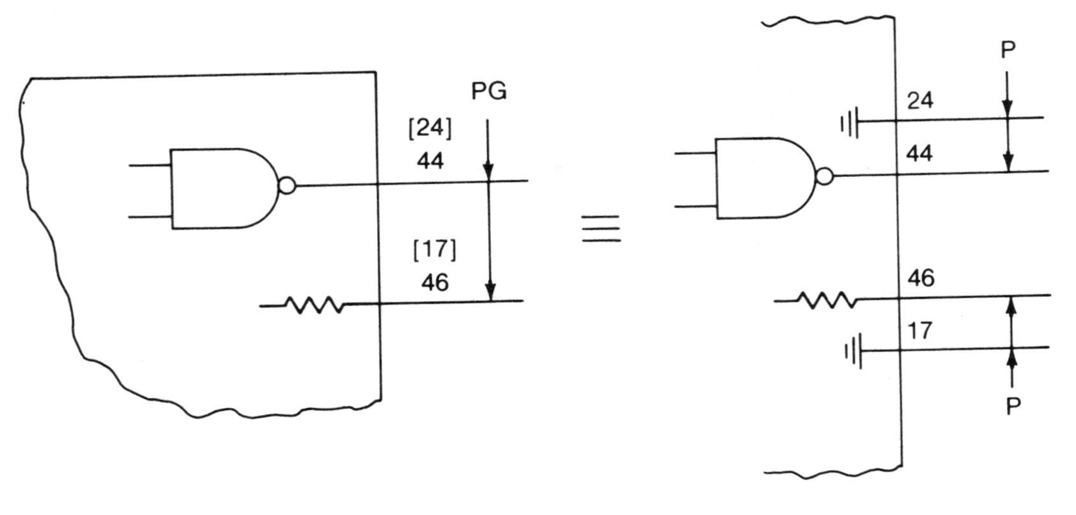

NOTES: (1) Terminal numbers shown in square brackets [ ] are ground pair termination pins.
(2) The letters PG indicate connections that are paired with ground.

Figure 9.33 Single line representation of connections that are paired with ground. (From ANSI/IEEE Std 991-1986.)

9.11.3 Connections Paired with Ground

Signal connections paired with a ground connection may be shown by a simplified diagram, as illustrated in Fig. 9.33.

Appendix A. List of Abbreviations and Acronyms

The following abbreviations and acronyms have been compiled from numerous sources, including *Modern Dictionary of Electronics,* sixth edition, by Rudolf F. Graf (Howard W. Sams & Co.), *The Illustrated Dictionary of Electronics,* third edition. (TAB Books, Inc.), *QST* magazine, and the *American Radio Relay League Handbook;* all by permission of the publishers.

A — ampere
  angstrom
AACK — advanced acknowledge
AAFC — antiaircraft fire control
ABC — automatic bandwidth control
  automatic bass compensation
  automatic brightness control
ABEL — advanced Boolean expression language
ABM — antiballistic missile
ABS — acrylonitrile butadiene-styrene
AC — alternating current
ACB — air circuit breaker
ACC — accumulate
  asynchronous communications controller
ACE — asynchronous communication element
  automatic checkout equipment
ACIA — asynchronous communications interface adapter
ACK — acknowledge
ACL — advanced CMOS logic
ACM — Association for Computing Machinery
ACS — add-compare-select
  address control shift
ACT — a CMOS logic family from VTC
ACTG — actuating
ACTR — actuator
ACU — automatic calling unit
A/D — analog to digital
A-D — analog to digital
ADC — analog-to-digital converter
ADCCP — advanced data communication control procedure
ADF — automatic direction finder
ADJ — adjust
ADL — automatic data link
ADLC — advanced data link controller
ADP — automatic data processing
ADPT — adapter

ADPTR adapter
ADRC advanced dynamic RAM controller
ADSSC analog data separator support circuit
ADU analog data acquisition unit
automatic dialing unit
AEW airborne early warning
AF audio frequency
AFC automatic frequency control
AFSK audio-frequency shift keying
AFT automatic fine tuning
AGC automatic gain control
AGM air-to-ground missile
Ah ampere-hour
AHCT a CMOS logic family from Integrated Device Technology
AHM ampere-hour meter
AI artificial intelligence
AILAS Automatic Instrument Landing Approach System
AL aluminum
ALBO automatic line buildout
ALC automatic level control
automatic load control
ALGOL algorithmic language
ALM alarm
ALS advanced low-power Schottky
ALT alternator
ALTE altitude transmitting equipment
ALU arithmetic and logical operation unit
arithmetic and logic unit
arithmetic-logic unit
AM amplitude modulation
A/m ampere per meter
AML automatic modulation limiting
AMM ammeter
AMNL amplitude-modulation noise level
AMP ampere
AMPL amplifier
AMPS advanced mobile phone service
AN anode
AND a logical operator
ANL automatic noise limiter
ANSI American National Standards Institute
ANT antenna
AOI AND-OR-Invert
area of interest
AOS acquisition of signal
APC automatic phase control
automatic picture control
automatic power control
API air-position indicator
APL approved products list
automatic phase lock
average picture level
APM analog panel meter
APPAR apparatus
APT automatically programmed tool
APU auxiliary power unit
AQL acceptable quality level
ARM armature
ARQ automatic repeat request
ARR arrestor
ARSR arrestor
ARY array
ASC automatic sensitivity control
ASCII American Standard Code for Information Interchange
ASCR asymmetrical SCR
ASIC application-specific IC
ASK amplitude-shift keying
ASMA asynchronous serial Manchester adapter
ASME American Society of Mechanical Engineers
ASPLD application-specific PLD
ASR airborne surveillance radar
automatic send receive
ASROC antisubmarine rocket
ASSC asynchronous serial communications controller
ASSY assembly
ASTRO asynchronous–synchronous receiver–transmitter
ASW antisubmarine warfare
ATC address translation chip
ATCE automatic test and checkout equipment
ATE automatic test equipment
ATR anti-transmit-receive
ATTEN attenuator, attenuation
ATV amateur television
AU arithmetic unit
AUD audio
AUX auxiliary
A/V audiovisual
AVC automatic volume control

AVE	automatic volume expansion
AVG	average
AVPC	adaptive vector predictive coding
AWG	American wire gauge
AZ	azimuth
B	base (electron device)
	bit
	byte
	symbol for magnetic flux
	symbol for susceptance
BARRITT	barrier injection transit time
BASIC	beginner's all-purpose symbolic instruction code
BAW	bulk acoustic wave
BBD	bucket brigade device
BBM	break-before-make
BC	back-connected
	broadcast
	bus controller
BCD	binary-coded decimal
BCN	beacon
BCP	byte-controlled protocol
BD ELIM	band elimination
BEP	burst error processor
BER	bit error rate
BF	beat frequency
BFL	buffered FET logic
BFO	beat-frequency oscillator
BI	blanking input
BIDFET	bipolar, double-diffused, N-channel MOS and P-channel MOSFET
BIIC	bus interconnect integrated circuit
BiMOS	bipolar MOSFET
BIN	binary
BIOS	basic input/output system
BIP	binary image processor
BITE	built-in test equipment
BIU	bus interface unit
BLO	blower
BM	ballistic missile
BMFET	biopolar-mode FET
BNC	baby N connector
BNZS	binary N zero substitution
BO	beat oscillator
	blocking oscillator
BOART	bus-oriented programmable asynchronous receiver/transmitter
BOP	byte-oriented protocols
BORSCHT	battery feed, overvoltage protection, ringing, supervision, coding, hybrid, and testing
BOT	beginning of tape
	bottom
BP	bandpass
BPF	bandpass filter
BPI	bits per inch
BPS	bits per second
BPT	bipolar transistor
BRKR	breaker
B&S	Brown and Sharpe gauge
BSC	binary synchronous communications
BST	beam-switching tube
Btu	British thermal unit
BU	bulk (substrate)
BUBMEM	bubble memory
BUS	basic utility systems
BV	breakdown voltage
BW	bandwidth
BWA	backward-wave amplifier
BWL	bandwidth loaded
BWO	backward-wave oscillator
C	capacitance
	capacitor
	Celsius
	collector (electron device)
	Coulomb
C^3	command, control, and communications
C^4	command, control, communications, and computer
CAB	cabinet
CAD	computer-aided design
CADAM	computer-augmented design and manufacturing system
	content-addressable data manager
CAE	computer-aided engineering
CAL	calibrator
CAM	content-addressable memory
CAP	capacitor
	capacity
CAS	conditional array scaling
CASE	computer-aided software engineering

CAT	computer-aided tomography
	computer-aided testing
	computerized axial tomography
CATH	cathode
CATT	controlled avalanche transit-time triode
CATV	cable television
CATVI	cable television interference
CB	circuit breaker
	citizens' band
	common base
	common battery
C-band	a radio-frequency band of 3.9 to 6.2 GHz
CBDS	circuit board design system
CC	color code
CCA	circuit card assembly
	current-controlled amplifier
CCC	ceramic chip carrier
	cube-connected cycle
CCD	charge-coupled device
	computer-controlled display
CCDH	cassette/cartridge data handler
CCIF	International Telephone Consultative Committee
CCIR	International Radio Consultative Committee
CCITT	Consultative Committee for International Telephone and Telegraph
CCM	counter countermeasures
CCTV	closed-circuit television
CCW	coherent CW
	counterclockwise
CD	capacitor diode
	carrier detect
CdS	cadmium sulfide
CE	common emitter
CEMF	counter electromotive force
CERDIP	ceramic dual-in-line package
CERMET	ceramic metal element
CERT	character error-rate testing
CF	carrier frequency
	cathode follower
CFR	carbon film resistor
CGA	color graphics adapter
CGC	color graphics controller
CGI	color graphics interface
CGM	computer graphics metafile
CHAN	channel
CHIL	current-hogging injection logic
C^3I	command, control, communications, and intelligence
CIA	complex interface adapter
CID	charge-injection device
CIF	Caltech intermediate format
CKT	circuit
C^3L	complementary contact-current logic
CLB	configurable logic block
CLCC	ceramic leadless chip carrier
CLM	compact LISP machine
CLR	clear
CLV	constant linear velocity
CM	countermeasures
CMAC	color/monochrome attributes controller
CML	current mode logic
CMOS	complementary metal-oxide semiconductor
CMPTR	computer
CMR	common-mode rejection
CMRR	common-mode rejection ratio
CMS	current-mode switching
CMV	common-mode voltage
C/N	carrier to noise
CNC	computer numerical control
CND	conduit
CNDCT, COND	conductor
CO	cavity oscillator
COAX	coaxial
COB	chip on board
	complementary offset binary
COBOL	common business-oriented language
COHO	coherent oscillator
COLL	collector
COMM	communication
	commutator
COMP	compress
COND	conductor
CONN	connector
CONT	contact
	control
CONV	converter
COR	carrier-operated relay
COS	contactor, starting
	Corporation for Open Systems

CP clock pulse
CPC central processor chip
CPCU computer control unit
CPE central processing element
CPL controlled products list
CPLG coupling
CPM continuous-phase modulation
CPS characters per second
 cycles per second
CPTD call progress tone detection
CPU central processing unit
CR cathode ray
 controlled rectifier
 crystal rectifier
CRC cyclic redundancy check
CRM counter-radar measures
CRO cathode-ray oscilloscope, oscillograph
CROM cathode read-only memory
CRT cathode-ray tube
CRTC CRT controller
CRYPTO cryptography
CS chip select
CSB complementary straight binary
CSCR complementary semiconductor controlled rectifier
CSMA carrier-sense multiple access
CSMA/CD CSMA collision-detection
CSN carrier sense
CSP codebook-search processor
CSR control shift register
CT center tap
 current transformer
CTC complementary two's complement
CTCSS continuous tone-coded squelch system
CTD charge transfer device
CTL complementary transistor logic
CTS clear to send
CU piezoelectric-crystal unit
CUJT complementary unijunction transistor
CUR current
CUTS computer user tape system
CVSD continuously variable slope delta modulator
 continuously variable slope delta modulator/demodulator
CW clockwise
 continuous wave
C/W carrier wave
CWP communication word processor
CX control transmitter
CXR carrier
D dissipation factor
 drain electrode
 electrostatic flux density
DA digital to analog
D/A digital to analog
D-A digital to analog
DAA direct access arrangement
DAC digital-to-analog converter
DACIA dual asynchronous communication interface adapter
DAGC delayed automatic gain control
DAM data-addressed memory
DAMA dynamic assignment multiple access
DAS data-acquisition system
DAT diffused alloy transistor
 dynamic address translator
DAVC delayed automatic volume control
dB decibel
dBa decibel adjusted
dBc decibels referenced to carrier level
dBd antenna gain referenced to a dipole
dBi antenna gain referenced to isotropic
dBj decibels relative to 1000 microvolts
dBk decibels referred to 1 kilowatt
DBLR doubler
dBm decibels referenced to 1 milliwatt
DBM doubly balanced mixer
DBMU disk buffer management unit
dBr decibels relative level
dBRAP decibels above reference acoustical power
dBrn decibels above reference noise
dBRN decibels above reference noise
DBRT directed beam refresh tube
dBV decibels referred to a standard of 1 volt
dBW decibels referred to 1 watt

dBx decibels above the reference coupling
DC direct current
 double contact
D-C direct conversion
DCE data-communication equipment
DCFL direct-coupled FET
DCGG display character and graphics generator
DCH data channel
DCM digital capacitance meter
 distributed control module
DCP data ciphering processor
DCSR digitally corrected subranging
DCT data circuit-terminating equipment
 discrete-cosine-transform
DCTL direct-coupled transistor logic
DCU decade counting unit
DCV direct-current volts
DCWV direct-current working volts
DDP distributed data processing
DDS dual data separator
DEG degree
DEM demodulator
DEMUX demultiplexer
DET detector
DEUCE dual enhanced universal communications element
DF direction finder
 dissipation factor
DFP digital filter processors
DFT discrete Fourier transform
DGTZR digitizer
DHG digital harmonic generation
DI dielectric isolation
DIA diameter
DIAC bidirectional diode-thyristor
 bidirectional trigger diode
DIC dielectrically isolated integrated circuit
DICE digital intercontinental conversion equipment
DIF decimation in frequency
DIIC dielectrically isolated integrated circuit
DIL dual-in-line
DIN Deutsche Industrie Normenausschuss (German standards institute)
DIO diode
DIP dual-in-line package
DISC disconnect
DISCR discriminator
DL data link
 data list
 delay line
DLIC digital line interface controller
DLM double-layer metal
DLT data loop transceiver
 direct loop transceiver
DM delta modulation
DMA direct memory access
DMAC direct memory access controller
DMAI direct memory access interface
DMAR DMA request
DMC dynamic memory controller
DMM digital multimeter
DMOS diffused MOS
DMP dialer modem peripheral
DNC direct numerical control
DNL dynamic noise limiter
DP dash pot (relay)
 double pole
DPDT double-pole, double-throw
DPFC double-pole, front-connected
DPM digital panel meter
DPST double-pole, single-throw
DQPSK differential quaternary phase-shift keying
DRAM dynamic random-access memory
DRAMC dynamic RAM controller
DRAW direct read after write
DRB double row buffer
DRC design rules checking
DRO destructive readout
 dielectric resonator oscillator
 digital readout
DSB double sideband
DSC double silk covered
DSCRM discriminator
DSP digital signal processing
DS&R data storage and retrieval
DT double throw
DTBRD daughter board
DTC DMA transfer controller
DTE data terminal equipment
DTL diode transistor logic
DTMF dual-tone multifrequency
DTR data terminal ready

DUART	dual universal asynchronous receiver and transmitter
DUF	diffusion under epitaxial film
DUSCC	dual universal serial communications controller
DUT	device under test
DVM	digital voltmeter
DVST	direct-view storage tube
DWG	drawing
DXE	data transmitting equipment
DYNM	dynamotor
DYNO	dynamometer
E	emitter
	voltage
EAC	extended arithmetic chip
EAROM	electrically alterable read-only memory
	electronically alterable read-only memory
EBMD	electron-beam mode discharge
EBR	electron-beam recording
EBS	electron-bombarded semiconductor
EC	enamel covered
E^2C^2	expandable error checker and corrector
ECC	electrochemichromic
	error check correction
ECCM	electronic counter-countermeasures
ECCSL	emitter-coupled current steered logic
ECDT	electrochemical diffused-collector transistor
ECG	electrocardiogram
ECL	emitter-coupled logic
ECM	electronic countermeasures
ECO	electron-coupled oscillator
ECTL	emitter-coupled transistor logic
E/D	enhancement/depletion
EDAC	error detection and correction
EDC	error detection and correction
EDFA	electronic differential analyzer
EDLC	Ethernet data link controller
EDP	electronic data processing
EDPM	electronic data processing machine
EDS	error detection and support
EDU	electronic display unit
EEG	electroencephalograph
EEPROM	electrically erasable programmable read-only memory
EFD	electronic funds transfer
EFTS	electronic funds transfer system
EGA	extended graphics adapter
EGC	electronic gain control
EHF	extremely high frequency
EHP	electric horsepower
EHS	extrahigh strength
EHT	extrahigh tension
EHV	extremely high voltage
EIA	Electronic Industries Association
EIRP	equivalent isotropically radiated power
EL	electroluminescent
ELANC	enhanced local area network controller
ELD	edge-lighted display
	electroluminescent display
ELECT	electrolytic
ELEK	electronic
ELEX	electronics
EMC	electromagnetic compatibility
EME	Earth–Moon–Earth (moonbounce)
EMF	electromotive force
EMI	electromagnetic interference
EMP	electromagnetic pulse
EMR	electromagnetic radiation
EMU	electromagnetic unit
EMV	emulation vehicle
ENGR	engineer
ENGRG	engineering
EOM	end of message
EOT	end of tape
EPA	enhanced-performance architecture
EPCI	enhanced programmable communication interface
EPLD	erasable programmable logic device
EPR	expander parameter register
EPROM	erasable programmable read-only memory
EPS	electric power supply
EQ	equal
EQL	equal
E^2 PROM	electrically erasable programmable read-only memory

ERP	effective radiated power
ERU	external register unit
ES	electronic switching
ESC	escutcheon
ESD	electrostatic discharge
ESDI	enhanced small-device interface
ESG	electronic sweep generator
ESM	electronic support measure
E&SP	equipment and spare parts
ESR	equivalent series resistance
ESU	electrostatic unit
E/TMS	EnMasse transaction management system
ETVM	electrostatic transistorized voltmeter
EU	execution unit
EUV	extreme ultraviolet radiation
EVM	electronic voltmeter
EVR	electronic video recording
EWB	embedded wiring board
EXC	exciter
EXCTR	exciter
F	Fahrenheit
	farad
	filament
	frequency
	fuse
FACT	Fairchild advanced CMOS technology
FAGC	fast automatic gain control
FAMOS	floating-gate avalanche-injection MOSFET
FAST	Fairchild advanced Schottky TTL
FAW	frame alignment word
FAX	facsimile
FBR	feedback resistance
FC	front-connected
FCC	Federal Communications Commission
FCS	fire control system
	frame check sequence
FCT	filament center tap
FD	frequency diversity
FDC	floppy disk controller
FDDS	floppy disk data separator
FDM	frequency-division multiplexing
FDR	frequency-domain reflectometry
FDX	full duplex
FEC	forward-error control
	forward-error correction
FEP	fluorinated ethylene propylene copolymer
	front-end processing
FET	field-effect transistor
FF	flip-flop
FFC	flat flexible cable
FFCT	fast Fourier-cosine transform
FFT	fast Fourier transform
FH	flat head
FHP	fractional horsepower
FIFO	first in, first out
FIL	filament
FIO	FIFO I/O
FIR	finite impulse response
FIRQ	fast interrupt request
FIT	failure in 10^9 device hours
FL	filter
FLA	flexible linear array
FLTR	filter
FLUOR	fluorescent
FM	frequency modulation
FMC	false memory cycle
FO	fiber optics
FORTRAN	formula translation
FPC	Federal Power Commission
FPGA	field-programmable gate array
FPLA	field-programmable logic array
FPLS	field-programmable logic sequencer
fps	feet per second
FPU	floating-point unit
FREQ	frequency
FSD	full-scale deflection
FSK	frequency-shift keying
FSR	FPU status register
ft	foot (feet)
FZ	fuze
g	gram
G	conductance
	gate of a field-effect transistor
	giga (10^9)
	grid
	ground
G/A	ground to air
GaAs	gallium arsenide
GaAsFET	gallium arsenide field-effect transistor
GAL	generic array logic
GBW	gain bandwidth product
Gbyte	gigabyte

GCA	ground-controlled approach
GCP	graphics control processor
GCS	gate-controlled switch
GCT	Greenwich civil time
GDO	gate-dip oscillator
	grid-dip oscillator
GDM	graphics data manager
GDS	graphic display system
Ge	germanium
GEN	generator
GFI	ground-fault interrupter
GFLOPS	giga floating operations per second
GHz	gigahertz
GIGO	garbage in, garbage out
GIP	graphics/imaging processor
GKS	graphical kernel system
GMS	graphics memory system
GMT	Greenwich mean time
GMV	guaranteed minimum value
GND	ground
GOS	graphics output system
GPA	general-purpose amplifier
GPC	general-purpose controller
GPI	ground-position indicator
GPIA	general-purpose interface adapter
GPIB	general-purpose interface bus
GPS	general problem solver
GR	general-purpose register
GROM	grommet
GSI	grand-scale integration
GSP	graphics system processor
GTO	gate-turn-off thyristor
GWEN	ground-wave emergency network
GYRO	gyroscope
h	hour
H	heater
	henry
	magnetic field strength
HAAT	height above average terrain
HAL	hard array logic
HCD	hot carrier diode
HDB	high-density bipolar
HDBK	handbook
HDC	hard disk controller
HDLC	high-level-data link control
HEMT	high-electron-mobility transistor
HEX	hexadecimal
HF	high frequency
HFO	heterodyne-frequency oscillator
	high-frequency oscillator
HI	hybrid integrated circuit
hi-fi	high fidelity
HIPOT	high potential
HIT	high-impedance transceiver
HLL	high-level language
	high-level logic
HLTL	high-level transistor logic
HNIL	high-noise immunity logic
HOL	higher-order language
HORIZ	horizon, horizontal
HOT	horizontal output transformer
hp	horsepower
HP	high pass
HPF	highest possible frequency
HPFL	high-pass filter
HRAM	hierarchical random access memory
HRC	horizontal redundancy checking
HSDH	hard sector data handler
HSL	high-speed shift-register time switch
HT	high tension
HTL	high-threshold logic
HTR	heater
HTSK	heat sink
HT TR	heat treat
HV	high voltage
HVDC	high-voltage dc
HVIC	high-voltage integrated circuit
HVPS	high-voltage power supply
HVR	high-vacuum rectifier
Hz	hertz
I	current
IARU	International Amateur Radio Union
IAS	immediate access storage
IAVC	instantaneous automatic volume control
IC	inductance–capacitance
	integrated circuit
	internal connection
ICC	intelligent communications controller
	interrupt control coprocessor
	ISDN communications controller
ICE ™	in-circuit emulator
ICR	inductance–capacitance–resistance
ICW	interrupted continuous wave

ID	inside diameter
I-D	identification, identifier
IDC	insulation displacement connector
IDP	industrial data processing
	integrated data processing
	intermodulation distortion percentage
IDT	interdigital transducer
IEC	integrated electronic component
	International Electrotechnical Commission, Geneva, Switzerland
IEEE	Institute of Electrical and Electronics Engineers
IER	interrupt enable register
IF	intermediate frequency
IFF	identification friend or foe
IFM	instantaneous frequency measurements
IFR	interrupt flag register
IFRU	interference rejection unit
IGFET	insulated-gate field-effect transistor
IGN	ignition
IGS	inertial guidance system
IGT	insulated-gate transistor
IHF	inhibit flip-flop
	Institute of High Fidelity
IHFM	Institute of High-Fidelity Manufacturers
IIR	infinite-duration, impulse response
	infinite-impulse response
ILD	injection laser diode
ILO	injection-locked oscillator
ILS	instrument landing system
IMD	intermodulation distortion
IMOS	ion-implanted MOS
IMPATT	impact avalanche and transit time
IMR	isolation-mode rejection
IMRR	IMR ratio
IMTS	improved mobile telephone system
in	inch
INCH	integrated chopper
IND	indicator
	induction
IN-LB	inch-pound
INST	instrument
INSTR	instrument
INV	inverter
I/O	input/output
IOB	I/O block
IOC	integrated optical circuit
IOCC	I/O channel converter/controller
IOP	input/output processor
IPA	intermediate power amplifier
IPC	intelligent peripheral controller
IR	current times resistance (voltage drop)
	infrared
	interrogator response
	insulation resistance
IRAC	Interdepartmental Radio Advisory Committee
IRED	infrared emitting diode
IRG	interrecord gaps
IRU	internal register unit
ISB	independent sideband
ISDN	integrated services digital network
ISHM	International Society of Hybrid Microelectronics
ISM	industrial, scientific, and medical
ISO	International Standardization Organization
ISV	instantaneous speed variations
ITDM	intelligent time-division multiplexer
ITU	International Telecommunication Union
ITV	industrial television
IVR	integrated voltage regulator
J	joule
	symbol for jack
JAN	joint Army–Navy
JB	junction box
JEDEC	Joint Electronic Device Engineering Council
JETEC	Joint Electron Tube Engineering Council
JFET	junction field-effect transistor
JK	jack
	two of the inputs of a flip-flop
JMOS	junction MOS
JSR	jump to subroutine
JTAC	Joint Technical Advisory Committee
k	kilo (1000)

K	degrees kelvin
	1024 when referring to memory storage capacity
	kilohm
KB	keyboard
	kilobit
KC	kilocycle (now kilohertz)
KCS	thousand characters per second
KEM	keyboard encoder memory
keV	kiloelectronvolt
kg	kilogram
kHz	kilohertz
KLIPS	thousand logical inferences per second
km	kilometer
km/h	kilometers per hour
KO	knockout
KOX	keyboard operated transmission
KSR	keyboard send–receive unit
kV	kilovolt
kVA	kilovolt-ampere
kVAR	kilovolt-ampere reactive
kVp	kilovolts peak
kW	kilowatt
kWh	kilowatthour
L	coil
	inductance (symbol for)
LAN	local-area network
LANC	local-area network controller
LANCE	local-area network controller for Ethernet
LAND	local-area network driver
LANT	local-area network transceiver
LAS	light-activated semiconductor
LASCR	light-activated silicon-controlled rectifier
LASCS	light-activated silicon-controlled switch
LAT	latitude
lb	pound
LC	induction–capacitance
	liquid crystal
LCA	logic cell array
LCC	leadless chip carrier
LCCC	leadless ceramic chip carrier
LCD	liquid-crystal display
LCP	left circular polarization
LCR	inductance–capacitance–resistance
LD	laser diode
LDM	LANCE development module
LDR	light-dependent resistor
LEC	layered elastomeric connectors
LED	light-emitting diode
LF	low frequency
LFO	low-frequency oscillator
LHCP	left-hand circular polarization
LIC	linear integrated circuit
LID	leadless inverted device
LIF	low insertion force
LIFO	last in, first out
LIM	limiter
	linear induction motor
LIMS	limit switch
LISP	list processing
LK	link
LLL	low-level logic
LLS	logic-level selectable
LMGR	left margin register
LMT	limit
LMTR	limiter
LNA	low-noise amplifier
LNB	low-noise band
LO	local oscillator
LOCOS	local oxidation of silicon
LP	long playing
	low pass
LPC	linear predictive coding
LPF	low-pass filter
lpm	lines per minute
LPTR	line printer
LRU	least recently used
LS	loudspeaker
LSA	limited space-charge accumulation diode
LSB	least-significant bit
	lower sideband
LSC	least-significant character
LSCFL	low-power source-coupled FET logic
LSD	least-significant decade
	least-significant digit
LSI	large-scale integration
LSTTL	low-power Schottky TTL
LTG	lighting
LTPD	lot tolerance percent defective
LTROM	linear-transformer read-only memory
LU	logic unit
LUF	lowest useful frequency
LUHF	lowest usable high frequency

LUT	look-up table
LV	low voltage
LVCD	least-voltage coincidence detection
LVDT	linear variable-differential transformer
LVP	low-voltage protection
LVDT	linear variable-differential transformer
	linear velocity displacement transformer
m	meter
	milli
M	mega
	mode
	symbol for mutual inductance
MA	magnetic amplifier
mA	milliampere
MAC	multiplier-accumulator
MAD	memory address director
MADT	microalloy diffused transistor
MAG	maximum available gain
MAGAMP	magnetic amplifier
MAG AMPL	magnetic amplifier
MAGMOD	magnetic modulator
MAGN	magnetron
MAN	manual
MAP	manufacturing automation protocol
MAR	memory address register
	multiplier-accumulator RAM
MAT	matrix
	microalloy transistor
MATV	master antenna television
MAU	math acceleration unit
MAX	maximum
MBB	make-before-break
MBM	magnetic bubble memory
MBO	monostable blocking oscillator
MC	megacycle (now MHz, megahertz)
	modulus control
	momentary contact
M^2CMOS	dual-layer metal CMOS
MCR	mode control register
MCU	microcomputer unit
MCW	modulated continuous wave
MDA	monochrome display adapter
MDC	memory disk controller
MDL	module
MDS	minimum discernible signal
MDU	multiply/divide unit
MEG	mega (10^6)
MEGO	megohm
MEM	memory
MF	medium frequency
	multifrequency
mF	millifarad
MFLOP	million floating operations per second
MFM	modified frequency modulation
MFSK	multiple frequency-shift keying
MG	motor generator
MGC	manual gain control
mH	millihenry
MHD	magnetohydrodynamic
MHF	medium high frequency
MHz	megahertz
mi	mile
MIB	microcomputer interface board
MIC	memory interface controller
	microphone
	microwave integrated circuit
	monolithic integrated circuit
MICMPTR	microcomputer
MICR	magnetic ink character recognition
micro	one millionth
MIDS	minimum discernible signal
mi/h	miles per hour
MIKE	microphone
MIMD	multiple-instruction multiple data
MIPS	million instructions per second
MIS	metal insulator silicon
mi/s	miles per second
MIT	master instruction tape
MLC	multilayer ceramic capacitor
MLM	multilayer metallization
MLSP	multilevel single-plane
mm	millimeter
MMF	magnetomotive force
MMIC	monolithic microwave integrated circuit
MMPU	memory management and protection unit
MMU	memory management unit
MNOS	metal-nitride-oxide semiconductor
MO	master oscillator

MOD	modification
	modulator
MOE	metal on elastomer
MOLELEX	molecular electronics
MOPA	master oscillator power amplifier
MOS	metal-oxide semiconductor
MOSFET	metal-oxide semiconductor field-effect transistor
MOSIGT	metal-oxide semiconductor insulated-gate transistor
MOST	metal-oxide semiconductor transistor
MOT	motor
MOV	metal-oxide varistor
MOX	manually operated switching
MPC	message-processing coprocessor
MPCC	multiprotocol communications controller
MPF	mask programmable filter
MPG	microwave pulse generator
MPO	maximum power output
MPTC	multiproject test chip
MPU	microprocessing unit
	microprocessor unit
MPX	multiplex
ms	millisecond
MS	master switch
m/s	meters per second
MSB	most-significant bit
MSD	most-significant decade
	most-significant digit
msec	millisecond
MSI	medium-scale integration
MSPS	megasample per second
MSR	machine status register
MSTV	medium-scan television
MSU	memory storage unit
MSW	master switch
MTBF	mean time between failures
MTC	memory timing controller
MTG	mounting
MTHBD	motherboard
MTI	moving-target indicator
MTNS	metal–thick nitride semiconductor
MTOS	metal–thick oxide semiconductor
MTRG	metering
MTS	multichannel television sound
MTSO	mobile telecommunications switching office
MTTF	mean time to failure
MTTR	mean time to repair
MUF	maximum usable frequency
MUL	multiplier
MUPO	maximum undistorted power output
MUT	module under test
MUX	multiplex
mV	millivolt
MV	megavolt
	multivibrator
MVB	multivibrator
MVC	manual volume control
mW	milliwatt
MW	megawatt
MWV	maximum working voltage
MX	multiplex
MΩ	megohm
N	number of bits
n	nano (10^{-9})
nA	nanoampere
NA	not applicable
	not available
NAB	National Association of Broadcasters
NAND	not AND (a logic operator)
NAPLPS	North American Presentation Level Protocol System
NAS	numerical aerodynamic simulation
NASA	National Aeronautics and Space Administration
NB	narrow band
NBFM	narrow-band frequency modulation
NBMU	network buffer management unit
NBS	National Bureau of Standards
NBTDR	narrow-band time-domain reflectometry
NBVM	narrow-band voice modulation
NBW	noise bandwidth
NC	no connection
	normally closed
	not connected
	numerical control
NDRO	nondestructive readout
NDT	nondestructive testing
NEB	noise equivalent bandwidth
NEC	*National Electrical Code®*
NEG	negative

NEMA	National Electrical Manufacturers Association
NEMP	nuclear electromagnetic pulse
NEP	noise equivalent power
NEUT	neutral
nF	nanofarad
NF	noise figure
NFET	N-channel JFET
NFM	narrowband frequency modulation
NFS	network file system
nH	nanohenry
NiCad	nickel cadmium
NIF	noise improvement factor
NIM	national instrumentation module
NIPO	negative input, positive output
NLR	nonlinear resistor
NMOS	N-channel MOSFET
NMRR	normal-mode rejection ratio
NO	normally open
	number
NOR	not OR (logical operator)
NOT	logical inverter
NPIN	negative–positive–intrinsic–negative
NPN	negative–positive–negative
NPNP	negative–positive–negative–positive
NRZ	nonreturn to zero
NRZI	nonreturn to zero interchange
	nonreturn to zero inverted
NRZL	nonreturn-to-zero level
NRZM	nonreturn-to-zero mark
ns	nanosecond
NTC	negative temperature coefficient
NTIA	National Telecommunications and Information Administration
NTS	not to scale
NTSC	National Television Systems Committee
NTWK	network
nV	nanovolt
NVR	no voltage release
NVRAM	nonvolatile random access memory
nW	nanowatt
OCB	oil circuit breaker
OCR	optical character reader (recognition)
OCTL	open-circuit transmission line
OD	outside diameter
ODT	octal debugging technique
OEIC	optoelectronic integrated circuit
OEM	original equipment manufacturer
OFC	offset control circuit
OLTP	on-line transaction processor
OMR	optical mark recognition
OMS	ovonic memory switch
OOK	on–off keying
OP AMP	operational amplifier
OR	logical operator
OROM	optical read-only memory
OS	one-shot
	operating system
OSC	oscillator
OSCP	oscilloscope
OSI	open systems interconnection
OSP	operating system processor
OSR	output shift register
OTA	operational transconductance amplifier
OTPROM	one-time PROM
OTS	ovionic threshold switch
OVLD	overload
OVP	overvoltage protector
oz	ounce(s)
P	plate
	power
	primary winding
pA	picoampere
PA	power amplifier
	public address
PACM	pulse-amplitude code modulation
PAD	pocket assembler/disassembler
PAL	phase alternation line
	programmable array logic
PAM	pulse-amplitude modulation
PAP	programmable algorithm processor
PAR	precision approach radar
PAX	private automatic exchange
PB	pushbutton
PBW	power bandwidth
PBX	private branch exchange
PC	phase corrector
	printed circuit
	program counter
	programmable controller
PCA	parts configuration analysis
PCB	printed-circuit board

PCC	plastic chip carrier
PCI	programmable communications interface
PCM	pulse-code modulation
PCM/FM	pulse-code modulation/frequency modulation
PD	photodetector
	plate dissipation
	potential difference
	pulse doppler
	pulse duration
PDM	pulse duration modulation
PDM-FM	pulse duration modulation–frequency modulation
PDVM	printing digital voltmeter
PE	processor element
PEG	programmable event generator
PENT	pentode
PEP	peak envelope power
pF	picofarad
PF	power factor
	pulse frequency
PFA	probabilistic fault analysis
PFN	page frame number
PG	power gain
PGA	pin grid array
PH	phase
PHONO	phonograph
PIC	power integrated circuit
PIM	pulse interval modulation
PIN	position indicator
	positive intrinsic negative
PINO	positive input, negative output
PIO	parallel input/output
	programmed I/O
PIT	programmable interval timer
PI/T	parallel interface/timer
PIV	peak inverse voltage
PK	peak
PK-PK	peak to peak
PL	plug
PLA	programmable logic array
	programmed logic array
PLC	programmable logic controller
PLCC	plastic lead chip carrier
PLD	programmable logic device
PLL	phase-locked loop
PLM	pulse-length modulation
PLRT	polarity
PLT	power-line transients
PM	permanent magnet
	phase modulation
PMC	pattern-matching chip
PMMC	permanent-magnet moving coil
PMMU	page-memory management unit
PMOS	P-channel MOS
PMU	parametric measurement unit
PNdB	perceived noise level expressed in decibels
PNIN	positive negative intrinsic positive
PNM	pulse number modulation
PNP	positive–negative–positive
PNPN	positive–negative–positive–negative
POS	positive
POT	potential
	potentiometer
PP	push-pull
p-p	peak to peak
PPI	plan position indicator
ppm	parts per million
PPM	pulse-position modulation
PPTR	page printer
PREAMP	preamplifier
PRF	pulse-repetition frequency
PRI	primary
PRL	parallel
PRM	pulse-rate modulation
PROM	programmable read-only memory
PRR	pulse-repetition rate
PRV	peak reverse voltage
ps	picosecond
PS	power supply
PSAR	programmable synchronous–asynchronous receiver
PSART	programmable synchronous–asynchronous receiver–transmitter
PSAT	programmable synchronous–asynchronous transmitter
PSB	parallel system bus
PSK	phase-shift keying
PSM	pulse spacing modulation
PSPS	planar silicon photoswitch
PSRR	power supply rejection ratio
PSVM	phase-sensitive voltmeter
PSW	process status word
PTC	positive temperature coefficient

PTD parallel transfer disk
PTM programmable timer module
pulse-time modulation
PTN private telecommunications network
PTO permeability-tuned oscillator
PTOUT printout
PTT press to talk
push to talk
PU pickup
power unit
PUT programmable unijunction transistor
PVC polyvinyl chloride
PVTC programmable video timing controller
PW printed wiring
PWA printed wiring assembly
PWM pulse-width modulation
PWR power
PWR SUP power supply
Q figure of merit for a capacitor, inductor, or LC circuit
symbol for quantity of electric charge
symbol for transistor
QAM quadrature amplitude modulation
QAVC quiet automatic volume control
QC quality control
QOD quick-opening device
QPL qualified products list
QPLTT qualified product list throughput time
QPSK quarternary phase-shift keying
QTY quantity
QUIL quad in-line
R designator for resistor
radar
reluctance
resistance
RAD radiation
radio
RADC Rome Air Development Center
RAIAM random-access indestructive advanced memory
RAM random-access memory
RBI ripple blanking input
RBO ripple blanking output
RC resistance–capacitance
R/C radio control
RC CPLD resistance–capacitance coupled
RCL resistance–capacitance–inductance
RCM radar countermeasures
RCP right circular polarization
RCPT receptacle
RCS remote-controlled systems
RCTL resistor–capacitor–transistor logic
RCVR receiver
RDF radio direction finder
RDOUT readout
REC recorder
RECP receptacle
RECT rectifier
REF reference
REF L reference line
REG regulator
RES resistor
RET return
RETFI return from interrupt
rev/min revolutions per minute
RF radio frequency
RFC radio-frequency choke
RFI radio-frequency interference
RGLTR regulator
RH roundhead
RHCP right-hand circular polarization
RHEO rheostat
RHI range-height indicator
RIAA Recording Industry Association of America
RIPS real-time image-processing package
RISC reduced instruction set computer
RIT receiver incremental tuning
RLS reference low sense
RLY relay
RM range marks
RMGR right-margin register
RMM read mainly memory
read mostly memory
RMOS refractory-metal oxide semiconductor
RMS root mean square
RNS residue number system
RO receive only
ROM read-only memory
ROTR receive-only typing reperforator
RPC remote procedure call

RPG	report program generator
rpm	revolutions per minute
RSLVR	resolver
RSU	register storage unit
RT	real time
RTC	real-time clock
RTD	resistance-temperature detector
RTL	resistor–transistor logic
RTS	request to send
RTTY	radio teletype
RTV	room temperature vulcanizing
RTZ	return to zero
RVDT	rotary-variable differential transformer
RVS	reverse
RW	read/write
RZ	return to zero
s	second
S	screen of an electron tube
	secondary
SADT	surface alloy diffused-base transistor
SAG MOS	self-aligning gate MOS
SAM	sequential access memory
SAP	second audio program
SAR	storage address register
	successive-approximation register
SAS	silicon asymmetrical switch
SASI	Shugart Associates Systems interface
SATO	self-aligned thick oxide
SAW	surface acoustic wave
SB	sideband
SBC	S-bus interface circuit
	single-board computer
SBDT	surface-barrier diffused transistor
SBS	semiconductor bilateral switch
SBT	surface-barrier transistor
SC	single contact
SCA	scale
	Secondary Communications Authorization
	subcarrier authorization
	Subsidiary Communications Authorization
SCAT	strip-chart architecture topology
SCC	serial communications controller
SCD	source control drawing
SCH	schedule
SCHEM	schematic
SCI	serial communications interface
SCO	subcarrier oscillator
SCOPE	oscilloscope
SCR	screw
	semiconductor-controlled rectifier
	silicon-controlled rectifier
SCS	semiconductor-controlled switch
	silicon-controlled switch
SCSI	small computer systems interface
SCT	surface-charge transistor
SDA	source and detector assemblies
SDFL	Schottky-diode FET
SDLC	synchronous data-link control
SDS	single-ended to differential mode selectable
Se	selenium
SEC	secondary
SEEDS	self-electro-optic-effect devices
SEF	soft-error filtering
SEL	selector
SER	series
SERVO	servomechanism
SF	single frequency
SFA	single-frequency amplifier
SGCS	silicon gate-controlled switch
SHE	socket head eliminator
SHF	superhigh frequency
SHLD	shield
S/I	signal-to-intermodulation ratio
SIA	serial interface adapter
SIC	semiconductor integrated circuit
	specific inductive capacity
	system interface controller
SICOS	sidewall base-contact structure
SID	sound interface device
SIDAC	(bidirectional voltage-triggered switch)
SIG	signal
SIMD	single instruction/multiple data
SIMM	single in-line memory module
SIO	serial input/output
SIP	single in-line package
SIR	system interface receiver
SIT	static induction transistor
	system interface transmitter
SKT	socket
SLB	side lobe blanking
SLC	straight-line capacitance
SLD	subscriber line data link
SLF	straight-line frequency

SLIC	subscriber loop interface circuit
	subscriber line interface card
SLR	side-looking radar
SLSI	super-large-scale integration
SM	silver mica (capacitor)
SMART®	synchronous mode avionic receiver–transmitter
SMD	storage module device
	surface-mount device
	surface-mounted device
SMI	standard memory interface
SMT	surface-mount technology
	surface-mounting technology
S/N	signal-to-noise ratio
SNAP	static nibble access path
SNOS	silicon nitride oxide silicon
SNR	signal-to-noise ratio
SO	slow operate (relay)
	small outline
SOA	safe operating area
SOC	socket
SOIC	small outline IC
SOL	solenoid
SOS	silicon on sapphire
SOT	small outline transistor
SP	single pole
	stack pointer
SPA	software performance analysis
SPC	stored-program control
SPCC	sync-protocol communications controller
SPDT	single-pole, double-throw
SPEC	specification
SPKR	speaker
SPL	sound pressure level
SPLY	supply
SPST	single-pole, single-throw
SQR	signal-to-quantizing noise ratio
SR	saturable reactor
	selenium rectifier
	set reset
	shift register
	silicon rectifier
	slew rate
	slow release (relay)
SRAM	static random-access memory
SRB	single row buffer
SRD	step recovery diode
SRE	series relay
SS	signal strength
	spread spectrum
SSB	single sideband
SSC	special service channels
SSDA	synchronous serial data adapter
SSI	small scale integration
SSP	subsatellite point
SSR	solid-state relay
SSS	solid-state switching
SSSC	single-sideband-suppressed carrier
SSTV	slow-scan television
STC	sensitivity time control
	system timing controller
STL	Schottky transistor logic
	simulation and test language
STN	station
STP	signal transfer point
SUP	supply
SUPPR	suppressor (ion)
SUS	semiconductor unilateral switch
	silicon unilateral switch
SVFC	synchronous voltage/frequency converter
SW	shortwave
	switch
SWBD	switchboard
SWGR	switchgear
SWL	shortwave listener
SWTL	surface-wave transmission line
SWR	standing-wave ratio (voltage)
SYM	symbol
SYN	synchronous
SYNC	synchronous, synchronizing
T	transformer
TAC	terminal access circuit
	terminal access controller
TACH	tachometer
TAIC	teleset audio interface circuit
TB	terminal board
TC	technical circular
	temperature coefficient
TCA	time of closest approach
TCCO	temperature-controlled crystal oscillator
TCE	thermal coefficient of expansion
TCL	transistor-coupled logic
TCP/IP	transmission control protocol/internet

TCR temperature coefficient of resistance
TCW translation control word
TD time delay
transmitter-distributor
TDA tunnel diode amplifier
TDL tunnel diode logic
TDM time-division multiplexing
TDMA time-division multiple access
TDR time-delay relay
time-domain reflectometry
TE transequatorial (propagation)
transverse electric
TED text editing device
transferred electron device
TEL telephone
TEM transverse electromagnetic
TER tertiary
TERM terminal
TFR terrain following radar
time full register
TGT target
TGTP tuned-grid, tuned-plate
THD total haromonic distortion
THz terahertz
TI target identification
TIA true instrumentation amplifier
TIC twinax interface circuit
TID touch information display
TIM transient intermodulation distortion
TLB translation lookaside buffer
TLI transport-level interface
TLM telemetry, telemeter
TLP transmission-level point
TLU table look-up
TM technical manual
transverse magnetic
TMGR top margin register
TMO thermomagnetic-optic
TMR triple modulation redundancy
TMTR thermistor
TNC terminal-node controller
TNLDIO tunnel diode
TOP technical office protocol
TP test point
TPI transport provider interface
turns per inch
TPTG tuned-plate, tuned grid
TR transmit receive
transmitter–receiver
TRF tuned radio frequency
TSAC time slot assigner circuit
TSS time-sharing system
TSW test switch
TTL transistor-transistor logic
TTY Teletypewriter
TU terminal unit
TUN tuning
TV television
TVI television interference
TVM transistor voltmeter
TW twisted
traveling wave
TWSB twin sideband
TWT traveling-wave tube
TWX Teletypewriter Exchange
TX transmitter
TXCO temperature-compensated crystal oscillator
TWA traveling-wave amplifier
U symbol for unrepairable unit
UART universal asynchronous receiver–transmitter
UDC universal disk controller
UDLT universal digital loop transceiver
UFET unipolar field-effect transistor
UHF ultrahigh frequency
UJT unijunction transistor
UL Underwriters' Laboratories
ULF ultralow frequency
UNDV undervoltage
UNF unfused
UPC universal product code
USART universal synchronous–asynchronous receiver–transmitter
USB upper sideband
USRT universal synchronous receiver–transmitter
USYNRT universal synchronous receiver–transmitter
UT universal time
UTC universal time, coordinated
UTS utility strength
UUT unit under test
UVPROM ultraviolet-light erasable read-only memory

UWS user workstation
V volt(s)
VA Viterbi algorithm
 volt-ampere
VAC video attributes controller
 volts, alternating current
Vac volts, alternating current
VAN value-added network
VAR volt-amperes, reactive
VCA voltage-controlled amplifier
VCCO voltage-controlled crystal oscillator
VCD variable-capacitance diode
VCO voltage-controlled oscillator
VCR videocassette recorder
 voltage-controlled resistor
VCSR voltage-controlled shift register
VCXO voltage-controlled crystal oscillator
VDAC™ video display attributes controller
VDC video display controller
 volts, direct current
Vdc volts, direct current
VDCT volts, dc test
VDI virtual device interface
VDM virtual device metafile
VDU video display unit
VF voice frequency
V-F voltage to frequency
V/F voltage to frequency
VFBO variable-frequency beat oscillator
VFC voltage-to-frequency converter
VFD vacuum fluorescent display
VFO variable-frequency oscillator
VGA variable-gain amplifier
VHDL VHSIC hardware description language
VHF very high frequency
VHLL very-high-level language
VHSIC very-high-speed integrated circuit
VIA versatile interface adapter
VID video
VIDF video frequency
VIR vertical interval reference
VIS video interface system
VLF very low frequency
VLT visible line trace
VLSI very-large-scale integration
VM voltmeter
VME virtual machine environment
VMOS vertical metal-oxide field-effect transistor
 vertical power FET
 V-groove metal-oxide silicon
VMPU virtual memory processor unit
VOL volume
VOM volt-ohm-milliameter
VOR VHF omnirange
VOX voice-operated control
 voice-operated switching
 voice-operated transmitter keyer
VPAC™ video processor and controller
VPCM vector PCM
VPN virtual page number
VPSP VHSIC programmable signal processor
VR voltage regulator
VRAM video random-access memory
VRH VAR-hour meter
VRM voice-recognition module
VSB vestigial sideband
VSC variable speech control
 video system controller
 voltage-saturated capacitor
VSM voice synthesis memories
VSP vector system processor
 voice synthesis processor
VSWR voltage standing-wave ratio
VT vacuum tube
VTAC® video timer and controller
VTL variable threshold logic
VTLC video terminal logic controller
VTM voltage-tunable magnetron
VTO voltage-tunable oscillator
VTOC volume table of contents
VTR videotape recording (recorder)
VTVM vacuum-tube voltmeter
VU volume unit
VVC voltage-variable capacitor
VVCD voltage-variable capacitor diode
VXO variable crystal oscillator
W watt
WARC World Administrative Radio Conference
WBFM wide-band frequency modulation
WFR wafer
WG waveguide
WHM watt-hour meter
WHR watt-hour
WM wattmeter

WORM	write-once read-many optical storage
WP	word processing
WPM	words per minute
WPS	word processing software
WS	weapon system
WSI	wafer-scale integration
WVDC	working voltage, direct current
wVdc	working voltage, direct current
WW	wire-wound
WWV	call letters of National Bureau of Standards radio station at Fort Collins, Colorado

Appendix B. Letter Symbols for Diodes and Transistors*

GENERAL

Symbol	Term	Symbol	Term
$\overline{F}$ or $\overline{NF}$	average noise figure average noise factor	T_A	free-air temperature ambient temperature
F or NF	spot noise figure spot noise factor	T_C	case temperature
I_F	forward current, dc	T_J	virtual junction temperature
I_n	noise current, equivalent input	T_{stg}	storage temperature
I_R	reverse current, dc	T_n	noise temperature
R_θ (formerly θ)	thermal resistance	T_0	reference noise temperature
$R_{\theta CA}$	thermal resistance, case-to-ambient	t_d	delay time
$R_{\theta JA}$ (formerly $\theta_{J\text{-}A}$)	thermal resistance, junction-to-ambient	t_f	fall time
		t_{off}	turn-off time
$R_{\theta JC}$ (formerly $\theta_{J\text{-}C}$)	thermal resistance, junction-to-case	t_{on}	turn-on time
		t_p	pulse time
s_f or s_{21}	forward transmission coefficient	t_r	rise time
		t_s	storage time
s_i or s_{11}	input reflection coefficient	t_w	pulse average time
		V_F	forward voltage, dc
s_o or s_{22}	output reflection coefficient	V_n	noise voltage, equivalent input
s_r or s_{12}	reverse transmission coefficient	V_R	reverse voltage, dc

* Courtesy of Texas Instruments Incorporated

SIGNAL DIODES AND RECTIFIERS

Symbol	Term	Symbol	Term
$I_{F(RMS)}$, I_f, I_F, $I_{F(AV)}$, i_F, I_{FM}	forward current	R_θ	thermal resistance
I_{FRM}	forward current, repetitive peak	T_J	junction temperature
I_{FSM}	forward current, surge peak	t_{fr}	forward recovery time
I_O	average rectified forward current	t_p	pulse time
$I_{R(RMS)}$, I_r, I_R, $I_{R(AV)}$, i_R, I_{RM}	reverse current	t_r	rise time
$i_{R(REC)}$, $I_{RM(REC)}$	reverse recovery current	t_{rr}	reverse recovery time
I_{RRM}	reverse current, repetitive peak	t_w	pulse average time
I_{RSM}	reverse current, surge peak	$V_{(BR)}$, $v_{(BR)}$	breakdown voltage (dc, instantaneous total value)
P_F, $P_{F(AV)}$, p_F, P_{FM}	forward power dissipation	$V_{F(RMS)}$, V_f, V_F, $V_{F(AV)}$, v_F, V_{FM}	forward voltage
P_R, $P_{R(AV)}$, p_R, P_{RM}	reverse power dissipation	$V_{R(RMS)}$, V_r, V_R, $V_{R(AV)}$, v_R, V_{RM}	reverse voltage
Q_S	stored charge	V_{RWM}	working peak reverse voltage
		V_{RRM}	repetitive peak reverse voltage
		V_{RSM}	nonrepetitive peak reverse voltage

VOLTAGE-REGULATOR AND VOLTAGE-REFERENCE DIODES

Symbol	Term	Symbol	Term
I_F	forward current, dc	V_Z, V_{ZM}	regulator voltage, reference voltage (dc, dc at maximum-rated current)
I_R	reverse current, dc	z_Z, z_{ZK}, z_{ZM}	regulator impedance, reference impedance, (small-signal, at I_Z, at I_{ZK}, at I_{ZM})
I_Z, I_{ZK}, I_{ZM}	regulator current, reference current (dc, dc near breakdown knee, dc maximum-rated current)		
T_J	junction temperature		
V_F	forward voltage, dc		
V_R	reverse voltage, dc		

VOLTAGE-VARIABLE-CAPACITANCE DIODES (Varactor Diodes)

Symbol	Term	Symbol	Term
α_C	temperature coefficient of capacitance	f_{co}	cut-off frequency
C_c	case capacitance	L_s	series inductance
C_j	junction capacitance	η	efficiency
C_t	total capacitance	Q	figure of merit
$\frac{C_{t1}}{C_{t2}}$	capacitance ratio	r_s	series resistance, small-signal
		T_J	junction temperature

MULTIJUNCTION TRANSISTORS

Symbol	Term
C_{cb}, C_{ce}, C_{eb}	interterminal capacitance (collector-to-base, collector-to-emitter, emitter-to-base)
C_{ibo}, C_{ieo}	open-circuit input capacitance (common-base, common-emitter)
C_{ibs}, C_{ies}	short-circuit input capacitance (common-base, common-emitter)
C_{obo}, C_{oeo}	open-circuit output capacitance (common-base, common-emitter)
C_{obs}, C_{oes}	short-circuit output capacitance (common-base, common-emitter)
C_{rbs}, C_{res}	short-circuit reverse transfer capacitance (common-base, common-emitter)
C_{tc}, C_{te}	depletion-layer capacitance (collector, emitter)
$\overline{F}$ or F	noise figure, average or spot
f_{hfb}, f_{hfe}	small-signal short-circuit forward current transfer ratio cutoff frequency (common-base, common-emitter)
f_{max}	maximum frequency of oscillation
f_T	transition frequency or frequency at which small-signal forward current transfer ratio (common-emitter) extrapolates to unity
f_1	frequency of unity current transfer ratio
G_{PB}, G_{PE}	large-signal insertion power gain (common-base, common-emitter)
G_{pb}, G_{pe}	small signal insertion power gain (common-base, common-emitter)
G_{TB}, G_{TE}	large-signal transducer power gain (common-base common-emitter)
G_{tb}, G_{te}	small-signal transducer power gain (common-base, common-emitter)
h_{FB}, h_{FE}	static forward current transfer ratio (common-base, common-emitter)
h_{fb}, h_{fe}	small-signal short-circuit forward current transfer ratio (common-base, common-emitter)
h_{ib}, h_{ie}	small-signal short-circuit input impedance (common-base, common-emitter)

MULTIJUNCTION TRANSISTORS

Symbol	Term
$h_{ie(imag)}$ or $Im(h_{ie})$	imaginary part of the small-signal short-circuit input impedance, (common-emitter)
$h_{ie(real)}$ or $Re(h_{ie})$	real part of the small-signal short-circuit input impedance, (common-emitter)
h_{ob}, h_{oe}	small-signal open-circuit output admittance (common-base, common-emitter)
$h_{oe(imag)}$ or $Im(h_{oe})$	imaginary part of the small-signal open-circuit output admittance, (common-emitter)
$h_{oe(real)}$ or $Re(h_{oe})$	real part of the small-signal open-circuit output admittance, (common-emitter)
h_{rb}, h_{re}	small-signal open-circuit reverse voltage transfer ratio (common-base, common-emitter)
I_B, I_C, I_E	current, dc (base-terminal collector-terminal, emitter-terminal)
I_b, I_c, I_e	current, rms value of alternating component (base-terminal, collector-terminal, emitter-terminal)
i_B, i_C, i_E	current instantaneous total value (base-terminal, collector-terminal, emitter-terminal)
I_{BEV}	base cutoff current, dc
I_{CBO}	collector cutoff current, dc, emitter open
I_{CEO}	collector cutoff current, dc, with (base open,
I_{CER},	resistance between base and emitter,
I_{CES}	base short-circuited to emitter,
I_{CEV}	voltage between base and emitter,
I_{CEX}	circuit between base and emitter)
$I_{E1E2(off)}$	emitter cutoff current
I_{EBO}	emitter cutoff current, dc, collector open
$I_{EC(ofs)}$	emitter-collector offset current
I_{ECS}	emitter cutoff current, dc, base short-circuited to collector
I_n	noise current, equivalent input
$\overline{NF}$ or NF[a]	noise figure, average or spot
P_{IB}, P_{IE}	large-signal input power (common-base, common-emitter)

MULTIJUNCTION TRANSISTORS

Symbol	Term	Symbol	Term
P_{ib}, P_{ie}	small-signal input power (common-base, common-emitter)	t_f	fall time
P_{OB}, P_{OE}	large-signal output power (common base, common-emitter)	t_{off}	turn-off time
P_{ob}, P_{oe}	small-signal output power (common-base, common-emitter)	t_{on}	turn-on time
P_T	total nonreactive power input to all terminals	t_p	pulse time
$r_b'C_c$	collector base time constant	t_r	rise time
$r_{CE(sat)}$	saturation resistance, collector-to-emitter	t_s	storage time
$r_{ele2(on)}$	small-signal emitter-emitter on-state resistance	t_w	pulse average time
R_θ	thermal resistance	V_{BB}, V_{CC}, V_{EE}	supply voltage, dc (base, collector, emitter)
s_{fb} or s_{21b}, s_{fe} or s_{21e}	forward transmission coefficient (common-base, common-emitter)	V_{BC}, V_{BE}, V_{CB}, V_{CE}, V_{EB}, V_{EC}	voltage, dc or average (base-to-collector, base-to-emitter, collector-to-base, collector-to-emitter, emitter-to-base, emitter-to-collector)
s_{ib} or s_{11b}, s_{ie} or s_{11e}	input reflection coefficient (common-base, common-emitter)	v_{bc}, v_{be}, v_{cb}, v_{ce}, v_{eb}, v_{ec}	voltage, instantaneous value of alternating component (base-to-collector, base-to-emitter, collector-to-base, collector-to-emitter, emitter-to-base, emitter-to-collector)
s_{ob} or s_{22b}, s_{oe} or s_{22e}	output reflection coefficient (common-base, common-emitter)	$V_{(BR)CBO}$ (formerly BV_{CBO})	breakdown voltage, collector-to-base, emitter open
s_{rb} or s_{12b}, s_{re} or s_{12e}	reverse transmission coefficient (common-base, common-emitter)	$V_{(BR)CEO}$ (formerly BV_{CEO})	breakdown voltage, collector-to-emitter with (base open,
T_J	junction temperature	$V_{(BR)CER}$ (formerly BV_{CER})	resistance between base and emitter,
t_d	delay time	$V_{(BR)CES}$ (formerly BV_{CES})	base short-circuited to emitter,

MULTIJUNCTION TRANSISTORS

Symbol	Term
$V_{(BR)CEV}$ (formerly BV_{CEV})	voltage between base and emitter,
$V_{(BR)CEX}$ (formerly BV_{CEX})	circuit between base and emitter)
$V_{(BR)E1E2}$	emitter-emitter breakdown voltage
$V_{(BR)EBO}$ (formerly BV_{EBO})	breakdown voltage, emitter-to-base, collector open
$V_{(BR)ECO}$ (formerly BV_{ECO})	breakdown voltage, emitter-to-collector, base open
$V_{CB(fl)}$, $V_{CE(fl)}$, $V_{EB(fl)}$, $V_{EC(fl)}$,	dc open-circuit voltage (floating potential) (collector-to-base, collector-to-emitter, emitter-to-base, emitter-to-collector)
V_{CBO}	collector-to-base voltage, dc, emitter open
$V_{CE(ofs)}$	collector-emitter offset voltage
$V_{CE(sat)}$	saturation voltage, collector-to-emitter
V_{CEO}	collector-to-emitter voltage, dc, with (base open,
V_{CER},	resistance between base and emitter,
V_{CES},	base short-circuited to emitter,
V_{CEV},	voltage between base and emitter,
V_{CEX}	circuit between base and emitter)
V_{EBO}	emitter-to-base voltage, dc, collector open
$V_{EC(ofs)}$	emitter-collector offset voltage
$\lvert V_{E1E2(ofs)} \rvert$	magnitude of the emitter-emitter offset voltage
$\lvert \Delta V_{E1E2(ofs)} \rvert_{\Delta I_B}$	magnitude of the change in offset voltage with base current
$\lvert \Delta V_{E1E2(ofs)} \rvert_{\Delta T_A}$	magnitude of the change in offset voltage with temperature
V_n	noise voltage, equivalent input
V_{RT}	reach-through (punch-through) voltage
y_{fb}, y_{fe}	small-signal short-circuit forward-transfer admittance (common-base, common-emitter)
y_{ib}, y_{ie}	small-signal short-circuit input admittance (common-base, common-emitter)
$y_{ie(imag)}$ or $Im(y_{ie})$	imaginary part of the small-signal short-circuit input admittance (common-emitter)
$y_{ie(real)}$ or $Re(y_{ie})$	real part of the small-signal short-circuit input admittance (common-emitter)

MULTIJUNCTION TRANSISTORS

Symbol	Term	Symbol	Term
y_{ob}, y_{oe}	small-signal short-circuit output admittance (common-base, common-emitter)	$y_{oe(real)}$ or $Re(y_{oe})$	real part of the small-signal short-circuit output admittance (common-emitter)
$y_{oe(imag)}$ or $Im(y_{oe})$	imaginary part of the small-signal short-circuit output admittance (common-emitter)	y_{rb}, y_{re}	small-signal short-circuit reverse transfer admittance (common-base, common-emitter)

UNIJUNCTION TRANSISTORS

Symbol	Term	Symbol	Term
η	intrinsic standoff ratio	t_w	pulse average time
		V_{B2B1}	interbase voltage
$I_{B2(mod)}$	interbase modulated current	$V_{EB1(sat)}$	emitter saturation voltage
I_{EB2O}	emitter reverse current	V_{OB1}	base-1 peak voltage
I_P	peak-point current	V_P	peak-point voltage
I_V	valley-point current		
r_{BB}	interbase resistance	V_V	valley-point voltage
T_J	junction temperature		
t_p	pulse time		

FIELD-EFFECT TRANSISTORS

Symbol	Term	Symbol	Term
b_{fs}, b_{is}, b_{os}, b_{rs}	common-source small-signal (forward transfer, input, output, reverse transfer) susceptance	$\overline{F}$ or F	noise figure, average or spot
		G_{pg}, G_{ps}	small-signal insertion power gain (common-gate, common-source)
C_{ds}	drain–source capacitance	G_{tg}, G_{ts}	small-signal transducer power gain (common-gate, common-source)
C_{du}	drain-substrate capacitance		
C_{iss}	short-circuit input capacitance, common-source	g_{fs}, g_{is}, g_{os}, g_{rs}	common-source small-signal (forward transfer, input, output, reverse transfer) conductance
C_{oss}	short-circuit output capacitance, common-source		
C_{rss}	short-circuit reverse transfer capacitance, common-source	I_D	drain current, dc
		$I_{D(off)}$	drain cutoff current

FIELD-EFFECT TRANSISTORS

Symbol	Term
$I_{D(on)}$	on-state drain current
I_{DSS}	zero-gate voltage drain current
I_G	gate current, dc
I_{GF}	forward gate current
I_{GR}	reverse gate current
I_{GSS}	reverse gate current, drain short-circuited to source
I_{GSSF}	forward gate current, drain short-circuited to source
I_{GSSR}	reverse gate current, drain short-circuited to source
I_n	noise current, equivalent input
I_S	source current, dc
$I_{S(off)}$	source cutoff current
I_{SDS}	zero-gate-voltage source current
$\overline{NF}$ or NF	noise figure, average or spot
$r_{DS(on)}$	static drain–source on-state resistance
$r_{ds(on)}$	small-signal drain–source on-state resistance
R_θ	thermal resistance
s_{fg} or s_{21g}, s_{fs} or s_{21s}	forward transmission coefficient (common-gate, common-source)
s_{ig} or s_{11g}, s_{is} or s_{11s}	input reflection coefficient (common-gate, common-source)
s_{og} or s_{22g}, s_{os} or s_{22s}	output reflection coefficient (common-gate, common-source)
s_{rg} or s_{12g}, s_{rs} or s_{12s}	reverse transmission coefficient (common-gate, common-source)
T_J	junction temperature
$t_{d(off)}$	turn-off delay time
$t_{d(on)}$	turn-on delay time
t_f	fall time
t_{off}	turn-off time
t_{on}	turn-on time
t_p	pulse time
t_r	rise time
t_w	pulse average time
$V_{(BR)GSS}$	gate–source breakdown voltage
$V_{(BR)GSSF}$	forward gate–source breakdown voltage
$V_{(BR)GSSR}$	reverse gate–source breakdown voltage
V_{DD}, V_{GG}, V_{SS}	supply voltage, dc (drain, gate, source)
V_{DG}	drain–gate voltage
V_{DS}	drain–source voltage
$V_{DS(on)}$	drain–source on-state voltage
V_{DU}	drain–substrate voltage
V_{GS}	gate–source voltage
V_{GSF}	forward gate–source voltage
V_{GSR}	reverse gate–source voltage
$V_{GS(off)}$	gate–source cutoff voltage
$V_{GS(th)}$	gate–source threshold voltage
V_{GU}	gate–substrate voltage
V_n	noise voltage, equivalent input
V_{SU}	source–substrate voltage
y_{fs}	common-source small-signal short-circuit forward transfer admittance
y_{is}	common-source small-signal short-circuit input admittance
y_{os}	common-source small-signal short-circuit output admittance
y_{rs}	common source small-signal short-circuit reverse transfer admittance
$y_{fs(imag)}$, $y_{is(imag)}$, $y_{os(imag)}$, $y_{rs(imag)}$	common-source small-signal (forward transfer, input, output, reverse transfer) susceptance
$y_{fs(real)}$, $y_{is(real)}$, $y_{os(real)}$, $y_{rs(real)}$	common-source small-signal (forward transfer, input, output, reverse transfer) conductance

Index